Ocean Optics: Methods and Applications

Ocean Optics: Methods and Applications

Edited by Harold Mason

SYRAWOOD
PUBLISHING HOUSE
New York

Published by Syrawood Publishing House,
750 Third Avenue, 9th Floor,
New York, NY 10017, USA
www.syrawoodpublishinghouse.com

Ocean Optics: Methods and Applications
Edited by Harold Mason

International Standard Book Number: 978-1-64740-406-2 (Hardback)

Cataloging-in-publication Data

Ocean optics : methods and applications / edited by Harold Mason.
p. cm.
Includes bibliographical references and index.
ISBN 978-1-64740-406-2
1. Optical oceanography. 2. Seawater--Optical properties. 3. Underwater photography. I. Mason, Harold.
GC178.2 .O24 2023
551.4601--dc23

TABLE OF CONTENTS

PREFACE

Ocean optics is the study of the transmission of light through the ocean and interaction of light with other bodies in water. The sun rays that fall on the ocean water interact with the different dissolved and particulate constituents of the seawater. This interaction is studied using the methods of ocean optics. The information about the interaction of light in water is important for understanding the underwater ecosystems and their functioning. Ocean optics is a branch of oceanography that is engaged in the research on a wide range of issues related with engineering and ocean science. In-situ study of ocean optics is facilitated by using satellite data. This book elucidates the various methods and applications of ocean optics in a multidisciplinary manner. It will also provide interesting topics for research, which interested readers can take up. This book is a resource guide for experts as well as students.

Significant researches are present in this book. Intensive efforts have been employed by authors to make this book an outstanding discourse. This book contains the enlightening chapters which have been written on the basis of significant researches done by the experts.

Finally, I would also like to thank all the members involved in this book for being a team and meeting all the deadlines for the submission of their respective works. I would also like to thank my friends and family for being supportive in my efforts.

Editor

1

A Virtual Geostationary Ocean Color Sensor to Analyze the Coastal Optical Variability

Marco Bracaglia [1,2,*], Rosalia Santoleri [1], Gianluca Volpe [1], Simone Colella [1], Mario Benincasa [1] and Vittorio Ernesto Brando [1]

1 Istituto di Scienze Marine (CNR-ISMAR), Via Fosso del Cavaliere 100, 00133 Rome, Italy; rosalia.santoleri@cnr.it (R.S.); gianluca.volpe@cnr.it (G.V.); simone.colella@cnr.it (S.C.); mario.benincasa@artov.ismar.cnr.it (M.B.); vittorio.brando@cnr.it (V.E.B.)

2 Dipartimento di Scienze e Tecnologie, Università degli Studi di Napoli Parthenope, Via Amm. F. Acton 38, 80133, Naples, Italy

* Correspondence: marco.bracaglia@artov.ismar.cnr.it

Abstract: In the coastal environment the optical properties can vary on temporal scales that are shorter than the near-polar orbiting satellite temporal resolution (~1 image per day), which does not allow capturing most of the coastal optical variability. The objective of this work is to fill the gap between the near-polar orbiting and geostationary sensor temporal resolutions, as the latter sensors provide multiple images of the same basin during the same day. To do that, a Level 3 hyper-temporal analysis-ready Ocean Color (OC) dataset, named Virtual Geostationary Ocean Color Sensor (VGOCS), has been created. This dataset contains the observations acquired over the North Adriatic Sea by the currently functioning near-polar orbiting sensors, allowing approaching the geostationary sensor temporal resolution. The problem in using data from different sensors is that they are characterized by different uncertainty sources that can introduce artifacts between different satellite images. Hence, the sensors have different spatial and spectral resolutions, their calibration procedures can have different accuracies, and their Level 2 data can be retrieved using different processing chains. Such differences were reduced here by adjusting the satellite data with a multi-linear regression algorithm that exploits the Fiducial Reference Measurements data stream of the AERONET-OC water-leaving radiance acquired at the Acqua Alta Oceanographic Tower, located in the Gulf of Venice. This work aims to prove the suitability of VGOCS in analyzing the coastal optical variability, presenting the improvement brought by the adjustment on the quality of the satellite data, the VGOCS spatial and temporal coverage, and the inter-sensor differences. Hence, the adjustment will strongly increase the agreement between the satellite and in situ data and between data from different near-polar orbiting OC imagers; moreover, the adjustment will make available data traditionally masked in the standard processing chains, increasing the VGOCS spatial and temporal coverage, fundamental to analyze the coastal optical variability. Finally, the fulfillment by VGOCS of the three conditions for a hyper-temporal dataset will be demonstrated in this work.

Keywords: fiducial reference measurements; hyper-temporal dataset; optical radiometry; coastal environment; observation geometry

---

1. Introduction

Ocean Color Radiometry (OCR) satellite data allow observing the variability of the inherent optical properties (IOPs) and the concentrations of the water components in the coastal and open ocean with a large spatial coverage [1].

Near-polar orbiting OC satellites usually have a spatial resolution that goes from 300 m to 1000 m and provides ~1 image per day of basins located at middle latitudes. These spatial and temporal resolutions are of the same order as the temporal and spatial scales of most of the open ocean bio-optical processes and, consequently, are sufficient to properly characterize such areas [2–4]. On the contrary, in coastal waters, meteorological, marine, and hydrological drivers set up shorter temporal scales for such processes, compared to the open ocean [4]; indeed, in those areas, rapid changes can be observed in the time scale of few hours or less [4–8]. Consequently, an improved temporal resolution than what is currently provided by existing near-polar orbiting sensors is needed to accurately analyze the coastal environment [2–4].

An improved temporal resolution can be provided by OCR geostationary sensors [9–15], which acquire multiple observations of the same area during the same day. Nevertheless, the only currently working OCR geostationary sensor is the Geostationary Ocean Color Imager (GOCI) [16] over the Korean Peninsula, while such sensors are absent over the European Seas.

The first step in improving the near-polar orbiting sensor temporal resolution has been accomplished in [5], where the overlapping scenes of the Visible Infrared Imaging Radiometer Suite (VIIRS), mounted on the SUOMI-NPP satellite, have been exploited to observe the short term variability of some OC quantities in the Gulf of Mexico. This approach has then been used in [6] to observe the particulate backscattering (b_{bp}) variability in the North Adriatic Sea (NAS, Figure 1), adjusting the remote sensing reflectance (R_{rs}) spectra following [17]. This adjustment, based on a multi-linear regression (MLR) procedure and the AERONET-OC Acqua Alta Oceanographic Tower (AAOT) [18,19] in situ radiometric data (Figure 1), was used to reduce the uncertainties introduced by the different viewing geometry of different VIIRS images and to use data generally masked in the standard processing chains. The AERONET-OC in situ data are part of the Fiducial Reference Measurements (FRM) data stream as they guarantee traceability, accuracy, long-term stability, and cross-site consistency [20]. Those aspects are central for system vicarious calibration, product validation, uncertainty estimation, and development and assessment of bio-optical models for the generation of derived high-level products [21–26].

The present work aims to reduce the gap between near-polar orbiting and geostationary mission temporal resolutions, creating a Level 3 hyper-temporal analysis-ready OC dataset [27,28], containing observations provided by the currently functioning near-polar orbiting OC imagers.

This dataset is defined as an analysis-ready dataset as it is conceived to provide to users, without the expertise in pre-processing procedures, "ready to use" OC data to analyze the coastal optical variability at the high temporal resolution, similarly to the CEOS Analysis Ready Data for Land (CARD4L) [27].

The definition of a hyper-temporal dataset is based on the three conditions provided in [29], for which a hyper temporal dataset must:

(a) Be univariate (i.e., with multiple images of the same parameter only).

(b) Contain a set of time slices, all of which must be precisely coregistered (with image-to-image pixels perfectly aligned spatially).

(c) Exhibit radiometric consistency between images (i.e., they are measured using the same sensors or inter-validated sensor systems, and exhibit a degree of normalization between time slices).

This dataset is named Virtual Geostationary Ocean Color Sensor (VGOCS). It contains multiple images acquired during the same day by the currently functioning near-polar orbiting OC imagers over the NAS study area, allowing approaching the temporal resolution of a geostationary satellite. For each sensor, the time series starts for the first day of acquisition until 31 October 2019.

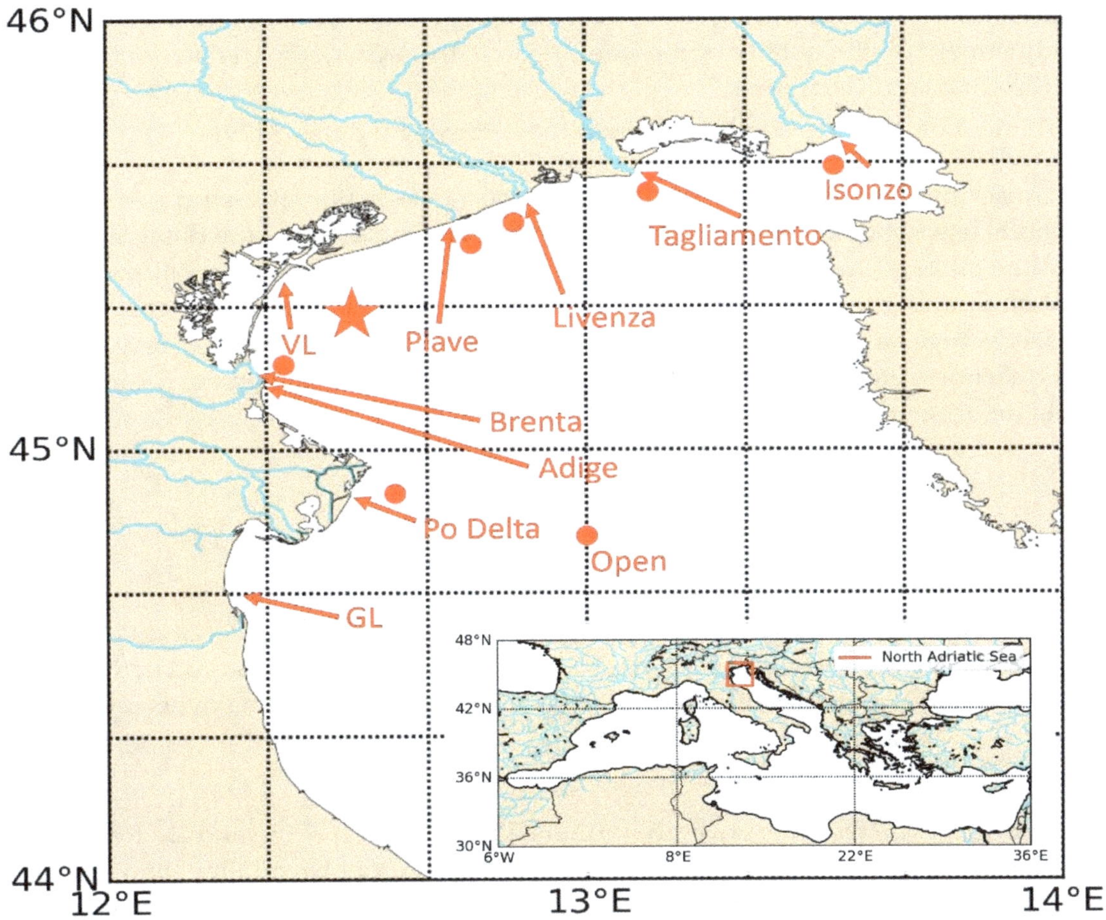

Figure 1. Map of the NAS with the rivers of the basin. The red star identifies the Acqua Alta Oceanographic Tower (AAOT), VL the Venice Lagoon, GL the Goro Lagoon, the red dots the virtual buoys (Section 2.7). In the box on the bottom right the Mediterranean Sea, with the NAS identified by the red square.

The creation of this dataset started from the Level 2 (L2) R_{rs} spectra provided by National Aeronautics and Space Administration (NASA) and the European Organisation for the Exploitation of Meteorological Satellites (EUMETSAT); the radiometric data of all sensors have been preserved at their native spectral resolution and re-projected on a common 1 km × 1 km equirectangular grid. The grid spatial resolution of 1 km is probably too coarse for the short spatial scale processes that can take place in the coastal environment [4,30–33]; nevertheless, this is the best that is possible to achieve for the creation of a hyper-temporal dataset created with several OCR sensors. After the re-projection, the R_{rs} spectra, adjusted following [17], have been used to calculate various IOPs by the Quasi Analytical Algorithm version 6 (QAA) [34,35]. This algorithm has been chosen as it has been demonstrated suitable for the b_{bp} retrieval in the NAS [6,36,37].

The adjustment presented in [17] is here exploited to use data from different sensors, as its application to the R_{rs} spectra can reduce the artifacts that can be observed between different satellite images. Indeed, the use of data from different sensors introduces an additional source of uncertainty, besides the one due to the different viewing geometry of each satellite image [6,38]. Hence, the sensors have different spatial and spectral resolutions, their calibration procedures can have different accuracies, and their L2 data can be retrieved using different processing chains. Consequently, looking at two scenes acquired during the same day by different sensors, it is not easy to discern how much of the observed difference is due to a real process or the artifacts, introduced by the different data uncertainties [8,39].

The objective of this work is to prove:

- The suitability of VGOCS in analyzing the coastal optical variability, by the study of the effect of the adjustment on the quality of the satellite data, the VGOCS spatial and temporal coverage, and the intersensor differences.
- The fulfillment of the three conditions for a hyper-temporal dataset [29].

The manuscript is structured as follows: in the next section the study area, the material, and the methods used in this work will be presented, while in Section 3 the obtained results will be analyzed. In the latter section, the results of the match-up analyses between different satellite and in situ data before and after the application of the adjustment will be presented. Then the spatial and temporal coverage and the inter-sensor differences of the adjusted VGOCS data will be analyzed. In the last two sections, the results will be discussed and the conclusions and the future perspective for this work will be presented.

2. Materials and Methods

2.1. Study Area

The NAS is characterized by a large number of rivers (listed in Figure 1) that can discharge waters with different biogeochemical and optical properties [40–42]. This basin is located in the North-Eastern Mediterranean (Figure 1) and its main rivers are shown in Figure 1. All those rivers are relevant for the optical dynamics of the basin [40,43,44], but Po and Adige are those that provide most of the NAS freshwater [45,46].

The NAS optical variability is strongly influenced by meteo-marine and hydrological forcings, and by the basin shallow bathymetry, which has an average depth smaller than 35 m [41,43–50]. Due to this complexity, several works studied the bio-optical properties and the dynamics of the NAS using several kinds of data and methods such as in situ, ship-board, and satellite observations and oceanographic models [51–56].

2.2. VGOCS Sensors

VGOCS is composed of the observations acquired over the NAS by the current functioning near-polar orbiting OC sensors. Those involved in the dataset are

- The NASA Moderate-Resolution Imaging Spectroradiometer (MODIS) sensors mounted on the AQUA and TERRA satellites, here simply referred to as AQUA and TERRA respectively.
- The National Oceanic and Atmospheric Administration (NOAA) VIIRS sensors mounted on the SUOMI-NPP and NOOA-20 (previously JPSS-1) satellites, here referred to as VIIRSN and VIIRSJ respectively.
- The European Space Agency (ESA)/EUMETSAT Ocean and Land Color Instrument (OLCI) mounted on Sentinel 3A, here referred simply as OLCI.

The two MODIS sensors, mounted on the AQUA and TERRA satellites, are the only sensors of the dataset that have the same technical characteristic and resolutions; indeed, the two VIIRS sensors are very similar to each other but differ for the spectral resolution [57]. The latter is depicted for each sensor in Figure 2 [57–60].

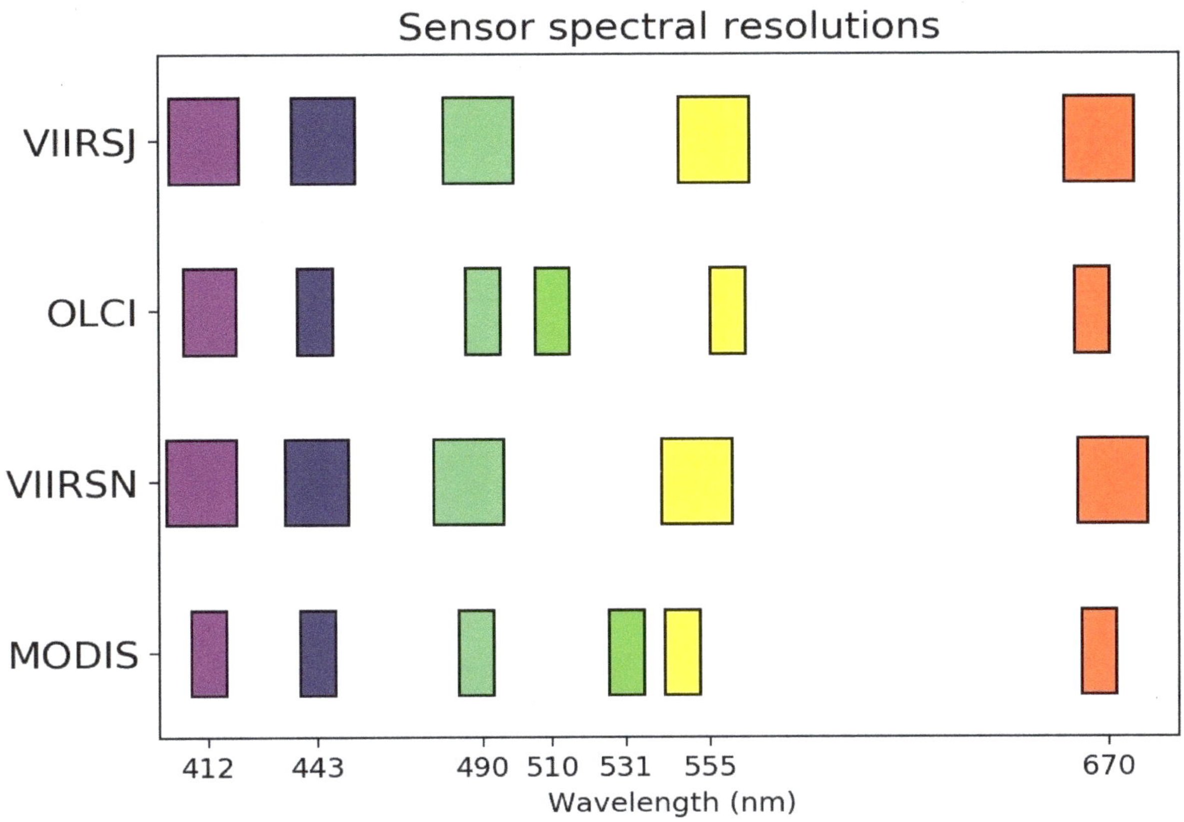

Figure 2. Spectral resolutions of the sensor involved in the Virtual Geostationary Ocean Color Sensor (VGOCS).

The MODIS OC bands have a 1000 m spatial resolution [60], the two VIIRS sensors a 750 m one [58], while the OLCI data are provided at two different resolutions: Full resolution (FR, 300 m) and Reduce Resolution (RR, 1200 m) [59]. In this study, the RR products are exploited, as the FR data suffer from a salt and pepper effect that is due to atmospheric correction problems (i.e., a lot of valid observations are surrounded by masked pixels, not shown).

OLCI, which is the only ESA/EUMETSAT sensor used in VGCOS, is the one with the lower revisiting frequency. This sensor has an orbital cycle of 27 days [59], while for the NASA/NOAA (N/N) sensors this is of 16 days [58,60].

OLCI has also a different viewing geometry and engineering than the N/N sensors. Indeed, the swath of the MODIS and VIIRS sensors is centered on the nadir and a rotating mirror reflects the radiation onto a set of Charge-Coupled Device (CCD) detectors. In OLCI, 5 CCD detectors are mounted in a fan arrangement on a common optical bench and the radiance arrives at them after passing a calibration assembly [59]. The sensor has a total field of view of 68.5°, while each detector has a field of view 14.1° with a slight overlap of 0.6° between each other [59]. Moreover, to reduce the sun-glint effect, which impacted a large number of the MEdium Resolution Imaging Spectrometer (MERIS) observations at sub-tropical latitudes [59,61], the OLCI swath is not centered at nadir but is tilted 12.6° westwards [59].

2.3. VGOCS Dataset and Processing Chain

The VGOCS dataset is created starting from the standard L2 R_{rs} spectra provided by EUTMESAT (OLCI, processing baseline v2.23) [62–64] and NASA (N/N sensors, processing version R2018.0) [65], that are stored in the archive of CNR ISMAR in Rome [39].

The VGOCS processing chain (Figure 3) is mostly the same as the one used in [6], which was based on the standard Copernicus Marine Environment Monitoring System (CMEMS) operational processing chain for the Mediterranean Sea [39], with some changes. The main difference is that CMEMS provides a single daily image, derived from the merging of data from different sensors [39], while in the VGOCS processing chain the merging procedure is not performed. Hence, all the images are retained to have multiple satellite observations of the basin during the same day.

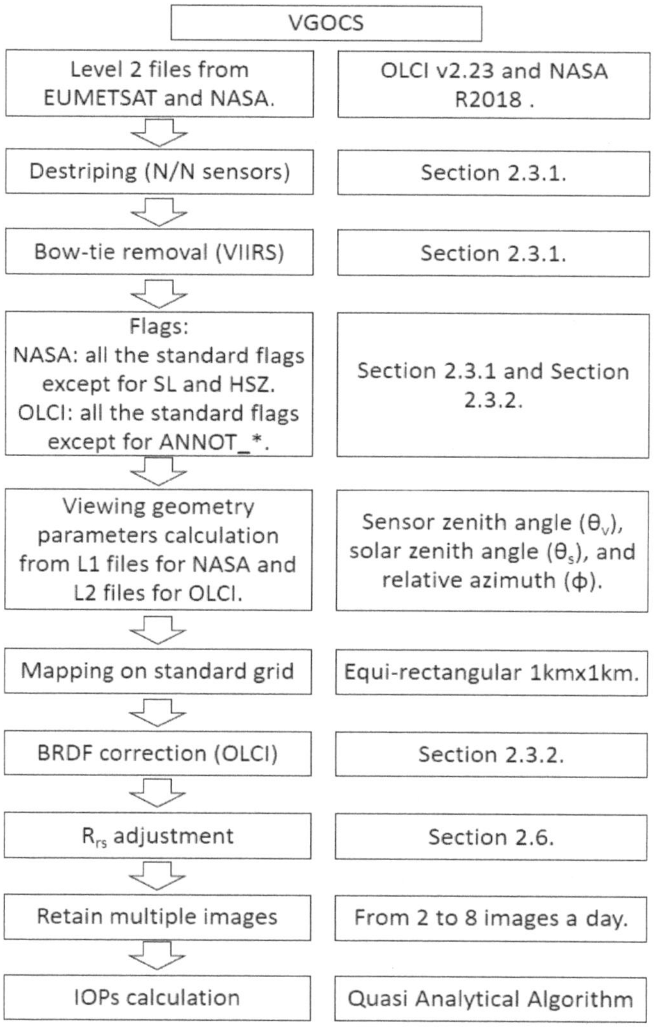

Figure 3. On the left the flow chart of the VGOCS processing chain, on the right some additional information for each step.

VGOCS contains observations of the NAS from 2002 (AQUA and TERRA operating) to 31 October 2019 (all sensors operating), providing 2 to 8 images a day depending on the number of functioning sensors and on the day of the satellite orbital cycles. The latter parameter is useful as for each sensor the number of images and their viewing geometry is the same for the same orbital cycle day, except for some minor changes in the ground-track and overpass times. The orbital cycle day is calculated in the same way for all the N/N sensors and it is a number that goes from 1 to 16. Hence, all the N/N sensors involved in VGOCS have an orbital cycle of 16 days [58,60]. For each file, defining as NDT the number of days between the satellite observation and the first TERRA acquisition (8 February 2000), the orbital cycle day can be defined as the remainder of NDT/16 plus one. The orbital cycle of OLCI is of 27 days, thus the orbital cycle day is defined as the remainder of NDO/27 plus one (where NDO is the number of days between the satellite observation and the first OLCI acquisition, 26 April 2016).

To use the same grid for each image of the dataset, all of them are re-projected on a standard 1 km $\times$ 1 km equirectangular grid by the nearest neighbor approach.

The name format of each file is VGOCS_YYYYJJJHHMMSS_X.nc, with Y, J, H, M, S (year, Julian day of the year, hour, minute, second) referring to the start time of the sensor acquisition, while X is the label which identifies the sensor (J: VIIRSJ, V: VIIRSN, A: AQUA, T: TERRA, O: OLCI).

The parameters that can be found in each VGOCS file are stored in different groups:

- The R_{rs}_data group, where the R_{rs} spectra at the native sensor resolution are stored.
- The IOP_data group, where all the parameters (listed below) calculated with the QAA can be found [34,35]:
    - reference wavelength (λ_0)
    - total absorption at the reference wavelength ($a(\lambda_0)$)
    - particulate backscattering at the reference wavelength ($b_{bp}(\lambda_0)$)
    - b_{bp} spectral slope (η)
    - b_{bp} at 443 nm ($b_{bp}(443)$)
    - absorption by nonalgal particles and dissolved matter at 443 nm ($a_{dg}(443)$)
    - a_{dg} spectral slope (s)
    - absorption by phytoplankton at 443 nm ($a_{phy}(443)$)
- The Atmospheric_data group contains some atmospheric parameters, such as aerosol optical thickness and angstrom coefficients, extracted from the L2 files.
- The Geo_data group contains information about the applied flags, extracted from the L2 files, and about some viewing geometry parameters, such as the sensor zenith angle (θ_v), the solar zenith angle (θ_s), and the relative angle between the solar and sensor azimuth angle (φ). Those parameters for the OLCI sensor are extracted from the L2 file, for the VIIRS sensors from the L1 GEO files, while for the MODIS sensors they are retrieved using the L1B files as the input of l2gen.

Moreover, latitude, longitude, and the satellite orbital cycle day are provided as stand-alone variables.

2.3.1. NASA/NOAA Sensors

The processing chain used for the N/N sensors is the same reported in [6] for the VIIRSN data. All the N/N sensor images have been destriped following [66,67], while, for the two VIIRS sensors, the bow-tie missing data have been filled by linear interpolation [58]. Not all the standard L2 flags [68] have been applied; hence, the data have been unmasked for the High Sensor Zenith (HISATZEN, here HSZ) and the stray-light flag (STRAYLIGHT, here SL). The first one masks data acquired at a $\theta_v > 60°$ [69,70], while the second one masks data that could be affected by the adjacency effect of clouds and lands [71–74].

The HSZ flag has not been applied because a large number of images are acquired at $\theta_v > 60^{\circ}$, and unmasking for this flag makes available a larger number of valid pixels. Using the HSZ data is fundamental for the aim for which VGOCS is conceived, as, to analyze the daily optical variability, it is important to have at disposal as many full images as possible. In this study, an image will be considered as "full" if the sensors detected at least 90% of the water pixels of the NAS, while it will be considered as "partial" if the sensor detected between 30% and 90% of the water pixels.

In Figure 4 the number of images for each day of the orbital cycle is presented for the N/N sensors considering as masked the HSZ observations. As can be seen, only for days 5, 8, 13, and 16 there are five full images (in green), while for the other days there are maximum one partial (in black) and four full images.

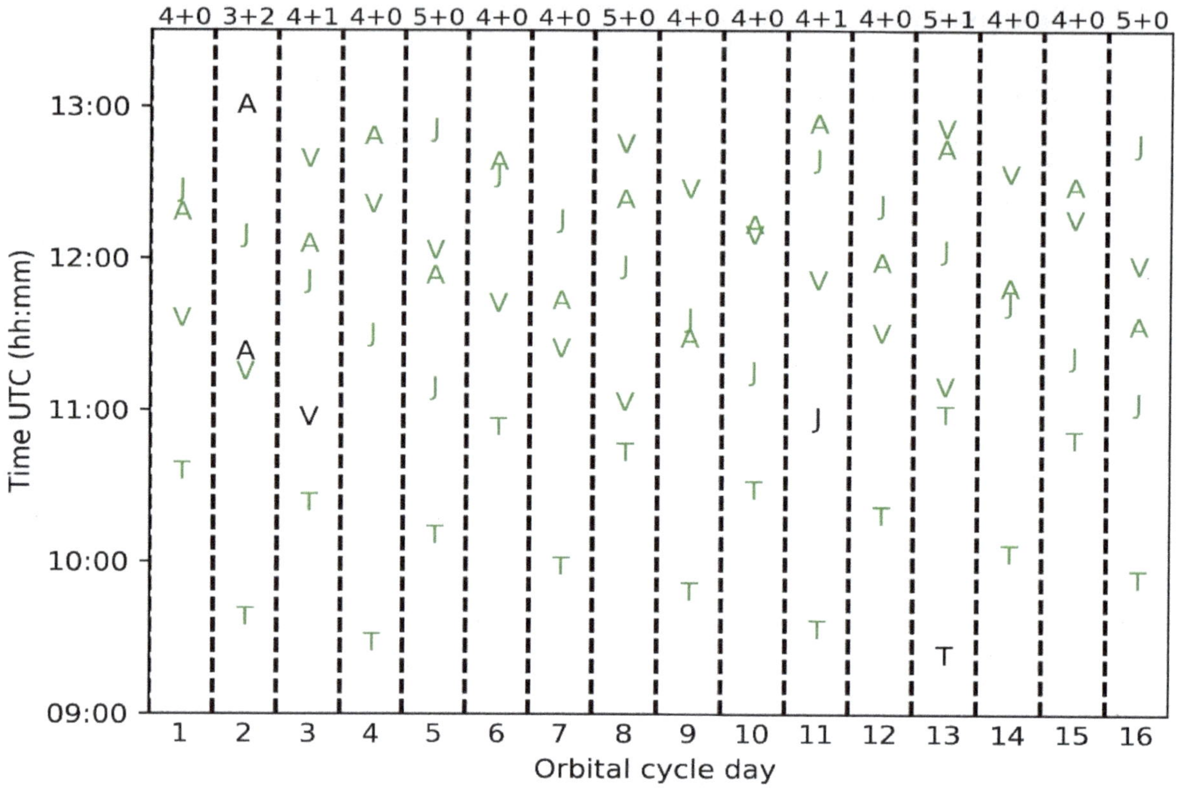

Figure 4. Available images for the N/N sensors for each day of the orbital cycle, with the application of the High Sensor Zenith (HSZ) flag. On the x-axes the orbital cycle day, on the y-axes the scan start time, with the different letters identifying the different sensors. In green the full images, in black the partial images. On the top for each orbital cycle day the number of full images plus the number of the partial ones.

Figure 5 is the same of Figure 4, but with the pixels unmasked for the HSZ flag. The HSZ pixels are 22.9% of the pixels scanned by the N/N sensors in the NAS, and unmasking them provides a larger number of full and partial images, in comparison with Figure 4. Now the minimum number of available images is reached on day 3 with 5 full images and the maximum on day 4 with 7 full and 1 partial image. Particularly:

- For VIIRSN there are 9 full orbit overlaps in comparison with the two available maskings for HSZ.
- For VIIRSJ there are 8 full and 2 partial overlaps in comparison with the two available maskings for HSZ.
- For AQUA and TERRA, there are 3 full orbit overlaps (previously zero), and some additional partial images.

Thus, the unmasking for this flag strongly improves the spatial and temporal coverage of VGOCS, making available up to 3 additional full images a day.

It is important to remind here that OLCI has not been taken into account in Figures 4 and 5 as it has a different orbital cycle and no HSZ flag. This sensor can provide one additional complete image of the NAS per day except for days 7, 11, 15, 19, 23, and 27 of its orbital cycle when the NAS is not overpassed by Sentinel-3A.

The SL flag masks all the pixels that could be affected by the light reflected by the land surface that can be forward scattered into the sensor field of view [71–74]. Nevertheless, the application of this flag results in the masking of a large part of the coastal area (here identified in red in Figure 6) [6,75], as can be noted in Figure 7. Here two $b_{bp}(443)$ maps retrieved from the original VIIRSN R_{rs} during 21 March 2013 12:48 UTC are presented: on the left the SL flag is active, and a large portion of the coastal area results masked; on the right, the SL flag is not applied and the majority of the coastal pixels are now available.

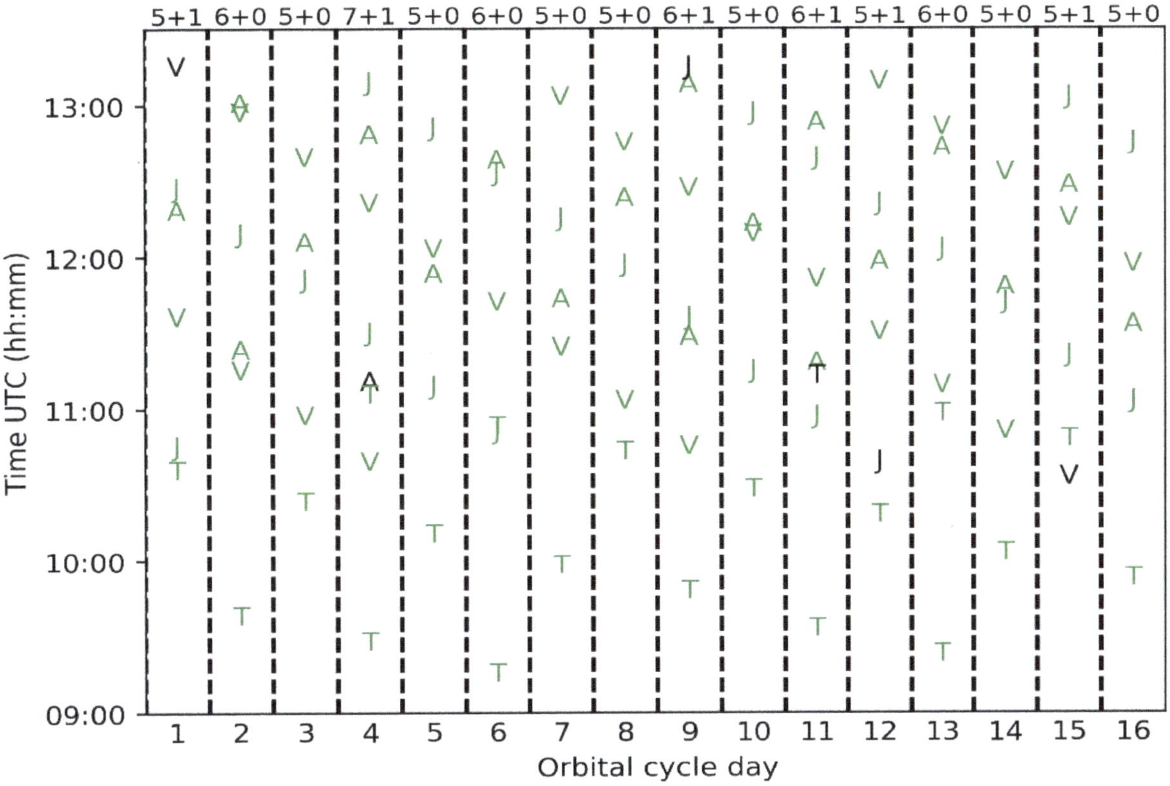

Figure 5. Available images for the N/N sensors for each day of the orbital cycle, without the application of the HSZ flag. On the x-axes the orbital cycle day, on the y-axes the scan start time, with the different letters identifying the different sensors. In green the full images, in black the partial images. On the top for each orbital cycle day the number of full images plus the number of the partial ones.

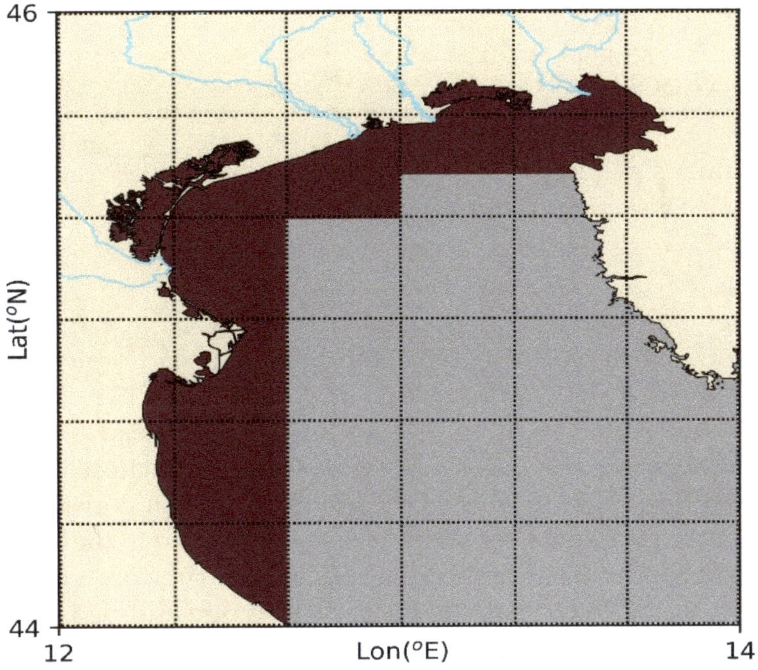

Figure 6. Map of the North Adriatic Sea (NAS), in red the coastal area used to quantify the stray-light flag (STRAYLIGHT, here SL) and ANNOT_* effect on the VGOCS spatial coverage.

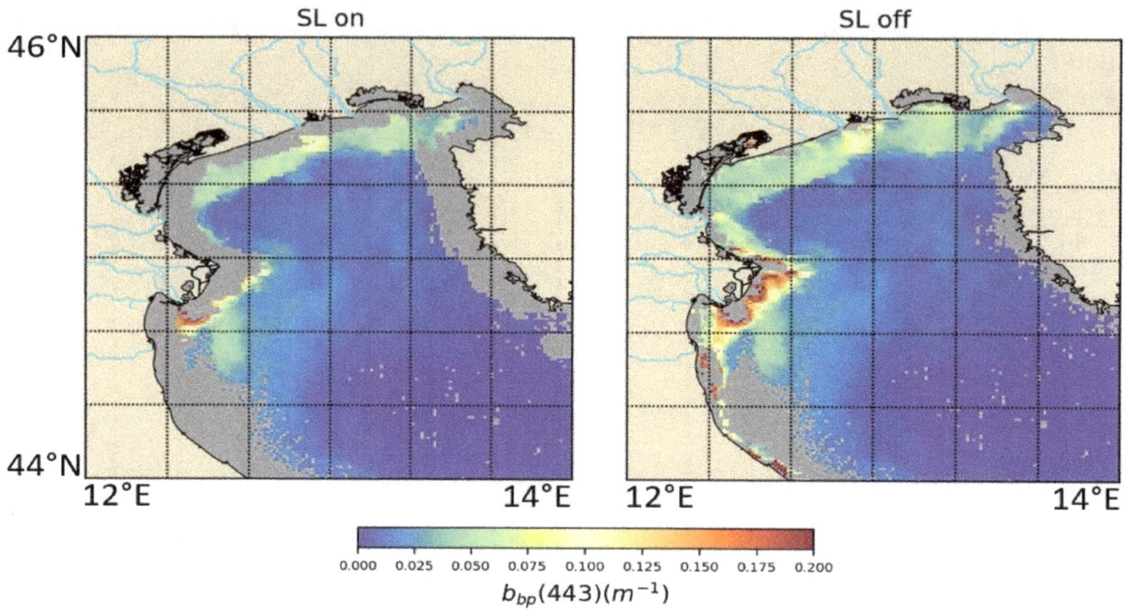

Figure 7. Maps of $b_{bp}(443)$ retrieved from the original VIIRSN R_{rs} for 21 March 2013 12:48 UTC. On the left, the SL flag is active (SL on), while the pixels are unmasked (SL off) on the right.

The behavior observed in Figure 7 is common to a large number of images of the dataset, as the mean percentage of the valid pixels flagged by SL in the coastal area is >40% for every single N/N sensor and the entire VGCOS N/N time series (Table 1). Hence, to analyze the coastal optical variability it is necessary to unmask such pixels.

Table 1. The mean percentage of SL flagged pixels in comparison with the total number of coastal valid pixels, for each N/N sensor and for the entire N/N VGOCS time series.

Sensor	Percentage of SL Masked Pixels
TERRA	42.0%
AQUA	41.2%
VIIRSN	45.1%
VIIRSJ	44.7%
VGOCS (no OLCI)	42.4%

The larger uncertainties generally observed for the HSZ and SL data can be reduced by the adjustment procedure [6]; consequently, the agreement of such data with the in situ observations will be analyzed in Section 3.2.

2.3.2. OLCI Sensor

The EUMETSAT OLCI data have different L2 processing, leading to files with structures and variables differing than those for the N/N sensors. Particularly, a different atmospheric correction algorithm is applied to the TOA radiances; hence, for OLCI, instead of the algorithm from [76], used in the L2 NASA processing, the Baseline Atmospheric Correction (BAC) algorithm [77,78] is exploited. Moreover, the R_{rs} provided in the OLCI L2 files are not corrected for the bi-directional effect. For this reason, for the OLCI data, an additional step is needed in the processing chain (Figure 3) to be coherent with the N/N data. Hence, the OLCI R_{rs} spectra are corrected with the bi-directional reflectance distribution function (BRDF), following the procedure outlined in the works of Morel and his co-workers from 1991 [69,78–81].

The EUMETSAT OLCI L2 flags are different than those used in the NASA L2 processing. In this work, in agreement with [82] and the latest EUMETSAT indication (I. Cazzaniga, pers. comm.) [83], the ANNOT_DROUT flag is not applied, as in the latest version of CMEMS [84]. This flag masks pixel where the value of the residual for the surface reflectance at 510 nm is above a certain threshold, defined in the climatology. Moreover, in this study, the ANNOTMIXR1, ANNOT_TAU6, and ANNOT_ABSO_D named annotation flags (here ANNOT_*), have not been applied in the match-up analysis to analyze their effect on the VGOCS spatial coverage and their agreement with in situ data. Hence, all those flags concern the quality of the atmospheric correction in open waters, and the application to optically complex waters could be questionable [82,83].

In Figure 8, the maps of $b_{bp}(443)$ retrieved from the original OLCI R_{rs} during 21 September 2017, with and without the application of the ANNOT_DROUT and ANNOT_* flags, are presented. The mean percentage of valid pixels masked by the ANNOT_DROUT flag in the coastal area (Figure 6) is around 40.9% for the OLCI time series, the ANNOT_MIXR1 flag masks the 20.1% of them, while the other two ANNOT_* flags mask the 0.8% of the coastal pixels.

Also for the ANNOT_DROUT and ANNOT_* data, the effect of the adjustment on their agreement with in situ observations will be analyzed in Section 3.2.3 to justify the decision for unmasking data traditionally masked in the standard processing chains.

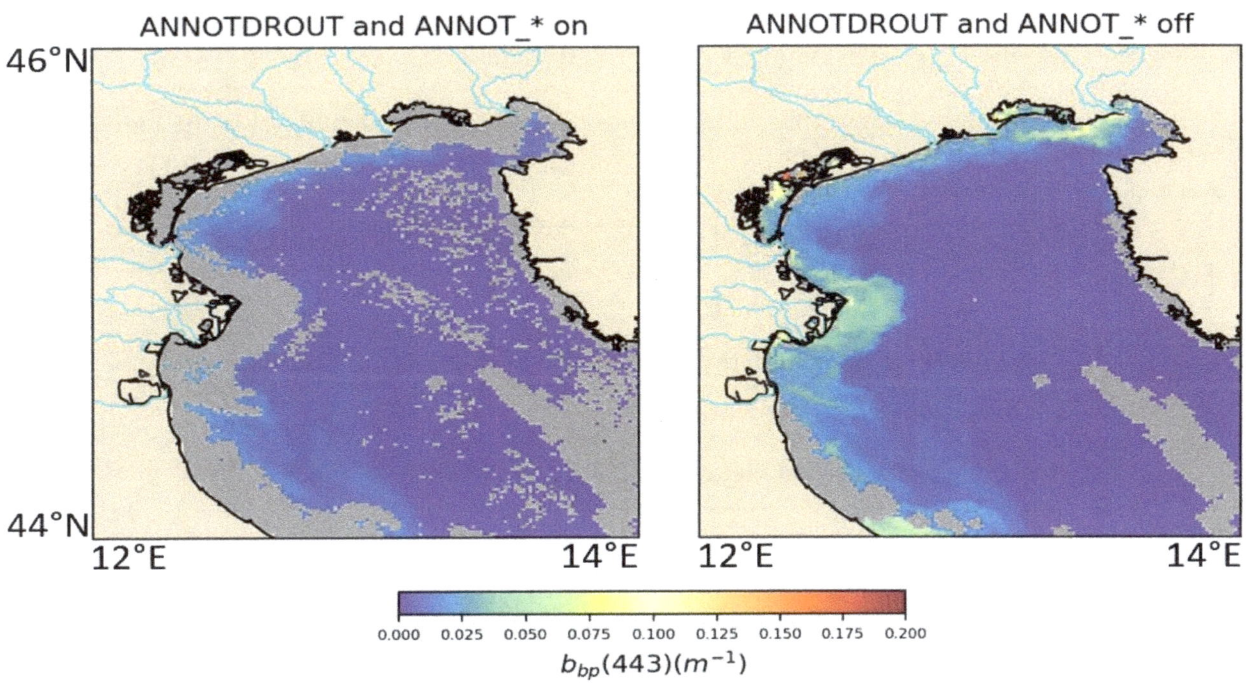

Figure 8. Maps of $b_{bp}(443)$ retrieved from the original OLCI R_{rs} for 21 September 2017 09:40 UTC. On the left, the ANNOT_ * flags are active (ANNOT_* on), while the pixels are unmasked on the right (ANNOT_* off).

2.4. In Situ Data

AAOT, part of the AERONET-OC network [18,19], is located in the NAS, 8 nautical miles south-east of the Venice Lagoon (12.509° E, 45.314° N, the red star in Figure 1). The in situ data here acquired has been used to validate and adjust the satellite R_{rs} spectra of each sensor.

AERONET-OC can be considered as the archetype FRM for the validation of satellite remote-sensing radiometric data over water [20,85]. Hence, to compare the AAOT data with the satellite R_{rs} spectra, the normalized water-leaving radiances here acquired have been divided by the extra-solar irradiance (F0) [86]. Particularly, in this study the Version 3 L2 AAOT data (cloud-screened and quality assured data) have been exploited and integrated with the L1.5 data (only cloud-screened) for periods where the L2 data were not still processed.

The exploitation of these data to adjust the satellite spectra is fundamental for this study; indeed, AAOT is located in an area with a large optical variability that can be considered as representative of the basin variability [19,87]. Consequently, the adjustment coefficients here retrieved can be used to adjust the R_{rs} spectra in the entire basin [6].

2.5. Match-up Analyses

To assess the quality of the satellite R_{rs} spectra, pairwise match-up analyses were performed using the AAOT in situ data as a reference. Those analyses were conducted for every sensor of VGOCS separately, because different sensors can have different sources of uncertainties and, consequently a different agreement with the in situ spectra.

The match-up procedure is consistent with the one of [6], where for each couple of satellite/in situ observations acquired at $\Delta t < 30$ min, a 3×3 box around the AAOT area has been chosen, keeping only boxes with at least 3 valid pixels. The value of the in situ R_{rs} is the mean of all the AAOT data acquired between t (time of the satellite observation)-Δt (30 min) and t+Δt, while the value of the satellite R_{rs} is the mean of the 3x3 box data For each valid couple of observations, the in situ R_{rs} spectrum has been band-shifted to the satellite wavelengths, following [88].

To test the agreement between the satellite and in situ data, different statistical parameters have been calculated and their formulas have been presented in Table 2.

Table 2. Formulations of the statistical parameters used in this study to assess the agreement between two datasets, where n is the number of concurrent observations of the match-up, x the in situ measurement, and y the satellite observation.

Statistical Parameter	Formulation
Determination coefficient (r^2)	$r^2 = \frac{\sum_{i=1}^{n}(x_i-\bar{x})(y_i-\bar{y})}{\sqrt{\sum_{i=1}^{n}(x_i-\bar{x})^2}\sqrt{\sum_{i=1}^{n}(y_i-\bar{y})^2}}$
Mean Absolute Difference (MAD)	$MAD = \frac{\sum_{i=1}^{n}\lvert y_i-x_i\rvert}{n}$
Root Mean Squared Difference (RMSD)	$\mathrm{RMSD} = \sqrt{\frac{\sum_{i=1}^{n}(y_i-x_i)^2}{n}}$
Mean absolute percentage difference (MAPD)	$\mathrm{MAPD}(\%) = \frac{100}{n}\sum_{i=1}^{n}\left\lvert\frac{y_i-x_i}{x_i}\right\rvert$

In the R_{rs} match-up analyses, the entire dataset (ED) of each sensor was divided into two subsets: training (TD) and validation dataset (VD). The first was used to retrieve the coefficients of the adjustment (presented in Section 2.6), while the second one was used to validate the satellite data before and after the adjustment procedure (Section 3). For each N/N sensor (except for VIIRSJ), VD is composed of all the data acquired between 1 September 2015 and 31 October 2019, while the remaining data are part of TD. The split of the dataset is not applied to the OLCI and VIIRSJ data, as they are the youngest sensors between those used in VGOCS and they have a lower number of match up points in comparison with the other sensors. Hence, the splitting may not allow capturing all the optical variability of the basin; consequently, for those two sensors, the data used to retrieve the adjustment coefficients are the same used to validate the dataset (TD=VD=ED).

Match-up analyses have been performed also for data traditionally masked in the standard processing chains. In Section 3.2 match-up analyses for the HSZ, SL, ANNOT_DROUT, and ANNOT_* data have been performed. For the HSZ flag, observation is considered as flagged if the pixel collocated with AAOT is scanned at $\theta v > 60°$; for SL, ANNOT_DROUT, and ANNOT_*, an observation is identified as flagged if the considered flag is active for at least 1/3 of the box valid pixels.

2.6. Satellite R_{rs} Adjustment

To reduce discrepancies and biases between in situ and satellite measurements, to exploit data usually masked in the standard processing chains, and to reduce the inter-sensor differences, a multi-linear regression algorithm (MLR), like the one presented in [17], is exploited to adjust the satellite R_{rs} spectra. As different sensors have different uncertainties, temporal coverages, and agreements with the in situ observations, a different set of adjustment coefficient is calculated for each of them.

Particularly, for each couple of observations of the TD the difference between the in situ and satellite measurement is evaluated for each band:

$$\Delta R_{rs}(\lambda) = R_{rs}^{is}(\lambda) - R_{rs}^{or}(\lambda), \tag{1}$$

where $R_{rs}^{is}(\lambda)$ is the in situ R_{rs} and $R_{rs}^{or}(\lambda)$ is the original satellite R_{rs}. The multilinear regression scheme is

$$\langle \Delta R_{rs}(\lambda) \rangle = a_0^{sat} + \sum_{i=1}^{n} a_i^{sat} R_{rs}^{or}(\lambda_i) + a_6^{sat}\theta_v + a_7^{sat}\theta_s + a_8^{sat}\varphi, \tag{2}$$

The input vectors of the adjustment are the original R_{rs} spectra, and the values of the three viewing geometry parameters (θ_v, θ_s, and φ) corresponding to the pixel collocated with AAOT.

The coefficients a_i^{sat} (i = 0, ... , 8) were calculated performing the MLR scheme between $\Delta R_{rs}(\lambda)$ and the input vectors. Then the adjusted R_{rs} can be retrieved with:

$$R_{rs}^{adj}(\lambda) = R_{rs}^{or}(\lambda) + \langle \Delta R_{rs}(\lambda) \rangle, \tag{3}$$

This set of fixed coefficients have been used to adjust the R_{rs} spectra of each image of the single sensor dataset, as the AAOT in situ data are representative of the optical variability of the basin. Finally, the adjusted R_{rs} have been exploited as the input of the QAA to calculate different IOPs [34,35].

2.7. *Inter-Sensor Differences*

As stated in the previous sections, the adjustment is applied to reduce the R_{rs} inter-sensor differences that are due to different observation geometries, calibration accuracies, data processing, and resolutions of the sensors. To quantify such effect r^2 and MAPD have been calculated in different locations of the NAS for images acquired in temporal proximity by different sensors ($\Delta t < 10$ min). This has been done starting from the hypothesis that in a time range of 10 min no considerable oceanographic processes can occur and that consequently most of the differences observed between two images are likely due to artifacts. The chosen locations are here named virtual buoys (red dots in Figure 1) and they are situated:

- At AAOT, to analyze the effect of the adjustment in an area with large optical variability.
- Close to the Livenza, Brenta-Adige, Piave, Tagliamento, and Isonzo river mouths to analyze the effect of the adjustment in optically complex waters.
- In an off-shore location (here named simply OPEN), to analyze the effect of the adjustment in open waters.
- Close to the Po river mouth to analyze the effect of the adjustment in the area where the largest amount of fresh-water is usually available [45,46], due to the high river discharges.

Hence, the last two virtual buoys represent the two optical extreme situations for open and optically complex waters respectively.

As stated earlier, to calculate r^2 and MAPD only couples of images acquired at $\Delta t < 10$ min have been selected and for each virtual buoy a 3×3 box around its location has been chosen. Then only the couples of images with both boxes with at least 3 valid R_{rs} pixels have been maintained, and r^2 and MAPD have been calculated for each band of the R_{rs} spectrum, similar to a match-up analysis. As the sensors have a different spectral resolution, in this analysis all the R_{rs} spectra have been band-shifted to the MODIS bands, using the approach from [88].

3. Results

3.1. Match-Up Analyses

In this section, the agreement between the satellite and in situ R_{rs} spectra before and after the application of the adjustment will be evaluated. In Table 3 the numbers of match-up points for each VGOCS sensor are shown.

Table 3. The number of match-up points for each VGOCS sensor. Between parenthesis, for the N/N sensors the number of match-up points for the ED SL affected data, for OLCI those flagged by the ANNOT_DROUT/ANNOT_* flags.

Sensor	Number of Match-Up Points
TERRA	218 (178)
AQUA	357 (230)
VIIRSN	449 (200)
VIIRSJ	216 (48)
OLCI	64 (37/20)

Previous studies demonstrated how the MLR adjustment strongly increases the agreement between the in situ and satellite spectra [6,17]. In this work the effect of the adjustment on all the VGOCS sensors will be evaluated but, for the safe of brevity, the scatterplots will be presented only for OLCI. For the N/N sensors, only the spectra of r^2 and MAPD before and after the application of adjustment are shown (Figures 9 and 10) and the other statistical parameters (RMSD and MAD) are presented from Tables 4–7).

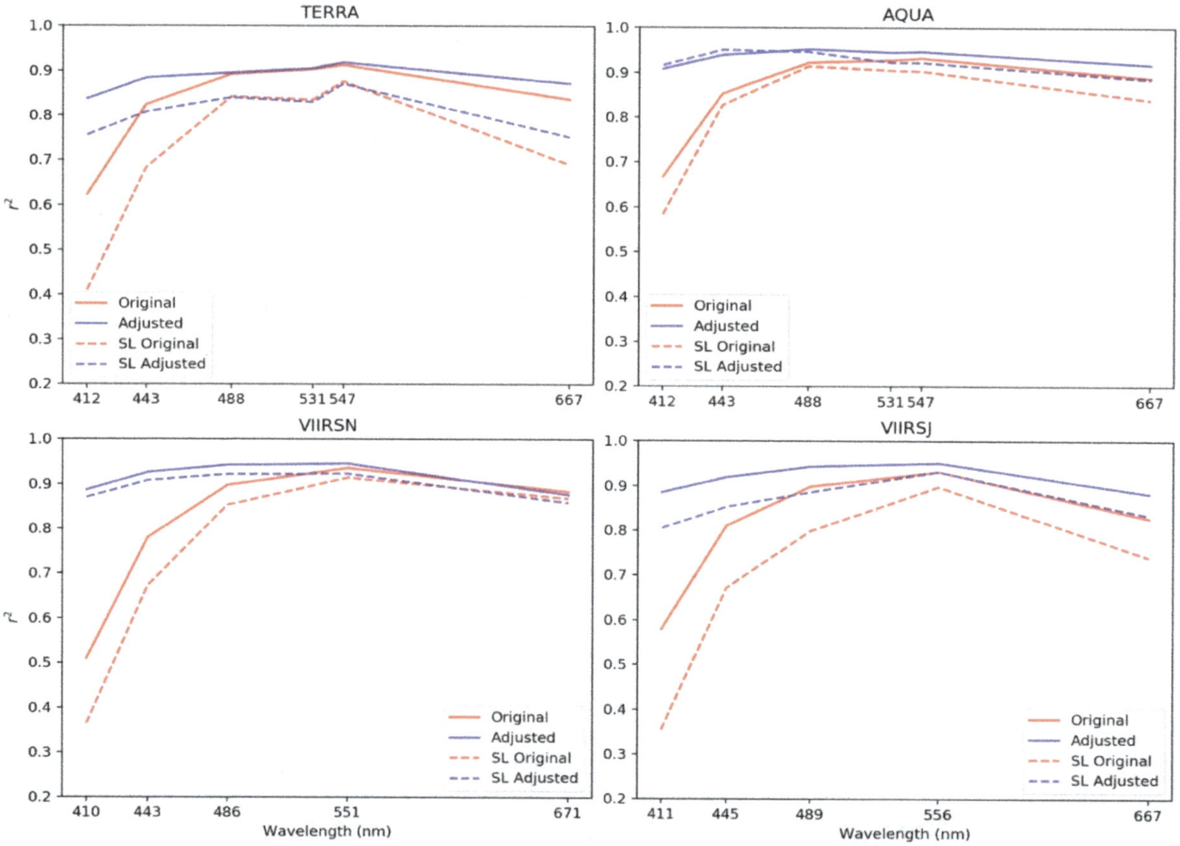

Figure 9. Spectra of r^2 before (red) and after (blue) the adjustment procedure for each N/N sensor. The continuous lines identified the results for VD, the dotted ones for the ED SL affected data.

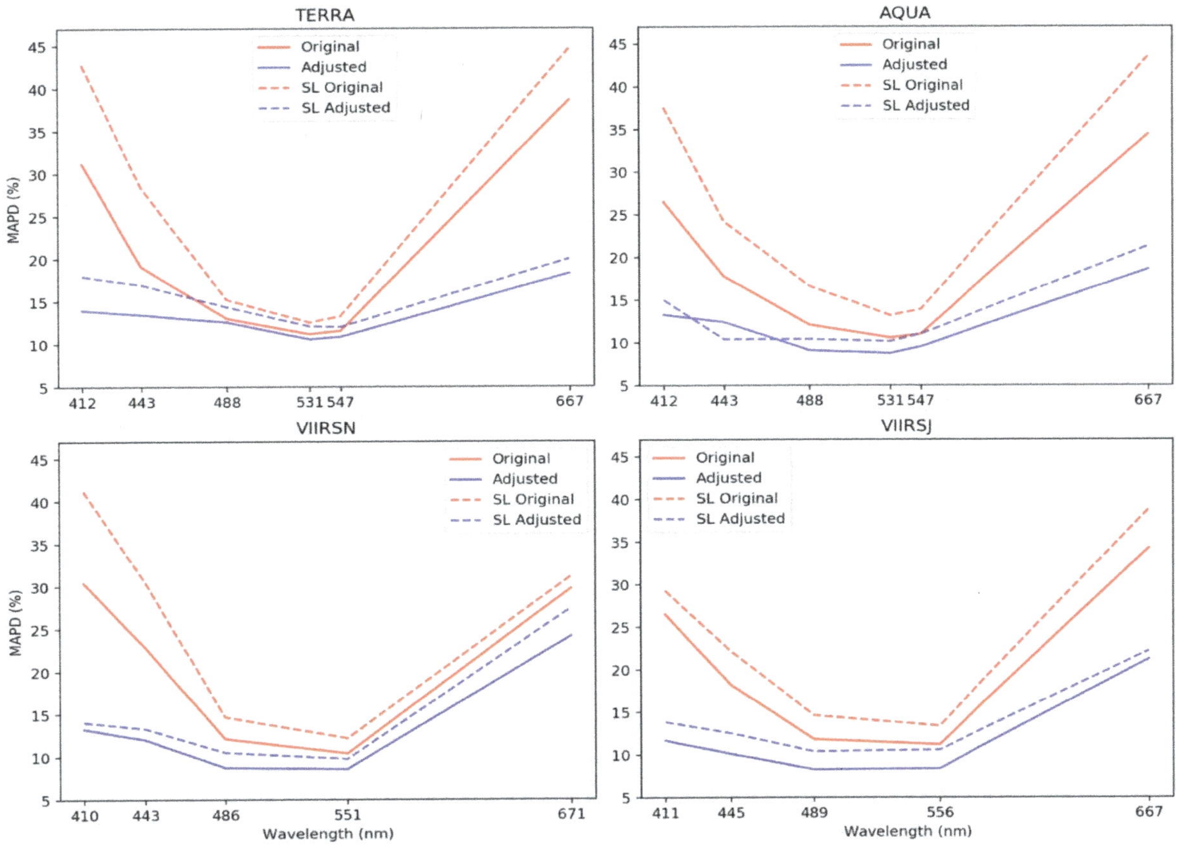

Figure 10. Spectra of MAPD before (red) and after (blue) the adjustment procedure for each N/N sensor. The continuous lines identified the results for VD, the dotted ones for the ED SL affected data.

Table 4. Statistical parameters for the match-up analysis between the AAOT in situ R_{rs} and the original (adjusted) VD TERRA R_{rs}.

TERRA Bands	r^2	MAD (sr^{-1})	RMSD (sr^{-1})	MAPD (%)
$R_{rs}(412)$	0.62 (0.84)	1.3×10^{-3} (6.6×10^{-4})	1.6×10^{-3} (1.0×10^{-3})	31.2 (14.0)
$R_{rs}(443)$	0.82 (0.89)	9.1×10^{-4} (6.3×10^{-4})	1.2×10^{-3} (9.0×10^{-4})	19.1 (13.5)
$R_{rs}(488)$	0.90 (0.90)	9.2×10^{-4} (8.1×10^{-4})	1.3×10^{-3} (1.2×10^{-3})	13.1 (12.6)
$R_{rs}(531)$	0.90 (0.90)	8.5×10^{-4} (7.6×10^{-4})	1.3×10^{-3} (1.2×10^{-3})	11.2 (10.6)
$R_{rs}(547)$	0.91 (0.92)	8.4×10^{-4} (7.6×10^{-4})	1.2×10^{-3} (1.2×10^{-3})	11.6 (10.9)
$R_{rs}(667)$	0.83 (0.87)	4.1×10^{-4} (2.3×10^{-4})	5.2×10^{-4} (3.5×10^{-4})	38.6 (18.3)

Table 5. Statistical parameters for the match-up analysis between the AAOT in situ R_{rs} and the original (adjusted) VD AQUA R_{rs}.

AQUA Bands	r^2	MAD (sr^{-1})	RMSD (sr^{-1})	MAPD (%)
$R_{rs}(412)$	0.67 (0.91)	1.1×10^{-3} (6.1×10^{-4})	1.6×10^{-3} (7.9×10^{-4})	26.5 (13.3)
$R_{rs}(443)$	0.85 (0.94)	8.7×10^{-4} (5.7×10^{-4})	1.2×10^{-3} (7.4×10^{-4})	17.8 (12.5)
$R_{rs}(488)$	0.92 (0.95)	8.8×10^{-4} (6.4×10^{-4})	1.2×10^{-3} (9.1×10^{-4})	12.1 (9.1)
$R_{rs}(531)$	0.93 (0.95)	8.1×10^{-4} (6.6×10^{-4})	1.2×10^{-3} (9.9×10^{-4})	10.6 (8.8)
$R_{rs}(547)$	0.93 (0.95)	8.0×10^{-4} (6.9×10^{-4})	1.1×10^{-3} (1.0×10^{-3})	11.0 (9.6)
$R_{rs}(667)$	0.89 (0.92)	3.5×10^{-4} (2.1×10^{-4})	4.4×10^{-4} (2.9×10^{-4})	34.5 (18.7)

Table 6. Statistical parameters for the match-up analysis between the AAOT in situ R_{rs} and the original (adjusted) VD VIIRSN R_{rs}.

VIIRSN Bands	r^2	MAD (sr^{-1})	RMSD (sr^{-1})	MAPD (%)
$R_{rs}(410)$	0.51 (0.89)	1.5×10^{-3} (6.6×10^{-4})	2.1×10^{-3} (8.8×10^{-4})	30.5 (13.3)
$R_{rs}(443)$	0.78 (0.93)	1.2×10^{-3} (6.0×10^{-4})	1.7×10^{-3} (8.3×10^{-4})	22.9 (12.1)
$R_{rs}(486)$	0.90 (0.94)	9.2×10^{-4} (6.6×10^{-4})	1.3×10^{-3} (1.0×10^{-3})	12.2 (8.8)
$R_{rs}(551)$	0.94 (0.95)	7.7×10^{-4} (6.1×10^{-4})	1.1×10^{-3} (9.7×10^{-4})	10.5 (8.6)
$R_{rs}(671)$	0.88 (0.88)	3.1×10^{-4} (2.6×10^{-4})	4.1×10^{-4} (3.7×10^{-4})	29.8 (24.2)

Table 7. Statistical parameters for the match-up analysis between the AAOT in situ R_{rs} and the original (adjusted) VD VIIRSJ R_{rs}.

VIIRSJ Bands	r^2	MAD (sr^{-1})	RMSD (sr^{-1})	MAPD (%)
$R_{rs}(411)$	0.58 (0.89)	1.4×10^{-3} (6.0×10^{-4})	1.9×10^{-3} (7.9×10^{-4})	26.6 (11.7)
$R_{rs}(445)$	0.81 (0.92)	1.0×10^{-3} (6.0×10^{-4})	1.3×10^{-3} (8.2×10^{-4})	18.3 (10.2)
$R_{rs}(489)$	0.90 (0.94)	9.5×10^{-4} (6.4×10^{-4})	1.3×10^{-3} (9.0×10^{-4})	11.9 (8.3)
$R_{rs}(556)$	0.93 (0.95)	7.9×10^{-4} (5.6×10^{-4})	1.1×10^{-3} (8.0×10^{-4})	11.3 (8.4)
$R_{rs}(667)$	0.83 (0.88)	3.3×10^{-4} (1.9×10^{-4})	4.2×10^{-4} (2.7×10^{-4})	34.3 (21.3)

After the adjustment MAPD is spectrally reduced for each N/N sensor and r^2 is spectrally increased except for the 671 nm band for VIIRSN (nearly constant, Figures 9 and 10). Particularly, r^2 is now close to or larger than 0.90 for each band of each sensor, with just the exception of the first TERRA band for which r^2 is 0.84. For MAPD, a lower decrease is observed in the green part of the spectrum, since the green bands were already in good agreement with the in situ observations before the adjustment. The blue and red bands show the larger improvement: for the first original blue band, MAPDs were larger than 25% for all sensors, while after the adjustment they are lower than 15%; for the red band MAPDs were between 25.0% and 30.0%, and after the adjustment, these decreased to values that go from 18.3% (TERRA) to 24.2% (VIIRSN). Hence, after the adjustment, the agreement between the satellite and in situ spectra is strongly improved, and this is confirmed also by the other statistical parameters (from Tables 4–7) that are spectrally reduced.

The agreement with the in situ spectra differs from one sensor to another. Particularly, TERRA is the one that shows the larger discrepancies with the in situ data both before and after the adjustment procedure. This can be due to the serious degradation of the TERRA mirror; indeed, for this sensor, not even the vicarious calibration has led to the retrieval of reliable OC products, and a calibration based on other functioning OC sensor observations is applied to the TERRA data [3,89]. This degradation can also explain the lower number of valid observations and match-up points for TERRA (218), in comparison with AQUA (357) and VIIRSN (449). Despite that, the adjustment has strongly improved the quality of the TERRA R_{rs} spectra, enabling a more reliable usage of the data from this sensor.

Some differences in the statistics for different sensors can also be partly due to the different spectral resolutions; indeed, in the band shifting procedure the spectral distances between the native in situ (AAOT) and target bands (satellite) are different for different sensors, and larger spectral distances can increase the uncertainty of the band-shifting [88]. This validates the choice to keep the spectral resolution of each VGOCS sensor and to do not band-shift all of them to a set of predefined wavelengths. Indeed, the AAOT central wavelengths varied during the years, and choosing a set of predefined wavelengths for the satellite sensors could lead to the need to band-shift both in situ and satellite data to this set, increasing the uncertainty.

In Figures 11 and 12, the scatterplots for the original and adjusted OLCI R_{rs} in comparison with the in situ data are shown, with the statistical parameters spectra shown in Figure 13 and listed in Table 8. The number of match-up points is only 67 due to the short OLCI time series, the larger revisiting time of Sentinel 3A in comparison with the other VGOCS satellites, and the lack of AAOT data from the end of June to the beginning of October of 2017. The uncertainty observed for $R_{rs}(412)$ is larger than those observed for the other OC sensors of the constellation. Despite that, the adjustment largely reduces this uncertainty, showing similar results to those of the previous analyses (Figure 13). Indeed, after the adjustment r^2 is 0.89 for the red band and larger than 0.9 for the other ones, while MAPD is ≈23% for the red band, 10.3% for the 412 nm band, and lower than 10.4% for the other ones.

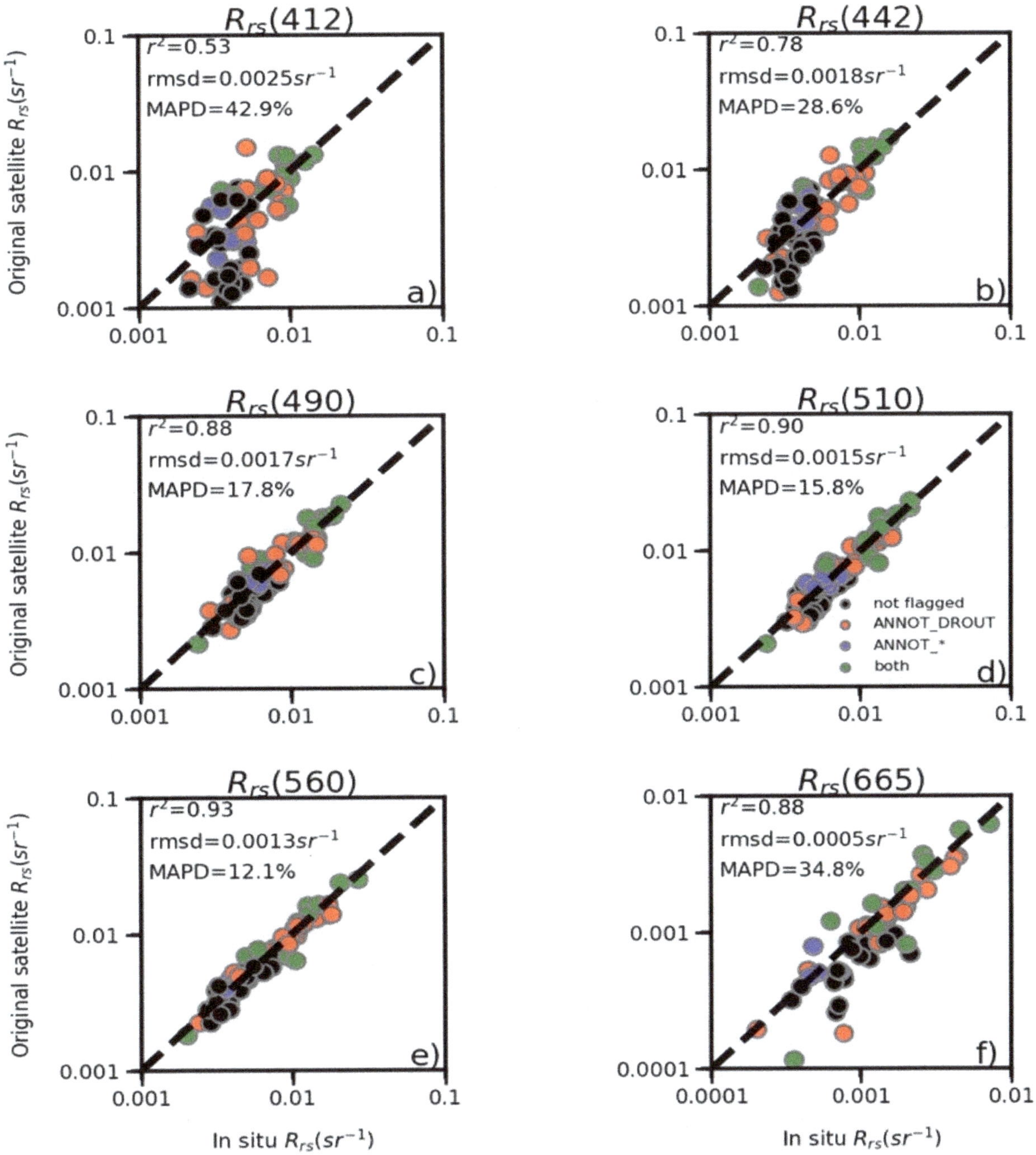

Figure 11. Match-up analysis between the in situ R_{rs} and the original VD OLCI R_{rs}. In the logarithmic scale on the x-axes the in situ R_{rs} and on the y-axes the original OLCI R_{rs} for each band (**a**–**f**). The red points are the observations where the ANNOT_DROUT flag is active, the blue points those where the ANNOT_* flags are active, the green ones those where both flags are active, and the black points where none of those flags is active.

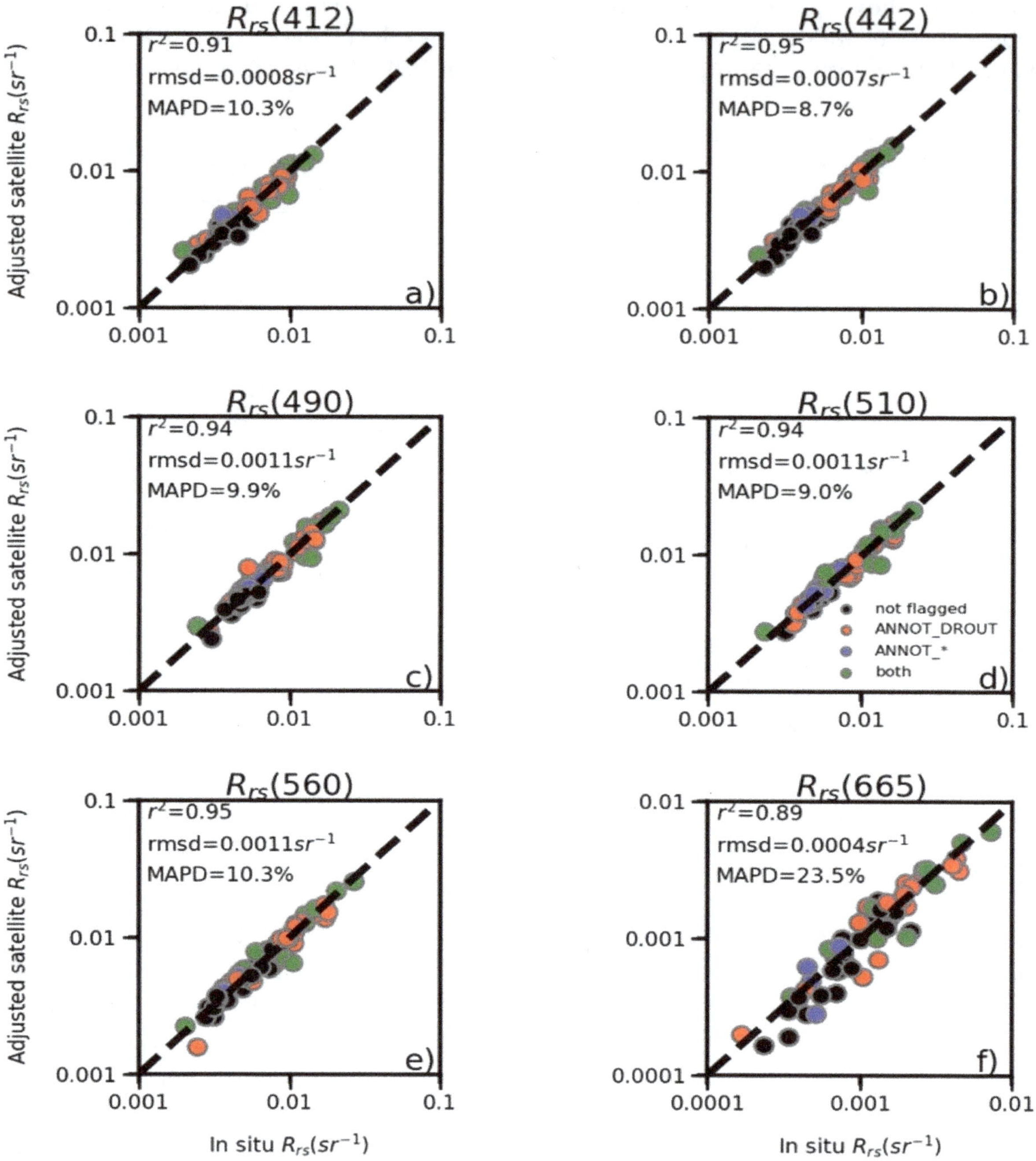

Figure 12. Match-up analysis between the in situ R_{rs} and the adjusted VD OLCI R_{rs}. In the logarithmic scale on the x-axes the in situ R_{rs} and on the y-axes the adjusted OLCI R_{rs} for each band (**a**–**f**). The red points are the observations where the ANNOT_DROUT flag is active, the blue points those where the ANNOT_* flags are active, the green ones those where both flags are active, and the black points where none of those flags is active.

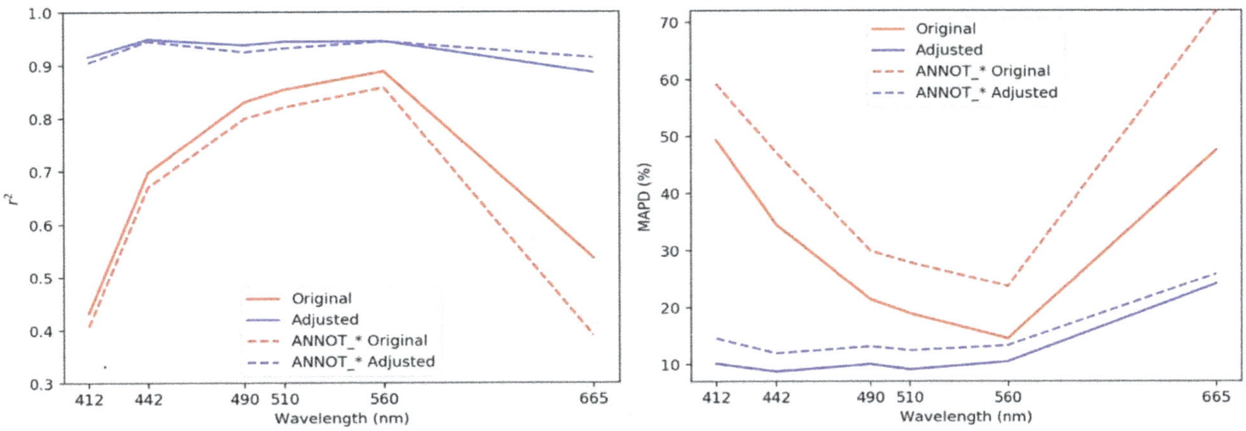

Figure 13. Spectra of r^2 (left) and MAPD (right) before (red) and after (blue) the adjustment procedure for OLCI. The continuous lines identified the results for VD, the dotted ones for the ANNOT_* data.

Table 8. Statistical parameters for the match-up analysis between the AAOT in situ R_{rs} and the original (adjusted) VD OLCI R_{rs}.

OLCI Bands	r^2	MAD (sr^{-1})	RMSD (sr^{-1})	MAPD (%)
$R_{rs}(412)$	0.43 (0.92)	2.1×10^{-3} (5.6×10^{-4})	2.8×10^{-3} (7.9×10^{-4})	49.4 (13.3)
$R_{rs}(442)$	0.70 (0.95)	1.6×10^{-3} (5.0×10^{-4})	2.2×10^{-3} (7.4×10^{-4})	34.6 (12.5)
$R_{rs}(490)$	0.83 (0.94)	1.4×10^{-3} (7.7×10^{-4})	1.9×10^{-3} (9.1×10^{-4})	21.4 (9.1)
$R_{rs}(510)$	0.85 (0.95)	1.3×10^{-3} (7.2×10^{-4})	1.8×10^{-3} (9.9×10^{-4})	19.0 (8.8)
$R_{rs}(560)$	0.89 (0.95)	1.0×10^{-3} (7.7×10^{-4})	1.6×10^{-3} (1.0×10^{-3})	14.6 (9.6)
$R_{rs}(665)$	0.54 (0.89)	4.9×10^{-4} (3.1×10^{-4})	1.0×10^{-3} (2.9×10^{-4})	47.6 (18.7)

The better agreement between the satellite and in situ spectra strongly increases also the quality of the IOPs retrieved by the QAA [6]. Indeed, the quality of the QAA outputs is dependent on the input R_{rs} spectrum uncertainty; consequently, the reduction of the discrepancies between the satellite and in situ R_{rs} bring to a better agreement between the satellite and in situ QAA parameters [6].

3.2. *VGCOS Spatial and Temporal Coverage*

As it is shown in Section 2.3, the unmasking of the HSZ, SL, ANNOT_DROUT, ANNOT_* data leads to an increase of the VGOCS spatial and temporal coverage. In this section, the agreement of those data with the in situ spectra will be analyzed, to evaluate if they can be unmasked in the VGOCS dataset.

3.2.1. HSZ Flag

The quality of the OC data can be dependent on the viewing geometry and, particularly, on θ_v. The HSZ flag masks data acquired at θ_v larger than 60° since they can be affected by larger uncertainties in the atmospheric correction and the normalization of the water leaving radiance [69,70]. Nevertheless, also data acquired at large θ_v, but lower than 60° (for example $50° < \theta v < 60°$) can be affected by larger discrepancies in comparison with in situ or other satellite data [6,38].

To evaluate this θ_v dependence for the N/N sensors, MAPD has been calculated for different ranges of θ_v with a step of 10° (Figure 14) for the match-up analyses of the previous section. Before the adjustment, the VIIRSN and VIIRSJ R_{rs} with $\theta_v > 50°$ show larger MAPDs in comparison of those acquired at $\theta_v < 50°$, especially for the 411/2 nm band and the 667/671 ones. After the adjustment MAPD is lower in each bin for each sensor, reducing a general bias with the in situ data, confirming the aggregated results presented in Section 3.1. MAPDs for data acquired at $\theta_v > 50°$ are still slightly larger than those acquired at lower θ_v, but the dependence is now consistently reduced. For example, for the original VIIRSJ 411 nm band MAPD is between 18.1% and 28.5% for the first five bins, while it is larger than 34% for the last two; after the adjustment, MAPD is between 10.4 and 13% for $\theta v < 50°$ and it is <14.0% for $\theta v > 50°$, confirming the reduced θv dependence.

AQUA and TERRA show similar behavior of the one observed for VIIRSJ/N, but for them also the original data acquired at $40° < \theta_V < 50°$ show larger discrepancies with the in situ data, but they are strongly reduced with the adjustment. This larger θ_V dependence for AQUA and TERRA can be due to the pixel growth effect [90,91], which can be more evident in the MODIS sensors, as this is reduced in VIIRS by its aggregation algorithm [91]. The main result of this analysis is that now most of the observations acquired at $\theta_V > 50°$ have MAPD lower of those acquired at $\theta_V < 50°$ before the adjustment.

Consequently, data previously masked by the HSZ flag ($\theta_V > 60°$) can now be exploited, making available a large number of data, usually discarded in standard applications.

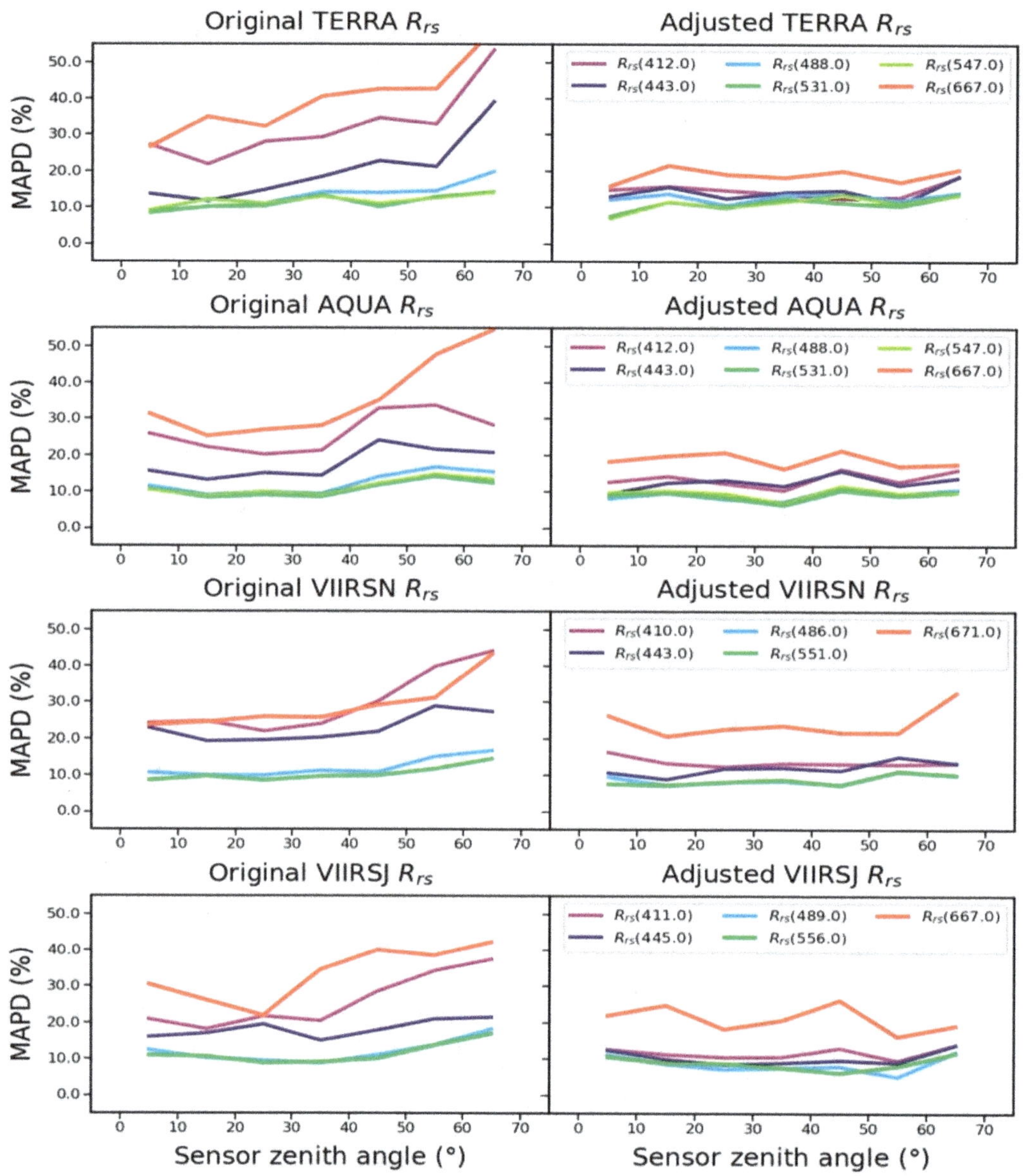

(a) (b)

Figure 14. Analysis of the MAPD dependence from θ_V before (left, column **a**) and after the adjustment (right, column **b**). On the x-axes the wavelength, on the y-axes MAPD, with the different colours of the lines identifying the different R_{rs} bands.

As already shown in Figures 4 and 5, unmasking for the HSZ flag provides up to 3 additional full images a day of the NAS, strongly increasing the VGOCS spatial and temporal coverage, which is fundamental for the aim for which VGOCS is conceived.

3.2.2. SL Flag

The SL affected pixels can have large sources of uncertainties [6,71–74], and, before using them, it is important to test their agreement with the in situ observations. For this reason, match-up analyses for the original and adjusted R_{rs} of the N/N sensors have been performed, considering only the SL affected data (spectra of r^2 and MAPD in Figures 9 and 10). For those analyses, for each sensor ED (TD+VD) has been used, to have a larger number of match-up points.

For the original SL affected data (red dotted line), the values of r^2 and MAPD are considerably lower and larger respectively than the values observed for VD (red continuous line), confirming that the stray-light effect can significantly lower the quality of the OC data. Nevertheless, after the adjustment r^2 and MAPD are considerably closer to 1 and 0 respectively. The behavior of each sensor is similar to the one observed in Section 3.1, with TERRA that still shows larger discrepancies in comparison with the in situ data, due to the degradation of its mirror [3,89]. There are still some residuals due to the stray-light effect, as the MLR algorithm can only minimize and not eliminate the larger uncertainties introduced by this effect. Nevertheless, the values of the statistical parameters (blue dotted lines) are now closer to those observed for VD after the adjustment (blue continuous line). Moreover, the values of MAPD and r^2 of the adjusted SL data are respectively lower and larger than those observed for the original R_{rs} in VD.

Hence, as for the HSZ flag (Section 3.2.1), the adjustment makes available such data, allowing to analyze the coastal optical variability and further improving the VGOCS spatial coverage close to shore (Section 2.3.1).

3.2.3. ANNOT_DROUT and ANNOT_* Flags

The ANNOT_DROUT and the ANNOT_* flags have not been applied in the OLCI processing chain, as they mask a large number of pixels close to the coast (Section 2.3.2).

As stated earlier, recently EUMETSAT has removed the ANNOT_DROUT flag from the standard L2 flags (I.Cazzaniga, pers.comm.) [83]. The validity of this choice is confirmed by the match-up analyses of Figures 11 and 12, where the comparison between the in situ and original/adjusted OLCI R_{rs} is presented. In these figures, the green and red data are those for which the ANNOT_DROUT flag is active. Not considering them, the number of match-up points is drastically reduced, as only 30 observations are now available in comparison with the 67 of Section 3.1. Masking for ANNOT_DROUT data, MAPD is larger and r^2 and RMSD lower (not shown) for all the bands in comparison with the match-up analyses accomplished considering the entire OLCI VD (Section 3.1). The RMSD reduction is not due to a better agreement of this dataset with the in situ data, but simply to the order of magnitude of the masked data. Indeed, only one observation with $R_{rs}(412) > 0.01\ sr^{-1}$ is not masked by ANNOT_DROUT (blue and black points, Figure 11), leading to a lower value of RMSD, which is an absolute difference, and it is dependent from the order of magnitude of the observations, while MAPD is a relative difference. Hence, the application of the ANNOT_DROUT flag does not allow capturing all the optical variability of the basin. For those reasons, this should not be applied in optically complex waters, as previously stated in [82].

In the OLCI match-up analysis, all the data flagged as ANNOT_* (blue and green points in Figures 11 and 12, with the r^2 and MAPD spectra in Figure 13) are due to the ANNOT_MIXR1 flag, as this is the only one which usually results to be applied in coastal waters (Section 2.3.2). The ANNOT_MIXR1 flag data show larger MAPDs (red dotted line) in comparison with those of the entire dataset (red continuous line), but those are strongly reduced after the adjustment (blue dotted line). Still slightly larger discrepancies are shown in comparison with those of the adjusted VD R_{rs}

(blue continuous line), but their agreement with in situ data is better than the one of the original R_{rs} of VD (red continuous line), similarly to the SL data.

Hence, the adjusted ANNOT_MIXR1 data should not be masked in the NAS coastal waters, for the OLCI processing baseline v2.23. On the contrary, the ANNOT_TAU6, and ANNOT_ABSO_D data have been finally masked in the VGOCS dataset. Hence, they are not usually applied in coastal waters and the use of such data could result in large uncertainties in open waters, as those data have not been characterized by a match-up analysis.

3.3. OLCI Camera Dependence

As stated earlier, the quality of the OLCI R_{rs} spectra can be dependent on the different cameras mounted on the optical bench. Indeed, the presence of different camera implies that different ranges of θ_v correspond to different detectors; consequently, a plot similar to Figure 14 is presented in Figure 15 but with MAPD calculated for the different camera observations (C1: Camera1, C2: Camera2 ...).

To distinguish between the different cameras, the detector_index field provided in the OLCI L2 files is exploited. Indeed, this field goes from 0 to 3699 (that are the total number of columns of a single L1 file) and each bin of 740 columns corresponds to a different camera (e.g., from 0 to 739 Camera1, from 740 to 1479 Camera 2 and so on ...).

The observations from each camera (from C1 to C5) are 17,19,17,8, and 6 respectively. Such numbers of observations are too low to characterize OLCI camera dependence. Nevertheless, using the available match-up points it is possible to observe that the value of MAPDs for the original R_{rs} are consistently different for different cameras. Those differences are strongly reduced after the adjustment, with MAPD showing similar values for different cameras. Hence, also if it is not possible to analyze the OLCI camera dependence, the observed reduction of this effect brought by our adjustment is sufficient for the aim of our study.

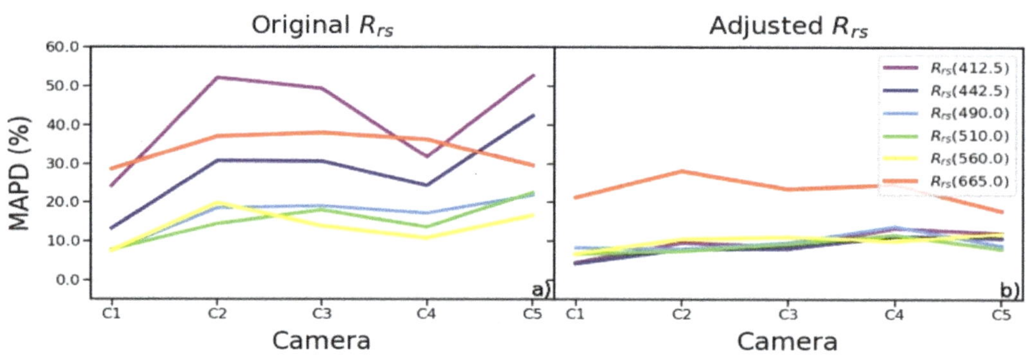

Figure 15. Analysis of the dependence of MAPD from the different OLCI cameras for the (**a**) original and (**b**) adjusted R_{rs}. On the x-axes the camera (from C1 to C5), on the y-axes MAPD, with the different colours of the lines identifying the different R_{rs} bands.

3.4. Inter-Sensor Differences

As described in Section 2.7, to qualify the residual inter-sensor differences, MAPD and r^2 have been calculated at the virtual buoys for couples of images acquired in temporal proximity.

In Figures 16 and 17, the spectra of r^2 and MAPD at the virtual buoys are presented for the original and adjusted R_{rs}, band-shifted to the MODIS bands. r^2 is everywhere increased for the two blue bands, while for the other bands it remains almost constant (or slightly reduced), except for the red band at OPEN, where this parameter is reduced. For what concerns MAPD, this is reduced for each band in each location, with a larger decrease observed for the two blue bands and the red band, and a slight increase in the green part of the spectrum at OPEN.

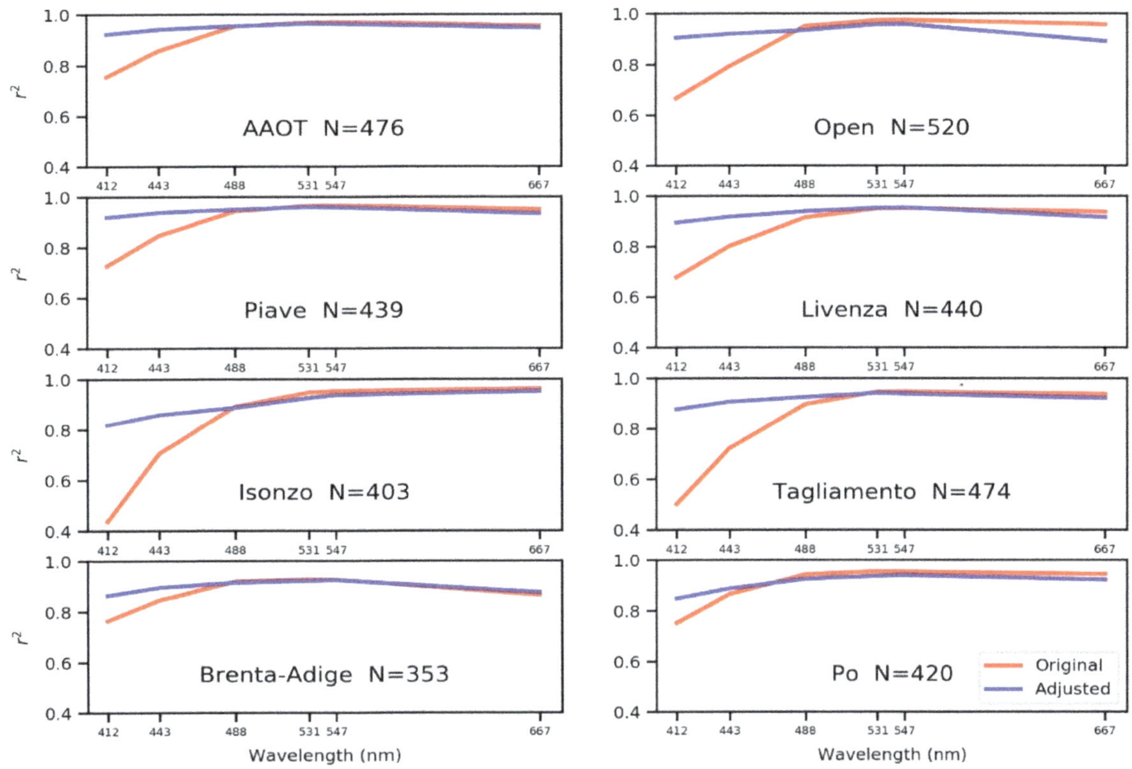

Figure 16. The r^2 calculated at the virtual buoys for the original (red) and adjusted R_{rs} (blue) for images acquired in temporal proximity by the different VGOCS sensors. Inside the box the virtual buoy name with the number of the couple of images used in the calculation of r^2.

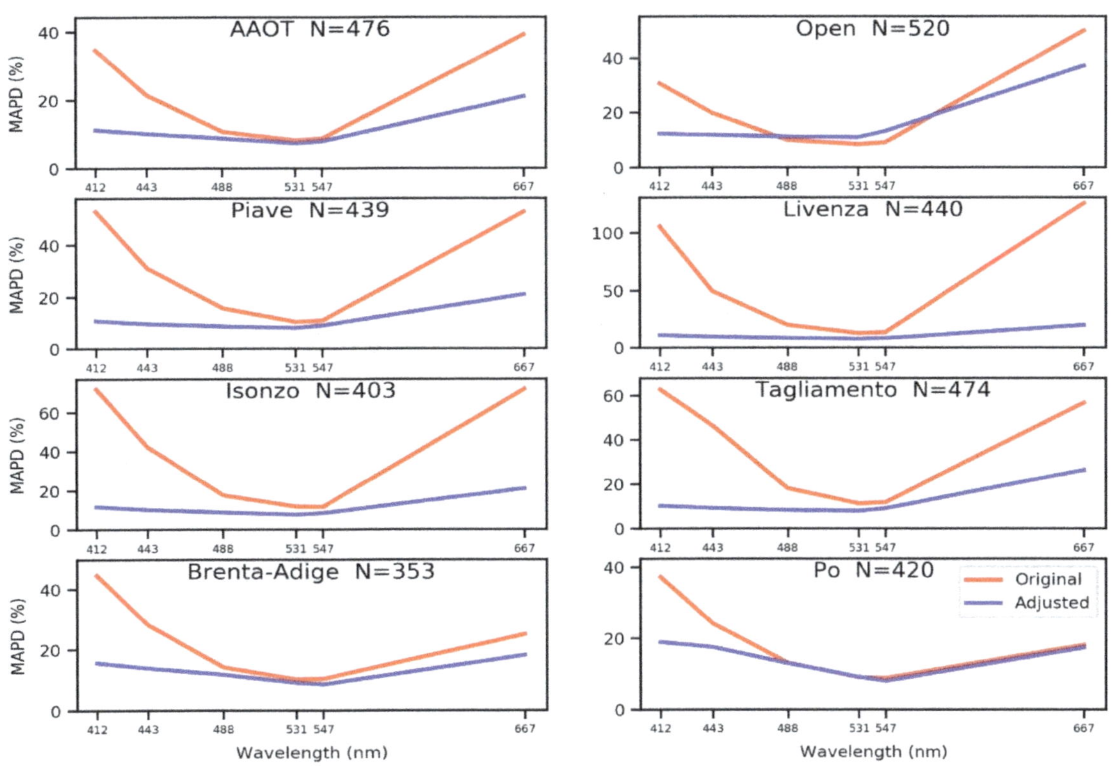

Figure 17. Mean absolute percentage difference (MAPD) calculated at the virtual buoys for the original (red) and adjusted R_{rs} (blue) for images acquired in temporal proximity by the different VGOCS sensors. Inside the box the virtual buoy name with the number of the couple of images used in the calculation of MAPD.

The lower reduction/increase of MAPD/ r^2 in the 488–547 nm range is due to the good agreement of these bands also for the original R_{rs}; thus, the adjustment does not influence too much the R_{rs} retrieval for them, as the original bands are already in good agreement with in situ observations.

The MAPD decrease observed at AAOT in Figure 17 after the adjustment is consistent with the behavior observed for the match-up analyses of Section 3.1, performed used the AAOT in situ data as a reference. For the original sensor match-up analyses, MAPD for the first blue band was between 26.5% and 42.9%, while for the second blue band this was between 17.8% to 28.6%; after the adjustment, those ranges decreased to 10.3–14.0% and 8.7–13.5% respectively. Similarly, for AAOT in the inter-sensor analysis, the values of MAPD decreased from 34.8% and 21.4% to 11.2% and 10.1% for the 412 and 443 nm bands respectively. Moreover, after the adjustment, the MAPD values for the green bands are slightly lower than 10.0%, while the one for the red band is around 20.0%, similarly to the results of Section 3.1.

At the other virtual buoys, the MAPD spectra after the adjustment are similar to the one observed for AAOT, while for the original R_{rs} very different spectra were observed for each of them. Indeed, for the original R_{rs}, the ranges of MAPD for the blue bands were between 30.7% (OPEN) and 105.8% (Livenza) and between 19.9% (OPEN) and 49.6% (Livenza). After the adjustment, for the first band, MAPD is between 10.4% (Tagliamento) and 19.0% (Po), while for the second one it is between 9.4% (Tagliamento) and 17.4% (Po). Not considering Po and Brenta-Adige, the superior limits of the two ranges reduce to 12.4% and 11.9% (OPEN), which are values similar to those observed for AAOT in both match-up and inter-sensor analysis. Similarly, for the green bands, all values of MAPD are around 10.0%, while for the red band MAPD is always around 20.0%, except for OPEN, where an MAPD of 37.3% is observed.

Different values of MAPD are observed for the blue bands at Po/Brenta-Adige and the red band at OPEN in comparison with the other virtual buoys since such locations represent the extreme situation for optically complex and open waters respectively, and consequently, the adjustment can be less effective for such type of waters at certain wavelengths.

Similar values of the statistical parameters for the match-up and inter-sensor analysis imply that the residual differences now observed between different satellite images are mostly due to the uncertainty in the atmospheric correction and the absolute calibration of the space sensors; consequently, those that are due to the different technical characteristics, resolutions, viewing geometry, and processing chains are notably reduced. Moreover, these findings confirm the underlying assumption of this study, i.e., that the AAOT in situ data can be exploited to adjust the R_{rs} spectra of the entire basin [6,87].

4. Discussion

This study tested the effect of an adjustment on the quality of the satellite data, the VGOCS spatial and temporal coverage, and the inter-sensor differences to demonstrate the suitability of VGOCS in analyzing the coastal optical variability.

Several match-up analyses between the original R_{rs} of different satellite sensors and the AAOT R_{rs} data were performed. The adjustment brought to a significant improvement of the satellite R_{rs} retrieval with a large spectral increase of r^2 and a spectral reduction of the other statistical parameters for all the VGOCS sensors. The better agreement between the satellite and in situ spectra strongly increases also the quality of the IOPs retrieved by the QAA, which are needed to analyze the coastal optical variability [6].

The adjustment was exploited also to make available data generally masked in the standard processing chains. Indeed, data acquired at $\theta v > 60°$ are usually masked by the application of the HSZ flag, due to larger uncertainties [38]. This is confirmed by the analysis of the dependence of the quality of the original R_{rs} spectra on θ_v; indeed, for the original R_{rs} larger discrepancies were observed for high θ_v, confirming the results of previous works [6–38]. Nevertheless, the adjustment reduced the biases between the satellite and in situ data for all ranges of θ_v, especially for those acquired at $\theta_v > 50°$, leading to a strong reduction of the θ_V dependence. This enabled to use also data usually masked by the HSZ flag, improving the VGOCS temporal and spatial coverage.

The VGOCS spatial coverage is further increased by the use of data usually masked by the SL flag for the N/N sensors and by the ANNOT_DROUT and ANNOT_MIXR1 flags for OLCI.

The adjustment strongly reduces the uncertainty for the SL affected pixels and it allows to have more valid pixels close to the shore, as the application of this flag leads to the masking of the 40% of the coastal pixels. In the future, to additionally reduce the SL data uncertainty, the calculation of a different set of adjustment coefficient could be considered, using in the TD only SL data.

The analysis of the ANNOT_DROUT data confirms the findings of [82] and the validity of the choice by EUMETSAT to remove this flag from the list of the standard L2 flags (I. Cazzaniga, pers.comm.) [83], as they result to be in good agreement with in situ data in the coastal area and they allow to analyze the coastal optical variability.

The behavior of the ANNOT_MIXR1 data, which are 20.1% of the OLCI coastal pixels, is similar to the SL ones, thus they have been unmasked. On the contrary, as it is not possible to characterize the ANNOT_TAU6 and ANNOT_ABSO_D data, they have been masked in our dataset, as they are not encountered in the coastal area and they can show large uncertainties in open waters.

The VGOCS spatial and temporal coverage will be further increased in the future. Hence, in the dataset, only the sensors that are currently functioning were included, as the present days are those with a larger number of sensors on orbit. Nevertheless, in the future also other sensors, that are not anymore functioning, such as the Sea-Viewing Wide Field-of-View Sensor (SeaWiFS) mounted on the satellite SeaStar and MERIS mounted on the Envisat satellite, will be included. Moreover, when the vicarious calibration process will be accomplished, also the OLCI Sentinel 3B sensor will be added to the constellation.

The analysis of the θ_V dependence was not performed for OLCI as different ranges of θ_V are scanned by different cameras. The low number of match-up points does not allow to deeply analyze the camera dependence, but different values of MAPD were observed for different cameras. After the adjustment, the difference between different cameras is strongly reduced and MAPD is similar for each of them. In the future, to deeply characterize this effect, the analysis can be performed simply enlarging the Δt of the match-up. Indeed, in this study Δt has been chosen as very narrow (30 min) to take into account the large temporal variability of the coastal area.

Starting from the hypothesis that no considerable oceanographic processes can occur in 10 min, the differences between R_{rs} spectra measured by different sensors have been estimated using images acquired in temporal proximity. The values of MAPD observed in this analysis after the adjustment, similar to those observed in the comparison with the in situ data, proved that the MLR scheme notably reduced the inter-sensor differences, which can lead to a misleading interpretation of the basin optical variability when two images from different sensors are compared.

In the introduction, the VGOCS has been defined as a hyper-temporal dataset, but to be coherent with this definition, the dataset must fulfill the three conditions defined in [29] and reported in Section 1:

(a) The variables stored in the dataset must be univariate. As the R_{rs} data are provided at their native spectral resolution to reduce the uncertainty in the band-shifting procedure, the variables stored in the R_{rs}_data group are different for each sensor and this condition seems to be not fulfilled. Nevertheless, the R_{rs} data group is mainly provided for post-processing procedures and it allows the users to calculate different IOPs and water component concentrations using different algorithms [92–95]. The parameters that are needed to analyze the coastal optical variability, the aim for which VGOCS is conceived, are the IOPs, that are the same for each sensor, fulfilling the first requirement.

(b) All the images must be time referenced and the same grid should be used for each image of the dataset. This is fulfilled because the name of each file gives information about the acquisition time, and each image of the dataset is reprojected on a standard 1 km × 1 km equirectangular grid.

(c) All the sensors must be inter-validated between each other. This condition is also reached, as they are all adjusted on the AERONET-OC AAOT in situ data and the adjustment significantly reduced the inter-sensor differences.

5. Conclusions

This work aims to fill the gap between the near-polar orbiting and geostationary sensor temporal resolutions over the European coastal basins by the creation of a hyper-temporal Ocean Color analysis-ready dataset, named Virtual Geostationary Ocean Color Sensor (VGOCS). The latter fulfills the conditions needed to be defined as a hyper-temporal dataset and contains observations acquired from several near polar-orbiting Ocean Color imagers over the North Adriatic Sea, allowing approaching the geostationary sensor temporal resolution.

The problem in using data from different sensors is that they are characterized by different uncertainty sources that can introduce artifacts between different satellite images. For this reason, a multi-linear regression adjustment, based on a Fiducial Reference Measurements data-stream, has been applied to the remote sensing reflectance spectra of the VGOCS sensors. Hence, this adjustment, besides improving the agreement between the satellite and in situ data, reduced the inter-sensor differences, allowing the use of data from different sensors in analyzing the coastal optical variability. Moreover, the adjustment made available data traditionally masked in the standard processing chains, strongly increasing the VGOCS spatial and temporal coverage in the coastal area, an improvement that is fundamental to properly capture most of the coastal optical variability.

The approach here presented can be extended to all the coastal areas where one or more automatic radiometric stations, that represent the entire basin optical properties, are present. In European waters, this approach could be extended to:

- The Western Black Sea where the Gloria, Section-7, and Galata Platform AERONET-OC stations are present [96].
- In the Northern Sea where the Zeebrugge-MOW1 and Thornton_C-Power, part of the AERONET-OC network, are located [97].
- In the Baltic Sea where the Helsinki Lighthouse, Irbe Lighthouse, and Gustav Dalen Tower AERONET-OC stations are present [19]. As this basin is located at higher latitudes than the NAS, the satellite overpasses above such area are more frequent, making the approach also more powerful. Nevertheless, due to a large amount of CDOM present in these waters [98], a different atmospheric correction is needed to retrieve reliable Rrs spectra [99–101].
- The North-Western Mediterranean where the BOUee pour l'aquiSition d'une Serie Optique a Long termE mooring (BOUSSOLE) [102] and the Casablanca AERONET-OC [103] station are located. Particularly, BOUSSOLE has been used for the system vicarious calibration of the OLCI sensor [104]; hence, it could be interesting to test the effect of the adjustment using such data as input of the MLR algorithm.
- The areas where the different WATERHYPERNET network sites are located [105]. This network, which will be developed in the next years, is based on the concept of AERONET-OC [26], but with the essential advantage of the exploitation of a hyperspectral radiometer [106]. The use of such an instrument will allow validating each band of each satellite mission [106], reducing the uncertainties that could be introduced by band-shifting procedures [88,107].

The use of the Fiducial Reference Measurements data stream for the creation of this dataset for different coastal basins can be a reliable and unique way to compensate for the lack of an OCR geostationary sensor. Hence, the VGOCS dataset, available upon request to the authors, allows non-expert users to have at the disposal "ready to use" OC data to analyze the coastal optical variability at high temporal resolution. Moreover, expert users can perform analysis of exploiting products usually not provided in standard L3 files, such as flags and viewing geometry parameters.

Author Contributions: The general conception of this work was developed by M.B. and V.E.B.; M.B., G.V., M.B.E, and S.C. designed and implemented the processing chain. M.B. and V.E.B. designed the data analysis. M.B. processed the satellite imagery and performed the data analysis. M.B. and V.E.B. wrote the manuscript. S.C., G.V., and R.S. contributed to the interpretation of the results. All co-authors provided critical comments to the manuscript. All authors have read and agreed to the published version of the manuscript.

Acknowledgments: The AERONET Team is acknowledged for the continuous effort in supporting the AERONET-OC sub-network. Giuseppe Zibordi from the Joint Research Center of the European Commission is acknowledged for establishing and maintaining the AAOT AERONET-OC site. Vega Forneris and Flavio Lapadula from ISMAR-CNR are acknowledged for advice on the implementation of the processing chain. Ilaria Cazzaniga provided relevant information enabling the OLCI flag and camera dependence analysis. Francesco Bignami, Davide D'Alimonte, Davide Dionisi, Federico Falcini, Michela Sammartino, Enrico Zambianchi, and Giuseppe Zibordi, provided useful comments on earlier versions of this manuscript.

References

1. IOCCG. *Why Ocean Colour? The Societal Benefits of Ocean-Colour Technology*; Reports of the International Ocean-Colour Coordinating Group, No. 7; IOCCG: Bedford, NS, Canada, 2008.
2. IOCCG. *Mission Requirements for Future Ocean-Colour Sensors*; Reports of the International Ocean-Colour Coordinating Group, No. 13; IOCCG: Bedford, NS, Canada, 2012.
3. McClain, C.R.; Meister, G.; Monosmith, B. Satellite Ocean Color Sensor Design Concepts and Performance Requirements. In *Experimental Methods in the Physical Sciences*; Academic Press: Cambridge, MA, USA, 2014; Volume 47, pp. 73–119.
4. Mouw, C.B.; Greb, S.; Aurin, D.; DiGiacomo, P.M.; Lee, Z.; Twardowski, M.; Binding, C.; Hu, C.; Ma, R.; Moore, T.; et al. Aquatic color radiometry remote sensing of coastal and inland waters: Challenges and recommendations for future satellite missions. *Remote Sens. Environ.* **2015**, *160*, 15–30. [CrossRef]
5. Arnone, R.A.; Vandermeulen, R.A.; Soto, I.M.; Ladner, S.D.; Ondrusek, M.E.; Yang, H. Diurnal changes in ocean color sensed in satellite imagery. *J. Appl. Remote Sens.* **2017**, *11*, 032406. [CrossRef]
6. Bracaglia, M.; Volpe, G.; Colella, S.; Santoleri, R.; Braga, F.; Brando, V.E. Using overlapping VIIRS scenes to observe short term variations in particulate matter in the coastal environment. *Remote Sens. Environ.* **2019**, *233*, 111367. [CrossRef]
7. Chen, Z.; Hu, C.; Muller-Karger, F.E.; Luther, M.E. Short-term variability of suspended sediment and phytoplankton in Tampa Bay, Florida: Observations from a coastal oceanographic tower and ocean color satellites. *Estuar. Coast. Shelf Sci.* **2010**, *89*, 62–72. [CrossRef]
8. Qi, L.; Hu, C.; Barnes, B.B.; Lee, Z. VIIRS captures phytoplankton vertical migration in the NE Gulf of Mexico. *Harmful Algae* **2017**, *66*, 40–46. [CrossRef]
9. Choi, J.K.; Park, Y.J.; Ahn, J.H.; Lim, H.S.; Eom, J.; Ryu, J.H. GOCI, the world's first geostationary ocean color observation satellite, for the monitoring of temporal variability in coastal water turbidity. *J. Geophys. Res. Ocean.* **2012**, 117. [CrossRef]
10. Lavigne, H.; Ruddick, K. The potential use of geostationary MTG/FCI to retrieve chlorophyll-a concentration at high temporal resolution for the open oceans. *Int. J. Remote Sens.* **2018**, *39*, 2399–2420. [CrossRef]
11. Neukermans, G.; Ruddick, K.; Bernard, E.; Ramon, D.; Nechad, B.; Deschamps, P.Y. Mapping total suspended matter from geostationary satellites: A feasibility study with SEVIRI in the Southern North Sea. *Opt. Express* **2009**, *17*, 14029–14052. [CrossRef]
12. Choi, J.K.; Park, Y.J.; Lee, B.R.; Eom, J.; Moon, J.E.; Ryu, J.H. Application of the Geostationary Ocean Color Imager (GOCI) to mapping the temporal dynamics of coastal water turbidity. *Remote Sens. Environ.* **2014**, *146*, 24–35. [CrossRef]
13. Concha, J.; Mannino, A.; Franz, B.; Kim, W. Uncertainties in the Geostationary Ocean Color Imager (GOCI) Remote Sensing Reflectance for Assessing Diurnal Variability of Biogeochemical Processes. *Remote Sens.* **2019**, *11*, 295. [CrossRef]

14. Ruddick, K.; Neukermans, G.; Vanhellemont, Q.; Jolivet, D. Challenges and opportunities for geostationary ocean colour remote sensing of regional seas: A review of recent results. *Remote Sens. Environ.* **2014**, *146*, 63–76. [CrossRef]
15. Wang, M.; Son, S.; Jiang, L.; Shi, W. Observations of ocean diurnal variations from the Korean geostationary ocean color imager (GOCI). In *Ocean Sensing and Monitoring VI*; International Society for Optics and Photonics: Bellingham, WA, USA, 2014; Volume 9111, p. 911102.
16. Ryu, J.H.; Han, H.J.; Cho, S.; Park, Y.J.; Ahn, Y.H. Overview of geostationary ocean color imager (GOCI) and GOCI data processing system (GDPS). *Ocean Sci. J.* **2012**, *47*, 223–233. [CrossRef]
17. D'Alimonte, D.; Zibordi, G.; Melin, F. A statistical method for generating cross-mission consistent normalized water-leaving radiances. *IEEE Trans. Geosci. Remote Sens.* **2008**, *46*, 4075–4093. [CrossRef]
18. Zibordi, G.; Holben, B.; Hooker, S.B.; Mélin, F.; Berthon, J.F.; Slutsker, I.; Giles, D.; Vandemark, D.; Feng, H.; Rutledge, K.; et al. A network for standardized ocean color validation measurements. *EOS Trans. Am. Geophys. Union* **2006**, *87*, 293–297. [CrossRef]
19. Zibordi, G.; Mélin, F.; Berthon, J.F.; Holben, B.; Slutsker, I.; Giles, D.; D'Alimonte, D.; Vandemark, D.; Feng, H.; Fabbri, B.E.; et al. AERONET-OC: A network for the validation of ocean color primary products. *J. Atmos. Ocean. Technol.* **2009**, *26*, 1634–1651. [CrossRef]
20. Zibordi, G.; Voss, K.J. In situ optical radiometry in the visible and near infrared. In *Experimental Methods in the Physical Sciences*; Academic Press: Cambridge, MA, USA, 2014; Volume 47, pp. 247–304.
21. Zibordi, G.; Voss, K.J. Requirements and strategies for in situ radiometry in support of satellite ocean color. In *Experimental Methods in the Physical Sciences*; Academic Press: Cambridge, MA, USA, 2014; Volume 47, pp. 531–556.
22. Banks, A.C.; Vendt, R.; Alikas, K.; Bialek, A.; Kuusk, J.; Lerebourg, C.; Ruddick, K.; Tilstone, G.; Vabson, V.; Donlon, C.; et al. Fiducial Reference Measurements for Satellite Ocean Colour (FRM4SOC). *Remote Sens.* **2020**, *12*, 1322. [CrossRef]
23. Alikas, K.; Ansko, I.; Vabson, V.; Ansper, A.; Kangro, K.; Uudeberg, K.; Ligi, M. Consistency of Radiometric Satellite Data over Lakes and Coastal Waters with Local Field Measurements. *Remote Sens.* **2020**, *12*, 616. [CrossRef]
24. Liberti, G.L.; D'Alimonte, D.; di Sarra, A.; Mazeran, C.; Voss, K.; Yarbrough, M.; Bozzano, R.; Cavaleri, L.; Colella, S.; Cesarini, C.; et al. European Radiometry Buoy and Infrastructure (EURYBIA): A Contribution to the Design of the European Copernicus Infrastructure for Ocean Colour System Vicarious Calibration. *Remote Sens.* **2020**, *12*, 1178. [CrossRef]
25. Vabson, V.; Kuusk, J.; Ansko, I.; Vendt, R.; Alikas, K.; Ruddick, K.; Ansper, A.; Bresciani, M.; Burmester, H.; Costa, M.; et al. Field intercomparison of radiometers used for satellite validation in the 400–900 nm range. *Remote Sens.* **2019**, *11*, 1129. [CrossRef]
26. Zibordi, G.; Mélin, F.; Voss, K.J.; Johnson, B.C.; Franz, B.A.; Kwiatkowska, E.; Huot, J.; Wang, M.; Antoine, D. System vicarious calibration for ocean color climate change applications: Requirements for in situ data. *Remote Sens. Environ.* **2015**, *159*, 361–369. [CrossRef]
27. Lewis, A.; Lacey, J.; Mecklenburg, S.; Ross, J.; Siqueira, A.; Killough, B.; Szantoi, Z.; Tadono, T.; Rosenavist, A.; Goryl, P.; et al. CEOS Analysis Ready Data for Land (CARD4L) Overview. In Proceedings of the IGARSS 2018 IEEE International Geoscience and Remote Sensing Symposium, Valencia, Spain, 22–27 July 2018; pp. 7407–7410.
28. Scarrott, R.G.; Cawkwell, F.; Jessopp, M.; O'Rourke, E.; Cusack, C.; de Bie, K. From land to sea, a review of hypertemporal remote sensing advances to support ocean surface science. *Water* **2019**, *11*, 2286. [CrossRef]
29. Piwowar, J.M.; Peddle, D.R.; LeDrew, E.F. Temporal mixture analysis of arctic sea ice imagery: A new approach for monitoring environmental change. *Remote Sens. Environ.* **1998**, *63*, 195–207. [CrossRef]
30. Bissett, W.P.; Arnone, R.A.; Davis, C.O.; Dickey, T.D.; Dye, D.; Kohler, D.D.; Gould, R.W., Jr. From meters to kilometers: A look at ocean-color scales of variability, spatial coherence, and the need for fine-scale remote sensing in coastal ocean optics. *Oceanography* **2004**, *17*, 32–43. [CrossRef]
31. Aurin, D.; Mannino, A.; Franz, B. Spatially resolving ocean color and sediment dispersion in river plumes, coastal systems, and continental shelf waters. *Remote Sens. Environ.* **2013**, *137*, 212–225. [CrossRef]
32. Lee, Z.; Hu, C.; Arnone, R.; Liu, Z. Impact of sub-pixel variations on ocean color remote sensing products. *Opt. Express* **2012**, *20*, 20844–20854. [CrossRef] [PubMed]

33. Kutser, T. Quantitative detection of chlorophyll in cyanobacterial blooms by satellite remote sensing. *Limnol. Oceanogr.* **2004**, *49*, 2179–2189. [CrossRef]
34. Lee, Z.; Carder, K.L.; Arnone, R.A. Deriving inherent optical properties from water color: A multiband quasi-analytical algorithm for optically deep waters. *Appl. Opt.* **2002**, *41*, 5755–5772. [CrossRef]
35. Lee, Z.; Carder, K.L.; Arnone, R.A. "Quasi-Analytical Algorithm v6 Update". 2014. Available online: http://www.ioccg.org/groups/Software_OCA/QAA_v6_2014209.pdf (accessed on 5 May 2020).
36. Pitarch, J.; Bellacicco, M.; Volpe, G.; Colella, S.; Santoleri, R. Use of the quasi-analytical algorithm to retrieve backscattering from in-situ data in the Mediterranean Sea. *Remote Sens. Lett.* **2016**, *7*, 591–600. [CrossRef]
37. Pitarch, J.; Bellacicco, M.; Organelli, E.; Volpe, G.; Colella, S.; Vellucci, V.; Marullo, S. Retrieval of Particulate Backscattering Using Field and Satellite Radiometry: Assessment of the QAA Algorithm. *Remote Sens.* **2020**, *12*, 77. [CrossRef]
38. Barnes, B.B.; Hu, C. Dependence of satellite ocean color data products on viewing angles: A comparison between SeaWiFS, MODIS, and VIIRS. *Remote Sens. Environ.* **2016**, *175*, 120–129. [CrossRef]
39. Volpe, G.; Colella, S.; Brando, V.E.; Forneris, V.; La Padula, F.; Di Cicco, A.; Sammartino, M.; Bracaglia, M.; Artuso, F.; Santoleri, R. Mediterranean ocean colour Level 3 operational multi-sensor processing. *Ocean Sci.* **2019**, *15*, 127–146. [CrossRef]
40. Solidoro, C.; Bastianini, M.; Bandelj, V.; Codermatz, R.; Cossarini, G.; Canu, D.M.; Ravagnan, E.; Salon, S.; Trevisani, S. Current state, scales of variability, and trends of biogeochemical properties in the northern Adriatic Sea. *J. Geophys. Res. Oceans* **2009**, 114. [CrossRef]
41. Bignami, F.; Sciarra, R.; Carniel, S.; Santoleri, R. Variability of Adriatic Sea coastal turbid waters from SeaWiFS imagery. *J. Geophys. Res. Oceans* **2007**, 112. [CrossRef]
42. Zavatarelli, M.; Raicich, F.; Bregant, D.; Russo, A.; Artegiani, A. Climatological biogeochemical characteristics of the Adriatic Sea. *J. Mar. Syst.* **1998**, *18*, 227–263. [CrossRef]
43. Brando, V.E.; Braga, F.; Zaggia, L.; Giardino, C.; Bresciani, M.; Matta, E.; Bellafiore, D.; Ferrarin, C.; Maicu, F.; Benetazzo, A.; et al. High-resolution satellite turbidity and sea surface temperature observations of river plume interactions during a significant flood event. *Ocean Sci.* **2015**, *11*, 909–920. [CrossRef]
44. Marini, M.; Jones, B.H.; Campanelli, A.; Grilli, F.; Lee, C.M. Seasonal variability and Po River plume influence on biochemical properties along western Adriatic coast. *J. Geophys. Res. Oceans* **2008**, 113. [CrossRef]
45. Cozzi, S.; Giani, M. River water and nutrient discharges in the Northern Adriatic Sea: Current importance and long term changes. *Cont. Shelf Res.* **2011**, *31*, 1881–1893. [CrossRef]
46. Falcieri, F.M.; Benetazzo, A.; Sclavo, M.; Russo, A.; Carniel, S. Po River plume pattern variability investigated from model data. *Cont. Shelf Res.* **2014**, *87*, 84–95. [CrossRef]
47. Degobbis, D.; Precali, R.; Ivancic, I.; Smodlaka, N.; Fuks, D.; Kveder, S. Long-term changes in the northern Adriatic ecosystem related to anthropogenic eutrophication. *Int. J. Environ. Pollut.* **2000**, *13*, 495–533. [CrossRef]
48. Tesi, T.; Miserocchi, S.; Acri, F.; Langone, L.; Boldrin, A.; Hatten, J.A.; Albertazzi, S. Flood-driven transport of sediment, particulate organic matter, and nutrients from the Po River watershed to the Mediterranean Sea. *J. Hydrol.* **2013**, *498*, 144–152. [CrossRef]
49. Malačič, V.; Viezzoli, D.; Cushman-Roisin, B. Tidal dynamics in the northern Adriatic Sea. *J. Geophys. Res. Oceans* **2000**, *105*, 26265–26280. [CrossRef]
50. Wang, X.H.; Pinardi, N.; Malacic, V. Sediment transport and resuspension due to combined motion of wave and current in the northern Adriatic Sea during a Bora event in January 2001: A numerical modelling study. *Cont. Shelf Res.* **2007**, *27*, 613–633. [CrossRef]
51. Barale, V.; Jaquet, J.M.; Ndiaye, M. Algal blooming patterns and anomalies in the Mediterranean Sea as derived from the SeaWiFS data set (1998–2003). *Remote Sens. Environ.* **2008**, *112*, 3300–3313. [CrossRef]
52. Bellafiore, D.; Ferrarin, C.; Braga, F.; Zaggia, L.; Maicu, F.; Lorenzetti, G.; Manfè, G.; Brando, V.E.; De Pascalis, F. Coastal mixing in multiple-mouth deltas: A case study in the Po delta, Italy. *Estuar. Coast. Shelf Sci.* **2019**, *226*, 106254. [CrossRef]
53. Benincasa, M.; Falcini, F.; Adduce, C.; Sannino, G.; Santoleri, R. Synergy of Satellite Remote Sensing and Numerical Ocean Modelling for Coastal Geomorphology Diagnosis. *Remote Sens.* **2019**, *11*, 2636. [CrossRef]
54. Ferrarin, C.; Davolio, S.; Bellafiore, D.; Ghezzo, M.; Maicu, F.; Mc Kiver, W.; Drofa, O.; Umgiesser, G.; Bajo, M.; De Pascalis, F.; et al. Cross-scale operational oceanography in the Adriatic Sea. *J. Oper. Oceanogr.* **2019**, *12*, 86–103. [CrossRef]

55. Braga, F.; Zaggia, L.; Bellafiore, D.; Bresciani, M.; Giardino, C.; Lorenzetti, G.; Maicu, F.; Manzo, C.; Riminucci, F.; Ravaioli, M.; et al. Mapping turbidity patterns in the Po river prodelta using multi-temporal Landsat 8 imagery. *Estuar. Coast. Shelf Sci.* **2017**, *198*, 555–567. [CrossRef]
56. Boldrin, A.; Carniel, S.; Giani, M.; Marini, M.; Bernardi Aubry, F.; Campanelli, A.; Grilli, F.; Russo, A. Effects of bora wind on physical and biogeochemical properties of stratified waters in the northern Adriatic. *J. Geophys. Res. Oceans* **2009**, 114. [CrossRef]
57. Liu, X.; Wang, M. Filling the gaps of missing data in the merged VIIRS SNPP/NOAA-20 ocean color product using the DINEOF method. *Remote Sens.* **2019**, *11*, 178. [CrossRef]
58. Cao, C.; Xiong, X.; Wolfe, R.; De Luccia, F.; Liu, Q.; Blonski, S.; Lin, G.G.; Nishihama, M.; Pogorzala, D.; Oudrari, H.; et al. *Visible Infrared Imaging Radiometer Suite (VIIRS) Sensor Data Record (SDR) User's Guide*; NOAA Technical Report NESDIS: College Park, MD, USA, 2013.
59. Donlon, C.; Berruti, B.; Buongiorno, A.; Ferreira, M.H.; Féménias, P.; Frerick, J.; Goryl, P.; Klein, U.; Laur, H.; Mavrocordatos, C.; et al. The global monitoring for environment and security (GMES) sentinel-3 mission. *Remote Sens. Environ.* **2012**, *120*, 37–57. [CrossRef]
60. Masuoka, E.; Fleig, A.; Wolfe, R.E.; Patt, F. Key characteristics of MODIS data products. *IEEE Trans. Geosci. Remote Sens.* **1998**, *36*, 1313–1323. [CrossRef]
61. Kay, S.; Hedley, J.; Lavender, S. Sun glint correction of high and low spatial resolution images of aquatic scenes: A review of methods for visible and near-infrared wavelengths. *Remote Sens.* **2009**, *1*, 697–730. [CrossRef]
62. ESA. 2020. Available online: https://scihub.copernicus.eu/userguide/BatchScripting (accessed on 5 May 2020).
63. EUMETSAT. 2020. Available online: https://coda.eumetsat.int/ (accessed on 5 May 2020).
64. EUMETSAT. 2020. Available online: https://codarep.eumetsat.int/ (accessed on 5 May 2020).
65. NASA. 2020. Available online: https://oceancolor.gsfc.nasa.gov/reprocessing/r2018/ (accessed on 5 May 2020).
66. Bouali, M.; Ignatov, A. Adaptive reduction of striping for improved sea surface temperature imagery from Suomi National Polar-Orbiting Partnership (S-NPP) visible infrared imaging radiometer suite (VIIRS). *J. Atmos. Ocean. Technol.* **2014**, *31*, 150–163. [CrossRef]
67. Mikelsons, K.; Wang, M.; Jiang, L.; Bouali, M. Destriping algorithm for improved satellite-derived ocean color product imagery. *Opt. Express* **2014**, *22*, 28058–28070. [CrossRef]
68. NASA. 2020. Available online: https://oceancolor.gsfc.nasa.gov/atbd/ocl2flags (accessed on 5 May 2020).
69. Morel, A.; Antoine, D.; Gentili, B. Bidirectional reflectance of oceanic waters: Accounting for Raman emission and varying particle scattering phase function. *Appl. Opt.* **2002**, *41*, 6289–6306. [CrossRef] [PubMed]
70. Morel, A.; Mueller, J.L. Normalized water-leaving radiance and remote sensing reflectance: Bidirectional reflectance and other factors. *Ocean Opt. Protoc. Satell. Ocean Color Sens. Valid.* **2002**, *2*, 183–210.
71. Kaufman, Y.J. Atmospheric effect on spatial resolution of surface imagery. *Appl. Opt.* **1984**, *23*, 3400–3408. [CrossRef]
72. Sterckx, S.; Knaeps, E.; Ruddick, K. Detection and correction of adjacency effects in hyperspectral airborne data of coastal and inland waters: The use of the near infrared similarity spectrum. *Int. J. Remote Sens.* **2011**, *32*, 6479–6505. [CrossRef]
73. Bulgarelli, B.; Zibordi, G.; Mélin, F. On the minimization of adjacency effects in SeaWiFS primary data products from coastal areas. *Opt. Express* **2018**, *26*, A709–A728. [CrossRef]
74. Van Mol, B.; Ruddick, K. Total Suspended Matter maps from CHRIS imagery of a small inland water body in Oostende (Belgium). In Proceedings of the 3rd ESA CHRIS/Proba Workshop, Frascati, Italy, 21–23 March 2005; pp. 21–23.
75. Barnes, B.B.; Cannizzaro, J.P.; English, D.C.; Hu, C. Validation of VIIRS and MODIS reflectance data in coastal and oceanic waters: An assessment of methods. *Remote Sens. Environ.* **2019**, *220*, 110–123. [CrossRef]
76. Gordon, H.R.; Wang, M. Retrieval of water-leaving radiance and aerosol optical thickness over the oceans with SeaWiFS: A preliminary algorithm. *Appl. Opt.* **1994**, *33*, 443–452. [CrossRef]
77. Lavender, S.J.; Pinkerton, M.H.; Moore, G.F.; Aiken, J.; Blondeau-Patissier, D. Modification to the atmospheric correction of SeaWiFS ocean colour images over turbid waters. *Cont. Shelf Res.* **2005**, *25*, 539–555. [CrossRef]
78. Moore, G.F.; Aiken, J.; Lavender, S.J. The atmospheric correction of water colour and the quantitative retrieval of suspended particulate matter in Case II waters: Application to MERIS. *Int. J. Remote Sens.* **1999**, *20*, 1713–1733. [CrossRef]

79. Morel, A.; Gentili, B. Diffuse reflectance of oceanic waters: Its dependence on Sun angle as influenced by the molecular scattering contribution. *Appl. Opt.* **1991**, *30*, 4427–4438. [CrossRef]
80. Morel, A.; Gentili, B. Diffuse reflectance of oceanic waters. II. Bidirectional aspects. *Appl. Opt.* **1993**, *32*, 6864–6879. [CrossRef]
81. Morel, A.; Gentili, B. Diffuse reflectance of oceanic waters. III. Implication of bidirectionality for the remote-sensing problem. *Appl. Opt.* **1996**, *35*, 4850–4862. [CrossRef]
82. Zibordi, G.; Mélin, F.; Berthon, J.F. A regional assessment of OLCI data products. *IEEE Geosci. Remote Sens. Lett.* **2018**, *15*, 1490–1494. [CrossRef]
83. *Sentinel-3 Product Notice—OLCI Level-2 Ocean Colour*; EUMETSAT: Darmstadt, Germany, 2019.
84. Volpe, G.; Colella, S.; Brando, V.E.; Benincasa, M.; Forneris, V.; Bracaglia, M.; Di Cicco, A.; D'Alimonte, D. *Ocean Colour Production Centre. Ocean Colour Mediterranean and Black Sea Observation Product*; Copernicus Publications: Göttingen, Germany, 2019.
85. Ruddick, K.G.; Voss, K.; Boss, E.; Castagna, A.; Frouin, R.; Gilerson, A.; Hieronymi, M.; Johnson, B.C.; Kuusk, J.; Lee, Z.; et al. A Review of Protocols for Fiducial Reference Measurements of Water-Leaving Radiance for Validation of Satellite Remote-Sensing Data over Water. *Remote Sens.* **2019**, *11*, 2198. [CrossRef]
86. Thuillier, G.; Floyd, L.; Woods, T.N.; Cebula, R.; Hilsenrath, E.; Hersé, M.; Labs, D. Solar irradiance reference spectra for two solar active levels. *Adv. Space Res.* **2004**, *34*, 256–261. [CrossRef]
87. Mélin, F.; Vantrepotte, V.; Clerici, M.; D'Alimonte, D.; Zibordi, G.; Berthon, J.F.; Canuti, E. Multi-sensor satellite time series of optical properties and chlorophyll-a concentration in the Adriatic Sea. *Prog. Oceanogr.* **2011**, *91*, 229–244. [CrossRef]
88. Mélin, F.; Sclep, G. Band shifting for ocean color multi-spectral reflectance data. *Opt. Express* **2015**, *23*, 2262–2279. [CrossRef]
89. Lee, S.; Meister, G.; Franz, B. MODIS Aqua Reflective Solar Band Calibration for NASA's R2018 Ocean Color Products. *Remote Sens.* **2019**, *11*, 2187. [CrossRef]
90. Cao, C.; De Luccia, F.J.; Xiong, X.; Wolfe, R.; Weng, F. Early on-orbit performance of the visible infrared imaging radiometer suite onboard the Suomi National Polar-Orbiting Partnership (S-NPP) satellite. *IEEE Trans. Geosci. Remote Sens.* **2014**, *52*, 1142–1156. [CrossRef]
91. Pahlevan, N.; Sarkar, S.; Franz, B.A. Uncertainties in coastal ocean color products: Impacts of spatial sampling. *Remote Sens. Environ.* **2016**, *181*, 14–26. [CrossRef]
92. Brando, V.E.; Dekker, A.G.; Park, Y.J.; Schroeder, T. Adaptive semianalytical inversion of ocean color radiometry in optically complex waters. *Appl. Opt.* **2012**, *51*, 2808–2833. [CrossRef]
93. Loisel, H.; Stramski, D.; Dessailly, D.; Jamet, C.; Li, L.; Reynolds, R.A. An Inverse Model for Estimating the Optical Absorption and Backscattering Coefficients of Seawater From Remote-Sensing Reflectance Over a Broad Range of Oceanic and Coastal Marine Environments. *J. Geophys. Res. Ocean.* **2018**, *123*, 2141–2171. [CrossRef]
94. Werdell, P.J.; Franz, B.A.; Bailey, S.W.; Feldman, G.C.; Boss, E.; Brando, V.E.; Dowell, M.; Hirata, T.; Lavender, S.J.; Lee, Z.; et al. Generalized ocean color inversion model for retrieving marine inherent optical properties. *Appl. Opt.* **2013**, *52*, 2019–2037. [CrossRef]
95. Werdell, P.J.; McKinna, L.I.; Boss, E.; Ackleson, S.G.; Craig, S.E.; Gregg, W.W.; Lee, Z.; Maritorena, S.; Roesler, C.S.; Rousseaux, C.S.; et al. An overview of approaches and challenges for retrieving marine inherent optical properties from ocean color remote sensing. *Prog. Oceanogr.* **2018**, *160*, 186–212. [CrossRef]
96. Zibordi, G.; Mélin, F.; Berthon, J.F.; Canuti, E. Assessment of MERIS ocean color data products for European seas. *Ocean Sci.* **2013**, *9*, 521–533. [CrossRef]
97. Van der Zande, D.; Vanhellemont, Q.; De Keukelaere, L.; Knaeps, E.; Ruddick, K. Validation of Landsat-8/OLI for ocean colour applications with AERONET-OC sites in Belgian coastal waters. In Proceedings of the Ocean Optics Conference, Victoria, BC, Canada, 23–28 October 2016.
98. Berthon, J.F.; Zibordi, G. Optically black waters in the northern Baltic Sea. *Geophys. Res. Lett.* **2010**, 37. [CrossRef]
99. Darecki, M.; Stramski, D. An evaluation of MODIS and SeaWiFS bio-optical algorithms in the Baltic Sea. *Remote Sens. Environ.* **2004**, *89*, 326–350. [CrossRef]
100. IOCCG. *Atmospheric Correction for Remotely-Sensed Ocean-Colour Products*; Reports of International Ocean-Color Coordinating Group; IOCCG: Bedford, NS, Canada, 2010; p. 10.

101. Zibordi, G.; Berthon, J.F.; Mélin, F.; D'Alimonte, D.; Kaitala, S. Validation of satellite ocean color primary products at optically complex coastal sites: Northern Adriatic Sea, Northern Baltic Proper and Gulf of Finland. *Remote Sens. Environ.* **2009**, *113*, 2574–2591. [CrossRef]
102. Antoine, D.; Chami, M.; Claustre, H.; d'Ortenzio, F.; Morel, A.; Bécu, G.; Gentili, B.; Louis, F.; Ras, J.; Roussier, E.; et al. *BOUSSOLE: A Joint CNRS-INSU, ESA, CNES, and NASA Ocean Color Calibration and Validation Activity*; National Aeronautics and Space Administration: Washington, DC, USA, 2006.
103. Talone, M.; García-Ladona, E. *AERONET-OC-Bioptical Survey of the Ebro's Shelf*; Instituto de Ciencias del Mar (ICM): Barcelona, Brazil, 2020.
104. Lamquin, N.; Bourg, L.; Lerebourg, C.; Martin-Lauzer, F.R.; Kwiatkowska, E.; Dransfeld, S. System Vicarious Calibration of Sentinel-3 OLCI. In Proceedings of the CALCON Technical Meeting 2017, Logan, UT, USA, 23 August 2017.
105. HYPERNETS. Available online: https://www.hypernets.eu/from_cms/test_sites (accessed on 5 May 2020).
106. Vansteenwegen, D.; Ruddick, K.; Cattrijsse, A.; Vanhellemont, Q.; Beck, M. The pan-and-tilt hyperspectral radiometer system (PANTHYR) for autonomous satellite validation measurements—Prototype design and testing. *Remote Sens.* **2019**, *11*, 1360. [CrossRef]
107. Pahlevan, N.; Smith, B.; Binding, C.; O'Donnell, D.M. Spectral band adjustments for remote sensing reflectance spectra in coastal/inland waters. *Opt. Express* **2017**, *25*, 28650–28667. [CrossRef]

2

Improving the Retrieval of Carbon-Based Phytoplankton Biomass from Satellite Ocean Colour Observations

Marco Bellacicco [1,2,*], Jaime Pitarch [2], Emanuele Organelli [2], Victor Martinez-Vicente [3], Gianluca Volpe [2] and Salvatore Marullo [1,2]

1 Climate Modelling Laboratory, Italian National Agency for New Technologies, Energy and Sustainable Economic Development (ENEA), Via E. Fermi 45, 00044 Frascati, Italy; salvatore.marullo@enea.it

2 Consiglio Nazionale delle Ricerche (CNR), Istituto di Scienze Marine (ISMAR), Via Fosso del Cavaliere, 100 Rome, Italy; jaime.pitarchportero@artov.ismar.cnr.it (J.P.); emanuele.organelli@cnr.it (E.O.); gianluca.volpe@cnr.it (G.V.)

3 Plymouth Marine Laboratory (PML), Prospect Place, The Hoe, Plymouth PL1 3DH, UK; vmv@pml.ac.uk

* Correspondence: marco.bellacicco@enea.it

Abstract: Phytoplankton is at the base of the marine food web and plays a fundamental role in the global carbon cycle. Ongoing climate change significantly impacts phytoplankton distribution in the ocean. Monitoring phytoplankton is crucial for a full understanding of changes in the marine ecosystem. To observe phytoplankton from space, chlorophyll-*a* concentration (Chl) has been widely used as a proxy of algal biomass, although it can be impacted by physiology. Therefore, there has been an increasing focus towards estimating phytoplankton biomass in units of carbon (C_{phyto}). Here, we developed an algorithm to quantify C_{phyto} from space-based observations that accounts for the spatio-temporal variations of the backscattering coefficient associated with the fraction of detrital particles that do not covary with Chl. The main findings are: (i) a spatial and temporal variation of the detritus component must be accounted for in the C_{phyto} algorithm; (ii) the refined C_{phyto} algorithm performs better (relative bias of 23.7%) than any previously existing model; and (iii) our algorithm shows the lowest error in C_{phyto} across areas where picophytoplankton dominates (relative bias of 14%). In other areas, it is currently not possible to accurately assess the performance of the refined algorithm due to the paucity of in situ carbon data associated with nano- and micro-phytoplankton size classes.

Keywords: phytoplankton carbon; optical backscattering; non-algal particles; ocean colour observations; QAA algorithm; ESA OC-CCI

---

1. Introduction

Phytoplankton is responsible for approximately half of global primary production and is at the base of the marine food web [1]. Phytoplankton is consequently a fundamental actor in the global carbon cycle [2]. Moreover, these organisms are regarded as sentinels of changes in the ocean because of their capacity to respond rapidly to environmental perturbations. Several factors, such as ocean circulation, anthropogenic activities, and climate affect phytoplankton abundance and distribution. In particular, phytoplankton spatio-temporal patterns are expected to vary with climate change [3]. Monitoring the distribution of phytoplankton is crucial for a full understanding of the oceanic biogeochemical cycles.

Ocean colour observations have significantly improved our capability to map phytoplankton global distribution since the 1970s. Chlorophyll-*a* concentration (Chl; in mg m^{-3}) is a consolidated proxy of algal biomass [4]. However, Chl variability may not always correspond to actual changes in algal biomass but rather to cellular physiological adjustments in response to environmental stressors such as light and nutrient limitation [5,6]. Therefore, there has been an increasing focus on the estimation of phytoplankton biomass in units of carbon (C_{phyto}; mg C m^{-3}). C_{phyto} has found application in studies of: (i) primary production [7,8]; (ii) phytoplankton physiology [9]; (iii) phytoplankton growth/loss rates [10,11]; (iv) comparison with marine ecosystem models [12]; (v) pools of carbon in the ocean (i.e., their stocks and turnover rates); and (vi) ocean–atmosphere interactions in marine phytoplankton feedback in atmospheric aerosols [13].

Martinez-Vicente et al. (2017) recently reviewed algorithms for deriving C_{phyto} from satellite ocean colour observations [14]. Three groups of algorithms are currently available: (i) backscatter-based models developed using satellite [7] or in situ data [14–16]; (ii) Chl-based models [17,18]; and (iii) allometric conversion-based models [19–21]. Among these algorithms, those relying on the particulate optical backscattering coefficient, b_{bp} (in m^{-1}), raised high interest because b_{bp} is related to the concentration and composition of suspended particles in seawater, and to their size and shape [22]. Although former research suggested that b_{bp} was mainly influenced by submicron detrital particles [22], it has recently been demonstrated that most of the b_{bp} signal in the surface oligotrophic ocean is due to particles with equivalent spherical diameters between 1 and 10 μm [23], thus reinforcing the role of phytoplankton as a source of the open-ocean b_{bp} [24], and thus the usefulness of b_{bp} to retrieve C_{phyto}. However, satellite b_{bp}-based algorithms show large errors in predicting C_{phyto} mainly because the current formulation does not take into account the spatio-temporal changes of non-algal particles (NAP) [25].

Here we refine the first and widely applied satellite b_{bp}-based C_{phyto} model [7] and make crucial arrangements to improve its performance. Beh05 uses the linear relationship between Chl and b_{bp} to estimate the background contribution of NAP to $b_{bp}(443)$, b_{bp}^{k}, which corresponds to b_{bp} when Chl is zero (i.e., the intercept of the linear fit). Beh05 is based on the following equation:

$$C_{phyto} = \left[b_{bp}(443) - b_{bp}^{k}\right] \cdot SF \tag{1}$$

In Beh05, SF is the scaling factor used to convert b_{bp} into C_{phyto} and is set to 13,000 mg C m^{-2} [7]. b_{bp}^{k} is assumed to represent a stable surface population of heterotrophic and detrital particles that is not strictly dependent on phytoplankton dynamics and physiology. Thus, in Beh05, b_{bp}^{k} was assumed as a constant value ($3.5 \cdot 10^{-4}$ m^{-1}). However, NAP can include various types of particles ranging from heterotrophic organisms such as bacteria, micrograzers, and viruses, to fecal pellets and cell debris, and mineral particles and bubbles [22] that vary both in space and time. The nature, definition, and evaluation of b_{bp}^{k} is still under debate and has been investigated using in situ [24–26] and satellite data [27,28]. Understanding the spatial and temporal dynamics of b_{bp}^{k} in the open ocean will improve estimations of C_{phyto} and, more generally, our knowledge on the fate of marine carbon.

Beh05 estimated b_{bp}^{k} as the offset (intercept) of the least-square regression between monthly satellite-derived b_{bp} and Chl computed via the Garver-Siegel-Maritorena (GSM) semi-analytical algorithm [29]. The regression analysis used only Chl values higher than 0.14 mg m^{-3} to separate the changes in Chl due to physiology (i.e., photoacclimation) from actual changes in biomass, and thus in C_{phyto}. Bellacicco et al. (2018) used satellite data from the Mediterranean Sea to estimate b_{bp}^{k} for various bioregions and seasons, and showed that b_{bp}^{k} varied regionally and seasonally, thus rejecting the hypothesis of invariance of the heterotrophic and detrital components at the sea surface [27]. Global variability in b_{bp}^{k} was observed [28], with a reported median global value of 9.5 10^{-4} m^{-1}, nearly threefold that found by Beh05. The newly evaluated b_{bp}^{k} resulted in C_{phyto} that were half those estimated by Beh05.

The Chl–b_{bp} relationship is influenced by phytoplankton specific composition and diversity (e.g., size, shape, internal structure), physiology (e.g., photoacclimation), and the nature of NAP itself [22,30,31]. For these reasons, Brewin et al. (2012) (hereafter Bre12; [26]) developed an analytical non-linear fit between b_{bp} and Chl that accounted for changes in phytoplankton cell size. In Bre12, the offset of the non-linear fit (i.e., b^{k}_{bp}) between b_{bp} and Chl in clear waters was interpreted as a constant background of NAP or also partly influenced by very small phytoplankton (e.g., prochlorophytes) in addition to NAP. In Bre12, the reported b^{k}_{bp} values were $7.0 \cdot 10^{-4}$ m^{-1} at the wavelength of 470 nm and $5.6 \cdot 10^{-4}$ m^{-1} at 526 nm. More recently, the Bre12 model was used with a large database of 0–1000 m depth profiles of Chl and b_{bp} acquired by the global Biogeochemical-Argo (aka BGC-Argo) float array. b^{k}_{bp} was demonstrated also to vary over depth and to account only for a small fraction of total b_{bp} in productive areas, while being the main source of b_{bp} in oligotrophic waters [25].

The former evidence suggests that if a realistic spatio-temporal variation of b^{k}_{bp} is introduced in the C_{phyto} algorithm its precision and accuracy can be largely improved. Specifically, the hypothesis that b^{k}_{bp} varies in space and time is here applied monthly at the scale of satellite pixel rather than using a unique b^{k}_{bp} value or pre-defined bioregions. Algorithm outputs are validated against the only available, to the best of our knowledge, in situ dataset and compared to the performances of Beh05, Bel18, and Bre12 with single b^{k}_{bp} values. The performance of the new algorithm is also compared to empirical approaches [14,16]. The analysis was developed globally by partitioning the data among Optical Water Class (OWC; [32]) to explore the efficiency and applicability of the different tested approaches.

2. Data and Methods

2.1. Ocean Colour Data

The full European Space Agency (ESA) Ocean Colour–Climate Change Initiative (OC-CCI) time series (1997–2019) of global daily R_{rs} and Chl version 4.2 data at 4 km resolution was downloaded from the ESA OC-CCI FTP server (https://esa-oceancolour-cci.org; ftp://oceancolour.org/occci-v4.2/geographic/netcdf/daily/rrs/). ESA OC-CCI data products are the result of the merging of SeaWiFS, MERIS, MODIS, and VIIRS observations in which the inter-sensor biases are removed. Version 4.2 includes the latest NASA reprocessing R2018.0, which mostly accounts for the ageing of the MODIS sensor. The ESA OC-CCI provides the daily R_{rs} data and associated uncertainty maps in terms of bias and root mean square error. In this work, for each daily R_{rs}, the bias was also extracted and then corrected pixel-by-pixel, as recommended in the Product User Guide [33].

Chl was estimated with a blending of the OCI (as implemented by NASA, itself a combination of CI and OC4), OC5, and OC3 algorithms. For each daily Chl value, the bias was also extracted and compensated at the pixel scale. Daily b_{bp} maps at 443 nm were produced from daily R_{rs} for the same period (1997–2019) by applying the Quasi-Analytical Algorithm (QAA; [34,35]) with prior correction for Raman scattering [27,36]. The accuracy of the QAA in retrieving b_{bp} was fully demonstrated [27,36–39]. In this work, the QAA algorithm was thus selected for its high efficiency in the b_{bp} retrievals, and to show consistency and coherency with the OC-CCI program in which the QAA is the designated algorithm [37,40].

The daily datasets were then under-sampled to 25 km resolution to resolve only the broader oceanographic scales of variability.

2.2. Computation of b^{k}_{bp}

b^{k}_{bp} was estimated for every pixel and month by using the following basic relationship [7,25,27,28]:

$$b^{k}_{bp} = b_{bp} - k \cdot \text{Chl} \qquad (2)$$

where k is the slope of the least-square regression fit between daily Chl and b_{bp}, and b^{k}_{bp} is the intercept of the fit [7]. The term $k \cdot$Chl thus identifies the fraction of b_{bp} that covaries with Chl [7,25,27,28]. A scheme of the algorithm implementation is shown in Figure S1.

We computed monthly climatological b^{k}_{bp} maps at 25 km by applying Equation (2) to daily Chl and b_{bp} data within each month of the year from 1998 to 2019. The estimation of b^{k}_{bp}, and consequently of C_{phyto}, relies on good relationships between Chl and b_{bp} which then constitutes the fundamental condition to exploit this method. To this aim, for each b^{k}_{bp} map, the significance S (using Student's t-test), the Pearson correlation coefficient (r; not shown), and the 1σ uncertainty maps (due to the linear fit between Chl and b_{bp}; see Section 15.2 of [41]) were computed to give an estimate of the robustness of the fit.

We then applied for each pixel of every resulting monthly b^{k}_{bp} map an additional moving average of 1000 km to remove any source of noise and smaller scale variability, while conserving the large-scale oceanic patterns [42].

Finally, the monthly b^{k}_{bp} maps computed with the new approach were interpolated to obtain daily climatological b^{k}_{bp} maps at 25 km resolution. Daily pixel-based estimates were then used to assess C_{phyto} by applying a revised Equation (1) [7,27,28], such as:

$$C_{phyto} = \left[b_{bp}\ (443) - b^{k}_{bp}\ (443)\right]\cdot SF \tag{3}$$

In some pixels, C_{phyto} may show values less of, or close to 0. This occurred for pixels where the Chl–b_{bp} relationship had a significance $S < 0.95$ and $r \leq 0$. In these pixels, b^{k}_{bp} may be higher than b_{bp} thus giving non-reliable, negative estimations of C_{phyto} (less than 0; e.g., in the subtropical gyres). For these areas, we applied a threshold of C_{phyto} equal to 0.13 mg C m^{-3} to highlight that C_{phyto} is expected to be low [14]. All match-up points falling in the areas where the Chl–b_{bp} relationship showed a significance $S < 0.95$ and $r \leq 0$ were not included in the validation and algorithm performance analysis (Figure S1). These areas thus need to be interpreted and managed with caution.

For comparative purposes, C_{phyto} was also computed following the approaches of Beh05, Bel18, and Bre12 with single b^{k}_{bp} values of $3.5\cdot10^{-4}$ m^{-1}, $9.5\cdot10^{-4}$ m^{-1}, and $7.0\cdot10^{-4}$ m^{-1}, respectively. In addition, the performance of the new algorithm was compared to the empirical algorithms of Martinez-Vicente et al. (2017; hereafter MV17; [14]) and Graff et al., (2015; hereafter Gra15; [16].

2.3. In Situ Reference Data

The in situ C_{phyto} database [14] is a compilation of data acquired from many sources (e.g., MAREDAT and Atlantic Meridional Transect, AMT, cruises) for a total of N = 557 data points and consists of carbon biomass of picophytoplankton organisms (i.e., cell size < 2 μm). For complete details about the in situ dataset see [14]. Only pixels with a good ($S \geq 0.95$ and $r > 0$) satellite relationship between Chl and b_{bp} were retained so that the original 557 data points decreased to a total of 396 matchups (Figure 1). The final matchup database encompassed from oligotrophic to mesotrophic waters (Chl from 0.035 to 3.13 mg m^{-3} and C_{phyto} from 1.80 to 60.25 mg C m^{-3}) and OWC from 1 to 13. OWC from 1 to 6, corresponding to less productive waters, represented 56% of the in situ data.

2.4. Statistical Assessment

Estimated satellite C_{phyto} (y_i) computed with Equation (3) was compared to reference in situ data (x_i) by computing the bias (δ; mg m^{-3}), the relative bias in percentage (∇; in %), and the standard deviation of the difference (σ_Δ; mg m^{-3}). The assessment was made for each OWC individually and for all matchups at once:

$$\delta = \frac{1}{N}\sum_{i=1}^{N}(y_i - x_i) \tag{4}$$

$$\nabla = 100 * \frac{1}{N}\sum_{i=1}^{N}\frac{(y_i - x_i)}{x_i} \tag{5}$$

$$\sigma_{\Delta} = \sqrt{\frac{1}{N-1}\sum_{i=1}^{N}[(y_i - x_i) - \overline{(y_i - x_i)}]^2} \quad (6)$$

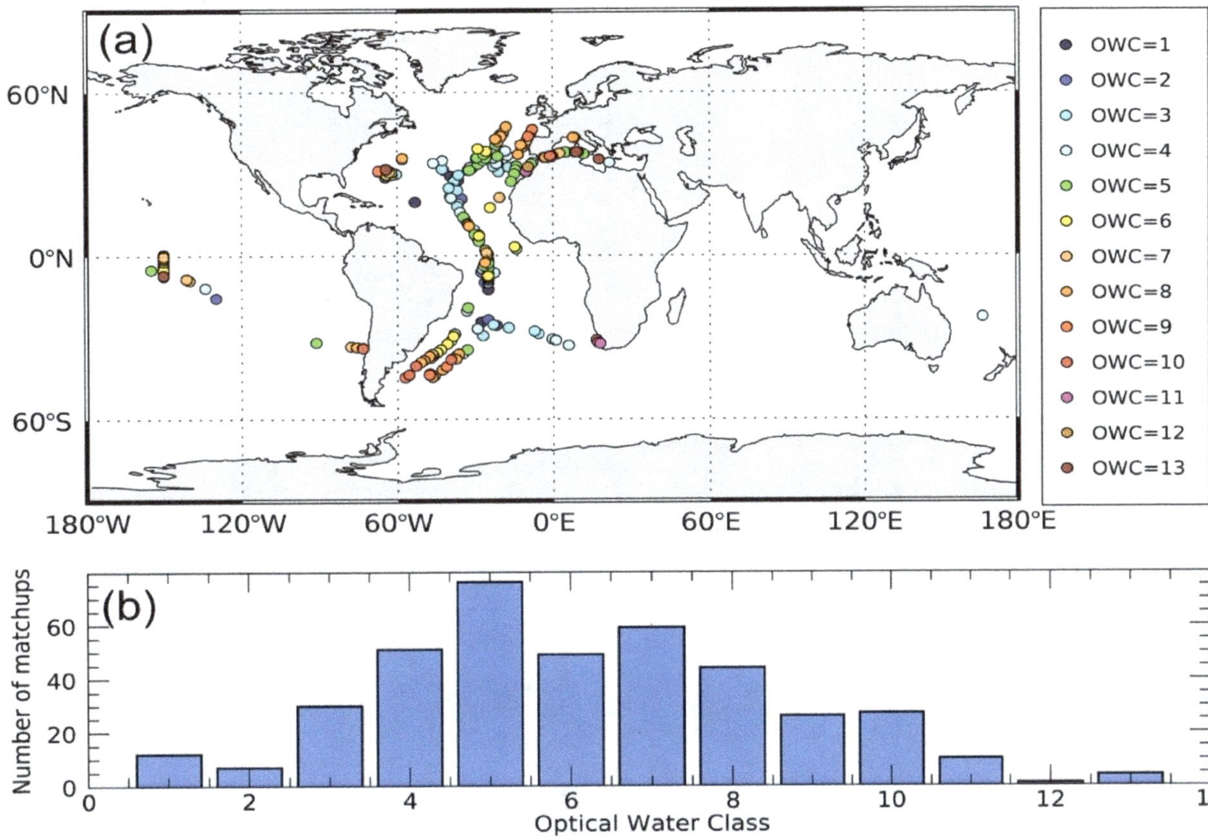

Figure 1. Geographical distribution of the matchup dataset (N = 396) used for the analysis. Each datapoint is also associated with the specific Optical Water Class (OWC) (**a**). Number of matchups per OWC (**b**).

3. Results

Algorithm Performance for C_{phyto} Retrievals

Results and statistics of the comparison between satellite C_{phyto} estimates obtained with our algorithm and the previously published values, and in situ C_{phyto}, are reported in Table 1 and Figure 2. Overall, algorithms perform with ∇ values spanning from 23.7% to 203%. The algorithm here developed shows the smallest bias (∇ of 23.7%). By comparison, the algorithms based on a constant b_{bp}^{k} value have lower performance, and Gra15, Beh05, and MV17 showed a systematic overestimation for low C_{phyto} values (Table 1 and Figure 2).

The b_{bp}-based algorithms that use a single b_{bp}^{k} constant value (Beh05, Bre12, and Bel18) show similar σ_{Δ}. Specifically, Bre12 has a b_{bp}^{k} value that is twice as high as that used in Beh05, and Bel18 has a value nearly three times larger than that of Beh05. ∇ ranges from 23.7% (Bel18) and 80.1% (Bre12), to 130.3% (Beh05), resulting in different performance in relation to the selection of the single value.

Table 1. Basic statistics for each single approach. For each approach is reported the bias (δ; in mg m^{-3}), the standard deviation (σ_Δ; in mg m^{-3}), and the relative percentage bias (∇; in %). Note that OWCs 1 and 2 are grouped in a single class to increase the number of observations available for the statistics. This is also replicated for OWCs 11, 12, and 13. The statistics using $\log_{10}$ transformed data for all OWCs are reported in Table S1.

OWCs	N. obs.	This Study			Bel18			Gra15			Beh05			Bre12			MV17		
		δ	σ_Δ	∇	δ	σ_Δ	∇	δ	σ_Δ	∇	δ	σ_Δ	∇	δ	σ_Δ	∇	δ	σ_Δ	∇
1:2	19	−1.9	5.0	5.0	−2.6	6.2	8.8	9.2	6.0	259.4	5.2	6.2	174.9	0.7	6.2	78.0	8.5	7.2	250.0
3	30	−1.8	3.4	−13.9	−2.8	3.4	−27.3	9.1	3.3	211.8	5.0	3.4	129.8	0.5	3.4	38.2	7.9	3.9	189.2
4	51	−1.7	7.6	18.2	−2.1	8.5	31.3	9.4	8.6	193.1	5.7	8.5	141.1	1.2	8.5	77.0	10.7	10.6	213.6
5	76	−1.6	6.6	8.1	−3.2	7.9	5.9	8.2	7.7	122.8	4.6	7.9	85.8	0.03	7.9	39.2	9.4	9.2	142.6
6	49	1.5	7.9	38.7	−0.2	9.3	26.9	11.0	9.0	131.4	7.6	9.3	99.7	3.1	9.3	57.2	14.4	11.4	163.0
7	59	−1.9	8.4	−0.4	0.2	9.5	13.7	11.3	9.0	102.9	8.0	9.5	76.0	3.5	9.5	39.7	15.2	12.4	132.3
8	44	−0.5	10.1	38.8	5.2	10.3	73.6	15.8	9.9	164.99	13.0	10.3	139.9	8.5	10.3	101.3	23.3	13.1	220.0
9	26	1.9	11.8	39.2	13.8	12.3	113.4	23.6	12.2	181.5	21.6	12.3	166.5	17.1	12.3	135.5	36.4	14.6	259.1
10	27	4.6	16.5	103.4	17.2	17.6	206.2	26.7	17.0	278.2	25.0	17.6	266.2	20.4	17.6	231.2	41.1	22.6	396.0
11:13	15	3.2	14.0	51.7	12.2	15.2	148.8	22.6	15.2	275.9	20.0	15.2	241.4	15.5	15.2	187.3	31.5	30.4	355.6
1:6	225	−1.0	6.8	14.0	−2.2	7.9	12.0	9.3	7.7	164.0	5.6	7.9	114.7	1.1	7.9	54.8	10.7	9.6	178.4
7:13	171	0.5	11.8	36.6	7.3	13.7	86.5	17.8	13.0	173.8	15.1	13.7	150.7	10.6	13.7	113.3	26.0	18.4	235.4
All	396	−0.4	9.2	23.7	1.9	11.8	44.2	13.0	11.1	168.2	9.7	11.8	130.3	5.2	11.8	80.1	17.3	16.0	203.0

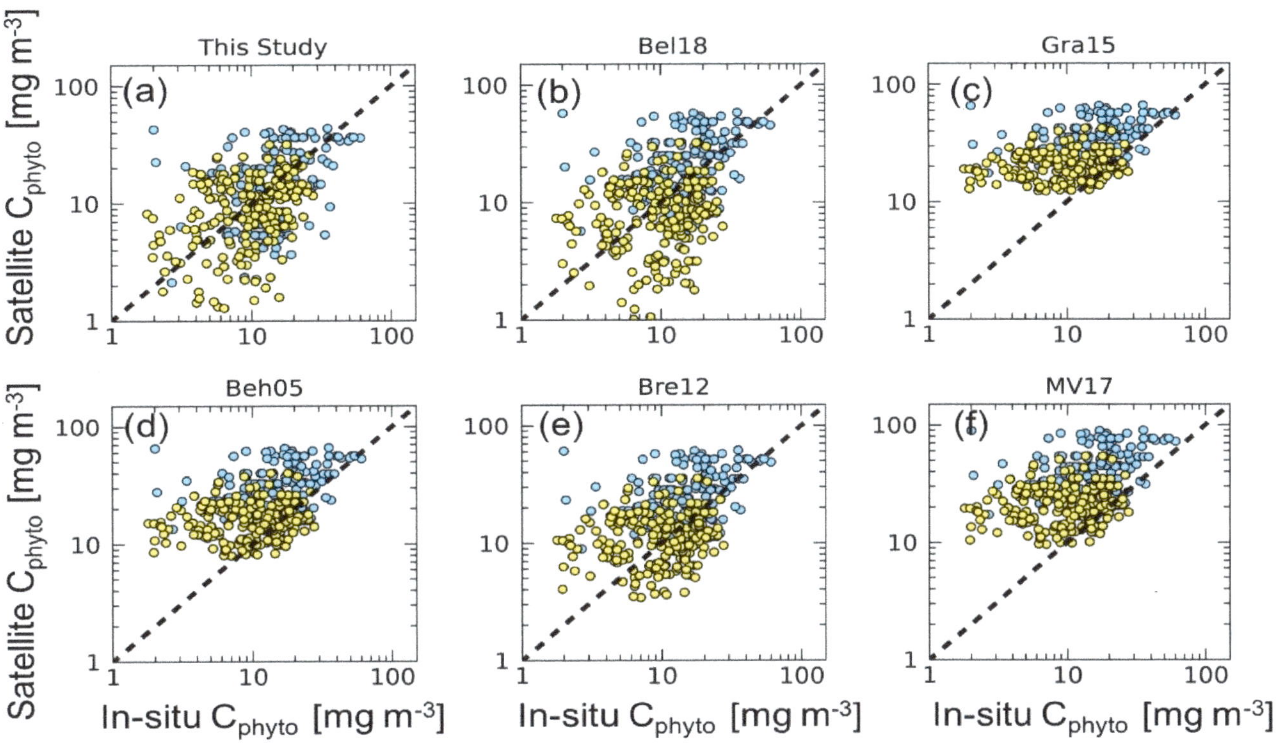

Figure 2. Matchups between satellite C_{phyto} versus in situ C_{phyto} data for each algorithm (N = 396). The dashed line represents the 1:1 ratio. Gold dots represent in situ C_{phyto} data corresponding to OWCs 1–6; blue dots represent in situ C_{phyto} data for OWCs 7–13. Statistics are reported in Table 1.

In the case of Beh05, a source of discrepancy leading to such a bias may be linked to the b_{bp} input from the algorithm as also highlighted in [27]. Indeed, the original equation was derived using the GSM algorithm [29] for obtaining a relationship between Chl and b_{bp} and thus for the b_{bp}^{k} estimation, instead of the QAA, applied here. In addition, it has recently been found that Raman scattering (here corrected prior to the QAA application) plays a fundamental role for the retrievals of b_{bp} in very clear waters because, if not corrected, it results in overestimation of b_{bp} [36,43].

However, these results highlight how the use of a single b_{bp}^{k} determines a strong overestimation of satellite C_{phyto}, in line with the results of MV17 (Table 1 and Figure 2). The comparison between our approach and previously published models underlines the necessity to account for the spatio-temporal variability of the b_{bp}^{k} for satellite-based C_{phyto} calculation. However, because Bre12 and Bel18 show relatively low bias, such approaches might be used alternatively to the new approach proposed here because it may occur in the most oligotrophic waters (e.g., subtropical gyre areas). Note that in Bre12, the b_{bp}^{k} was at 470 nm so that the overestimation of C_{phyto} can decrease if the b_{bp}^{k} is reported to the band of 443 nm.

The empirical algorithms of Gra15 and MV17 yield the worst results in terms of ∇. In the case of Gra15, this is the consequence of the few in situ measurements used for the algorithm definition, located in the Equatorial Pacific Ocean and in the Atlantic Ocean. Extrapolation of their algorithm, developed at 470 nm, to 443 nm, may also have had an effect. Regarding MV17, the results found are consistent with those reported in Tables 3 and 5 of their work.

MV17 reported that the geographical distribution of the in situ data, the median C_{phyto} concentrations, and the carbon-to-chlorophyll ratios are in agreement with previous observations in oligotrophic oceanic conditions [16,18]. The fact that the in situ C_{phyto} dataset used here contains only data from the picophytoplankton population must be taken into account for the interpretation of the algorithm evaluation. In most oligotrophic waters (OWCs from 1 to 6), δ shows the lowest values for all of the algorithms, with a maximum value of 14.4 mg m^{-3} (Table 1). However, when analyzing the ∇

index, the C_{phyto} computed by considering the spatio-temporal variability of b^{k}_{bp} is the most efficient (Table 1), with the exception of OWCs 5 and 6, for which Bel18 shows the best performance. Indeed, within OWCs in which picophytoplankton is expected to dominate (OWCs from 1 to 6), the performance of the majority of the tested algorithms generally improves (Table 1), with the exceptions of Gra15 and MV17. As expected, there is a decrease in the accuracy for all of the algorithms in the OWCs from 7 to 13. This could be influenced by the dominance of nano- and micro-phytoplankton (cell size > 2 μm) in these waters, whereas the picophytoplankton contribute in low relative proportions [14]. Because most of the b_{bp} signal is due to particles with equivalent diameters larger than that those associated with picophytoplankton [23], a decreasing performance of the b_{bp}-based algorithms in water dominated by larger cells (e.g., nano- and micro-phytoplankton) was expected (Table 1).

4. Discussion

4.1. Spatial and Temporal Distribution of C_{phyto}

The global mean climatology of C_{phyto} (Figure 3a) agrees with the expected geographical distribution [14]. C_{phyto} highs are found in the most productive regions such as the high latitude regions and coastal areas, and in upwelling systems such as those off Mauritania and Perù, the Benguela and Kuroshio currents, and along the equatorial belt of the Pacific Ocean (Figure 3a). C_{phyto} lows are found in mid-latitude areas such as the subtropical gyres or the Eastern Mediterranean Sea and Black Sea. C_{phyto} varies within the same order of magnitude of recent estimates made using satellite data [14] and biogeochemical models [44]. The two main oceanic regimes as defined by [45] are also highlighted from our analysis: "photoacclimation-dominance" and "biomass-dominance". The former is typical of oligotrophic areas (e.g., subtropical gyres) where the variability of Chl is uncoupled from biomass and the process of acclimation to light and nutrients drives Chl variations [45]. In these areas, C_{phyto} shows the lowest mean values (<10 mg C m^{-3}). Conversely, the latter is typical of most productive areas where Chl and b_{bp} strongly covary [31,46]. The high Chl-b_{bp} co-variability indicates that particles (and biomass) co-vary with phytoplankton abundance, whereas the physiological photoacclimation process plays a secondary role in determining the Chl variations; e.g., in the North Atlantic and Southern Oceans, C_{phyto} shows the highest mean values (>30 mg C m^{-3}).

The amplitude of seasonal variations (i.e., interannual variability) of C_{phyto} is shown in Figure 3b. Most of the variability is observed at high latitudes (e.g., Baltic Sea, Irish Sea, Norwegian Sea, North Sea), in the North Pacific and North Atlantic Oceans, and in the Southern Ocean, because σ_{std} values are higher than 12 mg m^{-3}. The North Atlantic and Southern Oceans are characterized by high seasonal variations due to intense productive months followed by unproductive periods [47–49] (Figures S2–S13). In the North Atlantic Ocean, C_{phyto} starts to increase in May, reaching the maximum values in June and July (>50 mg C m^{-3}) following the spring bloom. Then, C_{phyto} decreases to the lowest values for the remainder of the year [47]. In the Southern Ocean, C_{phyto} follows the typical seasonal cycle with high values during the most productive months which occur from November to March (>50 mg C m^{-3}). Highs are mostly observed in the Antarctic Circumpolar Current (ACC) and the Patagonia Shelf. Later, there is a decrease in C_{phyto} from May to September (Figures S2–S13). Additionally, the northwestern Indian Ocean, known to be an Oxygen Minimum Zone, as well as the Benguela upwelling system, also show significant variability in C_{phyto}, with σ_{std} greater than 12 mg m^{-3}. Conversely, mid-latitude areas (e.g., in the subtropical gyres), display much lower variability, with σ_{std} less than 2 mg C m^{-3}.

Figure S14 clearly shows how the single b^{k}_{bp} of Beh05 is located at the lowest limit of possible values, confirming that reported in [28], whereas the Bel18 and Bre12 single values are representative of a wider area. In the subtropical gyres, b^{k}_{bp} does not significantly deviate from the coefficient found by Beh05, but in the most productive regions (e.g., North Atlantic and Southern Oceans, high latitude seas, and coastal upwelling regions), they largely differ with great impact on C_{phyto} estimations. This discrepancy in C_{phyto} is still valid in the case of the use of Bel18 and Bre12 single values, but for other areas (e.g., in the subtropical gyres or in the Mediterranean Sea), the use of the single value of Bel18 or Bre12 yields an underestimation of C_{phyto}, whereas in the productive Arctic and Southern Oceans C_{phyto} is generally overestimated. This is also indicated in the seasonal b_{bp}^{k} variation as reported in the Figures S2–S13. In the North Atlantic Ocean, b_{bp}^{k} shows high values during the spring bloom and low values from December to April. However, in this area, the b_{bp}^{k} is only a small fraction of the entire b_{bp} signal as the consequence of the high variability in the b_{bp} and phytoplankton abundance [25,27] thus following the phytoplankton dynamics. In the Southern Ocean, and especially the region influenced by the ACC and Patagonia Shelf, b_{bp}^{k} shows highs from December to April (i.e., due to coccoliths) and lows in the period May–September in accordance also with the C_{phyto} seasonal cycle [25]. In less productive seas, the seasonal cycle of b_{bp}^{k} is smooth, suggesting low NAP variations. In these areas, the seasonal b_{bp}^{k} change is weak throughout the year but also dominates the b_{bp} signal [25]. These areas are characterized by limited nutrient availability determining low C_{phyto}, mostly of picophytoplankton class, high bacterial, and detrital concentrations [50–52]. It follows that the use of a varying b_{bp}^{k} enables accounting for its seasonal variations in the C_{phyto} computation, giving the model more flexibility in relation to the change of the biogeochemical and trophic conditions.

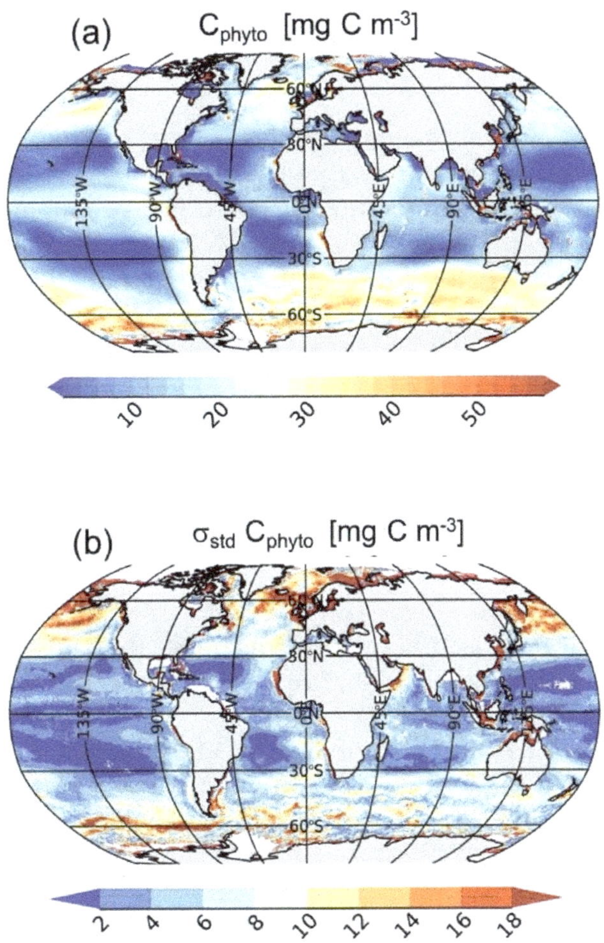

Figure 3. Annual global mean (**a**) and standard deviation (**b**) of C_{phyto}. Note that at high latitudes the number of satellite observations used for the mean and standard deviation computations are lower than those used at low or mid-latitudes due to the winter darkness (i.e., high sun zenith angle).

4.2. *Caveats of the C_{phyto} Algorithm*

Three main caveats of the algorithm should be noted. The first is that C_{phyto} assumes a constant scaling factor, SF, equal to 13,000 mg C m^{-2}, following [7]. This can be an additional possible source of error in the satellite retrievals of C_{phyto}. Therefore, one important future effort should be to investigate a refined scaling factor relating b_{bp} to C_{phyto} coupled with the b^{k}_{bp} space–time variability as presented in this work. With this perspective, additional laboratory work should be done to evaluate if change in SF values can affect the C_{phyto} estimations.

The second caveat of the algorithm is that it relies on a tight relationship between b_{bp} and Chl, which is influenced by the algorithms (including the atmospheric correction) used for Chl and b_{bp} retrievals, in addition to the environmental conditions. In some areas of the subtropical gyres, the Chl–b_{bp} relationship showed $S < 0.95$ (Figures S2–S13) and r smaller than or close to 0, as also reported by [28]. The main reason for this relationship may be the photoacclimation process, known to be dominant in such areas over the year, and which introduces variability in Chl that is unrelated to biomass. In these areas, our refined method must be applied with caution. With this perspective, one future challenge is to solve this limitation in b^{k}_{bp} computation and, consequently, in the satellite C_{phyto}. Indeed, as the subtropical gyres cover about 60% of the global ocean, it can be of a great importance to compute C_{phyto} and Chl:C_{phyto} to understand oceanic biogeochemical cycles. Conversely, the remaining areas of global ocean (e.g., the North Atlantic Ocean) present significant positive r and highly significant S (Figures S2–S13) throughout the year, indicating that either b_{bp} or Chl can be used for determining the phytoplankton dynamics and distribution, as expected in open-ocean waters [27]. In these areas, the algorithm performance is robust.

The third caveat is that the algorithm validation is restricted only to in situ C_{phyto} data associated with picophytoplankton carbon. It determines that the algorithm performance has to be interpreted with caution in those areas where other phytoplankton size classes dominate. This means that, currently, it is not possible to define the accuracy for OWCs from 7 to 13. Indeed, Table 1 shows that there is a decrease in the accuracy for all of the algorithms in the OWCs where nano- and micro-phytoplankton dominates. Thus, one future need is to improve the in situ C_{phyto} dataset with new measurements representative of all of the phytoplankton size classes.

5. Conclusions and Future Perspectives

In this work, a revisited version of the original b_{bp}-based algorithm for C_{phyto} retrieval from space [7] was proposed. The spatio-temporal variations of the background backscattering coefficient of non-algal particles (b^{k}_{bp}) is the crucial part of this refinement, and builds upon a series of recent studies [25,27,28]. The main findings are:

1. The new C_{phyto} algorithm proposed here performs better than any previously published model, with a relative error of 24% (Table 1 and Figure 2) with respect to a reference in situ dataset.

2. The new algorithm shows the lowest error in C_{phyto} (14.0%) across most of the OWCs in which the picophytoplankton population dominates (Table 1). On the contrary, the highest errors (36.6%) occur in OWCs 7–13 in which larger phytoplankton cells are supposed to dominate (Table 1).

However, we acknowledge that the refined algorithm presented here could be improved by addressing the caveats mentioned above. Enlarging the databases of quality-controlled and freely accessible in situ C_{phyto} at the different size-classes (pico-, nano-, and micro-), measured simultaneously with an array of optical variables (particularly b_{bp}) could help to reduce uncertainties in C_{phyto} retrieval from space. Improving the accuracy of satellite C_{phyto} could be of great importance to adding new knowledge regarding the contribution of phytoplankton to particulate organic carbon, and for the validation of ocean primary productivity and biogeochemical models at a wider scale [8,12].

Author Contributions: Conceptualization, M.B., J.P., V.M.-V. and S.M.; Methodology, M.B., J.P., E.O. and S.M.; Data curation, M.B., S.M., J.P.; Writing—review and editing, all authors contributed equally. All authors have read and agreed to the published version of the manuscript.

Acknowledgments: We thank ESA OC-CCI for providing publicly available ocean colour data (https://esa-oceancolour-cci.org). Giorgio Dall'Olmo (PML), Rosalia Santoleri (CNR-ISMAR), Vittorio Brando (CNR-ISMAR) and Marie-Hélène Rio (ESA) are thanked for all the inputs, suggestions and discussion done since the early version of the manuscript. We wish to thank the four anonymous reviewers for their criticisms and suggestions that helped the manuscript to be improved.

References

1. Siegel, D.A.; Buesseler, K.O.; Doney, S.C.; Sailley, S.F.; Behrenfeld, M.J.; Boyd, P.W. Global assessment of ocean carbon export by combining satellite observations and food-web models. *Glob. Biogeochem. Cycles* **2014**, *28*, 181–196. [CrossRef]
2. Boyce, D.G.; Lewis, M.R.; Worm, B. Global phytoplankton decline over the past century. *Nature* **2010**, *466*, 591–596. [CrossRef] [PubMed]
3. Dutkiewicz, S.; Hickman, A.E.; Jahn, O.; Henson, S.; Beaulieu, C.; Monier, E. Ocean colour signature of climate change. *Nat. Commun.* **2019**, *10*, 578. [CrossRef] [PubMed]
4. O'Reilly, J.E.; Werdell, P.J. Chlorophyll algorithms for ocean color sensors—OC4, OC5 & OC6. *Remote Sens. Environ.* **2019**, *229*, 32–47. [CrossRef]
5. Laws, E.A.; Bannister, T.T. Nutrient- and light-limited growth of Thalassiosira fluviatilis in continuous culture, with implications for phytoplankton growth in the ocean1. *Limnol. Oceanogr.* **1980**, *25*, 457–473. [CrossRef]
6. Behrenfeld, M.J.; Boss, E.S. Resurrecting the Ecological Underpinnings of Ocean Plankton Blooms. *Annu. Rev. Mar. Sci.* **2014**, *6*, 167–194. [CrossRef] [PubMed]
7. Behrenfeld, M.J.; Boss, E.; Siegel, D.A.; Shea, D.M. Carbon-based ocean productivity and phytoplankton physiology from space. *Glob. Biogeochem. Cycles* **2005**, *19*. [CrossRef]
8. Westberry, T.; Behrenfeld, M.J.; Siegel, D.A.; Boss, E. Carbon-based primary productivity modeling with vertically resolved photoacclimation. *Glob. Biogeochem. Cycles* **2008**, *22*. [CrossRef]
9. Behrenfeld, M.J.; Halsey, K.H.; Milligan, A.J. Evolved physiological responses of phytoplankton to their integrated growth environment. *Philos. Trans. R. Soc. B Biol. Sci.* **2008**, *363*, 2687–2703. [CrossRef] [PubMed]
10. Zhai, L.; Platt, T.; Tang, C.; Dowd, M.; Sathyendranath, S.; Forget, M.-H. Estimation of phytoplankton loss rate by remote sensing. *Geophys. Res. Lett.* **2008**, *35*. [CrossRef]
11. Zhai, L.; Platt, T.; Tang, C.; Sathyendranath, S.; Fuentes-Yaco, C.; Devred, E.; Wu, Y. Seasonal and geographic variations in phytoplankton losses from the mixed layer on the Northwest Atlantic Shelf. *J. Mar. Syst.* **2010**, *80*, 36–46. [CrossRef]
12. Dutkiewicz, S.; Hickman, A.E.; Jahn, O.; Gregg, W.W.; Mouw, C.B.; Follows, M.J. Capturing optically important constituents and properties in a marine biogeochemical and ecosystem model. *Biogeosciences* **2015**, *12*, 4447–4481. [CrossRef]
13. Fossum, K.N.; Ovadnevaite, J.; Ceburnis, D.; Dall'Osto, M.; Marullo, S.; Bellacicco, M.; Simó, R.; Liu, D.; Flynn, M.; Zuend, A.; et al. Summertime Primary and Secondary Contributions to Southern Ocean Cloud Condensation Nuclei. *Sci. Rep.* **2018**, *8*, 13844. [CrossRef]
14. Martínez-Vicente, V.; Evers-King, H.; Roy, S.; Kostadinov, T.S.; Tarran, G.A.; Graff, J.R.; Brewin, R.J.W.; Dall'Olmo, G.; Jackson, T.; Hickman, A.E.; et al. Intercomparison of Ocean Color Algorithms for Picophytoplankton Carbon in the Ocean. *Front. Mar. Sci.* **2017**, *4*. [CrossRef]
15. Martínez-Vicente, V.; Dall'Olmo, G.; Tarran, G.; Boss, E.; Sathyendranath, S. Optical backscattering is correlated with phytoplankton carbon across the Atlantic Ocean. *Geophys. Res. Lett.* **2013**, *40*, 1154–1158. [CrossRef]

16. Graff, J.R.; Westberry, T.K.; Milligan, A.J.; Brown, M.B.; Dall'Olmo, G.; Dongen-Vogels, V.V.; Reifel, K.M.; Behrenfeld, M.J. Analytical phytoplankton carbon measurements spanning diverse ecosystems. *Deep Sea Res. Part I Oceanogr. Res. Pap.* **2015**, *102*, 16–25. [CrossRef]
17. Sathyendranath, S.; Stuart, V.; Nair, A.; Oka, K.; Nakane, T.; Bouman, H.; Forget, M.H.; Maass, H.; Platt, T. Carbon-to-chlorophyll ratio and growth rate of phytoplankton in the sea. *Mar. Ecol. Prog. Ser.* **2009**, *383*, 73–84. [CrossRef]
18. Marañón, E.; Cermeño, P.; Huete-Ortega, M.; López-Sandoval, D.C.; Mouriño-Carballido, B.; Rodríguez-Ramos, T. Resource Supply Overrides Temperature as a Controlling Factor of Marine Phytoplankton Growth. *PLoS ONE* **2014**, *9*, e99312. [CrossRef]
19. Kostadinov, T.S.; Siegel, D.A.; Maritorena, S. Retrieval of the particle size distribution from satellite ocean color observations. *J. Geophys. Res. Ocean.* **2009**, *114*. [CrossRef]
20. Kostadinov, T.S.; Milutinović, S.; Marinov, I.; Cabré, A. Carbon-based phytoplankton size classes retrieved via ocean color estimates of the particle size distribution. *Ocean Sci.* **2016**, *12*, 561–575. [CrossRef]
21. Roy, S.; Sathyendranath, S.; Platt, T. Size-partitioned phytoplankton carbon and carbon-to-chlorophyll ratio from ocean colour by an absorption-based bio-optical algorithm. *Remote Sens. Environ.* **2017**, *194*, 177–189. [CrossRef]
22. Stramski, D.; Boss, E.; Bogucki, D.; Voss, K.J. The role of seawater constituents in light backscattering in the ocean. *Prog. Oceanogr.* **2004**, *61*, 27–56. [CrossRef]
23. Organelli, E.; Dall'Olmo, G.; Brewin, R.J.W.; Tarran, G.A.; Boss, E.; Bricaud, A. The open-ocean missing backscattering is in the structural complexity of particles. *Nat. Commun.* **2018**, *9*, 5439. [CrossRef] [PubMed]
24. Zhang, X.; Hu, L.; Xiong, Y.; Huot, Y.; Gray, D. Experimental Estimates of Optical Backscattering Associated with Submicron Particles in Clear Oceanic Waters. *Geophys. Res. Lett.* **2020**, *47*, e2020GL087100. [CrossRef]
25. Bellacicco, M.; Cornec, M.; Organelli, E.; Brewin, R.J.W.; Neukermans, G.; Volpe, G.; Barbieux, M.; Poteau, A.; Schmechtig, C.; D'Ortenzio, F.; et al. Global Variability of Optical Backscattering by Non-algal particles From a Biogeochemical-Argo Data Set. *Geophys. Res. Lett.* **2019**, *46*, 9767–9776. [CrossRef]
26. Brewin, R.J.W.; Dall'Olmo, G.; Sathyendranath, S.; Hardman-Mountford, N.J. Particle backscattering as a function of chlorophyll and phytoplankton size structure in the open-ocean. *Opt. Express* **2012**, *20*, 17632–17652. [CrossRef]
27. Bellacicco, M.; Volpe, G.; Colella, S.; Pitarch, J.; Santoleri, R. Influence of photoacclimation on the phytoplankton seasonal cycle in the Mediterranean Sea as seen by satellite. *Remote Sens. Environ.* **2016**, *184*, 595–604. [CrossRef]
28. Bellacicco, M.; Volpe, G.; Briggs, N.; Brando, V.; Pitarch, J.; Landolfi, A.; Colella, S.; Marullo, S.; Santoleri, R. Global Distribution of Non-algal Particles from Ocean Color Data and Implications for Phytoplankton Biomass Detection. *Geophys. Res. Lett.* **2018**, *45*, 7672–7682. [CrossRef]
29. Garver, S.A.; Siegel, D.A. Inherent optical property inversion of ocean color spectra and its biogeochemical interpretation: 1. Time series from the Sargasso Sea. *J. Geophys. Res. Ocean.* **1997**, *102*, 18607–18625. [CrossRef]
30. Dall'Olmo, G.; Westberry, T.K.; Behrenfeld, M.J.; Boss, E.; Slade, W.H. Significant contribution of large particles to optical backscattering in the open ocean. *Biogeosciences* **2009**, *6*, 947–967. [CrossRef]
31. Dall'Olmo, G.; Boss, E.; Behrenfeld, M.; Westberry, T. Particulate optical scattering coefficients along an Atlantic Meridional Transect. *Opt. Express* **2012**, *20*, 21532–21551. [CrossRef] [PubMed]
32. Jackson, T.; Sathyendranath, S.; Mélin, F. An improved optical classification scheme for the Ocean Colour Essential Climate Variable and its applications. *Remote Sens. Environ.* **2017**, *203*, 152–161. [CrossRef]
33. Jackson, T.; Chuprin, A.; Sathyendranath, S.; Grant, M.; Zühlke, M.; Dingle, J.; Storm, T.; Boettcher, M.; Fomferra, N. Ocean Colour Climate Change Initiative (OC_CCI)—Interim Phase. Product User Guide. D3.4 PUG. 2019. Available online: https://esa-oceancolour-cci.org/sites/esa-oceancolour-cci.org/alfresco.php?file=a68aa514-3668-4935-9235-fca10f7e8bee&name=OC-CCI-PUG-v4.1-v1.pdf (accessed on 22 December 2019).
34. Lee, Z.; Carder, K.L.; Arnone, R.A. Deriving inherent optical properties from water color: A multiband quasi-analytical algorithm for optically deep waters. *Appl. Opt.* **2002**, *41*, 5755–5772. [CrossRef]
35. Lee, Z. Update of the Quasi-Analytical Algorithm (QAA_v6). Available online: http://www.ioccg.org/groups/Software_OCA/QAA_v6_2014209.pdf (accessed on 22 December 2019).
36. Pitarch, J.; Bellacicco, M.; Organelli, E.; Volpe, G.; Colella, S.; Vellucci, V.; Marullo, S. Retrieval of Particulate

Backscattering Using Field and Satellite Radiometry: Assessment of the QAA Algorithm. *Remote Sens.* **2020**, *12*, 77. [CrossRef]

37. Brewin, R.J.W.; Sathyendranath, S.; Müller, D.; Brockmann, C.; Deschamps, P.-Y.; Devred, E.; Doerffer, R.; Fomferra, N.; Franz, B.; Grant, M.; et al. The Ocean Colour Climate Change Initiative: III. A round-robin comparison on in-water bio-optical algorithms. *Remote Sens. Environ.* **2015**, *162*, 271–294. [CrossRef]
38. Mélin, F.; Berthon, J.-F.; Zibordi, G. Assessment of apparent and inherent optical properties derived from SeaWiFS with field data. *Remote Sens. Environ.* **2005**, *97*, 540–553. [CrossRef]
39. Melin, F.; Zibordi, G.; Berthon, J. Uncertainties in Remote Sensing Reflectance from MODIS-Terra. *IEEE Geosci. Remote Sens. Lett.* **2012**, *9*, 432–436. [CrossRef]
40. Sathyendranath, S.; Brewin, R.J.W.; Brockmann, C.; Brotas, V.; Calton, B.; Chuprin, A.; Cipollini, P.; Couto, A.B.; Dingle, J.; Doerffer, R.; et al. An Ocean-Colour Time Series for Use in Climate Studies: The Experience of the Ocean-Colour Climate Change Initiative (OC-CCI). *Sensors* **2019**, *19*, 4285. [CrossRef]
41. Press, W.H.; Teukolsky, S.A.; Vetterling, W.T.; Flannery, B.P. *Numerical Recipes in C*; Cambridge University Press: Cambridge, UK, 1988.
42. Resplandy, L.; Lévy, M.; McGillicuddy, D.J., Jr. Effects of Eddy-Driven Subduction on Ocean Biological Carbon Pump. *Glob. Biogeochem. Cycles* **2019**, *33*, 1071–1084. [CrossRef]
43. Westberry, T.K.; Boss, E.; Lee, Z. Influence of Raman scattering on ocean color inversion models. *Appl. Opt.* **2013**, *52*, 5552–5561. [CrossRef]
44. Arteaga, L.; Pahlow, M.; Oschlies, A. Modeled Chl:C ratio and derived estimates of phytoplankton carbon biomass and its contribution to total particulate organic carbon in the global surface ocean. *Glob. Biogeochem. Cycles* **2016**, *30*, 1791–1810. [CrossRef]
45. Siegel, D.A.; Maritorena, S.; Nelson, N.B.; Behrenfeld, M.J. Independence and interdependencies among global ocean color properties: Reassessing the bio-optical assumption. *J. Geophys. Res. Ocean.* **2005**, *110*. [CrossRef]
46. Westberry, T.K.; Dall'Olmo, G.; Boss, E.; Behrenfeld, M.J.; Moutin, T. Coherence of particulate beam attenuation and backscattering coefficients in diverse open ocean environments. *Opt. Express* **2010**, *18*, 15419–15425. [CrossRef]
47. Westberry, T.K.; Schultz, P.; Behrenfeld, M.J.; Dunne, J.P.; Hiscock, M.R.; Maritorena, S.; Sarmiento, J.L.; Siegel, D.A. Annual cycles of phytoplankton biomass in the subarctic Atlantic and Pacific Ocean. *Glob. Biogeochem. Cycles* **2016**, *30*, 175–190. [CrossRef]
48. Balch, W.M. The Ecology, Biogeochemistry, and Optical Properties of Coccolithophores. *Annu. Rev. Mar. Sci.* **2018**, *10*, 71–98. [CrossRef]
49. Barbieux, M.; Uitz, J.; Bricaud, A.; Organelli, E.; Poteau, A.; Schmechtig, C.; Gentili, B.; Obolensky, G.; Leymarie, E.; Penkerc'h, C.; et al. Assessing the Variability in the Relationship between the Particulate Backscattering Coefficient and the Chlorophyll a Concentration from a Global Biogeochemical-Argo Database. *J. Geophys. Res. Ocean.* **2018**, *123*, 1229–1250. [CrossRef]
50. Heywood, J.L.; Zubkov, M.V.; Tarran, G.A.; Fuchs, B.M.; Holligan, P.M. Prokaryoplankton standing stocks in oligotrophic gyre and equatorial provinces of the Atlantic Ocean: Evaluation of inter-annual variability. *Deep Sea Res. Part II: Top. Stud. Oceanogr.* **2006**, *53*, 1530–1547. [CrossRef]
51. Grob, C.; Ulloa, O.; Claustre, H.; Huot, Y.; Alarcon, G.; Marie, D. Contribution of picoplankton to the total particulate organic carbon concentration in the eastern South Pacific. *Biogeosciences* **2007**, *4*, 837–852. [CrossRef]
52. Organelli, E.; Dall'Olmo, G.; Brewin, R.J.; Nencioli, F.; Tarran, G.A. Drivers of spectral optical scattering by particles in the upper 500 m of the Atlantic Ocean. *Opt. Express* **2020**, *28*, 34147–34166. [CrossRef]

3

Measuring and Modeling the Polarized Upwelling Radiance Distribution in Clear and Coastal Waters

Arthur C. R. Gleason [1], Kenneth J. Voss [1,*], Howard R. Gordon [1], Michael S. Twardowski [2] and Jean-François Berthon [3]

[1] Physics Department, University of Miami, Coral Gables, FL 33146, USA; agleason@physics.miami.edu (A.C.R.G.); gordon@physics.miami.edu (H.R.G.)
[2] Harbor Branch Oceanographic Institute, Fort Pierce, FL 34946, USA; mtwardowski@fau.edu
[3] Joint Research Centre of the European Commission, 21027 Ispra, Italy; Jean-Francois.berthon@ec.europa.eu
* Correspondence: voss@physics.miami.edu

Abstract: The upwelling spectral radiance distribution is polarized, and this polarization varies with the optical properties of the water body. Knowledge of the polarized, upwelling, bidirectional radiance distribution function (BRDF) is important for generating consistent, long-term data records for ocean color because the satellite sensors from which the data are derived are sensitive to polarization. In addition, various studies have indicated that measurement of the polarization of the radiance leaving the ocean can used to determine particle characteristics (Tonizzo et al., 2007; Ibrahim et al., 2016; Chami et al., 2001). Models for the unpolarized BRDF (Morel et al., 2002; Lee et al., 2011) have been validated (Voss et al., 2007; Gleason et al., 2012), but variations in the polarization of the upwelling radiance due to the sun angle, viewing geometry, dissolved material, and suspended particles have not been systematically documented. In this work, we simulated the upwelling radiance distribution using a Monte Carlo-based radiative transfer code and measured it using a set of fish-eye cameras with linear polarizing filters. The results of model-data comparisons from three field experiments in clear and turbid coastal conditions showed that the degree of linear polarization (*DOLP*) of the upwelling light field could be determined by the model with an absolute error of ± 0.05 (or 5% when the *DOLP* was expressed in %). This agreement was achieved even with a fixed scattering Mueller matrix, but did require in situ measurements of the other inherent optical properties, e.g., scattering coefficient, absorption coefficient, etc. This underscores the difficulty that is likely to be encountered using the particle scattering Mueller matrix (as indicated through the remote measurement of the polarized radiance) to provide a signature relating to the properties of marine particles beyond the attenuation/absorption coefficient.

Keywords: polarization; ocean optics; upwelling radiance distribution; remote sensing

---

1. Introduction

The in-water and water-leaving radiance in the ocean is partially polarized and this has implications for both biological activity [1] and for viewing the ocean from a satellite [2]. The biological implications of the polarization are a current topic of research [3] but the implications for ocean color remote sensing are just being exploited [2,4]. Various studies have indicated that measurement of the polarization of the radiance leaving the ocean can be used to determine particle characteristics [5–7]. In addition to using the polarization of the water leaving radiance for studies of the water properties, polarization may also be important for processing data from ocean color sensors that have unintended polarization sensitivities [8,9].

Measurements of the scalar (without regard to polarization) spectral upwelling radiance distribution have occurred more frequently since the development of the radiance distribution camera, RADS [10], and then the upwelling radiance distribution camera, NuRADS [11]. These instruments use electro-optic camera systems combined with filter changers and fisheye cameras to image the complete upwelling radiance distribution, for a specific wavelength, in one image. The CCD (charge coupled device) resolution and optics allow measurement of the radiance distribution with a 1° angular resolution. These systems have been used in studies of the in-water light field [12] and studies of the angular radiance distribution variations in ocean color algorithms [13,14]. Recently, two new cameras, polarized radiance camera (PolRADS) [15] and downwelling polarized camera (DPOL) [16], have been used to make measurements of the polarized spectral upwelling radiance distribution, as described below.

Measurements of the upwelling, in situ, polarized radiance have typically been done with variations of a Gershun-tube radiometer, obtaining the angular distribution by changing the viewing direction of the radiometer [17,18]. These instruments have advantages such as the simplicity of calibration and the ability to do hyperspectral measurements. Hyperspectral measurements of the polarized light field have been suggested as a method to measure in situ solar stimulated fluorescence [19]. Unfortunately, to obtain the full radiance distribution requires many measurements during which the illumination conditions can change, and spatially varying angular features resulting from the sea surface may not be captured. By making measurements using fisheye lenses, while limited to multi-spectral measurements rather than hyperspectral measurements, higher angular spatial resolution can be obtained nearly simultaneously (typical exposure times are less than 1 s).

Models for the unpolarized bi-directional reflectance distribution function (BRDF) [20,21] have been validated in Case I [13] and Case II [14] waters, but variations in the polarization of upwelling radiance due to sun angle, viewing geometry, dissolved material, and suspended particles have not been systematically documented.

2. Methods

The light field polarization is easily described by use of the four element Stokes vector, I, as in, for example, Bohren and Huffman [22]. The four components of I are prescribed relative to some reference plane, which here we choose as the plane defined by the viewing direction and the nadir direction. Consider an electromagnetic wave of angular frequency ω. If the unit vector $\hat{\ell}$ is parallel to, and the unit vector $\hat{r}$ is perpendicular to, the reference plane (such that $\hat{\ell} \times \hat{r}$ is in the direction of propagation), the electric field can be written as $\vec{E} = E_\ell \hat{\ell} + E_r \hat{r}$, with:

$$E_\ell = E_{\ell 0} \exp[i(\omega t + \delta_\ell)], \tag{1a}$$

$$E_r = E_{r0} \exp[i(\omega t + \delta_r)] \tag{1b}$$

where $E_{\ell 0}$ and E_{r0} are real. The four components of the Stokes vector are then defined as:

$$I \equiv E_\ell E_\ell^* + E_r E_r^* = E_{l0}^2 + E_{r0}^2 \tag{2a}$$

$$Q \equiv E_\ell E_\ell^* - E_r E_r^* = E_{l0}^2 - E_{r0}^2, \tag{2b}$$

$$U \equiv E_\ell E_r^* + E_r E_\ell^* = 2E_{l0} E_{r0} cos\delta \tag{2c}$$

$$V \equiv i(E_\ell E_r^* - E_r E_\ell^*) = 2E_{l0} E_{r0} sin\delta \tag{2d}$$

where $\delta \equiv \delta_r - \delta_\ell$. A derived parameter that we will use in our comparison is the degree of linear polarization, $DOLP$, defined as:

$$DOLP = \sqrt{\frac{Q^2 + U^2}{I^2}} \tag{3}$$

Another useful parameter is the angle of the plane of polarization, χ, which is defined as:

$$\chi = \frac{1}{2}\arctan\left(\frac{U}{Q}\right) \tag{4}$$

We measured the Stokes vector over the upwelling hemisphere using the PolRADS and DPOL cameras and simulated it using a Monte Carlo-based radiative transfer (RT) code. Data were available from three field campaigns: Hawaii, USA, in December 2005; the Ligurian Sea in March 2009; and the New York Bight in May 2009.

2.1. *Measurements of the Polarized Upwelling Radiance Distribution*

In Hawaii, measurements of the polarized upwelling radiance distribution were accomplished using the PolRADS camera system [15]. PolRADS is based on the NuRADS camera system [12], which is a compact (30 cm diameter, 30 cm length), multispectral camera that images the upwelling radiance distribution in six narrow ($\approx$10 nm full-width at half-maximum, FWHM) spectral bands centered at 412, 436, 486, 526, 548, and 615 nm. In the PolRADS instrument, three synchronized NuRADS cameras are used, each with a linear polarizer, to simultaneously acquire images. Combining the images from three NuRADS cameras, when in PolRADS configuration, allows for retrieval of three elements of the Stokes vector (I, Q, and U, but not V), as well as *DOLP*.

In the Ligurian Sea and New York Bight, measurements of the polarized upwelling radiance distribution were accomplished with the DPOL camera system [16]. DPOL is similar to PolRADS in the sense that simultaneous measurements are made using multiple fisheye lenses with different polarizing filters in the optical path. DPOL differs from PolRADS by using optical-fiber bundles to project images from all lenses through a single spectral filter and onto a single CCD array. Eliminating the redundant filters and cameras makes DPOL much smaller than PolRADS, thereby reducing the instrument shadow. Furthermore, DPOL has four lenses, rather than the three on PolRADS, enabling retrieval of all four Stokes vector elements. The spectral filters used on DPOL were also $\approx$10 nm FWHM and were centered at 442, 488, 520, 550, 589, and 650 nm. For both PolRADS and DPOL, the channels above 600 nm will not be used because of instrument self-shading. In addition, because of its absence on DPOL, the 412 nm channel on PolRads will also not be discussed.

Data acquisition for both the PolRADS and DPOL systems used filter changers to rotate interference filters and thus sequentially acquire images in each wavelength band. Typical exposure times were less than one second. However, acquiring a set of images from all wavelengths took about two minutes due to the time required to read and store the data from the CCD. With both systems, typical deployments lasted from one to several hours, enabling multiple acquisitions over a range of solar zenith angles.

Reduction of the raw images consisted of applying calibration factors [15,16] and averaging images in both space and time to reduce environmental noise. After calibration, but before averaging, every image was inspected to find the anti-solar point, to correct the geometry of the image, and to check for obstructions in the field of view such as fish, the power/data cable, the side of the ship, or other instruments. Calibrated images of the Stokes vectors were then averaged in 10-min bins, excluding those that had been flagged as unacceptable in the inspection stage. The symmetry of the images about the principal plane was exploited to further average both halves of each image. In addition, spatial binning of 3×3-pixel windows was performed to produce final average images at a $1^\circ \times 1^\circ$ resolution. Each pixel in the final, reduced image, therefore, could have been an average of up to 90 raw pixels (5 images $\times$ 2 image halves $\times$ 9-pixel window). The mean and coefficient of variation (CV = standard deviation divided by the mean) of the Stokes vector components were computed for each pixel in the reduced image using the up-to-90 raw pixels in the original images. The degree of linear polarization *DOLP* was then computed for each pixel from the mean I, Q, and U Stokes vector components (Equation (3)). Because V was not available from the PolRADS data, *DOLP* was calculated rather than the degree of polarization (*DOP*); however, in the upwelling lightfield, they are equivalent because the magnitude of *V* is negligible [23].

For in situ radiometric measurements, instrument self-shading must be considered [24]. In our case, instrument self-shading may also affect the *DOLP* measurements due to camera geometry. The physical displacement of the multiple lenses used means that the images through different polarizers will look into the instrument shadow in slightly different directions. When one lens views a portion of the radiance distribution with less shadow, it will appear brighter in that image. In the algorithm from which the polarization information is derived, for the scene to be unpolarized requires that region of the three camera images to be closely matched in intensity. If the region in one camera is less shaded, it will cause the algorithm to assume that the light field is polarized along the direction of the linear polarizer in that camera. Thus shadowing, which is not symmetric in all the images, will appear as an increase in the *DOLP*, and therefore, a negative model-data *DOLP* difference. This effect should be greater in turbid water and for the PolRADS instrument (because of its larger size) than in clear water or with the DPOL instrument. In our analysis, the area of direct shadow (the anti-solar point) was excluded from the data set. More subtle shadow areas may exist and could cause the measured *DOLP* to be larger than the modelled *DOLP* (as will be seen, this was not generally the case in our data set).

2.2. Modeling the Polarized Upwelling Radiance Distribution

A Monte Carlo model was used to solve the vector radiative transfer equation (Equation (5)) for the propagation of the Stokes vector, *I*, at wavelength λ,

$$cos(\theta)\frac{dI(\lambda,\tau,\theta,\phi)}{d\tau} = -I(\lambda,\tau,\theta,\phi) + \omega_0 \int_{4\pi} R(\alpha)P(\lambda,\tau,\theta',\phi' \to \theta,\phi)R'(\alpha')I(\lambda,\tau,\theta',\phi')d\Omega' \quad (5)$$

The four components of the column vector $I = [I, Q, U, V]^T$ were defined in Equation (2). The single-scattering albedo, ω_0, the optical depth of the medium, τ, the scattering phase matrix, P, and the rotation matrix, R, are defined below. The right-handed coordinate system in Equation (5) has its origin at the top of the medium, z-axis pointed downward, and x-axis directed away from the sun. The nadir angle, θ, relative to the z-axis, and the azimuth angle, ϕ, in the x-y plane, define the direction of photon propagation. Photons originate in the solar beam, i.e., with $\theta = \theta_0$, the solar zenith angle, and $\phi_0 = 0$.

The model simulated a horizontally homogenous, two-layer system with a non-absorbing, Rayleigh-scattering atmosphere over an ocean comprised of Rayleigh-scattering water molecules plus scattering and absorbing hydrosols ("particles"). Fresnel reflectance at the air–water interface used the boundary condition defined in Gordon et al. [8]. The single-scattering albedo, ω_0, and the optical depth of the medium, τ, were defined as per usual (e.g., Mobley [25]):

$$\omega_0 = \frac{b_t(\lambda)}{c_t(\lambda)}, \quad (6)$$

$$\text{and } \tau(\lambda) = \int_0^z c_t(\lambda)dz. \quad (7)$$

where the parameters $a_t(\lambda) + b_t(\lambda) = c_t(\lambda)$ are the total absorption, scattering, and attenuation coefficients (in m^{-1}), respectively. For the atmospheric layer, $a_t(\lambda) = 0$, therefore $\omega_0 = 1$, and τ was taken from Teillet [26] assuming standard atmospheric conditions. For the oceanic layer, absorption (a_w) and scattering (b_w) coefficients of seawater were interpolated to PolRADS and DPOL spectral bands from Table 1.1 in Pegau et al. [27]. In the Ligurian Sea and New York Bight experiments, the absorption coefficient of dissolved and particulate constituents (a_{pg}) and the particle scattering (b_p) coefficients were measured in situ, depth-weighted, and interpolated as necessary to the PolRADS and DPOL spectral bands as described by Gleason et al. [14]. No in situ measurements were available from

Hawaii, therefore for that dataset, a_{pg} and b_p were derived from Morel and Gentili [28] (Equations (7) and (8) in reference [28]) using an estimated total chlorophyll concentration of 0.1 g/m^3, the long-term average value at the measurement site. Summing the water and particle contributions gave the total coefficients required for Equations (6) and (7): $a_t = a_\omega(\lambda) + a_{pg}(\lambda)$ and $b_t = b_\omega(\lambda) + b_p(\lambda)$. For each image data set, the model was run at the corresponding solar zenith angle.

Scattering events in Equation (5) are represented by a 4×4 matrix called the Mueller matrix, M. For example, a photon travelling in direction ξ' with Stokes vector I' would be scattered in a new direction ξ with Stokes vector I by the linear transformation $I(\xi) = MI'(\xi')$. Note that M is defined relative to the scattering plane (as is traditional), the plane defined by vectors ξ' and ξ, but we have defined $I(\theta, \phi)$ relative to the viewing direction and the nadir direction. Pre- and post-multiplication by the rotation matrix $R(\alpha)$ is required to account for changes in reference frames: $I(\theta, \phi) = R(\alpha)MR(\alpha')I'(\theta', \phi')$, where

$$R(\alpha) = \begin{pmatrix} 1 & 0 & 0 & 0 \\ 0 & cos(2\alpha) & sin(2\alpha) & 0 \\ 0 & -sin(2\alpha) & cos(2\alpha) & 0 \\ 0 & 0 & 0 & 1 \end{pmatrix} \quad (8)$$

The rotation angle $\alpha = cos^{-1}(\hat{l}_l \cdot \hat{l}_r)$ is measured clockwise from the vector $\hat{l}_l$ in the initial reference frame to the vector $\hat{l}_r$ in the rotated reference frame [8].

The scattering phase matrix, P, is a Mueller matrix normalized to the integral of the M_{11} element over all solid angles:

$$P = M/b \quad (9)$$

where

$$b = 2\pi \int_0^\pi M_{11} sin(\theta) d\theta \quad (10)$$

M_{ij} represents the ith row and jth column of M, and Θ is the scattering angle.

It is advantageous, when investigating the polarization effects of the Muller matrix, to normalize in a different way, specifically to the M_{11} element at each angle Θ (Equation (11)).

$$S_{ij}(\Theta) = M_{ij}(\Theta)/M_{11}(\Theta) \quad (11)$$

Then,

$$P = \widetilde{\beta} S \quad (12)$$

where $\widetilde{\beta} = M_{11}/b$ is the scattering phase function typically used in scalar radiance transfer models.

For our modeling, P in the atmosphere and P_w, the seawater component of the ocean, were both set to P_r, given by Rayleigh scattering:

$$P_r = \frac{3}{16\pi} \begin{pmatrix} 1 + cos^2(\Theta) & -sin^2(\Theta) & 0 & 0 \\ -sin^2(\Theta) & 1 + cos^2(\Theta) & 0 & 0 \\ 0 & 0 & 2cos(\Theta) & 0 \\ 0 & 0 & 0 & 2cos(\Theta) \end{pmatrix} \quad (13)$$

(We examined a few cases using a code that included the (small) depolarization of Rayleigh scattering in the atmosphere and water by molecular anisotropy and concluded that its omission would not significantly affect the results.) Particle scattering in the ocean used $\widetilde{\beta}$ [25,29] and two alternative parameterizations of S. One set of model runs used S from Voss and Fry [30], which was experimentally determined from samples of seawater (referred to here after as V-F). Note that Voss and Fry found all off-diagonal elements of S equal to zero, within experimental error, except for S_{12} and S_{21}. Furthermore, Voss and Fry found $S_{12} \approx S_{21}$. Subsequent laboratory measurements of the Mueller matrix elements for phytoplankton [31–33] have been similar to V-F. Specifically, the $S_{12}(\Theta)$ element of phytoplankton samples tends to have a minimum value of between −0.6 to −0.8 at an angle between 90° and 100°. In contrast, experiments with solutions of suspended marine sediment have revealed smaller absolute values for $S_{12}(\Theta)$ element near Θ = 90° [32]. Because of this, we decided to carry out a second set of runs with a particle phase matrix having a smaller minimum in S_{12} near 90° (Figure 1). This was accomplished by replacing the V-F $S_{12}(\Theta)$ with that for Rayleigh scattering with a depolarization factor ρ. ρ is the ratio of light scattered with polarization parallel to the plane of incidence to that scattered perpendicular to the plane of incidence, when the incident radiation is polarized perpendicular to the plane of incidence. The element $S_{12}(\Theta)$ as a function of ρ is given by Equation (14) [34]. The other elements of S remained unchanged for this Mueller matrix (referred to as Mod-V-F). Setting ρ = 0.3 resulted in a minimum value of $S_{12}(90°)$ = −0.37, which is within the range of −0.38 to −0.25 observed by Volten et al. [32].

$$S_{12}(\Theta) = S_{21}(\Theta) = \frac{(1-\rho)\left(cos(\Theta)^2 - 1\right)}{\left(1 + cos(\Theta)^2\right) + \left(3 - cos(\Theta)^2\right)\rho} \tag{14}$$

Kuik et al. [35] presented four inequalities, originally derived by Fry and Kattawar [36] and Hovenier et al. [37], that must be satisfied by the scattering matrix elements of randomly oriented particles having a plane of symmetry. These inequalities are all satisfied by the modified Mueller matrix.

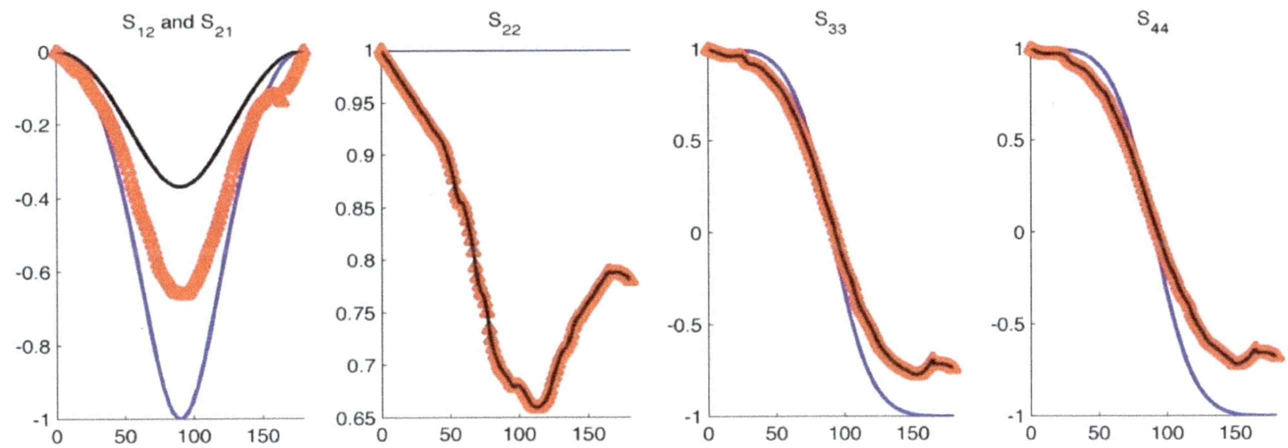

Figure 1. Non-zero, normalized Muller matrix, S, elements for Rayleigh scattering (blue), V-F (red), and Mod-V-F (black).

2.3. Model–Data Comparison

Visual comparisons provided a qualitative check that the Stokes vector components from our RT model were in agreement with the camera data. Quantitative comparisons between the model and data were performed by computing the model–data difference every 10° in nadir from 0° to 80° and every 30° in azimuth from 0° to 180°. Thus, the difference in $DOLP$, $DOLP_{diff}$, between each model output, $DOLP_{model}$, and the corresponding average PolRADS or DPOL image, $DOLP_{data}$, was computed at 63 angles using Equation (15).

$$DOLP_{diff}(\theta, \phi) = DOLP_{model}(\theta, \phi) - DOLP_{data}(\theta, \phi) \tag{15}$$

where $DOLP_{model}(\theta, \phi)$ and $DOLP_{data}(\theta, \phi)$ are the $DOLP$ predicted by the RT model and measured by the corresponding average PolRADS or DPOL image, respectively, at the same 63 nadir and azimuth angles.

3. Results

In total, approximately 528 individual images were selected and averaged in 10 min intervals to produce 219 data sets, which were compared with RT model runs (Table 1).

Table 1. Number, *N*, of reduced images used for model–data comparisons in each of the three datasets in each of four spectral bands. Note, the PolRADS and DPOL cameras used different filters with slightly different band centers; the center of each 10 nm-wide filter is listed in parenthesis.

Experiment	*N* (λ nm)	*N* (λ nm)	*N* (λ nm)	*N* (λ nm)
Hawaii	27 (436 nm)	31 (486 nm)	28 (526 nm)	24 (548 nm)
Ligurian Sea	9 (442 nm)	9 (488 nm)	13 (520 nm)	14 (550 nm)
New York Bight	7 (442 nm)	8 (488 nm)	9 (520 nm)	7 (550 nm)

3.1. Overall View of the Upwelling Polarization Signal

Figure 2 shows an example image for clear water. In the right column, the data parameters are shown (I, Q, U, $DOLP$, and χ), while in the center column, the results from the RT model are shown, where the model inputs are based on the measured parameters for the data (b_p, a_{pg}, solar zenith angle). As mentioned earlier, during processing of this data, the symmetry about the principal plane was used to average the left and right side of the data (with proper handling of the sign of U/I). Thus, to make these images, this symmetry was used to generate the left side of the data images. The left column is a single scattering calculation, where the Mueller matrix was assumed to be V-F, and the solar zenith angle was the value appropriate for the data. As can be seen, the same broad patterns were visible in the single scattering model, the RT model and data. The I component of the Stokes vector was largest near the horizon in the direction towards the sun. The Q/I and U/I patterns were very similar in all three cases. For the *DOLP*, the single scattering model was significantly different than the data and the RT model. With single scattering, the maximum *DOLP* was both larger than in the RT model and the data, and occurred for all nadir angles at the scattering angle matching the minimum in the S_{12} and S_{21} elements (90–100°). Interestingly, while it was a subtle effect in the clear water case, the maximum *DOLP* in the model and data occurred at the same scattering angle, but at an azimuth of 90° relative to the principal plane (the plane containing the anti-solar position and nadir). This can be contrasted with the case of the downwelling sky radiance distribution, where the maximum *DOLP* occurred in the principal plane, with a decrease in *DOLP* towards the horizon (for example, Liu and Voss [38]). The same cause was in effect in both cases: for downwelling radiance, the horizon had more multiple scattering than the principal plane at the 90° scattering angle, while for the upwelling radiance, in the water, there was more multiple scattering for the principal plane at a 90° scattering angle than for nadir angles closer to the horizon. The other feature in the data for this clear water case was the separation of the neutral points, areas of zero *DOLP*, on either side of the principal plane, as has been discussed previously [39,40]. Finally, the angle of the polarization plane, χ, was very similar in all three cases, perhaps less well-defined for the data in the area of the low polarization, around the anti-solar position.

An example with more turbid water is shown in Figure 3. Once again, the patterns were generally the same, however in this case the data are noisier. Note that the data shown in this case has not been averaged using the left/right symmetry across the principle plane, thus appears less symmetric than the example in Figure 2. As has been pointed out before [41], the most stable parameter was χ, with the exception of the area with very low *DOLP*, in which case the plane of polarization was not well defined. In general, the images appeared much noisier in the data from the two more turbid sites (New York Bight and Ligurian Sea) than in the clear water off Hawaii. The pattern of Q/I and U/I tended to shift more between each image that goes into these data averages. In the scale of Q/I and U/I ranging from -1 to 1, the standard deviation obtained during averaging the data was approximately 0.05 for the New York Bight and Ligurian Sea data, while it was 0.012 for the clearer Hawaii data. One can also see that in this more turbid case, the single scattering was a poor approximation of all parameters, with the exception of χ.

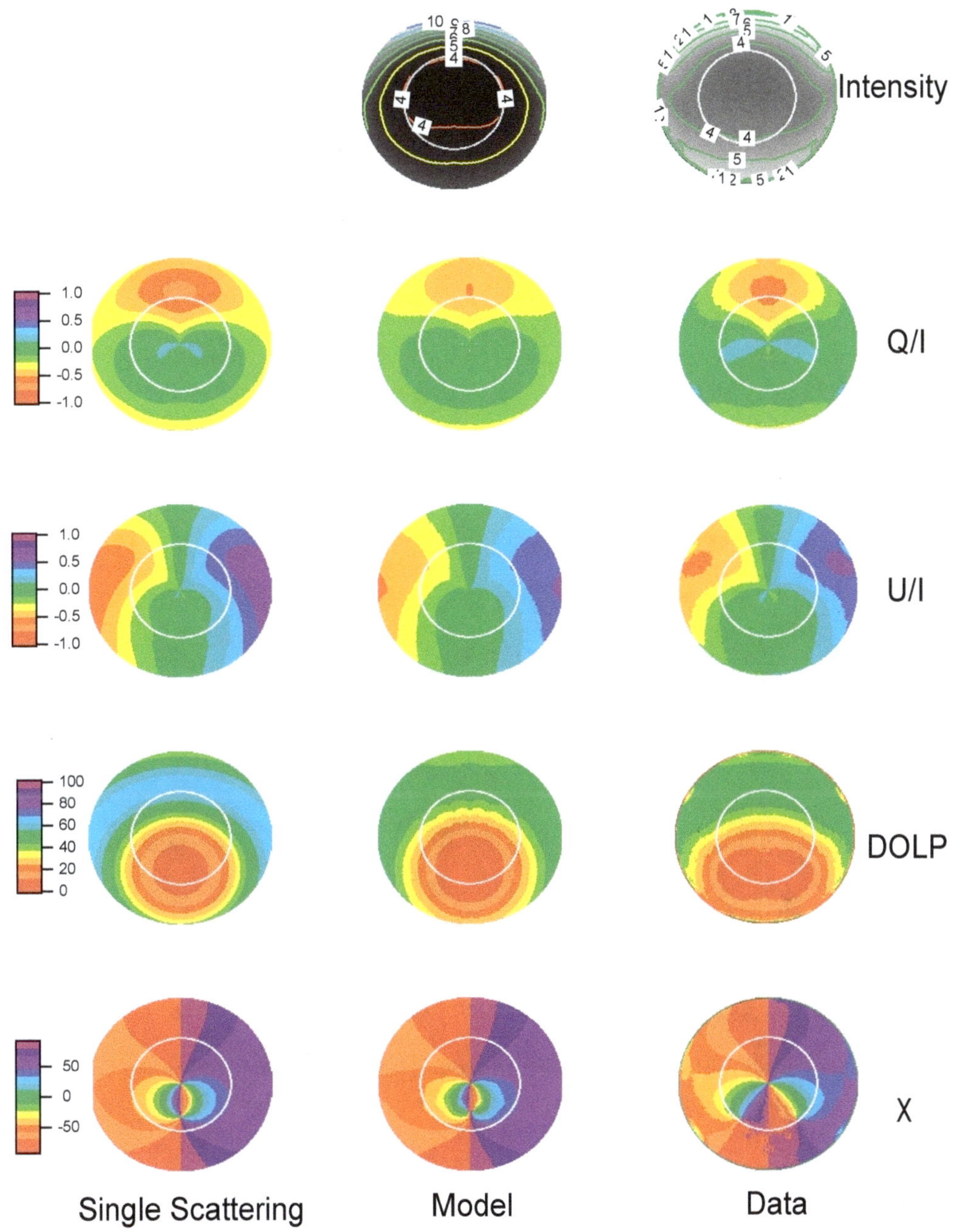

Figure 2. Comparison of the Stokes vector components I, Q/I, U/I, $DOLP$, and χ for clear water. Shown are the calculation for single scattering (left column), RT model (center column), and data (right column). The data were taken on 2 December 2005, at 20:46 UTC off of Oahu, Hawaii. The conditions were: SZA = 48°, 442 nm, Chl = 0.1 mg/m^3, c_t = 0.1 m^{-1} (calculated from Chl as described in text), ω_0 = 0.8, and clear skies. The intensity for the model was adjusted to match the data at nadir. Each image is a fisheye projection. Nadir is in the center of the circle, nadir angle is linearly proportional to radius from center. The angle of the Snell's circle (48° nadir angle) is shown in as the white circle. The principal plane (plane containing the anti-solar point and the nadir direction) is a vertical line through the center of the image. The anti-solar direction is towards the bottom of the image, and the sun direction is towards the top of the image.

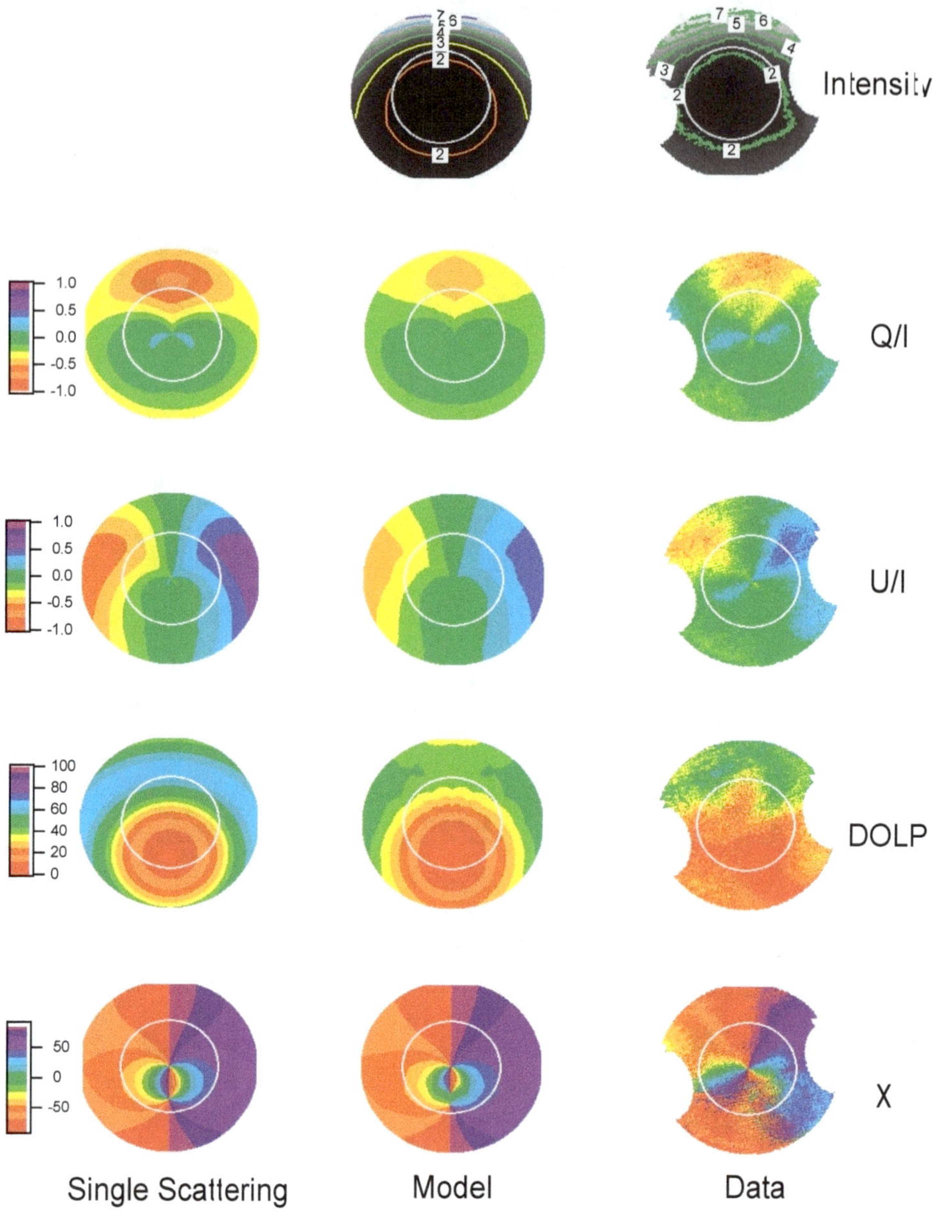

Figure 3. Similar to Figure 2, but for a case with more turbid water. The data were taken on 22 March 2009, at 9:40 UTC in the Ligurian Sea. The conditions were: SZA = 48°, 550 nm, $c_t = 0.44\ \mathrm{m}^{-1}$, $\omega_0 = 0.82$, and clear skies. The figure geometry is the same as Figure 2.

3.2. *Maximum DOLP and Nadir DOLP*

How large can the maximum *DOLP* be in the upwelling radiance? Figure 4 shows this as a function of c_t/a_t and c_t. We used the variation of the optical properties (c_t, b_t, b_{bt}, and a_t) with the wavelength to fill in the data, thus this figure includes all wavelengths with their appropriate optical properties. The quantity c_t/a_t can be shown to equal the mean number of scattering events in the medium taken as a whole [28]. As the water gets more turbid, there is more multiple scattering, which will decrease the *DOLP* [7,42]. The maximum *DOLP* is often outside the Snell's cone, thus not retrievable from above the surface, so we also show the maximum value of the *DOLP* inside the Snell's cone in Figure 4. As can be seen, this was less, sometimes significantly less, than the maximum *DOLP* in the total upwelling field. Another point was that the maximum *DOLP* in the total upwelling field occurred at an azimuth nearly perpendicular to the principal plane, as this position was always at a 90° scattering angle to the sun, and had the largest component of single scattered light due to a combination of light attenuation with depth and the relative smoothness of the scattering function in the backward direction. For the maximum *DOLP* inside the Snell's circle, as can be seen in Figures 2 and 3, this azimuthal angle will move towards the principal plane, and hence got closer to the position where, above the surface, it will be in the region of the solar glitter pattern.

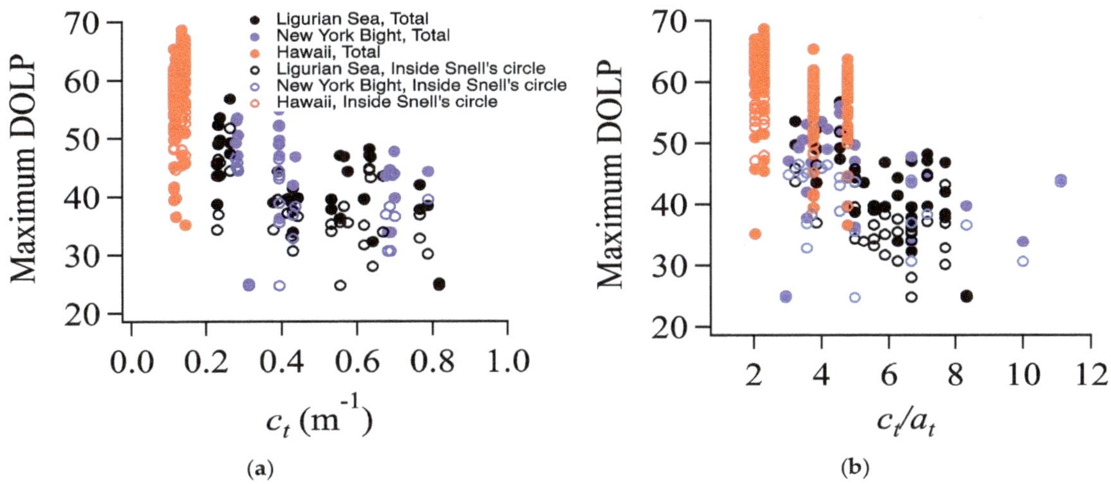

Figure 4. Maximum *DOLP* in upwelling radiance distribution vs c_t (**a**) and c_t/a_t (**b**). The symbol legend is the same for both graphs. Filled circles represent the maximum *DOLP* in the total upwelling field, while the open circles represent the maximum *DOLP* in the portion of the upwelling light field inside the Snell's circle (nadir angle less than 48°).

Another interesting parameter is the *DOLP* at the nadir angle, as this was the direction that many in situ radiometers make their measurement (Figure 5). Here the RT model results are presented rather than the data, as in our definition of the reference frame, the nadir direction is a point of singularity, and the data become excessively noisy at that point.

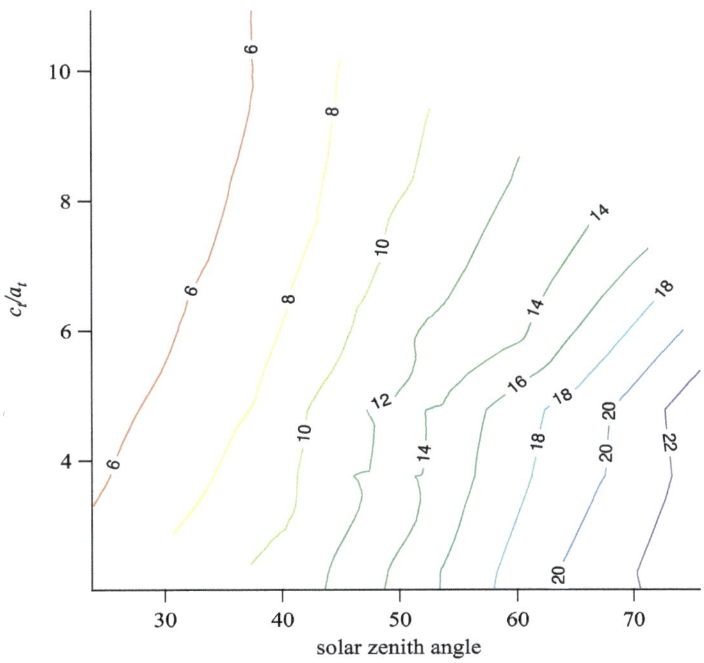

Figure 5. Nadir *DOLP* as a function of solar zenith angle and c_t/a_t, computed from the RT model. Contours are lines of constant *DOLP*.

Note that the nadir light field could be significantly polarized and the *DOLP* could be greater than 20%. Also, as can be seen, the largest values (22%) occurred at large solar zenith angles (when the scattering angle decreased to values closer to 90°), but there was also a trend of decreasing *DOLP* as c_t/a_t increased at a constant solar zenith angle due to increasing multiple scattering. The plane of polarization in all cases was perpendicular to the principal plane.

3.3. Comparison of DOLP Differences

For the first quantitative comparison of the RT model and the data, we looked at scatter plots comparing the measured and modeled *DOLP* (Figure 6). For each cruise, a best linear fit line was calculated, with the y-intercept = 0. The fit was calculated for all of the data, as well as the data inside the Snell's circle; however, the calculated slopes were not significantly different, so we will discuss only the fit for the total image. The slopes were 0.94 ($\pm$0.002, r^2 = 0.81), 0.88 ($\pm$0.006, r^2 = 0.84), and 0.82 ($\pm$0.008, r^2 = 0.76) for Hawaii, New York Bight, and Ligurian Sea, respectively.

Because the *DOLP* was significantly overestimated in the Ligurian Sea data, the alternative normalized Mueller matrix Mod-V-F (ρ = 0.30) was tried, with all other parameters the same, resulting in a slope of 1.20 ($\pm$ 0.009, r^2 = 0.656); we concluded that the depolarization of 0.3 was too strong. The true average Mueller matrix for this location could be somewhere between the two we used.

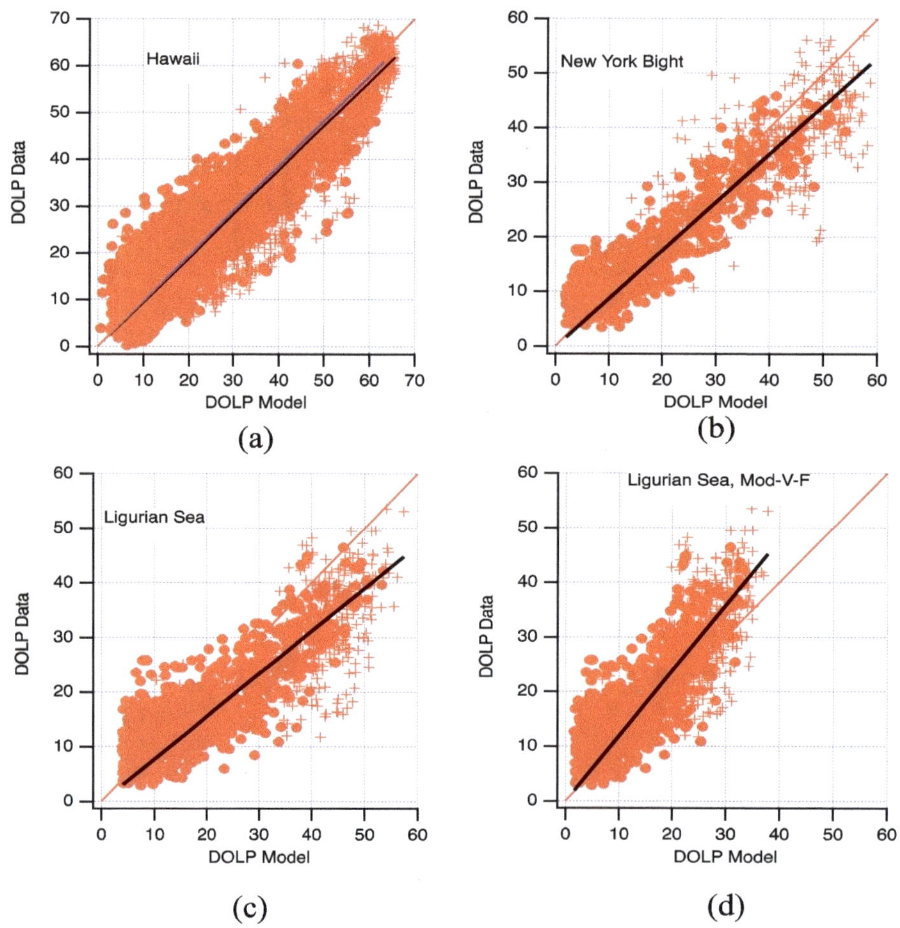

Figure 6. Scatter plots of *DOLP* for model versus data, separated by cruise. Also shown is the best fit line (with y-intercept = 0) (black line) and 1:1 line (red line). The crosses correspond to data outside the Snell's circle (nadir angle larger than 48°), and the dots to points inside the Snell circle. V-F was used in the model for (**a**–**c**), Mod-V-F was used for (**d**).

The correlation between $DOLP_{diff}$ and several environmental parameters was investigated. In Figure 7, we show the $DOLP_{diff}$ variation with solar zenith angle. In this figure, the mean $DOLP_{diff}$ was calculated for each data image, along with the standard deviation (shown as error bars). There was no trend in the data; however, there does appear to have been a slight increase in the magnitude of the deviation at solar zenith angles greater than 70°. In general though, in almost all cases, the mean $DOLP_{diff}$ was within one standard deviation of zero.

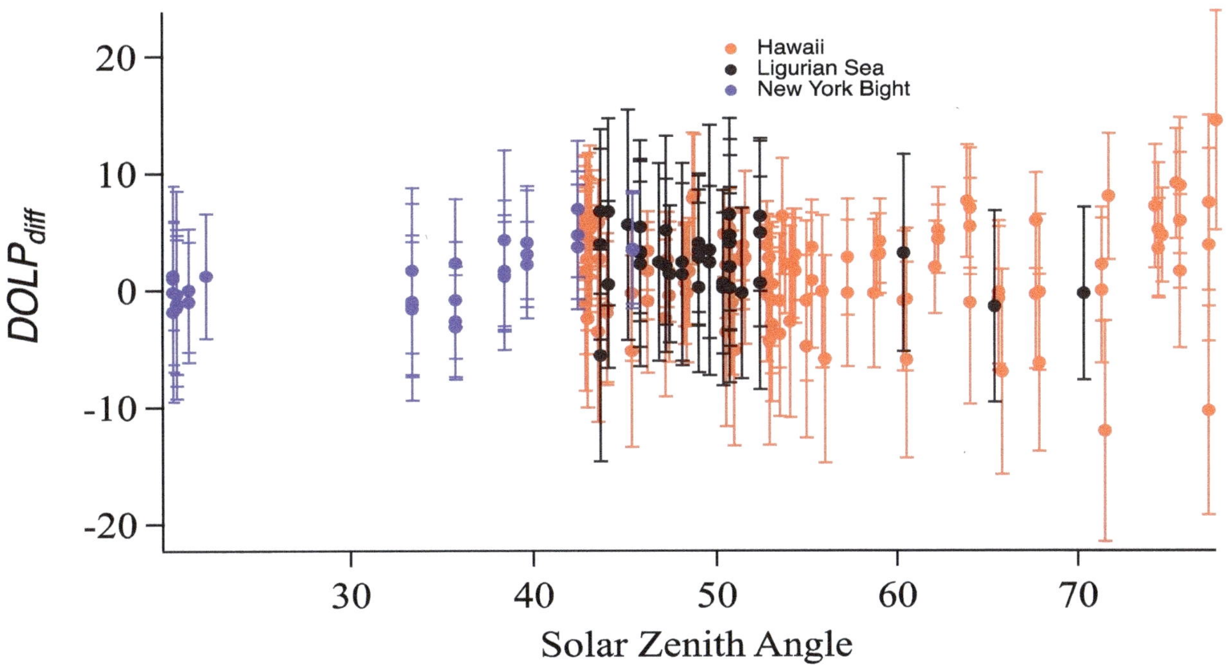

Figure 7. $DOLP_{diff}$ vs solar zenith angle for each data image. The error bars are ± the standard deviation of the mean.

In Figure 8, we look at the $DOLP_{diff}$ dependence on c_t and c_t/a_t. While there was no clear trend in $DOLP_{diff}$ with c_t, a slight trend occurred as a function of c_t/a_t. Low values of c_t/a_t (more absorption, less multiple scattering) had a slightly larger value of $DOLP_{diff}$, while $DOLP_{diff}$ tended towards 0 with larger values of c_t/a_t (more multiple scattering). However, the mean $DOLP_{diff}$ was less than 5% for almost all cases, and this was within the measurement uncertainty of these instruments in the field.

Finally, in Figure 9, we look at the $DOLP_{diff}$ as a function of b_{bt}/b_t (b_{bt}/b_t is the fraction of total scattering which was in the backwards direction, i.e., scattering angles from 90° to 180°). As b_{bt}/b_t increased, $DOLP_{diff}$ appeared slightly higher, indicating less agreement at higher values of b_{bt}/b_t. Low values of b_{bt}/b_t were indicative of a low refractive index, phytoplankton-dominated environment, while higher values of b_{bt}/b_t indicated an environment with other higher refractive index particles, perhaps sediment. V-F was based on measurements in a variety of environments (clear ocean water to more turbid coastal water); however, measurements of the Mueller matrix of phytoplankton cultures also agreed with V-F. It has often been stated that the polarization resulting from higher index particles is less than V-F (i.e., larger *depolarization* at a 90° scattering angle) [32]. It is probably the case in our data set, that as the b_{bt}/b_t ratio increases, the S_{12} and S_{21} elements of the Mueller matrix should also decrease in an absolute value sense, but not as much of a decrease as used in Mod-V-F, where the decrease in this matrix would have caused a significant negative $DOLP_{diff}$.

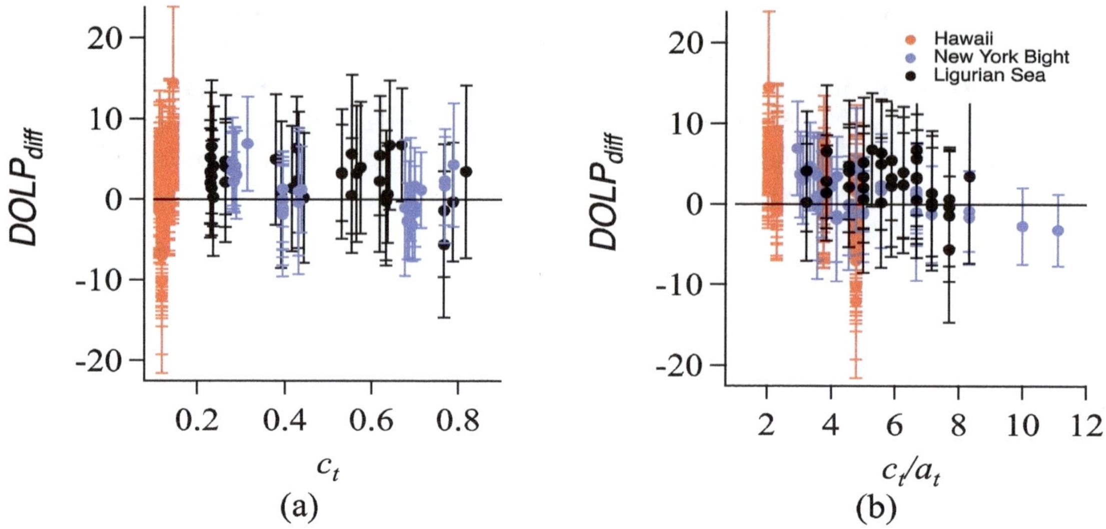

Figure 8. $DOLP_{diff}$ as a function of c_t and c_t/a_t. Color code of data is given in Figure 7. (**a**) shows $DOLP_{diff}$ vs c_t, while (**b**) is $DOLP_{diff}$ vs c_t/a_t.

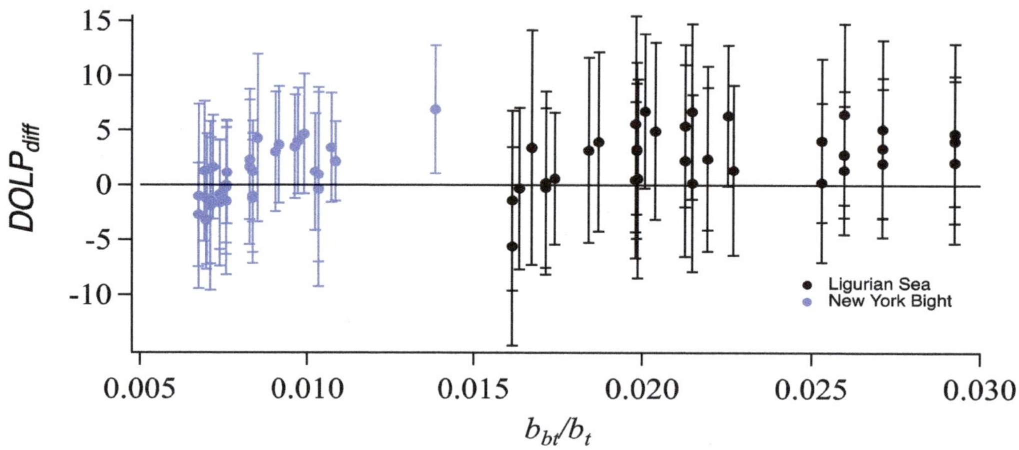

Figure 9. $DOLP_{diff}$ as a function of b_{bt}/b_t. There were no b_{bt} measurements during the Hawaii data collection.

3.4. Comparison of Q/I and U/I Differences

Rather than calculating *DOLP*, we could also show the difference between the model result for Q/I and U/I versus the data, shown in Figure 10 for the Ligurian Sea case. This case is shown because it was neither the best result, which was for Hawaii, nor the worse, which was New York Bight. The advantage of *DOLP*, and the reason most of the results are presented for this parameter, is that it is independent of the frame of reference, thus small errors in locating the solar plane in each image cause smaller differences in the comparison. Q/I and U/I comparisons are with respect to the frame of reference and depend on correctly locating the solar plane in the image, hence the comparison is not quite as good between the model and data result as with the DOLP. The line fits had a slope of 0.66 ($\pm$0.010, r^2 = 0.70) for Q/I and 0.71 ($\pm$0.013, r^2 = 0.67) for U/I. The fact that the scatter is qualitatively the same in these graphs as in the DOLP figure can be simply understood since Q/I and U/I can be derived from a combination of the *DOLP* and the plane of polarization, χ. As mentioned earlier, χ is a very stable parameter in the upwelling light field. The spread in the data is slightly larger for U/I versus Q/I because U/I has stronger spatial gradients, and small rotations affect this parameter more than Q/I.

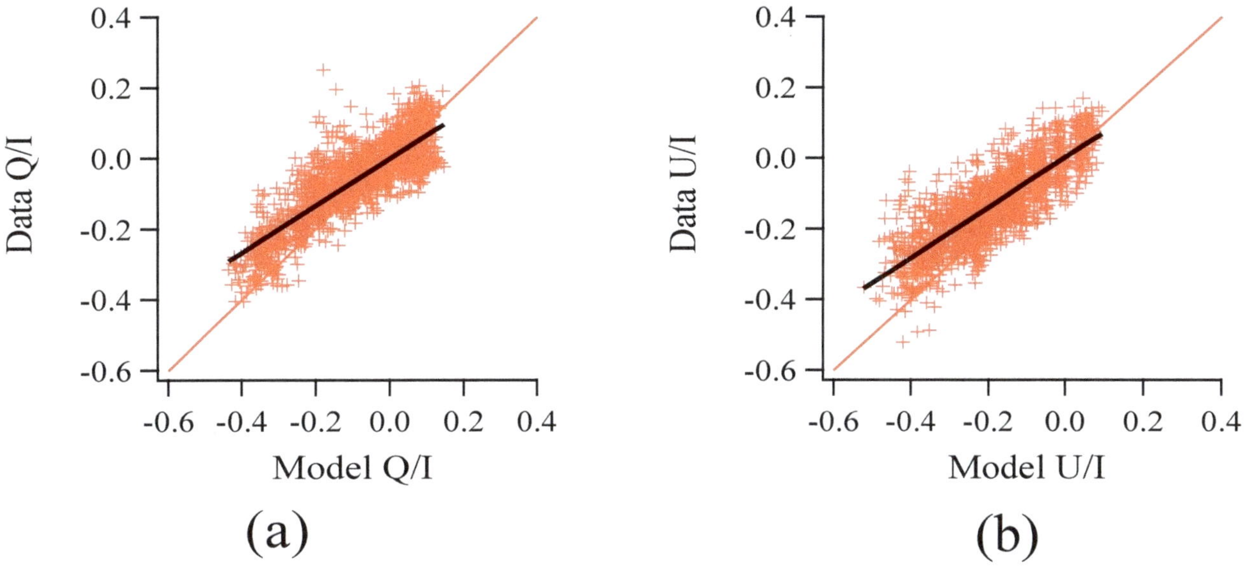

Figure 10. Scatter plots of Q/I (**a**) and U/I (**b**) for model versus data for the Ligurian Sea data set. Also shown is the best fit line between the model and data for each parameter (black line). The 1:1 line is shown in red.

4. Conclusions

We have compared measurements of the polarization properties of the upwelling light field in the marine environment to radiative transfer models based on the measured inherent optical properties (IOPs) and an average normalized Mueller matrix for the scattering. For the Ligurian Sea and New York Bight data sets, the base RT model used the same volume scattering phase function (Petzold) and Mueller matrix (V-F) for the standard model, with the scattering and absorption coefficients as measured coincidently on location. For Hawaii, the IOPs were not measured, but were estimated based on an average *Chl* value of 0.1 mg/m^3 at the measurement location. Even with the fixed scattering matrix parameters, on average, the $DOLP_{diff}$ was less than 5%, and in almost every case was within one standard deviation of zero. Thus, to get more information, or discrimination, from the *DOLP* measurements, the uncertainty of the measured *DOLP* must be smaller than 5%, and the other parameters (scattering phase function, Mueller matrix, and absorption coefficient) need to be known well enough to constrain the *DOLP* model to this accuracy.

Some of the variability in our data was due to instrument noise (on the order of 2%); however, other noise was due to the variability in the environment resulting from the incident light field interacting with the wavy air–sea interface. This was more of a problem for the downwelling light field but was still evident in the upwelling light field. For remote sensing, this radiance (Stokes vector) would have been propagated through the (wavy) air–sea interface, which would introduce more noise.

The 5% agreement of the model with the data shows that the model will provide sufficient accuracy for correcting residual polarization effects in ocean color instruments. However, more work must be done at improving the measurement accuracy of the instrumentation if the goal is to be able to use polarization effects to characterize (and differentiate with respect to) the physical properties of the suspended particles other than c/a.

Author Contributions: Data curation, K.J.V., M.S.T., and J.-F.B.; Formal analysis, A.C.R.G. and H.R.G.; Funding acquisition, K.J.V.; Methodology, A.C.R.G., K.J.V., and H.R.G.; Software, A.C.R.G. and H.R.G.; Supervision, K.J.V.; Writing—original draft, A.C.R.G., K.J.V., and H.R.G.; Writing—review and editing, A.C.R.G., K.J.V., H.R.G., and M.S.T.

Acknowledgments: Scott Freeman, Luke Logan, Puru Bhandari, and Nordia Souaidia are thanked for their technical assistance in the field and with the instrumentation and data reduction.

References

1. Waterman, T.H. Underwater light and orientation of animals. In *Optical Aspects of Oceanography*; Jerlov, N.G., Nielsen, E.S., Eds.; Academic Press: New York, NY, USA, 1974; pp. 415–443, ISBN 0123849500.
2. Loisel, H.; Duforet, L.; Dessailly, D.; Chami, M.; Dubuisson, P. Investigation of the variations in the water leaving polarized reflectance from the polder satellite data over two biogeochemical contrasted oceanic areas. *Opt. Express* **2008**, *16*, 12905–12918. [CrossRef] [PubMed]
3. Cronin, T.W.; Shashar, N.; Caldwell, R.L.; Marshall, N.; Cheroske, A.G.; Chiou, T.H. Polarization vision and its role in biological signaling. *Int. Comp. Biol.* **2003**, *43*, 549–558. [CrossRef] [PubMed]
4. Mishchenko, M.I.; Cairns, B.; Kopp, G.; Schueler, C.F.; Fafaul, B.A.; Hansen, J.E.; Hooker, R.J.; Itchkawich, T.; Maring, H.B.; Travis, L.D. Accurate monitoring of terrestrial aerosols and total solar irradiance: Introducing the glory mission. *Bull. Am. Meteorol. Soc.* **2007**, *88*, 677–691. [CrossRef]
5. Tonnizio, A.; Gilerson, A.; Harmel, T.; Ibrahhim, A.; Chowdhary, J.; Gross, B.; Moshary, F.; Ahmed, S.; Chami, M. Importance of the polarization in the retrieval of oceanic constituents from the remote sensing reflectance. *J. Geophys. Res.* **2007**, *112*, C05026. [CrossRef]
6. Ibrahim, A.; Gilerson, A.; Chowdhary, J.; Ahmed, S. Retrieval of macro- and micro-physical properties of oceanic hydrosols from polarimetric observations. *Remote Sens. Environ.* **2016**, *186*, 548–566. [CrossRef]
7. Chami, M.; Santer, R.; Dilligeard, E. Radiative transfer model for the computation of radiance and polarization in an ocean-atmosphere system: Polarization properties of suspended matter for remote sensing. *Appl. Opt.* **2001**, *40*, 2398–2416. [CrossRef] [PubMed]
8. Gordon, H.R.; Du, T.; Zhang, T. Atmospheric correction of ocean color sensors: Analysis of the effects of residual instrument polarization sensitivity. *Appl. Opt.* **1997**, *36*, 6938–6948. [CrossRef] [PubMed]
9. Meister, G.; Kwiatkowska, E.J.; Franz, B.A.; Patt, F.S.; Feldman, G.C.; McClain, C.R. Moderate-resolution imaging spectroradiometer ocean color polarization correction. *Appl. Opt.* **2005**, *44*, 5524–5535. [CrossRef] [PubMed]
10. Voss, K.J. Electro-optic camera system for measurement of the underwater radiance distribution. *Opt. Eng.* **1989**, *28*, 241–247. [CrossRef]
11. Voss, K.J.; Chapin, A.L. Upwelling radiance distribution camera system, NURADS. *Opt. Express* **2005**, *13*, 4250–4262. [CrossRef] [PubMed]
12. Voss, K.J. Use of the radiance distribution to measure the optical absorption coefficient in the ocean. *Limnol. Oceanogr.* **1989**, *34*, 1614–1622. [CrossRef]
13. Voss, K.J.; Morel, A.; Antoine, D. Detailed validation of the bidirectional effect in various case 1 waters for application to ocean color imagery. *Biogeosciences* **2007**, *4*, 781–789. [CrossRef]
14. Gleason AC, R.; Voss, K.J.; Gordon, H.R.; Twardowski, M.; Sullivan, J.; Trees, C.; Weidemann, A.; Berthon, J.-F.; Clark, D.; Lee, Z.-P. Detailed validation of the bidirectional effect in various case I and case II waters. *Opt. Express* **2012**, *20*, 7630–7645. [CrossRef] [PubMed]
15. Voss, K.J.; Souaidia, N. Polrads: Polarization radiance distribution measurement system. *Opt. Express* **2010**, *18*, 19672–19680. [CrossRef] [PubMed]
16. Bhandari, P.; Voss, K.J.; Logan, L. An instrument to measure the downwelling polarized radiance distribution in the ocean. *Opt. Express* **2011**, *19*, 17609–17620. [CrossRef] [PubMed]
17. Hoejerslev, N.K.; Aas, E. Spectral irradiance, radiance, and polarization in blue western Mediterranean waters. In Proceedings of the Ocean Optics XIII, Halifax, NS, Canada, 6 February 1997; Volume 2693. [CrossRef]
18. Tonizzo, A.; Zhou, J.; Gilerson, A.; Twardowski, M.S.; Gray, D.J.; Arnone, R.A.; Gross, B.M.; Moshary, F.; Ahmed, S.A. Polarized light in coast waters: Hyperspectral and multiangular analysis. *Opt. Express* **2009**, *17*, 5666–5682. [CrossRef] [PubMed]
19. Gilerson, A.; Zhou, J.; Oo, M.; Chowdhary, J.; Gross, B.M.; Moshary, F.; Ahmed, S. Retrieval of chlorophyll fluorescence from reflectance spectra through polarization discrimination: Modeling and experiments. *Appl. Opt.* **2006**, *45*, 5568–5581. [CrossRef] [PubMed]
20. Morel, A.; Antoine, D.; Gentili, B. Bidirectional reflectance of oceanic waters: Accounting for raman emission and varying particle scattering phase function. *Appl. Opt.* **2002**, *41*, 6289–6306. [CrossRef] [PubMed]

21. Lee, Z.P.; Du, K.; Voss, K.J.; Zibordi, G.; Lubac, B.; Arnone, R.; Weidemann, A. An inherent-optical-property-centered approach to correct the angular effects in water-leaving radiance. *Appl. Opt.* **2011**, *50*, 3155–3167. [CrossRef] [PubMed]
22. Bohren, C.F.; Huffman, D.R. *Absorption and Scattering of Light by Small Particles*; John Wiley and Sons: New York, NY, USA, 1983; ISBN 9783527618156.
23. Ivanoff, A.; Waterman, T.H. Elliptical polarization of submarine illumination. *J. Mar. Res.* **1958**, *16*, 255–282.
24. Gordon, H.R.; Ding, K. Self-shading of in-water optical instruments. *Limnol. Oceanogr.* **1992**, *37*, 491–500. [CrossRef]
25. Mobley, C.D.; Gentili, B.; Gordon, H.R.; Jin, Z.H.; Kattawar, G.W.; Morel, A.; Reinersman, P.; Stamnes, K.; Stavn, R.H. Comparison of numerical-models for computing underwater light fields. *Appl. Opt.* **1993**, *32*, 7484–7504. [CrossRef] [PubMed]
26. Teillet, P.M. Rayleigh optical depth comparisons from various sources. *Appl. Opt.* **1990**, *29*, 1897–1900. [CrossRef] [PubMed]
27. Pegau, S.; Zaneveld JR, V.; Mitchell, B.G.; Mueller, J.L.; Kahru, M.; Wieland, J.; Stramska, M. Inherent optical properties: Instruments, characterizations, field measurements, and data analysis protocols. In *Ocean Optics Protocols for Satellite Ocean Color Sensor Validation, Revision 4, Volume IV*; Mueller, J.L., Fargion, G.S., McClain, C.R., Eds.; NASA/TM-2003-21621/Rev4-Vol IV; Goddard Space Flight Center: Greenbelt, MD, USA, 2002.
28. Morel, A.; Gentili, B. Diffuse reflectance of oceanic waters: Its dependence on sun angle as influenced by the molecular scattering contribution. *Appl. Opt.* **1991**, *30*, 4427–4438. [CrossRef] [PubMed]
29. Petzold, T.J. *Volume Scattering Functions for Selected Ocean Waters*; Scripps Institution of Oceanography Ref. 72–78; University of California: San Diego, CA, USA, 1972; 79p.
30. Voss, K.J.; Fry, E.S. Measurement of the muller matrix for ocean water. *Appl. Opt.* **1984**, *23*, 4427–4439. [CrossRef] [PubMed]
31. Fry, E.S.; Voss, K.J. Measurement of the Mueller matrix for phytoplankton. *Limnol. Oceanogr.* **1985**, *30*, 1322–1326. [CrossRef]
32. Volten, H.; Haan, J.F.; Hovenier, J.W.; Schreurs, R.; Vassen, W.; Dekker, A.G.; Hoogenboom, H.J.; Charlton, F.; Wouts, R. Laboratory measurements of angular distributions of light scattered by phytoplankton and silt. *Limnol. Oceanogr.* **1998**, *43*, 1180–1197. [CrossRef]
33. Svensen, O.; Stamnes, J.J.; Kildemo, M.; Aas LM, S.; Erga, S.R.; Frette, O. Mueller matrix measurements of algae with different shape and size distributions. *Appl. Opt.* **2011**, *50*, 5149–5157. [CrossRef] [PubMed]
34. Adams, J.T.; Aas, E.; Hojerslev, N.K.; BLundgren, B. Comparison of radiance and polarization values observed in the mediterranean sea and simulated in a monte carlo model. *Appl. Opt.* **2002**, *41*, 2724–2733. [CrossRef] [PubMed]
35. Kuik, F.; Stammes, P.; Hovenier, J.W. Experimental determination of scattering matrices of water droplets and quartz particles. *Appl. Opt.* **1991**, *30*, 4872–4881. [CrossRef] [PubMed]
36. Fry, E.S.; Kattawar, G.W. Relationships between elements of the Stokes matrix. *Appl. Opt.* **1981**, *20*, 2811–2814. [CrossRef] [PubMed]
37. Hovenier, J.W.; van de Hulst, H.C.; van der Mee, C.V.M. Conditions for the elements of the scattering matrix. *Astron. Astrophys.* **1986**, *157*, 301–310.
38. Liu, Y.; Voss, K.J. Polarized radiance distribution measurements of skylight: II. Experiment and data. *Appl. Opt.* **1997**, *36*, 8753–8764. [CrossRef] [PubMed]
39. Chowdhary, J.; Cairns, B.; Travis, L.D. Contribution of water-leaving radiances to multiangle, multispectral polarimetric observations over the open ocean: Bio-optical model results for case 1 waters. *Appl. Opt.* **2006**, *45*, 5542–5567. [CrossRef] [PubMed]
40. Voss, K.J.; Gleason, A.C.R.; Gordon, H.R.; Kattawar, G.W.; You, Y. Observation of non-principal plane neutral points in the in-water upwelling polarized light field. *Opt. Express* **2011**, *19*, 5942–5952. [CrossRef] [PubMed]
41. You, Y.; Tonizzo, A.; Gilerson, A.A.; Cummings, M.E.; Brady, P.; Sullivan, J.M.; Twardowski, M.S.; Dierssen, H.M.; Ahmed, S.A.; Kattawar, G. Measurements and simulations of polarization states of understeer light in clear oceanic waters. *Appl. Opt.* **2011**, *50*, 4873–4893. [CrossRef]
42. Timofeyeva, V.A. Degree of polarization of light in turbid media. *Isvestiya Akademii Nauk Sssr Fisika Atmosfery I. Okeana* **1970**, *6*, 513–522.

4

Anomalous Light Scattering by Pure Seawater

Xiaodong Zhang [1,*] and Lianbo Hu [1,2]

1 Department of Earth System Science and Policy, University of North Dakota, Grand Forks, ND 58202, USA; hulb@ouc.edu.cn
2 Ocean Remote Sensing Institute, Ocean University of China, Qingdao 266001, China
* Correspondence: zhang@aero.und.edu

Abstract: The latest model for light scattering by pure seawater was used to investigate the anomalous behavior of pure water. The results showed that water exhibits a minimum scattering at 24.6 °C, as compared to the previously reported values of minimum scattering at 22 °C or maximum scattering at 15 °C. The temperature corresponding to the minimum scattering also increases with the salinity, reaching 27.5 °C for S = 40 psu.

Keywords: light scattering; light scattering by pure water; light scattering by pure seawater; anomalous properties of water

1. Introduction

Light scattering by pure water or pure seawater is a fundamental quantity in aquatic optics. Because of hydrogen bonding, many bulk properties of water exhibit anomalous behavior with temperature that is unlike any other liquids [1,2]. For example, liquid water has a maximum density near 4 °C [3], a minimum isothermal compressibility near 46 °C [4] and a maximum refractive index near 0 °C [5]. The scattering seems to behave "anomalously" too [6]. Cohen and Eisenberg [6] measured the scattering at 436 and 546 nm by pure water at temperatures from 5 to 65 °C, and found a scattering minimum at approximately 22 °C that is consistent with their theoretical estimate using the Einstein–Smoluchowski equation and the temperature variation of the isothermal compressibility. Using the same Einstein-Smoluchowski equation, with inputs re-evaluated using newer experimental results, Buiteveld et al. [7] improved the estimate of light scattering by pure water, which showed a better agreement with the spectral values measured by Morel [8]. However, their model predicts a maximum scattering at 15 °C, which differs from Cohen and Eisenberg [6] not only in value but in the behavior as well. To the best of our knowledge, few other studies have explored the temperature behavior of scattering by water. In addition, it is still unknown whether and how this temperature dependence of scattering by water would vary in the presence of sea salts.

2. Methods

Recently, Zhang and coworkers refined the models for light scattering by pure water [9], by pure seawater [10,11], and by simple sea salt solutions [12] using the improved measurements of the key thermodynamic parameters, and their models agree with the spectral scattering measurements [13,14] for both pure water and seawater within the experimental errors (2%). Localized fluctuation in density for pure water, as well as additional fluctuations in the mixing ratio of salt ions and water for pure seawater, lead to microscopic inhomogeneities in the refractive index (n) [15], which in turn cause scattering of light. Since the fluctuations in density and mixing ratio are independent, the scattering coefficient of seawater, b (m^{-1}) can be expressed as

$$b = b_d + b_c, \tag{1}$$

where b_d represents the scattering due to density fluctuation, and b_c the scattering due to fluctuation of mixing ratio (concentration). Following Zhang and Hu [9],

$$b_d = \frac{8\pi^3}{\lambda^4}\left(\rho\frac{\partial n^2}{\partial \rho}\right)_T^2 kT\beta_T h(\delta), \quad (2)$$

and following Zhang et al. [11]

$$b_c = \frac{8\pi^3}{\lambda^4 N_A}\left(\frac{\partial n^2}{\partial S}\right)^2 \frac{M_w}{\rho}\frac{S}{-\partial \ln a_w/\partial S}h(\delta), \quad (3)$$

where, respectively, λ, k (=1.38064852 $\times$ 10^{-23} $m^2 \cdot kg \cdot s^{-2} \cdot K^{-1}$), and N_A (=6.022 $\times$ 10^{23} mol^{-1}) are the wavelength of light, the Boltzmann constant, and Avogadro's number; ρ, n, T, β_T, S, and δ are the density, the absolute refractive index, the absolute temperature, the isothermal compressibility, the mass concentration of salts, and the depolarization ratio of the seawater; and a_w and M_w (=18.01528 g mol^{-1}) are the activity and molecular weight of pure water. Also, $h(\delta) = (2+\delta)/(6-7\delta)$.

In Equation (1), b_c vanishes for pure water, and the scattering of light is due entirely to density fluctuation. Replacing the density derivative in Equation (2) with pressure derivative, i.e., $(\rho\frac{\partial n^2}{\partial \rho})_T = \frac{2n}{\beta_T}(\frac{\partial n}{\partial P})_T$, Equation (2) becomes the Einstein–Smoluchowski equation

$$b_d = \frac{32\pi^3}{\lambda^4}\frac{n^2}{\beta_T}\left(\frac{\partial n}{\partial P}\right)_T^2 kTh(\delta) \quad (4)$$

which was used by Cohen and Eisenberg [6] and Buiteveld et al. [7] in evaluating the temperature dependence of scattering by pure water. Historically, Equation (4) was often used because the isothermal piezo-optic coefficient $(\partial n/\partial P)_T$ was relatively easier to measure, even though the uncertainty was high as compared to $n(T)$, $n(\lambda)$, or $n(S)$ [16]. However, recent theoretical development [17] has greatly improved our knowledge in $(\rho\partial n/\partial \rho)_T$. This, together with the development of Equation (3) to explicitly account for the effect of salinity on scattering [11], has advanced our capability for modeling scattering by seawater [9–11], and our confidence in using Equations (1)–(3) to evaluate its temperature effect. The formulae used in the equations to estimate n, ρ, β_T, and a_w can be found in Zhang and Hu [9] and Zhang et al. [11], and the Matlab code for the model can be accessed at https://goo.gl/jKAZgT. Light scattering by seawater is a function of salinity, temperature, and pressure. In this study, we focus on the temperature and salinity ranges of 0–60 °C and 0–40 psu under one atmospheric pressure, which cover the majority of natural inland, coastal, and oceanic surface water bodies. The presence of sea salts is expected to modify the value of δ through two contrasting effects: isotropic ions would decrease δ; whereas their electrostatic field would increase anisotropy, and hence the value of δ [18,19]. Both effects have been observed in pure salt solutions: δ for KNO_3 solution increases and δ for KCl solution decreases, with their respective concentrations [20]. To the best of our knowledge, however, no studies have been reported on how the δ of seawater would vary with salinity. For this study, we assumed a constant value of 0.039 for the depolarization ratio δ for pure water [21] and for seawater [8,9,22,23].

3. Results and Discussion

Light scattering by pure water at 436 and 546 nm was estimated using Equation (2) for temperatures 0–60 °C, and the values normalized to the scattering at 25 °C were compared with the measurements by Cohen and Eisenberg [6] in Figure 1. Between the two wavelengths, the temperature variations of the scattering are almost identical, showing a minimum scattering at 24.7 °C $\pm$ 0.2%. The scattering increases by 4.3% towards 0 °C and increases by 4.7% towards 60 °C at both wavelengths. The root mean square difference between our model and the measurements by Cohen and Eisenberg [6] is approximately 1.3% at both wavelengths. The refractive index model [24] used in Equations (2) and (3) were developed using the Austin and Halikas [16] measurements, which had a temperature precision of 0.1 °C. Also, the Cohen and Eisenberg [6] data we used for comparison in Figure 1 had a temperature precision of 0.1 °C as well. Therefore, we report the temperature in this study at a precision of 0.1 °C. We denote the temperature at which the scattering reaches the minimum as T_{min} hereafter. The predicted values of T_{min} are close to the value of 22 °C measured by Cohen and Eidenberg [6], but differ significantly in both value and trend from the Buiteveld, et al. [7] model, which predicts a maximum near 15 °C. We believe the difference is largely due to the uncertainty in modeling $(\partial n/\partial P)_T$ in Equation (4) that was used by Buiteveld, et al. [7]. Austin and Halikas [16] pointed out that the measurements of the refractive index of water as a function of the pressure, i.e., $n(P)$, were of worse quality when compared to those of $n(T)$, $n(\lambda)$, or $n(S)$. Also, it is well-known that to numerically approximate a derivative, such as $\partial n/\partial P$, as a ratio of measured values is very sensitive to the uncertainties in the measurements of $n(P)$. In addition, Buiteveld, et al. [7] derived the temperature dependency of $(\partial n/\partial P)_T$ by fitting the measurements [25] between 5 and 35 °C, which also explains the relatively large deviation as shown in Figure 1 when extrapolating their model beyond 35 °C.

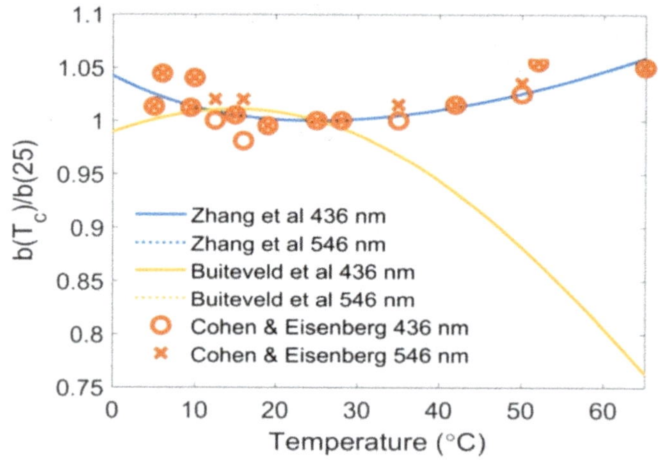

Figure 1. The temperature variations of light scattering by pure water, calculated using the Zhang and Hu [9] model (i.e., Equation (2)) at 436 and 546 nm and normalized by their respective values at 25 °C, are compared with the estimates using the Buiteveld, et al. [7] model and with the measurements by Cohen and Eisenberg [6]. Note that the normalized variations estimated by the Zhang and Hu model overlap with each other at the two wavelengths.

Several bulk properties of pure water needed to estimate the scattering coefficient behave "anomalously": density has the maximum near 4 °C [3]; isothermal compressibility has the minimum near 46 °C [4]; and the refractive index has the maximum near 0 °C [5]. Also, less apparent but indirectly relevant is that $(\partial n/\partial P)_T$, as in Equation (4), has its minimum near 50 °C [5]. Clearly, anomalous light scattering by pure water results from the combination of all of these anomalous properties, as well as its direct proportionality with the temperature (Equation (2) or (4)). Even though scattering by pure water varies strongly with wavelength, with a spectral slope of -4.28 [9], the anomalous temperature behavior of scattering varies little with wavelength (Figure 1).

The scattering coefficient at 546 nm as a function of temperature for different salinities is shown in Figure 2a for b_d (due to density fluctuation) and Figure 2b for b_c (due to concentration fluctuation). Both b_d and b_c vary with temperature in the same anomalous way, all exhibiting a minimum. Also, both T_{min} for b_d and T_{min} for b_c change with salinity—however, with differing patterns. T_{min} for b_d decreases about 20% from 24.6 °C to 19.1 °C for salinity varying from 0 to 40 psu, whereas over the same salinity range T_{min} for b_c increases slightly by ~3%, from 32.2 °C to 33.2 °C. In terms of absolute magnitude, b_d is about 2–10 times greater than b_c (Figure 2a), but in terms of change with respect to salinity, b_c is about 10 times greater than b_d (Figure 2b). As a result, the change of T_{min} for the total scattering coefficient, b is dominated by b_c, and increases from 24.6 °C to 27.5 °C for S from 0–40 psu (Figure 3). It is well known that T_{max} for density [26] and T_{min} for isothermal compressibility [27] decrease with the salinity. Here, we show for the first time that T_{min} for light scattering increases with salinity, which is largely due to the temperature variation of scattering introduced by sea salts. Table 1 lists the variations of T_{min} for b_d, b_c, and b at different salinities.

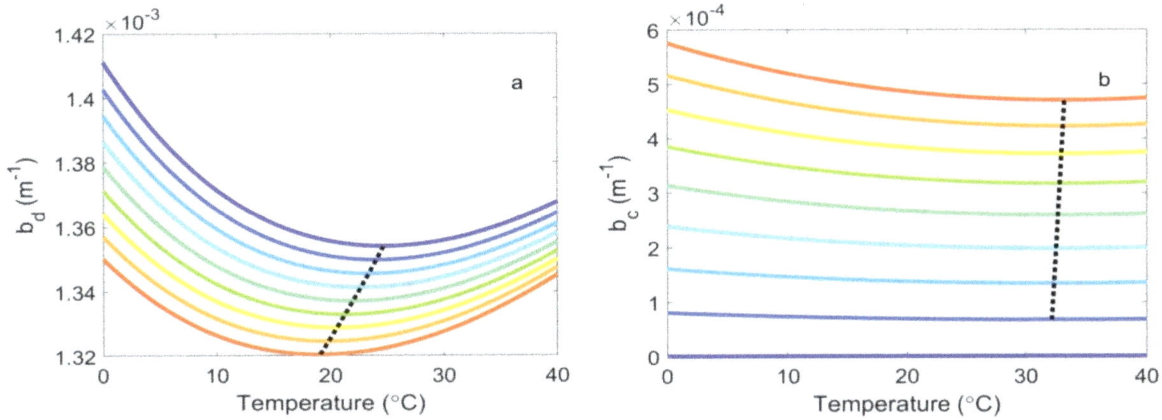

Figure 2. Light scattering by pure seawater at 546 nm as a function of temperature and salinity. (**a**) b_d, the scattering due to density fluctuation; and (**b**) b_c, the scattering due to concentration fluctuation. Lines of progressive colors from blue to red correspond to different salinities from 0 to 40 psu, at 5 psu increments. The dotted line in each plot connects T_{min} at different salinities.

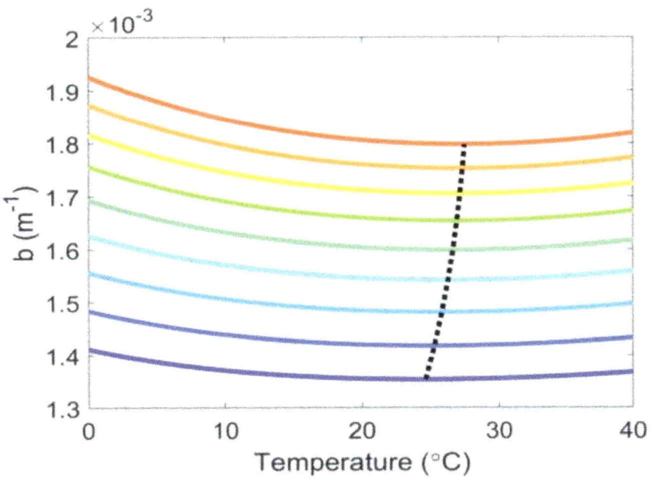

Figure 3. Total scattering coefficient by pure seawater at 546 nm as a function of temperature and salinity. Lines of progressive colors from blue to red correspond to different salinities from 0 to 40 psu at 5 psu increments. The dotted line connects T_{min} at different salinities.

Table 1. Temperatures (T_{min} in °C) at which the scattering of light at 546 nm by pure seawater due to density fluctuations (b_d), concentration fluctuations (b_c), and their total (b) reach the minimum for various salinities (S).

S (psu)	0	5	10	15	20	25	30	35	40
b_d (m^{-1})	24.6	24.0	23.4	22.7	22.0	21.2	20.5	19.8	19.1
b_c (m^{-1})		32.2	32.3	32.5	32.7	32.8	33.0	33.1	33.2
b (m^{-1})	24.6	25.3	25.9	26.3	26.6	26.9	27.2	27.3	27.5

In conclusion, using the latest scattering models for water and seawater, we find that water exhibits an anomalous light scattering behavior, with a minimum occurring at 24.6 °C for pure water, and that this minimum increases with the salinity, reaching 27.5 °C at 40 psu. This temperature behavior changes little spectrally. Caution should be exercised when using the Buiteveld, et al. [7] model, which predicts a temperature behavior of scattering that is inconsistent with the measurements [6] or the results of this study.

Author Contributions: X.Z. and L.H. conceived and designed the experiments; X.Z. wrote the paper.

Acknowledgments: We thank three anonymous reviewers and Dariusz Stramski for their comments, which have improved the manuscript.

References

1. Brovchenko, I.; Oleinikova, A. Multiple Phases of Liquid Water. *ChemPhysChem* **2008**, *9*, 2660–2675. [CrossRef] [PubMed]
2. Ball, P. Water—An enduring mystery. *Nature* **2008**, *452*, 291. [CrossRef] [PubMed]
3. Vedamuthu, M.; Singh, S.; Robinson, G.W. Properties of Liquid Water: Origin of the Density Anomalies. *J. Phys. Chem.* **1994**, *98*, 2222–2230. [CrossRef]
4. Vedamuthu, M.; Singh, S.; Robinson, G.W. Properties of Liquid Water. 4. The Isothermal Compressibility Minimum near 50 °C. *J. Phys. Chem.* **1995**, *99*, 9263–9267. [CrossRef]
5. Cho, C.H.; Urquidi, J.; Gellene, G.I.; Robinson, G.W. Mixture model description of the *T*-, *P* dependence of the refractive index of water. *J. Chem. Phys.* **2001**, *114*, 3157–3162. [CrossRef]
6. Cohen, G.; Eisenberg, H. Light Scattering of Water, Deuterium Oxide, and Other Pure Liquids. *J. Chem. Phys.* **1965**, *43*, 3881–3887. [CrossRef]
7. Buiteveld, H.; Hakvoort, J.H.M.; Donze, M. The optical properties of pure water. *SPIE* **1994**, *2258*, 174–183.
8. Twardowski, M.S.; Claustre, H.; Freeman, S.A.; Stramski, D.; Huot, Y. Optical backscattering properties of the "clearest" natural waters. *Biogeosciences* **2007**, *4*, 1041–1058. [CrossRef]
9. Zhang, X.; Hu, L. Estimating scattering of pure water from density fluctuation of the refractive index. *Opt. Express* **2009**, *17*, 1671–1678. [CrossRef] [PubMed]
10. Zhang, X.; Hu, L. Scattering by pure seawater at high salinity. *Opt. Express* **2009**, *17*, 12685–12691. [CrossRef] [PubMed]
11. Zhang, X.; Hu, L.; He, M.-X. Scattering by pure seawater: Effect of salinity. *Opt. Express* **2009**, *17*, 5698–5710. [CrossRef] [PubMed]
12. Zhang, X.; Hu, L.; Twardowski, M.S.; Sullivan, J.M. Scattering by solutions of major sea salts. *Opt. Express* **2009**, *17*, 19580–19585. [CrossRef] [PubMed]
13. Morel, A. Etude Experimentale de la diffusion de la lumiere par l'eau, les solutions de chlorure de sodium et l'eau de mer optiquement pures. *J. Chim. Phys.* **1966**, *10*, 1359–1366. [CrossRef]
14. Morel, A. Note au sujet des constantes de diffusion de la lumiere pour l'eau et l'eau de mer optiquement pures. *Cahiers Oceanogr.* **1968**, *20*, 157–162.
15. Einstein, A. Theorie der Opaleszenz von homogenen Flüssigkeiten und Flüssigkeitsgemischen in der Nähe des kritischen Zustandes. *Annalen der Physik* **1910**, *338*, 1275–1298. [CrossRef]
16. Austin, R.W.; Halikas, G. *The Index of Refraction of Seawater*; SIO Ref. No. 76-1; Scripps Institute of Oceanography: La Jolla, CA, USA, 1976; p. 121.
17. Proutiere, A.; Megnassan, E.; Hucteau, H. Refractive index and density variations in pure liquids: A new theoretical relation. *J. Phys. Chem.* **1992**, *96*, 3485–3489. [CrossRef]
18. Morel, A. Optical Properties of Pure Water and Pure Sea Water. In *Optical Aspects of Oceanography*; Jerlov, N.G., Nielsen, E.S., Eds.; Academic Press: New York, NY, USA, 1974; pp. 1–24.
19. Shifrin, K.S. *Physical Optics of Ocean Water*; American Institute of Physics: New York, NY, USA, 1988; p. 285.
20. Pethica, B.A.; Smart, C. Light scattering of electrolyte solutions. *Trans. Faraday Soc.* **1966**, *62*, 1890–1899. [CrossRef]
21. Farinato, R.S.; Rowell, R.L. New values of the light scattering depolarization and anisotropy of water. *J. Chem. Phys.* **1976**, *65*, 593–595. [CrossRef]

22. Zhang, X. Molecular Light Scattering by Pure Seawater. In *Light Scattering Reviews 7*; Kokhanovsky, A., Ed.; Springer: Heidelberg, Germany, 2013; pp. 225–243.
23. Werdell, P.J.; McKinna, L.I.; Boss, E.; Ackleson, S.G.; Craig, S.E.; Gregg, W.W.; Lee, Z.; Maritorena, S.; Roesler, C.S.; Rousseaux, C.S.; et al. An overview of approaches and challenges for retrieving marine inherent optical properties from ocean color remote sensing. *Prog. Oceanogr.* **2018**, *160*, 186–212. [CrossRef]
24. Quan, X.; Fry, E.S. Empirical equation for the index of refraction of seawater. *Appl. Opt.* **1995**, *34*, 3477–3480. [CrossRef] [PubMed]
25. O'Connor, C.L.; Schlupf, J.P. Brillouin Scattering in Water: The Landau—Placzek Ratio. *J. Chem. Phys.* **1967**, *47*, 31–38. [CrossRef]
26. Millero, F.J.; Chen, C.-T.; Bradshaw, A.; Schleicher, K. A new high pressure equation of state for seawater. *Deep-Sea Res.* **1980**, *27*, 255–264. [CrossRef]
27. Lepple, F.K.; Millero, F.J. The isothermal compressibility of seawater near one atmosphere. *Deep-Sea Res.* **1971**, *18*, 1233–1254. [CrossRef]

5

Remote Sensing of CDOM, CDOM Spectral Slope and Dissolved Organic Carbon in the Global Ocean

Dirk Aurin [1,*], Antonio Mannino [2] and David J. Lary [3]

[1] NASA Goddard Space Flight Center (USRA), Greenbelt, MD 20771, USA
[2] NASA Goddard Space Flight Center, Greenbelt, MD 20771, USA; antonio.mannino-1@nasa.gov
[3] University of Texas at Dallas, Richardson, TX 75080, USA; david.lary@utdallas.edu
* Correspondence: dirk.a.aurin@nasa.gov

Featured Application: Methods and algorithms developed in this manuscript may be applied to ocean color satellite or aircraft imagery for the remote sensing of oceanic CDOM spectral absorption, CDOM spectral slope, and DOC.

Abstract: A Global Ocean Carbon Algorithm Database (GOCAD) has been developed from over 500 oceanographic field campaigns conducted worldwide over the past 30 years including in situ reflectances and coincident satellite imagery, multi- and hyperspectral Chromophoric Dissolved Organic Matter (CDOM) absorption coefficients from 245–715 nm, CDOM spectral slopes in eight visible and ultraviolet wavebands, dissolved and particulate organic carbon (DOC and POC, respectively), and inherent optical, physical, and biogeochemical properties. From field optical and radiometric data and satellite measurements, several semi-analytical, empirical, and machine learning algorithms for retrieving global DOC, CDOM, and CDOM slope were developed, optimized for global retrieval, and validated. Global climatologies of satellite-retrieved CDOM absorption coefficient and spectral slope based on the most robust of these algorithms lag seasonal patterns of phytoplankton biomass belying Case 1 assumptions, and track terrestrial runoff on ocean basin scales. Variability in satellite retrievals of CDOM absorption and spectral slope anomalies are tightly coupled to changes in atmospheric and oceanographic conditions associated with El Niño Southern Oscillation (ENSO), strongly covary with the multivariate ENSO index in a large region of the tropical Pacific, and provide insights into the potential evolution and feedbacks related to sea surface dissolved carbon in a warming climate. Further validation of the DOC algorithm developed here is warranted to better characterize its limitations, particularly in mid-ocean gyres and the southern oceans.

Keywords: ocean color database; oceanic carbon; chromophoric dissolved organic matter; dissolved organic carbon; CDOM spectral slope; ocean color remote sensing; algorithm development; ocean color algorithm validation; ocean optics; CDOM climatology; CDOM and ENSO; machine learning

1. Introduction

1.1. Background

In 1896, Svante Arrhenius introduced the theory that adding carbon dioxide (CO_2) to the atmosphere enhances the planetary greenhouse effect. Over the intervening century, it became clear that the marine dissolved organic carbon (DOC) pool comprised the vast majority of the organic carbon in the oceans, and was nearly equivalent to the atmospheric pool of CO_2 [1]. In fact, remineralization of just 1% of the DOC in the oceans (e.g., by microbial metabolism and photo-oxidation) would generate a flux of CO_2 into the atmosphere greater than that resulting from all the fossil fuel burned in a year [2]. Recently, Belanger et al. [3] estimated that photoproduction of CO_2 from Chromophoric Dissolved Organic Matter (CDOM) has already increased by ~15% in Arctic waters due to an increase in ultraviolet radiation and the decrease in sea ice associated with global warming. Positive feedbacks such as this have potentially serious consequences for humans and ecosystems alike, and emphasize the urgency to develop robust, global algorithms for retrieving oceanic carbon products remotely and synoptically.

CDOM (refer to Table 1 for terms and abbreviations) is used to describe an often difficult to define fraction of the DOC pool (see Section 1.4) which has historically been called gilvin, *gelbstoff*, or simply "yellow substance". As its name suggests, the presence of CDOM imparts color to the water column through absorption of light by various chromophores, thereby providing an effective means of detecting CDOM remotely from ocean color reflection. Found in all natural waters and generally in highest concentration near shore, CDOM results from the breakdown products of plants and other organic matter into humic materials, and plays a significant role in aquatic photochemistry, photobiology, and as a tracer of the origins of oceanic water masses, e.g., [4,5]. DOC and CDOM can be terrigenous or autochthonous (i.e., deriving from in situ primary and bacterial production in river to ocean waters), with the DOC variously composed of high molecular weight (HMW) humic substances (which tend to be more labile) and low molecular weight (LMW) humics (such as fulvic acids), depending on its origin, labile fraction, age, and whether it has transitioned from fresh waters to marine [6–12]. Most estuarine and nearshore CDOM is terrigenous, and as it mixes in rivers on its transit to marine waters, the amount of HMW material declines from flocculation, photo-oxidation and microbial decomposition leaving marine waters dominated by LMW CDOM (e.g., [6]), a condition imparting a characteristic spectral shape to inherent light absorption by CDOM ($a_g(\lambda)$, where λ is wavelength) [7]. Inherent optical properties (IOPs) of the water column, such as the absorption and backscattering coefficients, depend on the composition and concentration of the dissolved and suspended material present, as well as the size and structure of the particles, and water itself. CDOM concentration—for which $a_g(\lambda)$ is the common proxy following Beer's law—varies widely in the ocean, tending to be highest near river outflows, but may also be high in upwelling regions and other regions of autochthonous, plankton-based production through exudation, excretion, and microbial breakdown of detritus [8]. It is degraded over time both by microbial activity, photooxidation, and other abiotic processes, ultimately resulting in remineralization of the carbon, and release from the ocean as CO and CO_2. In the case of the CDOM fraction of DOC, degradation over time scales of days to millennia can significantly change the magnitude and spectral characteristics of $a_g(\lambda)$.

Table 1. Definition of terms, units, and abbreviations.

	Units	Definition
$a_g(\lambda)$	m^{-1}	CDOM absorption coefficient
$a_d(\lambda)$	m^{-1}	NAP absorption coefficient
$a_{dg}(\lambda)$	m^{-1}	NAP and CDOM absorption coefficient
$a_p(\lambda)$	m^{-1}	Particulate absorption coefficient
$b_{bp}(\lambda)$	m^{-1}	Particle backscattering coefficient
$b_{bt}(\lambda)$	m^{-1}	Total backscattering coefficient
CDOM		Colored Dissolved Organic Matter
Chl	$mg\ m^{-3}$	Chlorophyll concentration
DOC, DOM	$\mu mol\ L^{-1}$	Dissolved Organic Carbon, -Material
$E_s(\lambda)$	$W\ m^{-2}\ nm^{-1}$	Downwelling surface irradiance
$L_w(\lambda)$	$W\ m^{-2}\ nm^{-1}\ sr^{-1}$	Water leaving radiance
$L_{wn}(\lambda)$	$W\ m^{-2}\ nm^{-1}\ sr^{-1}$	Normalized water leaving radiance
POC	$\mu mol\ L^{-1}$	Particulate Organic Carbon
$R_{rs}(\lambda)$	sr^{-1}	Remote sensing reflectance

Table 1. *Cont.*

	Units	Definition
$S_g(\lambda_1\text{-}\lambda_2)$	nm^{-1}	Exponential slope of CDOM and in select spectral range
SPM	$mg\ m^{-3}$	Suspended Particulate Material
TOC	$\mu mol\ L^{-1}$	Total Organic Carbon
AOP		Apparent Optical Properties
GIOP		Generalized IOP Algorithm
GOCAD		Global Ocean Carbon Algorithm Database
HMW		High Molecular Weight
IOCCG		International Ocean-Colour Coordinating Group
IOP		Inherent Optical Properties
LMW		Low Molecular Weight
MLR		Multiple Linear Regression Algorithm
MODIS		Moderate Resolution Imaging Spectroradiometer
NAP		Non-Algal Particulate
NOMAD		NASA bio-Optical Algorithm Dataset
QAA		Quasi-Analytical Algorithm
RFTB		Random Forest Tree Bagger Algorithm
SAA		Semi-Analytical Algorithm
SeaBASS		SeaWiFS Bio-optical Archive and Storage System
SeaWiFS		Sea-viewing Wide Field-of-view Sensor
UV, UVA, UVB		Ultraviolet spectrum, 315–400 nm, 280–315 nm
VIS		Visible spectrum

CDOM absorption is a superposition of the spectral absorption by its varied chromophores, and increases roughly exponentially (or hyperbolically [9]) with decreasing wavelength in the visible (VIS) and ultraviolet (UV) spectral ranges, as described in the next section. CDOM tends to dominate the blue and UV spectrum in many coastal and estuarine environments (e.g., [7,10–12]), and is the most important factor controlling UV and blue light penetration even in the open ocean [13] despite its generally lower concentration and distance from land. Within the visible spectrum, $a_g(\lambda)$ reduces the photosynthetically available radiation supporting phytoplankton and macrophytic growth, and generates heat in the surface layer of the water column, thus affecting mixing [14]. In the UV, CDOM causes surface heating as well, but also acts to protectively shade aquatic organisms, thus reducing the amount of damaging high frequency radiation reaching vulnerable cell structures.

From the passive remote sensing perspective, CDOM reduces the amount of blue light available for reflection out of the water column, and can therefore have a significant impact on ocean color algorithms, for example increasing uncertainty in blue-green band-ratio algorithms designed to estimate chlorophyll-*a* concentration (Chl) from its absorption peak near 443 nm [13,15]. These types of Chl algorithms assume covariance in Chl, CDOM, and other water column constituents (i.e., the "Case 1 waters" assumption [16]). By contrast, semi-analytical algorithms (SAAs) that invert the ocean color signal to retrieve individual component absorption spectra (particles, CDOM, water) are stymied by the presence of non-algal particulates (NAPs), which have a similar spectral shape to CDOM. As a result, these approaches tend to retrieve only the sum of these two elements [17] (and references therein).

1.2. Spectral Shape of CDOM

The CDOM absorption coefficient is generally modeled with an exponentially decaying function with increasing wavelength, λ.

$$a_g(\lambda) = a_g(\lambda_0)\, e^{-Sg\,(\lambda-\lambda_0)} \quad (1)$$

where S_g is the spectral slope parameter and λ_0 is a reference wavelength. S_g in various spectral ranges in the UV and VIS contains information about CDOM's photoreactive state, chemical composition, molecular weight distribution, and origin [4,7,18–21].

While the single exponential model in Equation (1) is accurate within limited wavebands, CDOM spectral slope, S_g, is not constant across the UV and VIS and depends on the wavelength range used, spectral resolution, and reference wavelength λ_0. Furthermore, comparative analysis of CDOM spectral shape as reported in the literature has been confounded by the multitude of methodologies and reference wavebands used historically to calculate S_g [9]. For instance, a linear fit to logarithmically transformed a_g data yields results for S_g biased by higher wavelength absorption, whereas a least-squared difference minimization fitting favors the lower wavelengths where the magnitude of a_g is higher, and is considered more accurate [7,22]. Changes to S_g resulting from photodegradation are wavelength dependent, i.e., increasing below 460 nm and decreasing above 510 nm [23], although when calculated across the VIS from 412–555, slope is expected to increase through the destruction of large humic complexes resulting in lower molecular weight CDOM [24]. This effect appears to reverse over time as more refractory, low-molecular weight compounds are also degraded, thereby reducing CDOM absorption at shorter wavelengths relative to longer, and decreasing spectral slope across the VIS.

All these factors lead to challenges in comparing CDOM spectral slope between studies, and a more standardized approach to CDOM spectral shape measurement still seems warranted [19]. The concept of the spectral slope curve, $S_g(\lambda)$—analogous to the first derivative of S_g with respect to λ—was explored by Loiselle et al. in 2009 [23]. Calculating S_g from natural waters, cultures, and laboratory standards at 20 nm waveband intervals between 200–700 nm, they showed that $S_g(\lambda)$ had complex spectral characteristics including peaks near 390 nm likely indicating a prevalence of autochthonous production of fulvic acid-type CDOM, and near 280 nm possibly due to the release of proteins or phenols by phytoplankton. While the spectral slope curve approach of Loiselle et al. 2009 represents an elegant method for quantifying many subtle characteristics of CDOM spectral shape when compared to, for example, using a single slope parameter across the UV and VIS, it does require relatively high spectral resolution data collection. Historically, this was not always available or reported, and here we focus on a set of eight different spectral ranges commonly seen in the literature and described in detail below.

1.3. Remotely Sensing CDOM and S_g

As interest in CDOM has grown in recent years, numerous empirical ocean color algorithms for retrieving CDOM within limited geographic regions have emerged, e.g., [25–30]. A smaller number of more generally applicable, global empirical algorithms have also been developed, including one for retrieving a unitless index of CDOM prevalence, though it does not retrieve $a_g(\lambda)$ or S_g and depends upon Case 1 assumptions. More recently, Tiwari and Shanmugam published global empirical algorithms for both $a_g(\lambda)$ and S_g [31,32]. These were optimized and tested using field data aggregated in NOMAD (the NASA bio-Optical Marine Algorithm Dataset version 2 [33]) and the synthetic ocean color dataset developed by the International Ocean Colour Coordinating Group (IOCCG) for the purpose of algorithm development and validation [34].

Other approaches to retrieving CDOM remotely depend on the premise that sea-surface reflectance is approximately inversely proportional to the total absorption coefficient [16,35,36], which can be linearly separated into various contributions by particulate and dissolved constituents. This forms the basis to semi-analytical ocean color algorithms (SAAs) for retrieving constituent absorption, e.g., [37–39], but, as already mentioned, owing to the similarity in spectral shape of non-algal particulate (i.e., detrital, microbial, and sedimentary) absorption and $a_g(\lambda)$, SAAs generally retrieve only their sum, a_{dg} [17]. To circumvent this difficulty, empirical methods are sometimes added to SAAs to help distinguish non-algal from dissolved absorption [40–45].

1.4. Remotely Sensing DOC

One of the most challenging aspects of developing robust, global ocean color algorithms for DOC is that the relationship between DOC and CDOM (i.e., the DOC-specific absorption) is highly variable, in some cases negatively correlated (e.g., Southern Ocean, [46]) and often poorly defined, particularly in open ocean areas such as the Sargasso Sea [47,48]. In some cases, the relationship is better constrained within a particular region and season, as shown by measurements made in the Mid-Atlantic Bight on the eastern shelf of North America [28]. Because absorption by CDOM is the only way in which ocean color is impacted by DOC, some other independent knowledge of water type is needed for retrieval of DOC from space.

1.5. Algorithm Development Data

One of the most confounding challenges in the development of both empirical and semi-analytical algorithms is the lack of a large, comprehensive database containing a broad enough dynamic range in optical characteristics to be representative of the majority of the world's oceans, while also having realistic combinations of inherent optical properties, which are not guaranteed in large, synthetic, modeled datasets. NOMAD represents the first (and most recent, as of this writing) major effort to provide the ocean color community with such a dataset. It was aggregated and selected from all of the relevant field data submitted to the NASA SeaBASS archive (http://seabass.gsfc.nasa.gov/), and has been extremely useful to those in the ocean color algorithm community since its original publication in 2005 and update in 2008. However, NOMAD was not focused on CDOM. For example, while it contains about 3700 coincident radiometric and phytoplankton pigment observations, coincident radiometric and CDOM observations number just ~1200. In part, this is because CDOM data collected using in situ instrumentation were excluded for various reasons discussed at greater length below. The remaining CDOM records—those measured from discrete water samples—were modeled spectrally at the preselected NOMAD wavebands after fitting field data to Equation (1), and do not extend into the UV where spectral shape can provide useful insights into the origin and photooxidation state of CDOM. NOMAD does not contain any DOC data observations.

Using the methodology described in the next section, we extend the NOMAD approach to create a global ocean color algorithm development database better suited to DOC and its optical components, CDOM and CDOM spectral slope, ultimately including over 51,000 field observations of surface-averaged inherent optical properties. These are matched to between ~8000 and ~11,000 coincident estimates of sea surface reflectance made from in situ measurements as well as satellite imagery from SeaWiFS and MODIS Aqua and Terra instruments. The global ocean carbon algorithm database (hereafter Global Ocean Carbon Algorithm Database (GOCAD) records are split into independent sets of field stations for training/optimization (i.e., with in situ radiometry) and validation (i.e., with satellite imagery) of algorithms, as described in the Section 2. A basic overview of the most relevant aspects of the global dataset is presented in Section 3.1. In Section 3.2, empirical and SAA approaches to retrieval of DOC, CDOM, and CDOM slope are developed and discussed. Finally, algorithms are applied to global climatological satellite imagery and discussed in Section 3.3.

2. Methodology

2.1. Database Assembly Overview

Field measurements of CDOM, DOC, remote sensing reflectance, $R_{rs}(\lambda)$, and ancillary data and metadata were downloaded from SeaBASS and the Hansell/Carlson collection (https://hansell-lab.rsmas.miami.edu/research/data-collection/index.html) in April 2013. Coincident, Level 2 (L2) SeaWiFS and MODIS Aqua and Terra satellite imagery at all field stations were downloaded from the NASA Ocean Color website (http://oceancolor.gsfc.nasa.gov). Due to the size of aggregated datasets for each of the key constituents (e.g., 117,291 raw CDOM records, 31,474 raw DOC records, 115,773 in situ reflectance records, and ~177,000 matching satellite scenes), extensive automation in the processing, quality control, and merging of the databases was a necessity. A station-by-station analysis (or field experiment-specific analysis, as in [33]) of the data for establishing the customized spatial and temporal thresholds for matching coincident inherent and apparent optical properties and satellite imagery was not feasible. Relatively broad guidelines conducive to automation were established, as described in detail in the following sections. We assume, for example, that geospatial and temporal variability of CDOM and DOC is higher in coastal and shelf waters (defined here as samples collected in waters of 1000 m depth or less) than in the pelagic.

2.2. Field Data

Targeted searches of SeaBASS were conducted for all records containing a_g, DOC, or in situ reflectances (see 2.2.3). Resultant data from the following physical, bio-optical, and biogeochemical fields were also retained where they happened to be present in SeaBASS files: depth, temperature, salinity, a_{nap}, a_p, a_{pg}, a_{dg}, b_{bt}, particulate organic carbon (POC), total organic carbon (TOC), and Chl. Ancillary data including time and date, latitude, longitude, and bottom depth were also retained, as well as complete SeaBASS metadata for each record. Carbon data were downloaded from each of the data repository resources linked in the Hansell/Carlson DOM Data Collection (http://yyy.rsmas.miami.edu/groups/biogeochem/Data.html). These were also queried for all the parameters above and assigned metadata for each cruise. Table 2 provides a complete listing and overview of all the field experiments retained in the final, quality-controlled database.

Table 2. Summary of field data collection campaigns.

Experiment	Principal Investigators	Numbers of Stations							Min. Lat	Max. Lat	Min. Lon	Max. Lon	Year(s)
		Cruises	CDOM	CDOM & IS*	CDOM & SAT**	DOC	DOC & IS*	DOC & SAT**					
MURI	A. Neeley, S. Freeman, J. Chaves, C. McClain	1	0	0	0	6	0	0	19.126	20.692	−157.36	−156.32	2012
EGEE3	A. Subramaniam	1	9	0	0	0	0	0	−6.003	3.327	−10.008	7.992	2006
EGEE5	A. Subramaniam	1	2	0	0	0	0	0	−5.977	−5.62	5.85	7.997	2007
MANTRA PIRANA	A. Subramaniam	5	41	2	2	0	0	0	3.386	25.002	−158.02	−42.276	2001–2003
MASS BAY	A. Subramaniam	7	39	0	13	0	0	0	41.85	42.619	−70.895	−70.228	2002–2005
IOFFE	A. Khrapko, S. Ershova	1	164	56	51	0	0	0	−66.46	48.59	−67.98	−5.54	2001–2002
B01	A. Mannino	2	15	5	0	16	5	0	36.713	37.786	−76.018	−74.644	2005
B02	A. Mannino	2	28	18	6	30	17	6	36.685	38.918	−76.069	−74.502	2005
B03	A. Mannino	2	26	0	14	29	0	16	36.413	38.87	−76.022	−74.499	2006
B04	A. Mannino	2	30	19	6	46	19	9	36.502	38.908	−76.019	−74.299	2006
B05	A. Mannino	1	15	0	5	15	0	5	36.431	38.586	−76.017	−73.517	2006
BIOD01	A. Mannino	1	15	11	1	15	11	1	42.361	43.574	−70.696	−69.863	2007
BIOD02	A. Mannino	1	17	14	6	17	14	6	42.593	43.708	−70.78	−69.691	2007
BIOD03	A. Mannino	1	13	12	0	12	11	0	41.201	42.812	−70.771	−70.445	2007
CBM01	A. Mannino	1	2	0	0	0	0	0	36.965	37.17	−76.172	−76.029	2004
CBM02	A. Mannino	1	4	0	3	0	0	0	36.987	37.182	−76.163	−76.018	2004
CBM03	A. Mannino	1	4	0	2	0	0	0	36.987	37.182	−76.163	−76.018	2004
CBM04	A. Mannino	1	4	0	0	0	0	0	36.987	37.182	−76.163	−76.018	2004
CBM05	A. Mannino	1	4	0	3	0	0	0	36.987	37.182	−76.163	−76.018	2004
CBM06	A. Mannino	1	4	0	3	0	0	0	36.987	37.182	−76.163	−76.018	2005
CBM07	A. Mannino	1	3	0	2	0	0	0	37.046	37.182	−76.138	−76.018	2005
CBM08	A. Mannino	1	3	0	0	0	0	0	37.047	37.182	−76.137	−76.019	2005
CBM09	A. Mannino	1	5	0	0	0	0	0	36.987	37.181	−76.161	−76.017	2005
CBM10	A. Mannino	1	4	0	3	0	0	0	36.987	37.181	−76.161	−76.017	2005
CBM11	A. Mannino	1	4	0	0	0	0	0	37.046	37.182	−76.136	−76.019	2006
CBM12	A. Mannino	1	4	0	0	0	0	0	36.987	37.182	−76.161	−76.018	2006
COI1	A. Mannino	1	4	2	3	4	2	3	36.69	36.969	−76.017	−75.713	2007
COI2	A. Mannino	1	4	1	3	4	1	3	36.69	36.969	−76.017	−75.71	2007
COI3	A. Mannino	1	4	1	0	3	0	0	36.69	36.969	−76.017	−75.713	2007
CV1	A. Mannino	1	53	0	24	51	0	23	35.745	42.498	−75.706	−65.736	2009
CV2	A. Mannino	1	69	0	35	69	0	35	36.475	43.062	−75.785	−66.088	2009
CV3	A. Mannino	1	43	0	9	43	0	9	37.089	43.112	−75.677	−65.779	2010
CV4	A. Mannino	1	79	0	30	78	0	30	36.073	44.233	−75.911	−65.772	2010
CV5	A. Mannino	1	67	0	24	63	0	24	36.142	44.299	−75.859	−65.775	2010
CV6	A. Mannino	1	92	0	18	92	0	18	36.187	43.928	−75.789	−65.768	2011
D01	A. Mannino	1	4	2	2	4	2	2	36.797	36.966	−76.02	−75.719	2005
D02	A. Mannino	1	6	6	6	6	6	6	36.805	36.973	−76.02	−75.712	2005
D03	A. Mannino	1	5	2	0	5	2	0	36.803	36.968	−76.018	−75.718	2006
D04	A. Mannino	2	6	3	0	6	3	0	36.801	36.966	−76.015	−75.64	2006
OCV1	A. Mannino	1	26	26	18	26	26	18	40.218	40.731	−74.153	−73.479	2007
OCV2	A. Mannino	1	22	18	2	22	18	2	40.208	40.724	−74.151	−73.451	2007–2009
OCV3	A. Mannino	1	8	8	0	8	8	0	40.392	40.739	−74.156	−73.554	2008

Table 2. *Cont.*

Experiment	Principal Investigators	Numbers of Stations							Min. Lat	Max. Lat	Min. Lon	Max. Lon	Year(s)
		Cruises	CDOM	CDOM & IS*	CDOM & SAT**	DOC	DOC & IS*	DOC & SAT**					
OCV5	A. Mannino	1	11	11	3	11	11	3	39.584	41.028	−73.901	−71.749	2009
PL6	A. Mannino	1	4	0	3	4	0	3	36.802	36.968	−76.017	−75.713	2007
GOMECC-2	A. Mannino, J. Salisbury	1	67	0	1	92	0	2	25.999	43.032	−90.809	−68.01	2012
GEO-CAPE	A. Mannino, M. Mulholland	1	59	53	6	54	51	6	38.098	39.17	−76.491	−76.084	2011
MONTEREY BAY	B. Arnone, R. Gould	1	57	51	11	0	0	0	36.271	36.988	−123.13	−121.81	2003
WOCE P14S P15S	B. Tilbrook	1	0	0	0	107	0	0	−66.99	−0.003	−174.79	173.982	1996
Carbon Transport MS R.	C. Del Castillo	2	9	0	6	0	0	0	28.295	28.925	−89.743	−89.411	2001–2003
GasEx	C. Del Castillo	1	44	10	3	0	0	0	−53.75	−50.14	−38.554	−36.622	2008
Big Bend	C. Hu	10	156	61	36	0	0	0	29.168	29.67	−83.635	−83.201	2010–2011
GEO-CAPE CBODAQ	C. Hu	2	16	12	0	23	11	2	38.098	39.17	−76.487	−76.083	2011
GOM Oil Spill	C. Hu	3	10	0	1	0	0	0	28.584	29.1	−88.42	−87.323	2010
Glider calibration	C. Hu	1	8	0	4	0	0	0	27.452	28.465	−83.992	−83.072	2009–2011
Glider validation	C. Hu	1	13	0	6	0	0	0	27.356	27.48	−83.115	−83.051	2009
SWFL	C. Hu	3	21	3	2	0	0	0	24.827	26.49	−82.318	−81.141	2010–2011
Tampa Bay	C. Hu	5	85	71	5	0	0	0	27.578	27.991	−82.783	−82.408	2008–2012
West Florida Shelf	C. Hu	5	134	67	53	0	0	0	25.057	28.439	−83.785	−81.143	2005–2008
MOCE	C. Trees, D. Clark	2	40	0	17	0	0	0	21.793	36.875	−122	−105.75	1992–1999
BowdoinBuoy	C. Roesler	1	2	0	0	0	0	0	43.762	43.795	−69.988	−69.947	2011
PenBaySurvey	C. Roesler	1	0	0	0	0	0	0	44.26	44.26	−68.983	−68.983	2008
San Diego Coastal Project	D. Stramski, M. Stramska	1	15	0	0	0	0	0	32.558	32.758	−117.26	−117.13	2004–2006
NASA Gulf of Maine	D. Phinney, D. Phinney, J. Brown	4	46	0	6	0	0	0	41	44.245	−70.567	−67.162	1998–1999
NOAA Gulf of Maine	D. Phinney, D. Phinney, J. Brown	4	37	0	4	0	0	0	40.209	44.344	−70.056	−65.549	1996–1998
Panama City Florida	D. Phinney, D. Phinney, J. Brown	1	0	0	0	0	0	0	30.167	30.172	−85.857	−85.852	2001
Plumes and Blooms	D. Siegel	5	88	1	22	0	0	0	34.024	34.464	−120.56	−119.28	2001–2003
CLIVAR A13.5 2010	D. Hansell	1	0	0	0	64	0	4	−54	4.62	−3.002	1.835	2010
CLIVAR I05 2009	D. Hansell	1	0	0	0	1	0	0	−31.19	−31.19	82.564	82.564	2009
CLIVAR I08S 2007	D. Hansell	1	0	0	0	28	0	2	−65.71	−28.32	81.962	95.014	2007
CLIVAR P02 2004	D. Hansell	1	0	0	0	56	0	7	29.991	32.644	−177.99	179.545	2004
CLIVAR P16N 2006	D. Hansell	1	0	0	0	78	0	2	−17	56.28	−153.22	−150	2006
CLIVAR P16S 2005	D. Hansell	1	0	0	0	57	0	1	−71	−16	−150.04	−149.91	2005
CLIVAR P18 2007	D. Hansell	1	0	0	0	72	0	4	−68.91	22.7	−110.04	−102.54	2007
HLY-02-01	D. Hansell	1	0	0	0	21	0	0	64.98	73.431	−169.14	−154.4	2002
HLY-02-03	D. Hansell	1	0	0	0	38	0	1	65.668	73.698	−168.86	−151.94	2002
HLY-0403	D. Hansell	1	0	0	0	36	0	0	65.661	73.827	−168.9	−152.02	2004
SR03	D. Hansell	1	0	0	0	24	0	0	−65.57	−44.38	139.658	146.189	2008

Table 2. *Cont.*

Experiment	Principal Investigators	Numbers of Stations: Cruises	CDOM	CDOM & IS*	CDOM & SAT**	DOC	DOC & IS*	DOC & SAT**	Min. Lat	Max. Lat	Min. Lon	Max. Lon	Year(s)
WOCE AR01 A05	D. Hansell	1	0	0	0	45	0	6	24.499	27.622	−79.937	−14.224	1998
ACE-ASIA	G. Mitchell, M. Kahru	1	45	22	10	0	0	0	28.207	38.905	−177	178.05	2001
AMLR	G. Mitchell, M. Kahru	7	47	0	7	0	0	0	−63.01	−57.5	−68.186	−53.296	2000–2007
Aerosols Index	G. Mitchell, M. Kahru	1	21	3	9	0	0	0	−34.53	27.368	−60.615	85.166	1999
CALCOFI	G. Mitchell, M. Kahru	16	179	8	29	0	0	0	29.847	36.057	−124.33	−117.3	1996–2002
Sea_of_Japan	G. Mitchell, M. Kahru	1	17	1	1	0	0	0	34.503	43.302	128.883	139.883	1999
Arc00	G. Cota	1	14	0	0	0	0	0	70.328	72.412	−167.59	−144.63	2000
LAB97	G. Cota	1	10	0	0	0	0	0	44.137	60.38	−58.19	−43.999	1997
Lab2000	G. Cota	1	6	0	2	0	0	0	49.518	60.047	−58.749	−48.899	2000
Lab96	G. Cota	1	10	0	0	0	0	0	52.08	60.999	−58.006	−47.908	1996
ORCA Ches. Light Tower	G. Cota	1	3	0	2	0	0	0	36.9	36.9	−75.71	−75.71	2000
Res95	G. Cota	1	1	0	0	0	0	0	74.645	74.645	−95.91	−95.91	1995
Res96	G. Cota	1	6	0	0	0	0	0	74.644	74.646	−94.915	−94.905	1996
NSF-BWZ	G. Mitchell	2	0	0	0	0	0	0	−62.26	−60.63	−58.375	−55.6	2004–2006
Benthic Ecol. from Space	H. Dierssen, R. Zimmerman	4	8	0	0	0	0	0	24.723	29.847	−85.382	−80.705	2005–2006
Kieber Photochemistry 03	H. Sosik	1	17	0	5	0	0	0	35.278	41.075	−75.218	−71.127	2003
MVCO	H. Sosik	34	129	0	23	0	0	0	41.143	41.342	−70.638	−70.415	2005–2011
GOCAL	J.R.V Zaneveld, W.S. Pegau	6	140	0	48	0	0	0	22.914	31.116	−114.64	−107.75	1996–1999
PREPP	J. Chen	5	47	0	0	0	0	0	22.15	22.555	113.673	114.43	2001
SAB Mapping	J. Nelson, A. Subramaniam	2	18	0	10	0	0	0	30.823	31.993	−81.024	−80.221	2005
GEOTRACES	J. Chaves	1	0	0	0	9	0	0	17.35	36.766	−24.496	−12.825	2010
COOA	J. Salisbury, D. Vandemark, C. Hunt	3	0	0	0	14	0	4	42.861	43.757	−70.66	−69.782	2008
NOAA CSC	J. Brock, A. Subramanian, K. Waters	1	10	0	0	0	0	0	31.335	31.965	−81.128	−80.454	1996
BOA	K. Carder	1	62	62	0	0	0	0	27.579	59.841	−91.768	−15.49	1991–1993
EcoHAB	K. Carder	19	398	208	126	0	0	0	25.3	27.572	−84.394	−81.259	1999–2003
Okeechobee	K. Carder	1	4	4	0	0	0	0	27.149	27.199	−80.794	−80.788	1997
Redtide	K. Carder	2	13	11	7	0	0	0	27.289	28.098	−83.253	−82.866	2005
TOTO	K. Carder	3	86	75	40	0	0	0	24.884	27.5	−82.776	−77.587	1998–2000
ACE-INC	L.W. Harding_Jr.,M. Mallonee, A. Magnuson	6	21	0	0	0	0	0	38.303	38.754	−76.62	−76	2002–2003
BIOCOMPLEXITY	L.W. Harding_Jr.,M. Mallonee, A. Magnuson	11	55	0	15	0	0	0	36.863	39.349	−76.451	−75.878	2001–2004
LMER-TIES	L.W. Harding_Jr.,M. Mallonee, A. Magnuson	17	220	0	22	0	0	0	36.866	39.421	−76.517	−75.749	1996–2000
SGER	L.W. Harding_Jr.,M. Mallonee, A. Magnuson	1	11	0	5	0	0	0	36.95	38.5	−76.481	−75.998	2003

Table 2. *Cont.*

Experiment	Principal Investigators	Numbers of Stations							Min. Lat	Max. Lat	Min. Lon	Max. Lon	Year(s)
		Cruises	CDOM	CDOM & IS*	CDOM & SAT**	DOC	DOC & IS*	DOC & SAT**					
ONR-MAB	L.W. Harding_Jr.,M. Mallonee, A. Magnuson	2	31	0	0	0	0	0	36.4	39.134	−75.949	−71.993	1996–1997
Ocean Color Cal Val	M.S. Twardowski, A.H. Barnard, J.R.V. Zaneveld	1	14	13	0	0	0	0	40.208	40.511	−74.054	−73.448	2007
Tokyo Bay	M. Kishino	1	1	0	0	0	0	0	35.223	35.223	139.718	139.718	1984
Global_CDOM	N. Nelson, D. Siegel	2	19	0	3	0	0	0	−8.458	7.004	−140.07	−124.35	2005–2006
CLIVAR	N. Nelson, D. Siegel, C. Carlson	9	80	17	4	45	8	4	−68.36	59.5	−150	95.028	2003–2008
TAO 2005	N. Nelson, D. Siegel, C. Carlson	1	61	0	9	0	0	0	−8.89	12	−140.2	−124.35	2005
TAO 2006	N. Nelson, D. Siegel, C. Carlson	2	78	0	7	0	0	0	−8.458	10.012	−140.17	−123.55	2006
BBOP	N. Nelson, D. Siegel	116	78	7	9	40	0	6	31.446	31.815	−64.991	−64.019	1994–2011
Active Fluorescence 2001	R. Morrison, H. Sosik	1	4	0	0	0	0	0	31.919	40.097	−70.528	−69.784	2001
Kieber Photochemistry 02	R. Morrison, H. Sosik	1	69	0	7	0	0	0	38.709	42.511	−75.564	−67.599	2002
GLOBEC	R. Morrison, H. Sosik	5	23	0	9	0	0	0	41.753	43.799	−70.445	−65.685	1997–1999
FRONT	R. Morrison, H. Sosik	3	9	0	4	0	0	0	42.245	40.985	−70.558	−71.75	2000–2002
CLIVAR A16N 2003	R. Freely	1	0	0	0	69	0	8	−6.004	63.295	−29.001	−19.994	2003
CLIVAR A16S 2005	R. Freely	1	0	0	0	49	0	3	−60.01	−2.334	−36.21	−24.997	2005
CLIVAR A20 2003	R. Freely	1	0	0	0	27	0	3	7.064	42.637	−53.51	−51.12	2003
CLIVAR A22 2003	R. Freely	1	0	0	0	37	0	6	11.001	39.857	−69.932	−64.161	2003
CLIVAR I09N 2007	R. Freely	1	0	0	0	51	0	2	−28.31	18.004	86.782	95.013	2007
North Carolina 2005	R. Stumpf, P. Tester	5	32	4	20	0	0	0	34.096	35.433	−76.693	−75.756	2005
North Carolina 2006	R. Stumpf, P. Tester	2	12	0	6	0	0	0	34.014	35.228	−76.388	−76.028	2006
Chesapeake Light Tower	R. Zimmerman, G. Cota	3	71	59	32	0	0	0	36.803	36.969	−76.101	−75.551	2005–2007
North Sea	R. Doerffer	1	32	0	0	0	0	0	52.226	55.367	0.591	8.123	1994
ICESCAPE	S.B. Hooker, A. Neeley	3	1609	31	31	85	28	6	56.211	73.828	−168.98	−150.44	2001–2011
B07	S.B. Hooker, M.E. Russ	1	11	0	3	20	0	0	42.65	43.18	−70.868	−70.616	2009
MALINA	S.B. Hooker, V. Wright	1	25	22	0	28	1	0	69.246	72.054	−140.83	−126.5	2009
B08	S.B. Hooker, J. Chaves	1	0	0	0	2	0	1	31.667	31.698	−64.169	−64.164	2009
COASTAL	S.B. Hooker, M.E. Russ	1	13	0	10	0	0	0	42.708	43.434	−70.794	−69.865	2008
USM pCO2	S. Lohrenz	1	1	0	0	0	0	0	28.858	28.858	−89.47	−89.47	2005
Catlin Arctic Survey	V. Hill	1	8	0	0	0	0	0	78.771	78.771	−104.72	−104.72	2011
AMT	W. Balch	5	23211	7703	1918	0	0	0	−47.27	49.716	−55.455	18.611	2005–2011
Gulf of Maine	W. Balch	33	8077	0	3251	0	0	0	42.683	44.058	−70.267	−66.172	2005–2008
Scotia Prince Ferry	W. Balch	47	11521	51	8397	0	0	0	43.604	43.798	−70.026	−66.164	2001–2004
2009oct Chesapeake	W.J. Rhea	1	13	10	4	0	0	0	38.136	39.062	−76.448	−76.229	2009
Totals		535	48574	8857	14568	1957	255	302	−71	78.771	−177.99	179.545	1984–2012

* IS is in situ $R_{rs}(\lambda)$, **SAT is satellite $R_{rs}(\lambda)$

2.2.1. CDOM

CDOM absorption was measured in field experiments using a variety of instruments and protocols. Examples include in-line filtered (generally 0.2 μm) flow-through systems outfitted with ac-9 or ac-S absorption and attenuation meters (Wet Labs) and processed to $a_g(\lambda)$ [49–51], as well as discrete sampling and filtration for bench-top spectrophotometry [52], or in liquid capillary waveguides [53]. Unfortunately, SeaBASS metadata did not historically specify which methods or protocols were used in data collection or processing, but more recently (since approximately 2012), investigators have been required to submit ancillary documentation, such as instrument calibration records, and encouraged to submit documentation retroactively.

CDOM data measured in situ (i.e., with ac-9 or ac-S instruments; 33.5% of the preliminary CDOM dataset) were subject to particulate and bubble contamination, especially in experiments in which an automated in-line flow valve switched between filtered and unfiltered water presenting the opportunity for unfiltered water to reside in the plumbing during CDOM data collection. To identify and eliminate particle contamination, any CDOM records with a notable (i.e., $\geq$0.006 m^{-1}) increase in absorption at 676 nm (a phytoplankton absorption peak) above the absorption curve from 650–715 nm were considered contaminated and removed (109 records).

Nonlinear, least squares minimization was used to fit $a_g(\lambda)$ to Equation (1) for calculating slopes of all hyperspectral a_g into seven spectral ranges: 275–295 nm, 290–600 nm, 300–600 nm, 350–400 nm, 350–600 nm, 380–600 nm, and 412–600 nm. Multispectral ac-9 data were fitted for slope using the six wavebands in the 412–555 nm range. To reduce outliers and noisy data, any CDOM slope data found to be outside of the range 0.005–0.05 nm^{-1} were considered unrealistic and eliminated, together with the $a_g(\lambda)$ data used to calculate them. This accounted for only 125 hyperspectral records in the 300–600 nm range (spectrophotometric), but nearly 6,000 records in the 412–600 nm range (predominantly flow-through). To further reduce outliers and noisy records, S_g and a_g data were eliminated where S_g in any slope range was greater than two standard deviations from the median for the entire database, or where they were outside the 2nd and 98th percentiles. This reduced the database of CDOM by nearly 11,000 records. In addition, 460 records were removed for $a_g(676) > 0.1\ m^{-1}$, $a_g(715) > 0.05\ m^{-1}$, or an average $a_g(\lambda > 680\ nm) > 0.05\ m^{-1}$, and an additional 1,057 CDOM records with extreme outliers (>4 standard deviations from the median) in the red ($\lambda > 620$) were eliminated.

2.2.2. DOC

While DOC was included in about 850 SeaBASS records, the majority of the carbon data retained after surface and spatial binning (see 2.2.4) were from the Hansell/Carlson datasets. In total, 1957, 625, and 45 stations included DOC, POC, and TOC, respectively. Outliers (1st and 99th percentiles) were eliminated, and stations were merged with the CDOM records after surface and spatial binning. Specifically, Hansell/Carlson data were matched to CDOM field stations if they were within 1 h, 2.5 m depth, and 1 km in continental shelf waters (bottom depth $\leq$ 1000 m) and within 3 h, 5 m depth, and 5 km off the shelf. Multiple matches within these criteria were averaged and retained if individual measurements were with 1.5 standard deviations of the mean and the coefficient of variability of the ensemble was $\leq$0.25.

2.2.3. In situ Reflectances

SeaBASS searches for field radiometry targeted R_{rs} (or equivalently L_w and E_s, where $R_{rs} = L_w/E_s$, or L_{wn}, where R_{rs} is L_{wn} divided by the top of atmosphere solar irradiance [54]). A total of 135,966 independent field observations of R_{rs} were binned as described in 2.2.4, quality controlled as described in 2.2.5, and then matched to the CDOM database using the same spatial, temporal, and outlier elimination criteria used for DOC (2.2.2).

2.2.4. Bathymetry, Surface Averaging, and Spatial and Spectral Binning

Records with no reported bottom depth (~85% of the database) were matched to the nearest pixel in the UNESCO GEBCO 08 0.05 degree bathymetry grid (http://www.gebco.net/data_and_products/gridded_bathymetry_data/documents/gebco_08.pdf). The purpose of GOCAD is the development of satellite algorithms for surface retrievals, so data collected at depth were discarded as follows: on the continental shelf (defined here as bottom depth 1000 m or less) samples from deeper than 5 m were discarded, as were data from deeper than 10 m off the shelf (~34% of the database combined). 57,127 surface records remained. Samples collected in profile within the surface layer (top 5 m on-shelf, top 10 m off-shelf) were averaged. Samples collected in transect were additionally binned to a 0.5 km grid and averaged.

All absorption related IOPs were matched to the following wavebands with a 2.5 nm tolerance: 245 nm, 1 nm resolution between 250 and 555 nm, 560, 620, 630, 645, 650, 665, 670, 676, 680, 705, and 715 nm. Backscattering data were similarly matched to 1 nm bands from 400 to 700 nm. Hyperspectral in situ $R_{rs}(\lambda)$ were matched to both SeaWiFS bands (412, 443, 490, 510, 555, and 670 nm) and MODIS bands (412, 443 488, 531, 547, 667) by weighting the data to the instrument-specific spectral response functions for SeaWiFS, Aqua, and Terra (https://oceancolor.gsfc.nasa.gov/docs/rsr/rsr_tables/). Multispectral in situ $R_{rs}(\lambda)$ were matched to satellite bands to within a 2.5 nm tolerance.

2.2.5. Additional Quality Controls

In addition to those measures already discussed for CDOM and CDOM slope outliers in Section 2.2.1, $a_d(\lambda)$ records were considered unrealistic and removed at all wavelengths if they exceeded 12 m^{-1} anywhere within the spectral range reported. Similarly, $a_p(\lambda)$ was removed if it exceeded 20 m^{-1}, $b_{bt}(\lambda)$ if it exceeded 0.15 m^{-1}. $R_{rs}(\lambda)$ were eliminated if they exceeded 0.075 sr^{-1} or were less than -0.001 sr^{-1} in any band, or if they were outside the 95th percentile for any given band.

2.3. Satellite Imagery and Matching

Ocean color satellite imagery from SeaWiFS, MODIS-Aqua, and MODIS-Terra that matched the field observations were selected and processed for further analysis. Scripted calls to the NASA GSFC Ocean Color browser (http://oceancolor.gsfc.nasa.gov/cgi/browse.pl) after the 2012.0 MODIS-Aqua reprocessing (http://oceancolor.gsfc.nasa.gov/WIKI/OCReproc20120MA.html) were used to identify and download 1 km nominal nadir resolution L2 SeaWiFS, Aqua, and Terra satellite scenes within 0.05 degrees of field observations on the same day. These were spatially extracted for a 5 × 5 pixel array around the station location. In <1% of stations, high resolution SeaWiFS imagery was not available, and Global Area Coverage (GAC; nominally 4.4 km spacing) scenes were substituted. By default, data were masked based on standard L2 flags using the criteria described in [55]: land, high solar or satellite zenith angle, clouds, sea ice, high light, stray light, glint, low water leaving radiance, and atmospheric correction failure. Extracted satellite data were then evaluated for coincidence with field sampling stations. Criteria were principally based on those of Bailey and Werdell (2006). Specifically, extracted satellite pixel arrays were retained in the database if the overpass occurred within 8 h of field sampling. For each waveband of R_{rs}, negative and outlier pixels within each array (>1.5 standard deviations from the mean) were set to null values. Data were only retained in each waveband if greater than 50% of non-land pixels were still valid, with no fewer than five valid pixels in total. Finally, the mean R_{rs} values for each array were calculated and retained in the database only if those pixels had a coefficient of variation (CV) < 0.25 (rather than <0.15 applied in Bailey and Werdell (2006)). Of the 50,127 field stations with spatially gridded, depth binned and quality controlled CDOM data, 8252 stations had matching quality controlled Aqua imagery, 11,156 matched Terra imagery, and 11,818 matched SeaWiFS imagery.

For the purpose of further quality assurance, several match-up metrics were retained in the final database, including the time difference, CV, the number of matched satellite pixel arrays for each R_{rs}

channel, the areal extent of the matched pixels (nominally ~25 km^2), and the distance between the field sampling location and the central pixel (nominally < ~1 km). Sensor viewing angle, which can significantly increase error in estimates of satellite R_{rs} due to increased uncertainties in the atmospheric correction, is not available on a pixel-by-pixel basis in standard L2 products. On the other hand, ground sample area, which can be approximated from the geographic coordinates of pixel arrays, is a good proxy for sensor viewing angle, with larger areas representing larger viewing angles. Area is also a reasonable metric for accuracy of the geographic collocation, wherein R_{rs} averaged over larger areas of the ocean may not be representative of those measured at the sampling location, depending on the degree of spatial variability of ocean color within the sampled region. These match-up metrics were described in greater detail in [56].

2.4. Statistical Methods

Various metrics and visualization techniques are employed below to gauge the performance of algorithms. Retrieval parameters are compared with the same parameter collected in situ for both the optimization/tuning dataset (i.e., using in situ reflectances) and satellite validation. In addition to common metrics such as the number of samples (N), the standard deviation (*STD*), and the squared correlation coefficient (r^2), we evaluate the adjusted r^2 (r$^{2'}$):

$$r^{2'} = r^2 - \left(1 - r^2\right)\left(\frac{\beta_n}{N - \beta_n - 1}\right), \tag{2}$$

(where β_n is the number of regressors) which adjusts the r^2 downward to correct for the number of predictive values relative to the number of samples in, for example, multiple linear regression. The root mean square difference (*RMSD*) was also calculated:

$$RMSD = \sqrt{\frac{\sum_i^N (mod_i - ref_i)^2}{N}}, \tag{3}$$

where "mod" is the model retrieved parameter and "ref" is the field measurement. The centered-unsigned (or unbiased) *RMSD* (*RMSD*$^{*\prime}$) was defined as follows:

$$RMSD^{*\prime} = \sqrt{\frac{\sum_i^N (mod_i - mean(mod_i)) - (ref_i - mean(ref_i))^2}{N}}, \tag{4}$$

and the signed *RMSD*$^{*\prime}$ is simply the *RMSD*$^{*\prime}$ multiplied by the sign of the difference between the *STD* of the model retrieval and the *STD* of the field data (*RMSD*$^{*\prime}(\sigma_d)$). The bias and the normalized bias (*Bias**) are also employed:

$$Bias^* = \frac{\sum_i^N (mod_i - ref_i)}{N \times STD(ref)} \tag{5}$$

as well as the percent bias (%*Bias*),

$$\%Bias = 100 \times \frac{mean(mod - ref)}{mean(ref)}, \tag{6}$$

and the mean average percent difference,

$$MAPD = 100 \times mean\left[abs\left(\frac{mod - ref}{ref}\right)\right]. \tag{7}$$

While most of these metrics are fairly straightforward, a few warrant further explanation and context. A powerful graphical tool for assessing the skill of model performance—and comparing one model to another—is the Taylor diagram [57], which combines the *RMSD*$^{*\prime}$, *STD*, and correlation into a single figure in which proximity to the field data indicates how well the pattern of the modeled data matches the observations. This is made possible in two dimensions because of the relationship between the $RMSD^{*\prime}$, the correlation, and the variances of model and reference.

Because the means of the model and reference are removed prior to calculating higher statistics shown in Taylor diagrams, they represent the comparisons between the patterns with any bias removed. For this reason, we have added color here to Taylor diagrams to include *%Bias*. Another graphical assessment used here which accounts for the bias (*Bias**) and adds a sign to the $RMSD^{*\prime}$ is the target diagram [58], in which the y-axis represents normalized bias of the model, the x-axis is signed-centered *RMSD*, and distance in any direction from the origin to the model is the total *RMSD*. Here, we also include color in our target diagrams to help visualize the *MAPD*.

3. Results and Discussion

3.1. Database Characteristics

GOCAD has over 40 times more CDOM records than NOMAD, and nearly 100 times as many spectra as IOCCG. It contains $a_g(412)$ data that is more normally distributed than either NOMAD or IOCCG, and a Chl distribution similar to NOMAD (Figure 1). IOCCG model data, while covering the same dynamic range as the two field databases, have unrealistically flat distributions of both CDOM and Chl, raising concerns for introducing bias when the dataset is used for algorithm development and optimization.

The dynamic range in $a_g(\lambda)$ is larger in GOCAD than NOMAD (e.g., $a_g(412)$ from 0.005 to 2.457 m^{-1}, and from 0.0013 to 1.923m^{-1}, respectively), but the data distribution of GOCAD is narrower than NOMAD and IOCCG ($a_g(412)$ 75th minus 25th percentiles of 0.095 m^{-1}, 0.204 m^{-1}, 0.627 m^{-1}, respectively) around a lower mean absorption level (mean $a_g(412)$ = 0.120 $\pm$ 0.133 m^{-1}, 0.194 $\pm$ 0.266 m^{-1}, 0.514 $\pm$ 0.745 m^{-1}, respectively), reflecting the predominance of low CDOM, offshore data in the database. The large number of field records of CDOM in GOCAD, its range and mean value, all indicate that it is suitable for developing global retrieval algorithms.

The data distributions shown along the bottom row of Figure 1 (with the exception of Chl—see figure caption) show data used to optimize algorithms developed in this study (i.e., from field stations with matching in situ radiometry and IOPs), versus data used for validation—in this case with SeaWiFS wave bands and satellite imagery. For each parameter, distributions of optimization and validation data were compared for similarity to test by analysis of variance (ANOVA) whether the populations share a common mean. Optimization and validation dataset were found to differ ($p << 0.01$) for CDOM absorption and spectral slope, but not for DOC and salinity.

The difference between the CDOM and DOC match-up datasets results from availability of the data (i.e., stations may not have both CDOM and DOC measurements in addition to in situ radiometry). Based on the distributions shown in Figure 1, as well as the geographic distributions highlighted in Figure 3, differences between the optimization and validation data populations for CDOM absorption and slope appear to derive from slightly fewer near-shore stations being present in the optimization dataset compared to the validation set, although there is clearly some endmember representation in the optimization set for near-shore conditions. We may conclude from this, however, that algorithms for CDOM absorption and spectral slope developed using these optimization data would perform best in oceanic conditions, while regional algorithms may be more accurate in coastal waters, or waters with very high CDOM absorption and low CDOM spectral slope.

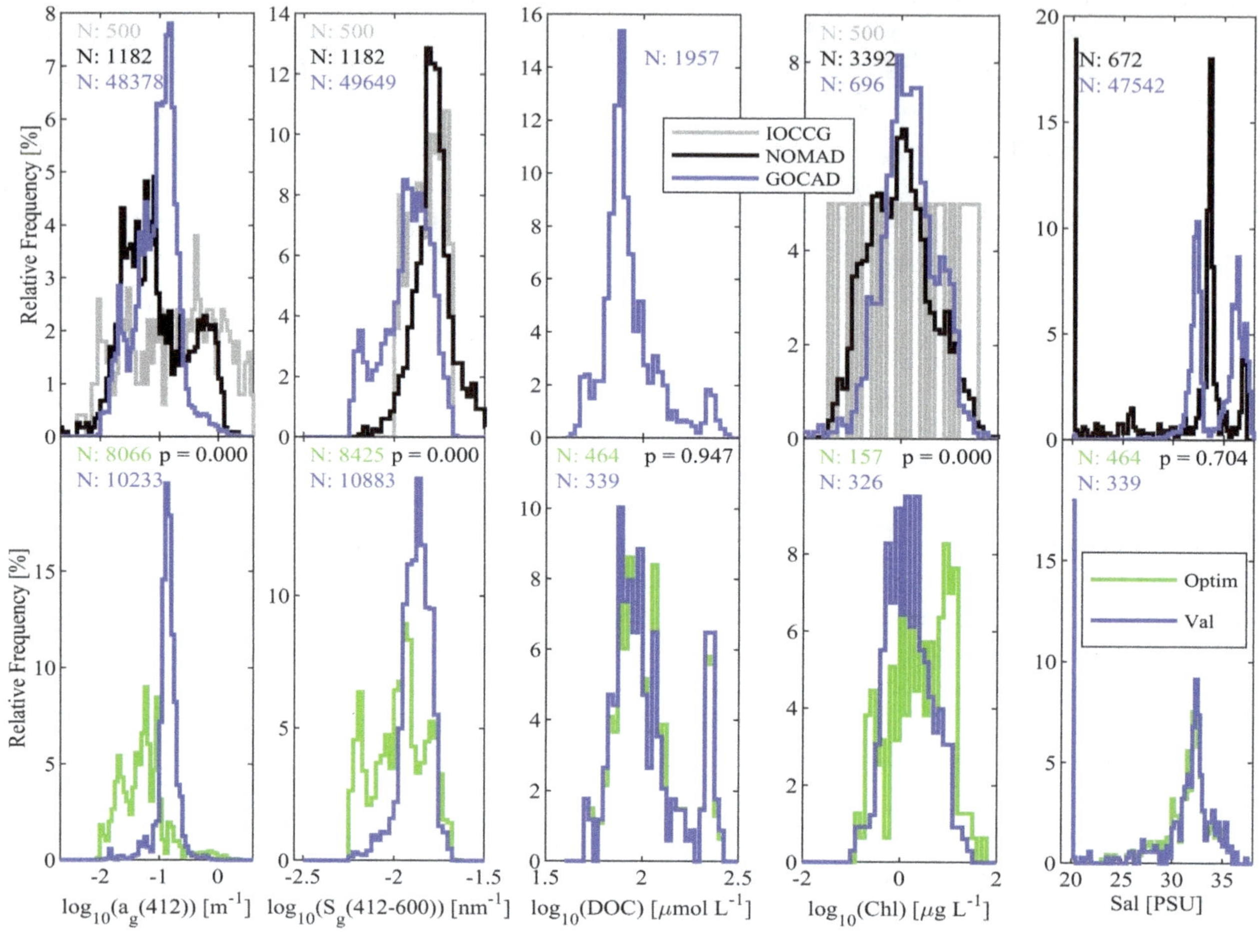

Figure 1. Top row: data distributions and counts (N) of relevant parameters and Chl (for context only) in Global Ocean Carbon Algorithm Database (GOCAD), NASA bio-Optical Marine Algorithm Dataset (NOMAD), and the synthetic ocean color dataset developed by the International Ocean Colour Coordinating Group (IOCCG). Bottom row: comparisons between the subset of GOCAD parameters used in optimization/tuning (Optim) and validation (Val) of algorithms (shown here with SeaWiFS match-ups, but also evaluated for MODIS Terra and Aqua with similar results). Populations of salinity and DOC share a common mean between optimization and validation datasets (ANOVA, $p > 0.01$).

IOCCG and NOMAD contain no UV CDOM data, so direct comparison of spectral slope is only possible in the VIS (Figure 2). The median S_g(412–600) is lower for GOCAD, demonstrating again the predominantly oceanic characteristics (i.e., photodegraded, primarily of marine origin, and presumably refractory) of the CDOM in the database. Slope decreases significantly ($p << 0.01$) as the reference wavelength (i.e., the shortest wavelength in the spectral range) increases from 275–412 nm. Overall, the variability in spectral slope for each range is quite low—generally no more than a factor of 2–3. This narrow dynamic range in slope within each waveband presents a challenge for retrieving fine scale differences in CDOM slope by limiting the sensitivity of algorithms built from inherently uncertain ocean color. However, errors in the retrievals should be small relative to the absolute magnitude of the slope even if the algorithm sensitivity (e.g., correlation between retrievals and field measurements) is low.

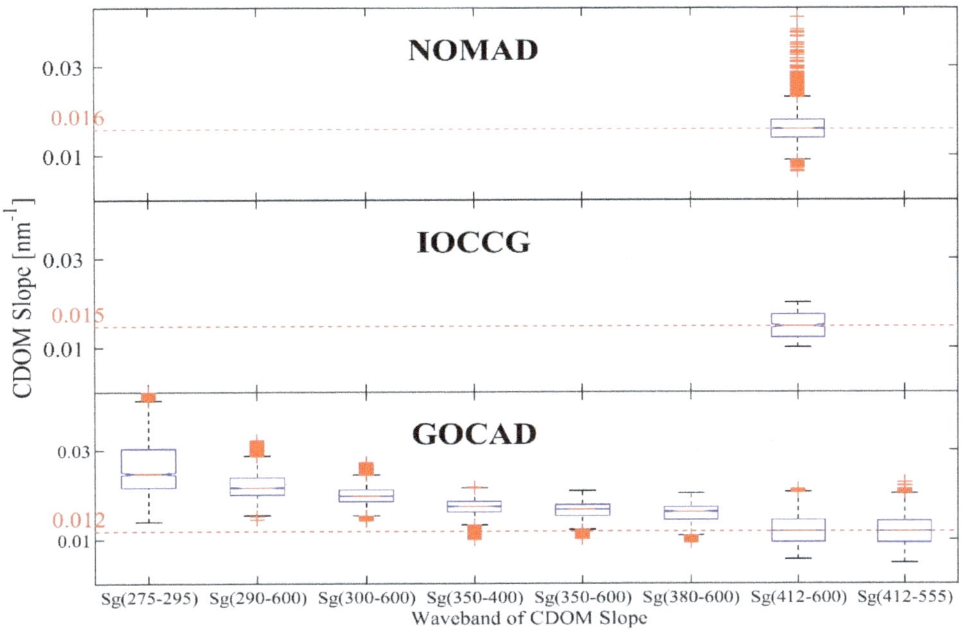

Figure 2. Exponential slope of CDOM in NOMAD, IOCCG, and GOCAD. Median values for $S_{412-600}$ are highlighted in red for comparison. NOMAD and IOCCG lack UV CDOM.

Geographic locations of GOCAD field stations overlap with NOMAD stations (Figure 3). We can see that many of the NOMAD stations were ultimately excluded from GOCAD during the quality assurance analysis described in Section 2.2. Highlighted in the central panel of Figure 3 are those stations with high-quality in situ radiometry, which were set aside for tuning, training, and optimization of ocean color algorithms. The geographic distributions of both the training data and the validation data show a representational combination of stations from both offshore and nearshore waters, which theoretically improves the odds of being able to retrieve a broad dynamic range of bio-optical properties, although as pointed out above the optimization data appears to be slightly dominated by oceanic stations. It is clear from Figure 3 that while there is significant overlap in the CDOM and DOC datasets in certain regions such as northern Alaska and the mid-Atlantic Bight in the Northeastern U.S., globally they follow a somewhat different pattern, and many DOC field stations are not obviously represented in the CDOM dataset.

The dense concentrations of field stations sampled in relatively smaller regions such as the Northeastern U.S. are difficult to resolve at the small scale in Figure 3. Figure 4 shows three-dimensional maps of select sub-regions with $a_g(412)$, $S_g(275–295)$, and $S_g(412–555)$, including the Northeast US and coastal Alaska between the Chukchi Sea and the Beaufort Sea. These are set in broad continental shelves with numerous nearby river outflows. Not surprisingly, CDOM is high throughout the regions shown in Figure 4 with low spectral slope in the UV. CDOM and $S_g(275–295)$ increase and decrease, respectively, in close relation to distance from shore, as expected given the considerations discussed in Section 1.2 and elsewhere. Variability is higher for spectral slope in the VIS ($S_g(412–555)$), but it generally follows the opposite pattern from that in the UV, i.e., decreasing with distance from new sources of CDOM. This is indicative of aging processes as the newly mobilized, near-shore CDOM mixes seaward and photo- and microbial degradation reduce absorption in the UVA relative to the UVB (thus increasing $S_g(275–295)$ and relative to the VIS (thus decreasing $S_g(412–555)$). It may indicate marine sources of CDOM with chromophores that absorb in the UVA and blue rather than terrestrial sources that also absorb in the UVB. These patterns are perhaps clearest at the outflows of the Colville River (~135° W and 70° N) and the Chesapeake Bay (~77° W and 37° N). An interesting exception for $S_g(412–555)$ can be found in the Gulf Stream transect (~70° W and 37°–40° N; GOMECC-2 experiment, Table 2), where slope increases upon entering the productive waters at the edge of the Gulf Stream despite a lack of CDOM increase, and then rapidly declines upon entering the oligotrophic waters south of the Gulf Stream.

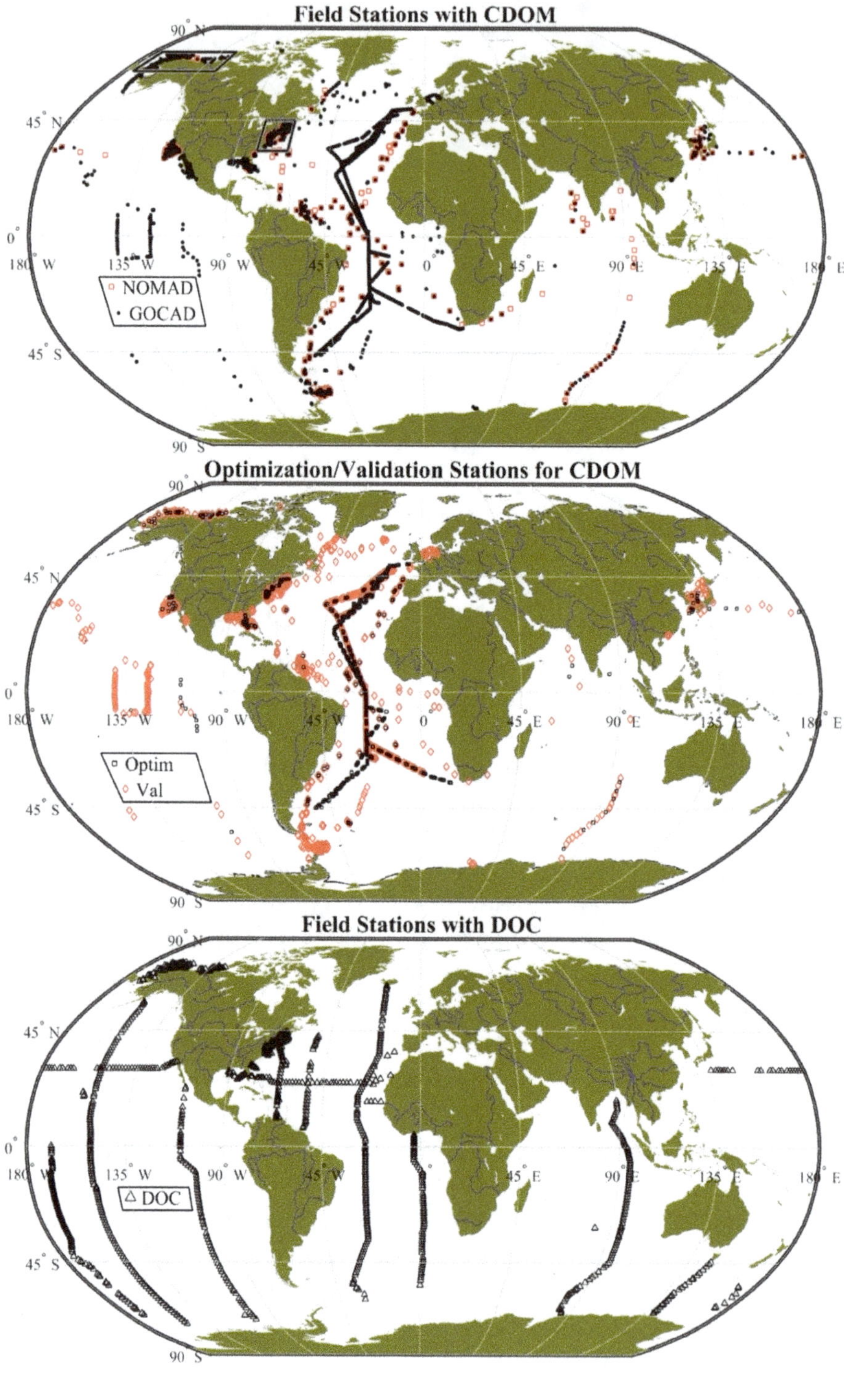

Figure 3. Global distribution of GOCAD and NOMAD field stations for CDOM (upper) and DOC (lower). The central panel shows the distributions of data within GOCAD separated into optimization (Optim) and validation (Val) dataset. Stations used in algorithm tuning are shown as red circles, the remainder of stations were available for satellite validation. The boxed subregions in the upper panel are shown in greater detail in Figure 4.

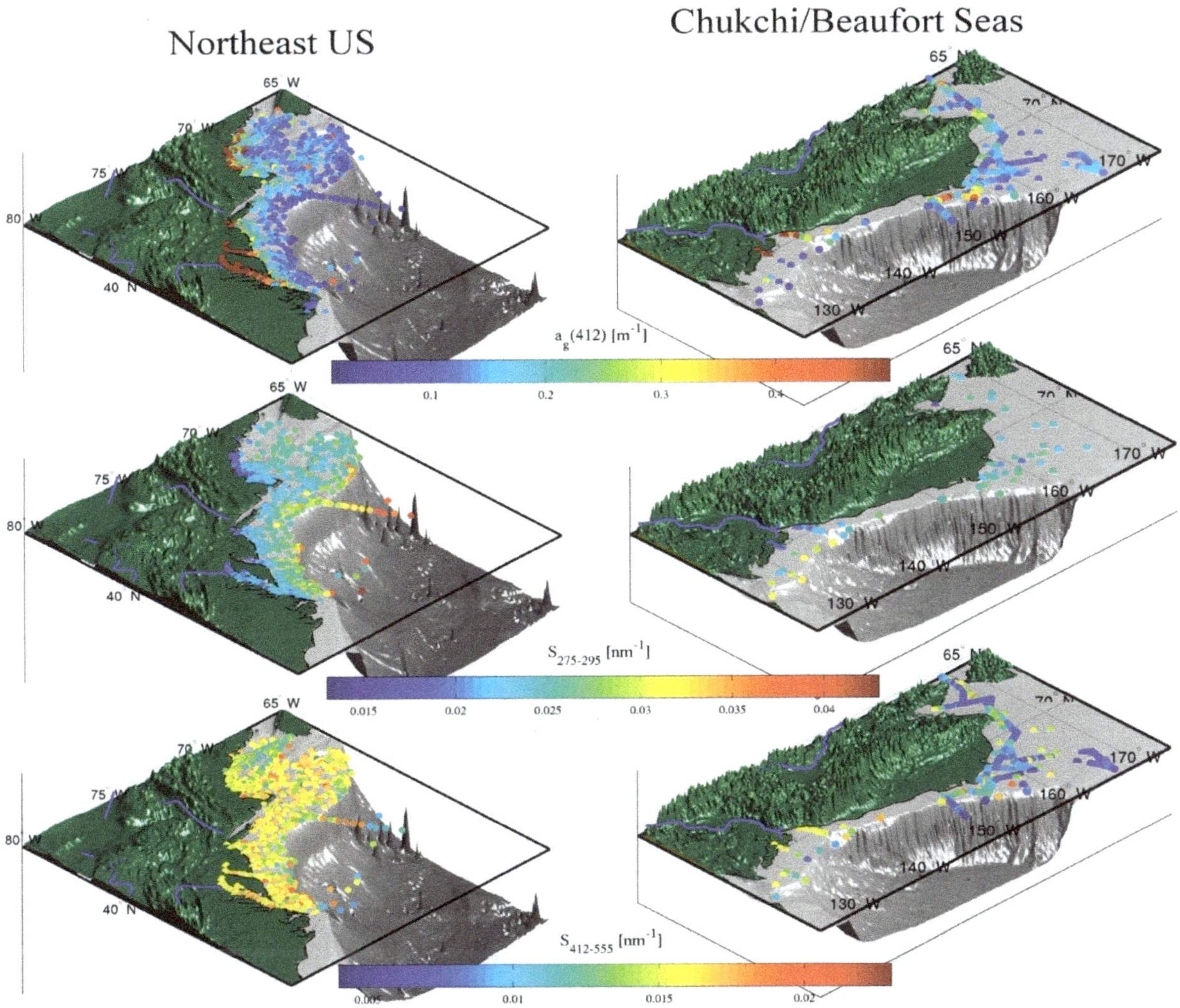

Figure 4. Examples of CDOM absorption at 412 nm (top row), and CDOM spectral slope in the UVB (middle row) and VIS (bottom row) from GOCAD show patterns which reflect the sources and age of CDOM in environments stretching from estuarine, such as the Chesapeake Bay in the eastern U.S., to stations sampled well offshore.

3.2. Algorithm Tuning and Validation

3.2.1. Algorithm Structures and Optimization

Both empirical and semi-analytical approaches to ocean color retrievals of CDOM, S_g, and DOC were explored using the GOCAD dataset. Of the former, a band ratio, single exponential decay model similar to that presented by Mannino (2008) was tested, but found to be better suited for the continental shelf waters for which it was derived rather than for the deep ocean, and will not be presented here. A multiple linear regression (MLR) approach was tested matching the natural logarithm of R_{rs} in four ocean color bands with the logarithm of $a_g(\lambda)$ and S_g at each waveband described in Section 2.2.1, and DOC. The least square difference minimization regression, performed using Matlab's *regstats* function (www.mathworks.com), follows the form:

$$\ln(Y) = \beta_0 + \beta_1 \times \ln(R_{rs}(\lambda_1)) + \beta_2 \times \ln(R_{rs}(\lambda_2)) + \beta_3 \times \ln(R_{rs}(\lambda_3)) + \beta_4 \times \ln(R_{rs}(\lambda_4)) \quad (8)$$

where β_0–β_4 are the regression coefficients, Y is the retrieval parameter, and λ_1–λ_4 are the sensor-specific wavelengths (i.e., 443, 488, 531, and 547 nm for MODIS, 443, 490, 510, and 555 nm for

SeaWiFS). Using monthly, binned L3 Aqua imagery for 2010, MLR retrievals were used to establish the 99th percentiles for each retrieval waveband of a_g. Retrievals above these values were considered outside the scope of this global algorithm, and eliminated. Regression coefficients, statistics, and thresholds are presented in Figure 5, and Tables 3–5. Model retrievals plotted against field data are well organized about the 1:1 line with low scatter, particularly in the UVA, which is reflected in high correlation coefficients and low error and bias. *MAPD* is below 30% for all bands below 488 nm; from this band to higher wavelengths, the CDOM signal becomes very weak in most of the global ocean.

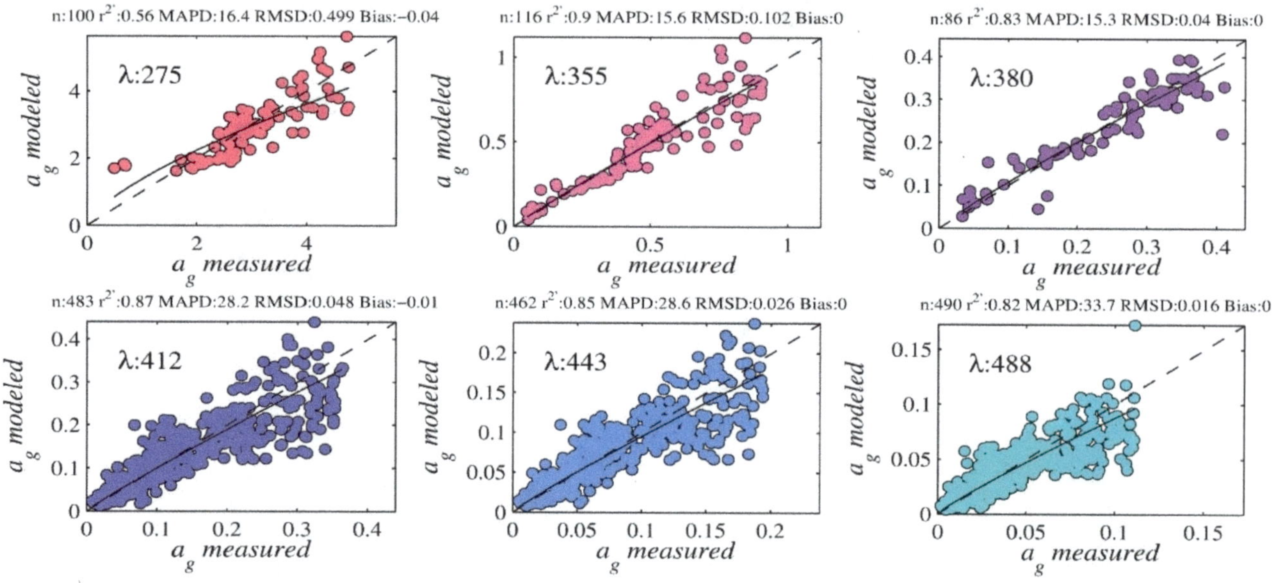

Figure 5. MLR retrievals of CDOM plotted against field data for the tuning dataset (i.e., in situ $R_{rs}(\lambda)$). The solid line shows the fit through the data, and the 1:1 line is dashed.

For the reasons outlined in Section 1.4 (i.e., a large and variable portion of DOC is unpigmented), DOC derived directly from ocean color alone using MLR was not robust (Table 5; MLR1). However, satellite retrievals of sea surface salinity are now available thanks to the Aquarius mission (http://aquarius.nasa.gov/), and for CDOM, salinity was a reasonable choice as an additional proxy for water type considering it will generally reflect proximity to sources of fresh water and CDOM as well as distinguishing water masses (e.g., Gulf Stream). Using GOCAD, a multiple linear regression approach was developed for retrieving DOC from $a_g(355)$ (in place of $R_{rs}(\lambda_1)$ in Equation (8)) and salinity (in place of $R_{rs}(\lambda_2)$), and proved very robust (e.g., $r^2 = 0.91$, *%Bias* = 0). Using CDOM and salinity as predictors significantly improved retrievals of DOC (Table 5; MLR2), with $r^{2'}$ increasing from 0.76 to 0.91, and *MAPD* dropping by about three percentage points. The strength of the correlation between field and retrieved DOC to CDOM and salinity is stronger than expected, considering the many ways in which changes in DOC, CDOM, and salinity may diverge across seasons or from region to region. It is worth pointing out that other factors may be contributing to the stronger statistical performance of MLR2 over MLR1, such as the higher number of coincident predictors and retrievals, as well as the absence of uncertainties associated with reflectance data in the tuning dataset (i.e., DOC is derived directly from CDOM absorption and salinity). Caution is therefore advised when applying this DOC algorithm in regions in which DOC is known to change without commensurate changes in CDOM and/or salinity. For example, the accumulation of DOC in surface subtropical waters including the BATS field station [59,60] appears to be decoupled from CDOM (Norman Nelson and Craig Carlson, personal communication).

Table 3. Coefficients of the MLR algorithm for retrieving CDOM absorption ($ag(\lambda)$) following Equation (8) and metrics of fit for the optimization data set.

	β_0	β_1	β_2	β_3	β_4	N	$r^{2'}$	*RMSD* [m^{-1}]	*MAPD* [%]	*%Bias* [%]	Threshold [m^{-1}]
						MODIS					
[nm]		443	488	531	547						
275	0.089	−0.540	−1.142	3.444	−1.875	100	0.56	0.499	16	−1.4	4.825
355	−2.246	−1.186	−0.558	2.912	−1.336	116	0.90	0.102	16	−0.9	0.9104
380	−2.263	−0.300	−1.882	3.831	−1.787	86	0.83	0.040	15	−1.9	0.4341
412	−2.535	−0.563	−1.294	1.606	0.170	483	0.87	0.048	28	−4.3	0.36419
443	−3.287	−0.727	−0.922	1.278	0.261	462	0.85	0.026	29	−4.3	0.1984
488	−3.722	−0.377	−1.429	1.424	0.300	490	0.82	0.016	34	−5.9	0.1114
						SeaWiFS					
[nm]		443	490	510	555						
275	−2.477	−2.880	2.225	0.480	−0.252	174	0.76	0.659	25	−2.2	4.825
355	−4.199	−2.563	1.214	0.955	−0.040	189	0.87	0.118	26	−2.2	0.9104
380	−4.544	−1.808	0.175	1.181	0.001	150	0.80	0.055	26	−2.4	0.4341
412	−6.004	−0.861	−0.006	−0.346	0.515	8066	0.37	0.035	56	−13.0	0.36419
443	−6.410	−0.743	−0.145	−0.367	0.547	8037	0.33	0.026	58	−13.6	0.1984
490	−7.014	−0.736	0.142	−0.796	0.678	7978	0.28	0.016	65	−15.5	0.1114

Another empirical approach tested here was the machine learning approach known as Random Forests [61,62], which is a method for multivariate, non-linear, non-parametric regression designed to help minimize over-fitting of the training dataset. The method improves on standard decision tree regression performance by using an ensemble of independent decision trees; bootstrapping for the regression is achieved by repeatedly, randomly resampling the original dataset to provide an ensemble of smaller independent datasets, which are each used to grow a decision tree (hence the term random forest tree-bagger, or RFTB). Here, 200 independent decision trees were used, and each tree is trained on approximately 66% of the training dataset. The inputs (i.e., reflectances) and retrievals of the regression (i.e., CDOM, CDOM slope, and DOC) were the same as in the MLR. Model performance and statistics for select bands in the UV and VIS with the training dataset are presented in Figure 6. Comparisons of model retrievals to field data are fairly well correlated, but error and bias are quite high, with *MAPD* reaching several hundred percent.

Table 4. Coefficients of the MLR algorithm for retrieving CDOM spectral slope ($S_g(\lambda)$) following Equation (8) and metrics of fit for the optimization data set.

λ	β_0	β_1	β_2	β_3	β_4	N	$r^{2'}$	*RMSD* [nm^{-1}]	*MAPD* [%]	*%Bias* [%]
					MODIS					
[nm]		443	488	531	547					
275–295	−3.289	0.270	−0.335	1.051	−0.921	322	0.61	0.002	6.4	−0.3
290–600	−3.471	0.127	−0.251	1.025	−0.843	322	0.38	0.002	6.2	−0.3
300–600	−3.607	0.044	−0.153	0.881	−0.722	324	0.30	0.001	5.7	−0.3
350–400	−3.924	−0.242	0.055	0.935	−0.710	331	0.26	0.001	6.8	−0.3
350–600	−3.908	−0.204	0.098	0.609	−0.463	331	0.22	0.001	6.3	−0.3
380–600	−3.912	−0.152	0.127	0.236	−0.173	340	0.14	0.001	6.3	−0.3
412–600	−4.219	−0.180	0.137	0.168	−0.131	782	0.16	0.002	7.4	−0.5
412–555	4.195	−0.162	0.147	0.096	−0.084	760	0.10	0.002	7.6	−0.5
					SeaWiFS					
[nm]		443	490	510	555					
275–295	−3.012	0.427	−0.459	0.357	−0.228	424	0.77	0.002	6.8	−0.4
290–600	−3.425	0.131	−0.085	0.145	−0.130	424	0.46	0.002	6.6	−0.4
300–600	−3.615	0.004	0.014	0.160	−0.129	426	0.29	0.002	6.0	−0.3
350–400	−3.968	−0.298	0.178	0.301	−0.150	433	0.23	0.002	7.4	−0.4
350–600	−4.058	−0.288	0.091	0.356	−0.138	433	0.33	0.001	6.9	−0.4
380–600	−4.072	−0.226	0.088	0.208	−0.051	445	0.32	0.002	7.2	−0.3
412–600	−4.498	−0.466	0.690	−0.202	−0.015	8550	0.06	0.004	28.5	−5.2
412–555	−4.533	−0.455	0.683	−0.214	−0.012	8425	0.05	0.004	28.2	−5.1

Table 5. Coefficients of the MLR algorithm for retrieving DOC following Equation (8) and metrics of fit for the optimization data set.

Algorithm	ß$_0$	ß$_1$	ß$_2$	ß$_3$	ß$_4$	N	$r^{2'}$	*RMSD* [µmol L^{-1}]	*MAPD* [%]	*%Bias* [%]
					MODIS					
[nm]		443	488	531	547					
MLR1	4.923	0.641	−2.424	3.503	1.692	183	0.76	23.9	13.9	−1.5
					SeaWiFS					
[nm]		443	490	510	555					
MLR1	5.272	0.526	−2.982	2.623	0.089	246	0.68	30.3	18.8	−3.0
					a_g(355) and Salinity					
		a_g(355)	*Sal*							
MLR2	192.718	26.790	−3.558	-	-	464	0.91	15.2	10.6	0.0

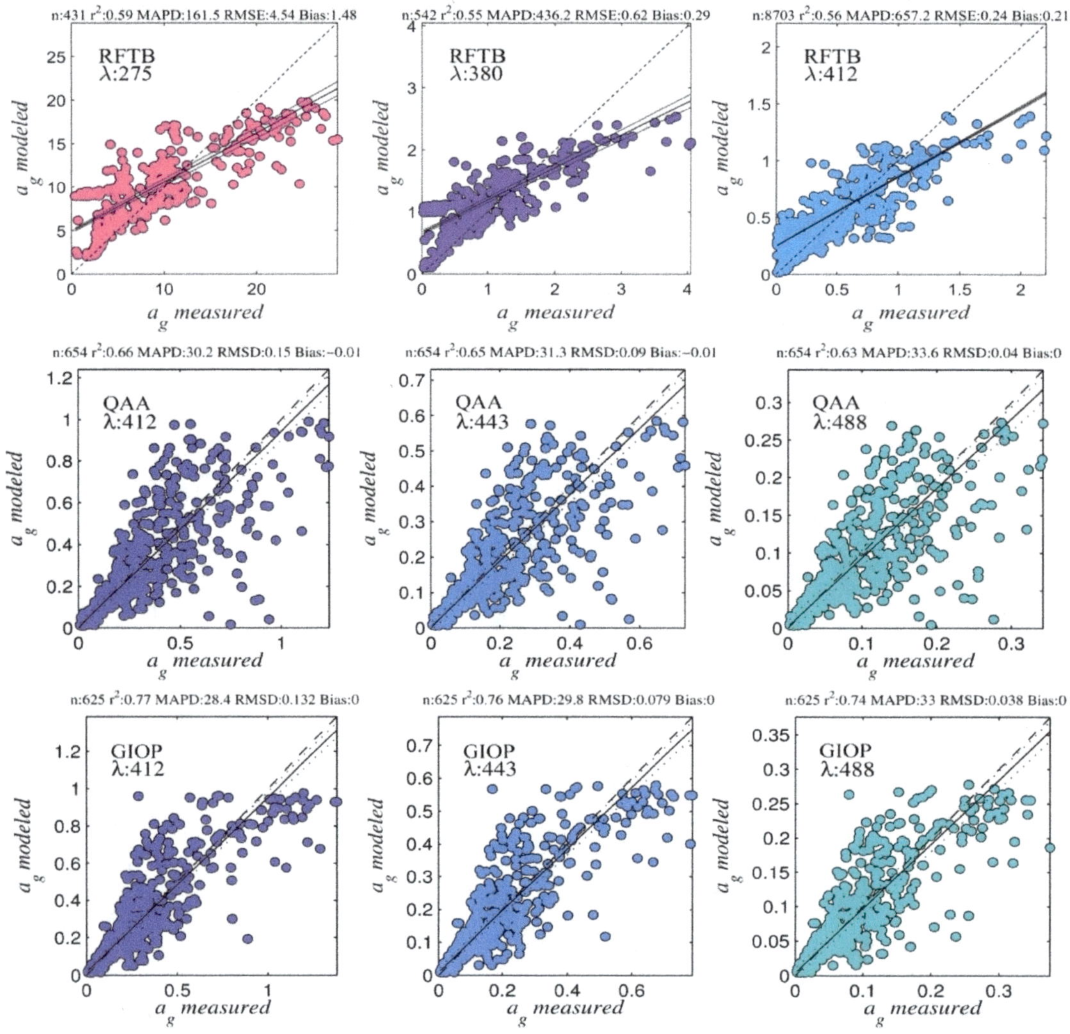

Figure 6. Random forest tree-bagger (RFTB), quasi-analytical algorithm (QAA), and generalized inherent optical property (GIOP) retrievals for tuning datasets.

Semi-analytical approaches included the Quasi-Analytical Algorithm of (QAA) [5,63,64], and the Generalized Inherent Optical Property (GIOP) algorithm [65]. These have the advantage that they are based on theoretical models for how the light field is affected by the inherent properties of the water, but can only retrieve IOPs at those wavebands for which $R_{rs}(\lambda)$ is measured (i.e., they do not extend into the UV for the current and historical suite of satellite sensors). In the future, however, data from GOCAD and elsewhere could be used in the development of linear matrix inversion-type semi-analytical algorithms including basis vector models extending into the UV for CDOM, thereby potentially enabling their retrieval directly using SAAs. In fact, using GOCAD to build more globally representational basis models extending CDOM into the UV may not only provide better retrievals of CDOM, but also of the other concurrently retrieved optical properties from linear matrix inversion. Both the GIOP and QAA invert the $R_{rs}(\lambda)$ to retrieve the water column IOPs following the theory that sea surface reflectance at a given wavelength is proportional to the backscattering coefficient, and inversely proportional to the absorption coefficient [16,66]. Each uses various assumptions, empirical parameterizations, and mathematical inversion techniques to solve for the IOPs and partition them into their constituents. These include the total backscattering coefficient, $b_{bt}(\lambda)$, backscattering by particles, $b_{bp}(\lambda)$, absorption by total particles, by phytoplankton, and by the combination of non-algal particles and CDOM, $a_{dg}(\lambda) = a_d(\lambda) + a_g(\lambda)$, where $a_d(\lambda)$ is non-algal (or detrital) absorption). These latter properties are similar in spectral shape, and therefore difficult to partition, which presents a challenge if we wish to compare the retrievals of SAAs to the other algorithms presented here. Therefore, while we do not re-develop or re-optimize the SAAs here—using them as published—we do utilize GOCAD to facilitate the separation of dissolved and detrital absorption components. Specifically, we solve for $a_g(\lambda)$ by assuming that $a_d(\lambda)$ is a function of the combined backscattering by water, non-algal particles, and the dissolved absorption by CDOM, which we assume does not backscatter, although there is some evidence supporting backscattering by colloids [46]. These SAAs retrieve only the combined $b_{bp}(\lambda)$ from phytoplankton and non-algal particles, but the latter tend to have a higher refractive index, and therefore contribute far more strongly to the backscattering signature, e.g., [67] and references therein. An empirical relationship was developed between $a_d(410)$, $b_{bt}(550)$ and $a_{dg}(410)$, and then $a_g(410)$ was found by subtracting $a_d(410)$ from SAA retrievals of $a_{dg}(410)$:

$$a_d(410) = 0.06822 \times a_{dg}(410) + 1.623 \times b_{bt}(550) + 0.0002123 \tag{9}$$

Due to a paucity of $a_d(\lambda)$ and $b_{bt}(\lambda)$ observations in GOCAD, this relationship was tuned for multiple linear regression using the IOCCG synthetic dataset ($r^{2'} = 0.76$, $RMSD = 0.07$, bias $= -0.004$ m^{-1}, $MAPD = 75\%$, N = 464). $a_g(410)$ was expanded using Equation (1) to other wavebands with the empirical retrieval for $S_g(412–555)$ (derived as per Equation (8) and Table 4). Regression statics for the optimization data are shown for the QAA and GIOP in Figure 6, with slightly better results in the GIOP. Although the current version of GOCAD is less well populated with some optical properties than others (i.e., data collection targeted carbon-related properties and only included others if they happened to be in the same SeaBASS file), the digital structures for each property mentioned in this section are included in the database, and future algorithm investigation (particularly using SAAs) would greatly benefit from incorporation of these data into GOCAD or a similar, climate-scale, global database.

3.2.2. Algorithm Validation

This work represents the most rigorously validated set of global CDOM and DOC algorithms to date. Optimization/training of algorithms as described in the previous section was conducted on GOCAD field stations with coincident in situ radiometry. These stations were then set aside from validation, which was performed only on those remaining stations in GOCAD that had coincident satellite imagery (i.e., MODIS Aqua, Terra, and SeaWiFS). In addition to the algorithms already mentioned, two other empirical algorithms based on band ratio approaches were included in validation analysis. The approach of Shanmugam (2011) [31] (hereafter Shan11) used a power-law relationship between the ratio of $R_{rs}(443)/R_{rs}(555)$ and $a_g(350)$ and $a_g(412)$, and performed well using the NOMAD dataset. The ratio of these was then used in another power-law function to solve for $S_g(350–412)$. Tiwari and Shanmugam (2011) [32] (hereafter TS11) used linear functions to relate the ratio of $R_{rs}(670)/R_{rs}(490)$ to $a_g(412)$ and $a_g(443)$, and solved for $S_g(412–670)$ analytically by inverting Equation (1). As these two algorithms were tuned using SeaWiFS bands, a slight adjustment was made to MODIS input reflectances to obtain the SeaWiFS reflectances required (only MODIS validation is shown here graphically).

The performance of all algorithms in independent validation is weaker than for optimization (Tables 6–8, Figures 7–9). This should not be surprising considering satellite imagery is subject to higher uncertainty associated with atmospheric correction, where the atmosphere comprises ~90% or more of the signal received by the satellite sensor. Furthermore, satellite match-ups exacerbate the issue of temporal and geographic coincidence with in situ measurements. Any regions of moderate to high variability in surface properties will likely not be well captured by the average of a nominally 5 km × 5 km pixel array. Nevertheless, results are encouraging, particularly for the MLR approach and particularly in the UV, where the CDOM signal is strongest (in terms of in situ data) and the SAAs are not currently useful.

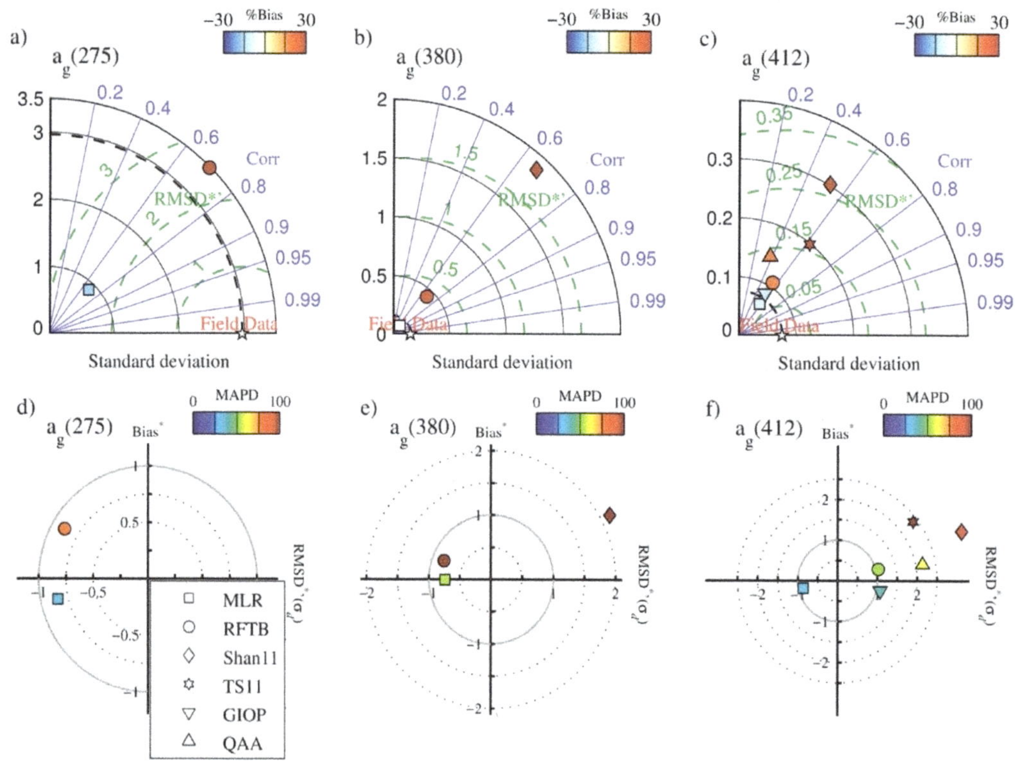

Figure 7. Taylor diagrams (top row) and target plots (bottom row) depicting comparative algorithm performance for retrieving CDOM absorption at 275 nm, 380 nm, and 412 nm from MODIS Aqua.

Figure 7 shows Taylor and target diagrams comparing the CDOM absorption retrieval metrics for various algorithms as described in Section 2.4. In the UVB (275 nm), and UVA (380 nm), only the empirical approaches were feasible, while SAAs (i.e., QAA and GIOP) are also shown at 412 nm. MLR and RFTB perform comparably with respect to correlation between the models and measurements at 275 nm, although MLR does have significantly lower *MAPD* and bias (Table 6; MLR highlighted in bold), and outperforms the RFTB at 380 nm in all but correlation for all sensors. MLR also outperforms all other CDOM absorption algorithms at 412 nm, although GIOP does not appear significantly worse as seen by its proximity to field data in the Taylor plots and the origin in the target diagrams. MLR shows a relatively strong negative bias in most sensors and channels, which is the result of underestimation in high CDOM waters (data not shown).

Table 6. Validation of algorithms for retrieving CDOM absorption (ag(λ)).

Algorithm	λ [nm]	N	r^2	*RMSE* [m^{-1}]	*MAPD* [%]	*%Bias* [%]
			MODIS-Aqua			
MLR	**275**	**186**	**0.47**	**2.499**	**33**	**−17**
MLR	**380**	**188**	**0.45**	**0.117**	**54**	**−1**
MLR	**412**	**7626**	**0.33**	**0.068**	**33**	**−10**
RFTB	275	191	0.50	3.100	78	37
RFTB	380	243	0.47	0.400	102	28
RFTB	412	6820	0.30	0.100	52	18
Shan11	350	237	0.45	1.669	134	108
Shan11	412	7748	0.28	0.287	92	72
TS11	412	7299	0.39	0.201	102	86
GIOP	412	6116	0.30	0.077	40	−12
QAA	412	6133	0.14	0.137	59	18
			MODIS-Terra			
MLR	**275**	**291**	**0.43**	**2.746**	**48**	**−26**
MLR	**380**	**326**	**0.32**	**0.337**	**45**	**−29**
MLR	**412**	**10612**	**0.19**	**0.081**	**35**	**−1**
RFTB	275	171	0.35	2.950	72	22
RFTB	380	269	0.35	0.380	125	21
RFTB	412	6962	0.20	0.110	63	34
Shan11	350	345	0.35	1.308	97	75
Shan11	412	10734	0.16	0.311	129	108
TS11	412	9607	0.20	0.202	109	87
GIOP	412	7976	0.12	0.108	51	11
QAA	412	8048	0.05	0.200	91	52
			SeaWiFS			
MLR	**275**	**342**	**0.25**	**2.976**	**49**	**−57**
MLR	**380**	**418**	**0.32**	**0.318**	**58**	**−53**
MLR	**412**	**10233**	**0.10**	**0.081**	**47**	**23**
RFTB	275	199	0.38	2.660	90	40
RFTB	380	423	0.36	0.370	209	41
RFTB	412	11594	0.06	0.110	55	27
Shan11	350	444	0.51	1.425	108	86
Shan11	412	10451	0.26	0.229	100	79
TS11	412	8890	0.20	0.250	152	128
GIOP	412	7863	0.11	0.085	45	11
QAA	412	7904	0.07	0.162	80	53

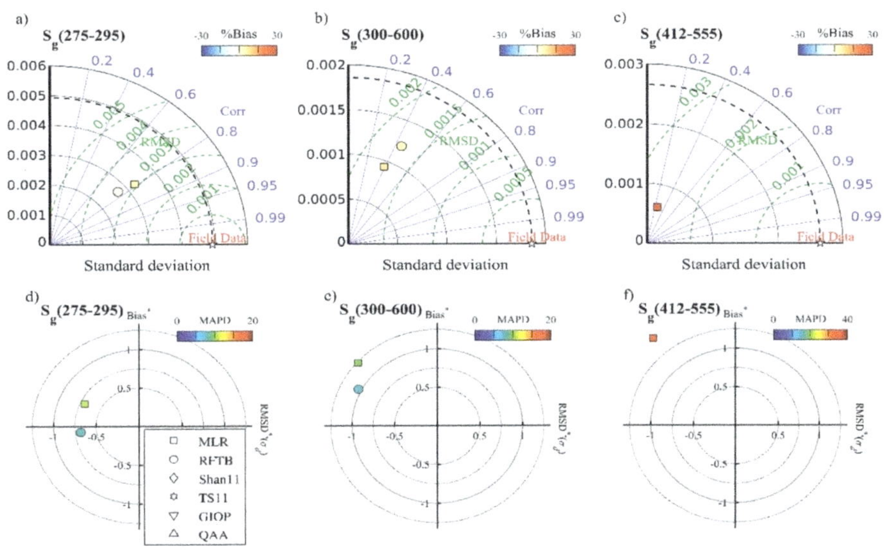

Figure 8. Taylor diagrams (top row) and target plots (bottom row) depicting comparative algorithm performance for retrieving CDOM slope at 275–295 nm, 300–600 nm, and 412–600 nm from MODIS Aqua. Results from Shan11 and TS11 were suppressed to preserve scale.

Table 7. Validation of algorithms for retrieving CDOM spectral slope (S_g).

Algorithm	Waveband	N	r^2	*RMSE*	*MAPD*	*%Bias*
	[nm]			[nm^{-1}]	[%]	[%]
			MODIS-Aqua			
MLR	**275–295**	**187**	**0.62**	**0.0034**	**11**	**6**
MLR	**300–600**	**213**	**0.15**	**0.0023**	**10**	**8**
MLR	**412–555**	**7825**	**0.06**	**0.0040**	**32**	**23**
RFTB	275–295	214	0.58	0.0033	8	−1
RFTB	300–600	244	0.20	0.0019	8	4
Shan11	350–600	188	0.01	0.2750	427	366
TS11	412–555	6223	0.02	0.0290	217	−209
			MODIS-Terra			
MLR	**275–295**	**284**	**0.41**	**0.0039**	**12**	**3**
MLR	**300–600**	**318**	**0.16**	**0.0024**	**10**	**9**
MLR	**412–555**	**10719**	**0.06**	**0.0036**	**27**	**20**
RFTB	275–295	242	0.43	0.0000	10	−3
RFTB	300–600	271	0.11	0.0000	8	5
Shan11	350–600	297	0.03	0.1250	255	176
TS11	412–555	8078	0.00	0.0280	208	−203
			SeaWiFS			
MLR	**275–295**	**350**	**0.44**	**0.0047**	**14**	**6**
MLR	**300–600**	**417**	**0.11**	**0.0021**	**8**	**4**
MLR	**412–555**	**10883**	**0.03**	**0.0029**	**17**	**−12**
RFTB	275–295	350	0.37	0.0053	11	−7
RFTB	300–600	418	0.09	0.0022	9	5
Shan11	350–600	372	0.01	0.1320	202	140
TS11	412–555	4716	0.03	0.0290	203	−198

CDOM spectral slope was only retrievable with empirical approaches. MLR and RFTB performed comparably to each other, although RFTB was not tested at S_g(412–555). In the application of retrieval algorithms for S_g below (Section 3.3), the MLR is used mainly for its simplicity, but we would expect RTFB retrievals to yield nearly equally accurate results. Shan11 and TS11 performed poorly (Table 7; MLR highlighted in bold). Correlations between modeled and measured CDOM slope were weak in the UVA and VIS, but as the dynamic range of the field data is quite low (Figure 2), error and bias were still low in the retrievals (Table 7). In all sensors and bands for the MLR and RFTB, S_g tends to be slightly overestimated in waters with low S_g, and slightly underestimated in waters with high S_g, indicating the weak sensitivity of these empirical approaches also reflected in the low correlation coefficients.

Table 8. Validation of algorithms for retrieving DOC.

Algorithm	N	r^2	*RMSE*	*MAPD*	*%Bias*
			[$\mu mol\ L^{-1}$]	[%]	[%]
			MODIS-Aqua		
MLR1	164	0.23	40.8	41	24.7
MLR2	**382**	**0.89**	**27.8**	**16**	**−13.0**
RFTB	161	0.57	27.3	26	13.8
			MODIS-Terra		
MLR1	158	0.23	40.2	32	13.9
MLR2	**369**	**0.90**	**26.7**	**15**	**−12.3**
RFTB	114	0.47	29.4	27	12.1
			SeaWiFS		
MLR1	274	0.29	34.2	34	4.5
MLR2	**339**	**0.89**	**28.9**	**15**	**−14.2**
RFTB	182	0.30	25.3	23	6.7

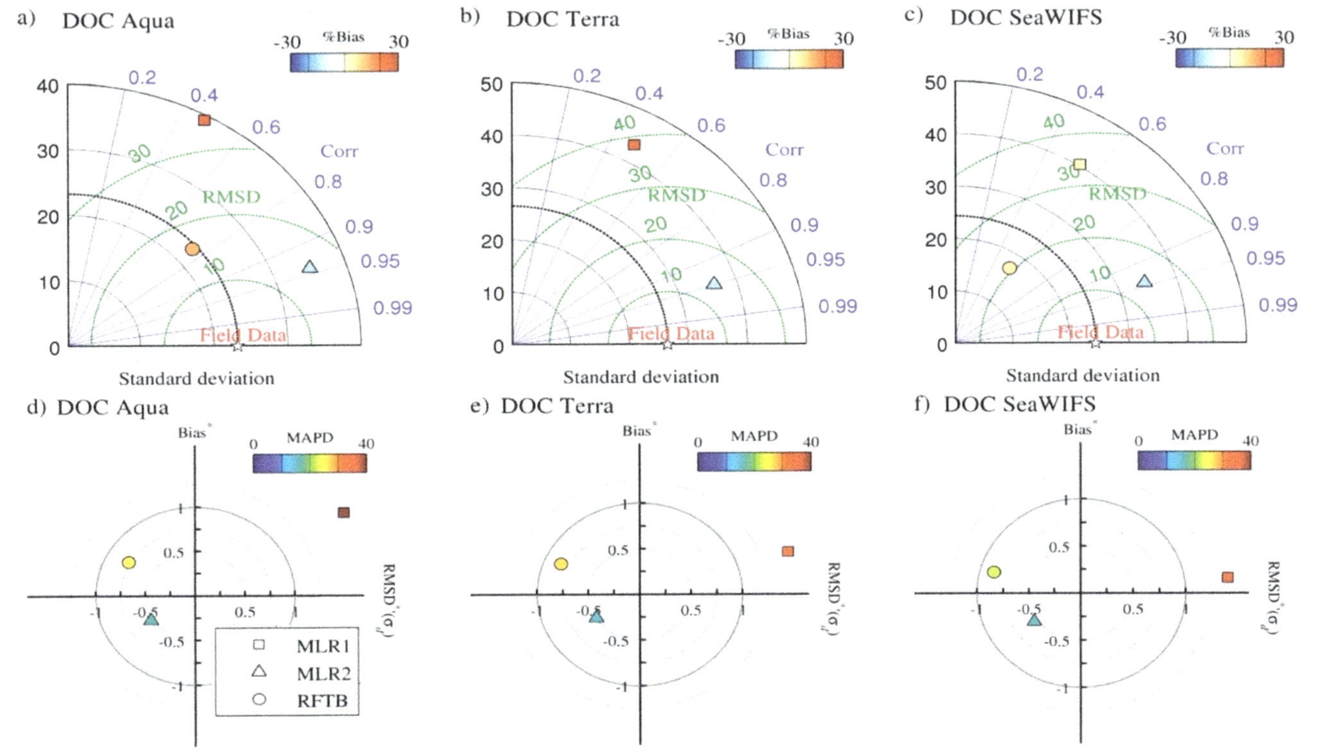

Figure 9. Taylor diagrams (top row) and target plots (bottom row) depicting comparative algorithm performance for retrieving DOC from MODIS Aqua and Terra, and SeaWiFS.

As anticipated, MLR retrievals of DOC using ocean color alone were only weakly correlated with field data (i.e., $r^2 < 0.3$, Table 8). RFTB performed considerably better, but was unable to match the performance of MLR2 (i.e., regression with retrievals of $a_g(355)$ and known salinity; highlighted in bold in Table 8). Due to the newness of the Aquarius mission, there were too few retrievals available for incorporation in these validation results, and further validation of this approach is encouraged based on these results.

We speculated above (Section 3.1) that small differences between the optimization and validation records may bias algorithm performance to favor oceanic waters. To test this hypothesis, a sensitivity analysis was evaluated for CDOM absorption retrieval by MLR to test correlations between algorithm error (percent error between retrievals and field data) and salinity, water column depth, and $a_g(412)$ measured in the field. We found no sensitivity to these factors ($r^2 < 0.04$ in each case, n = 29,757 for Aqua, Terra, and SeaWiFS combined), indicating that the algorithm is not optimized in a way that would limit its performance in, for example, high salinity, offshore waters, or fresher waters with high inputs of fresh CDOM. A geographic distribution of algorithm retrieval error (percent error) for $a_g(412)$ and $S_g(412–600)$ is shown in Figure 10.

A similar sensitivity analysis was evaluated for MLR2 (DOC retrieval) performance at validation stations to help identify limitations of the algorithm. We tested the correlation between the DOC retrieval error (percent difference between retrieved and measured DOC) and salinity, water column depth, and DOC concentration, but found no strong trends in the distribution of error (r^2 = 0.17, 0.19, 0.43, respectively, see included figures below), although it could be argued that absolute retrieval error increases somewhat (overestimates) at the extremely high salinity stations, and at extremely low DOC stations. In general, it appears that shallow stations underestimate DOC, and deeper stations tend to overestimate. The geographic distribution of error in algorithm retrievals (Figure 10) revealed no patterns with respect to distance from shore or nearby fluvial sources, but MODIS Aqua retrievals did overestimate DOC in southern oceans (south of 40° S, 41% ± 16%, n = 18) compared to minor underestimates from other sensors and at latitudes north of 40° S (−8% ± 18%, n = 1054).

Care should therefore be taken when evaluating algorithm retrievals in these areas. Only 1,090 stations (all sensors combined) were available for validation of the MLR2 and this analysis of sensitivity, and their distribution is not uniform across the world's oceans, but, as shown in Figure 1, a broad spectrum of water types with a large dynamic range of DOC were represented in both the optimization and validation datasets. Unfortunately, no validation stations for MLR2 were identified for mid-ocean gyres, and therefore the performance of the MLR2 in those waters remains poorly defined, and caution is advised in the interpretation of DOC retrievals in those areas.

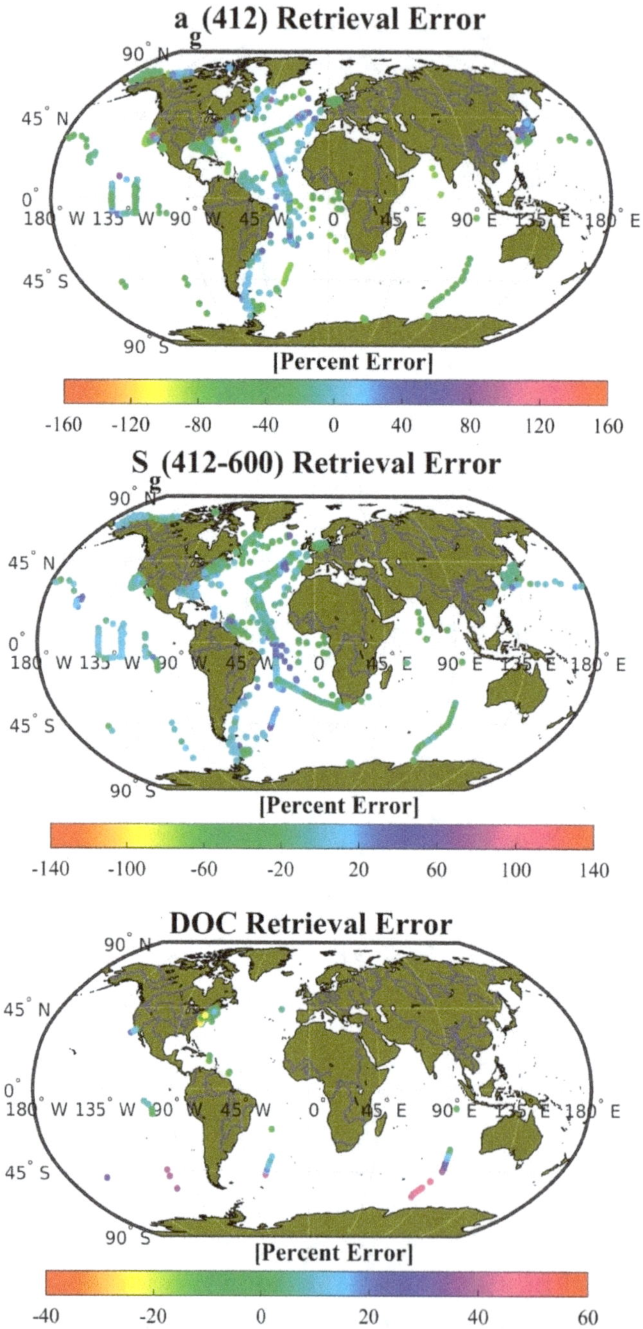

Figure 10. Geographic distribution of error in MLR algorithm retrievals of CDOM absorption and slope in the VIS (top and center), and MLR2 retrievals of DOC (bottom) using validation stations and satellite imagery.

Differences between retrieval statistics across satellite platforms using the MLR approaches were generally small, with Aqua and Terra outperforming SeaWiFS for CDOM absorption (Table 6). All three sensors performed comparably for S_g and DOC (Tables 7 and 8).

3.3. CDOM, S_g, and DOC Climatology

Because the Aquarius mission (providing sea surface salinity) was limited to <4 year data record (~August 2011–June 2015), climatologies for retrieved DOC similar to those presented below for a_g and S_g are not possible. Instead, three years (2011–2013) of coincident MODIS Aqua and Aquarius data were used to generate a three-year mean 9 km global DOC product (Figure 11). An overlay of in situ surface DOC from GOCAD was examined, but not included here because with no temporal coincidence in this comparison, strong biases likely to occur in the field data (e.g., field sampling of high latitudes is proscribed during winter for obvious practical reasons) will not be reflected in the mean DOC satellite product. Nevertheless, the relatively large (±~50%–~100%) disparities apparent in several regions—including high latitudes and the Atlantic subtropical gyre—indicate fundamental weaknesses in the global DOC algorithm. For instance, as mentioned in 3.2.1, the subtropical Atlantic gyre is characterized by an accumulation of DOM not reflected in the CDOM nor apparently traceable with changes in salinity, and is therefore overlooked by the DOC algorithm presented here (MLR2). Based on the tuning statistics, there appears to be merit in the approach, but more study will be required to establish when and where the algorithm works best, and what (if anything) can be done for remotely sensing DOC in regions where no robust optical proxies exist.

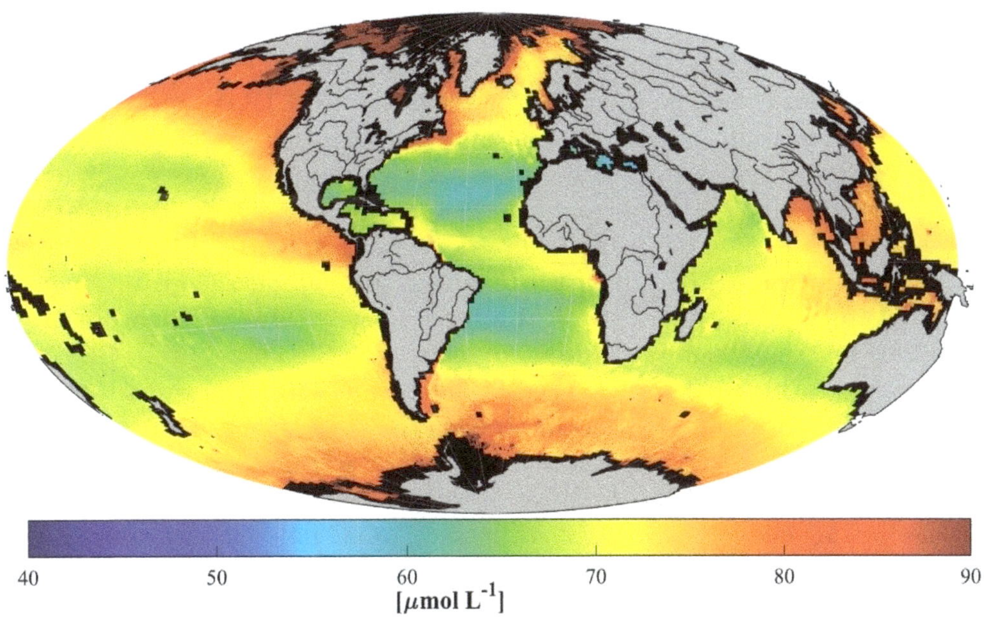

Figure 11. Retrieved three-year mean, 9 km nominal resolution DOC from Aquarius and MODIS Aqua using the MLR2 inversion. Validation statistics are reasonably good for the MLR2 (Figure 9, Table 8), but a larger number and wider geographic distribution of validation stations than are currently available is required to thoroughly evaluate the geographic and water-type limitations for MLR2, particularly in the mid-ocean gyres (see text Section 3.2.2). Overestimates of DOC (~41%) retrieved with the MLR2 were found in the southern oceans (S of 40° S)), but only for MODIS Aqua (i.e., not Terra, and no SeaWiFS stations were identified). Elsewhere (i.e., north of 40° S), retrievals tend to slightly underestimate DOC (<10%). Caution is therefore advised in interpreting MLR2 retrievals in mid-ocean gyres, and in the southern oceans using Aqua.

As outlined in the introductory sections, a common assumption made in ocean color remote sensing on a global scale is that CDOM and other water-borne pigmented material covary with Chl.

A valid concern with empirical approaches to retrieve CDOM from ocean color—particularly for those that use some of the same spectral bands as Chl algorithms like OC3M—is that they are essentially tuning themselves to Chl, and not CDOM. While performance metrics for MLR are quite robust (e.g., Figure 5, Table 3), there remains the possibility that this is true in part because these spectral bands are sensitive to Chl, and CDOM is simply covarying with Chl as per Case 1 assumptions. In fact, this was not found to be the case generally from global CDOM and Chl field data in GOCAD ($r^2 = 0.00$, N = 19,446, $\lambda = 412$), nor in the investigation by Siegel et al. (2002), although there exist areas of open ocean outside of the strong influence of terrestrial run-off and upwelling zones where fluctuations in CDOM are clearly driven by local productivity.

To quantify the distinction between retrievals of CDOM using MLR and Chl using OC3M, we calculate retrievals of each over the course of the entire MODIS Aqua mission, and then calculate a residual of the normalized properties, as defined by:

$$\overline{Chl} - \overline{CDOM} = \frac{Chl_i}{median(Chl)} - \frac{a_{g_i}(\lambda)}{median(a_g(\lambda))} \tag{10}$$

where the median is taken over an entire composite image to scale each property by its magnitude. For example, MODIS imagery was separated into seasonal composites for the entire Aqua era, then processed to CDOM and Chl, and the residual calculated for each (Figure 12). Bear in mind in this analysis that OC3M-like algorithms for retrieving Chl have been shown to be strongly influenced by not only phytoplankton biomass, but also physiology (particularly in tropical and subtropical regions) as well as the presence of significant absorption by CDOM and non-algal particulates (a_{dg}) [68].

The results show that over much of the world's oceans—particularly at high latitudes, upwelling zones, and regions influenced by large river plumes—normalized CDOM and Chl diverge by as much as a factor of three. Interestingly, the regions shown by Siegel et al. (2013) to be most negatively impacted by a_{dg} in terms of empirical Chl retrievals are the same regions which are shown here to diverge most strongly in terms of the normalized CDOM and Chl residual, indicating that a similar pattern would be expected even when using Chl retrieval algorithms less susceptible to error induced by a_{dg}. The pattern that emerges is that in open ocean regions characterized by strong seasonal blooms such as the North Atlantic and North Pacific, high primary productivity in the presence of lower CDOM (i.e., high residual) is followed after approximately a season by higher CDOM and a collapse in Chl (i.e., low residual). This can be seen in the boreal Spring–Summer transition in the N. Atlantic and Pacific, in the bloom and collapse associated with the reversal of the monsoons in the Arabian Sea between Summer and the following Winter/Spring, and in the Congo and Amazon river plumes over the same period where the residual often shifts by approximately a factor of six between seasons. This observed seasonal lag between peak Chl and peak CDOM helps explain why the two properties rarely covary, as described above for GOCAD, and in [13]. The lag may be explained by the time required for microbial degradation of the bloom's less labile particulate detrital material after the bloom has collapsed.

Application of these algorithms also shows the Spring–Summer transition in the CDOM absorption (left column, Figure 12) as an increase in CDOM from major river outflows such as the Amazon and Congo following peak runoff [69], and in the upwelling region of the Arabian Sea induced by the southwesterly monsoon. In the case of the Amazon River, the CDOM in the distal plume can be seen well into the following season as it drifts slowly eastward across the Atlantic from the retroflection of the North Brazil Current [70], indicating that satellite retrievals of CDOM using the MLR can successfully track surface DOM as it evolves over time scales of weeks to years and over very long distances. The results shown in Figure 12 are broadly similar to those described in [13] for CDM at 440 nm retrieved using the GSM algorithm [71], and to the empirical algorithm of Shanmugam [31] for $a_g(350)$ (their Figure 12), although we show generally higher absorption across the equatorial regions and some parts of the Southern Ocean.

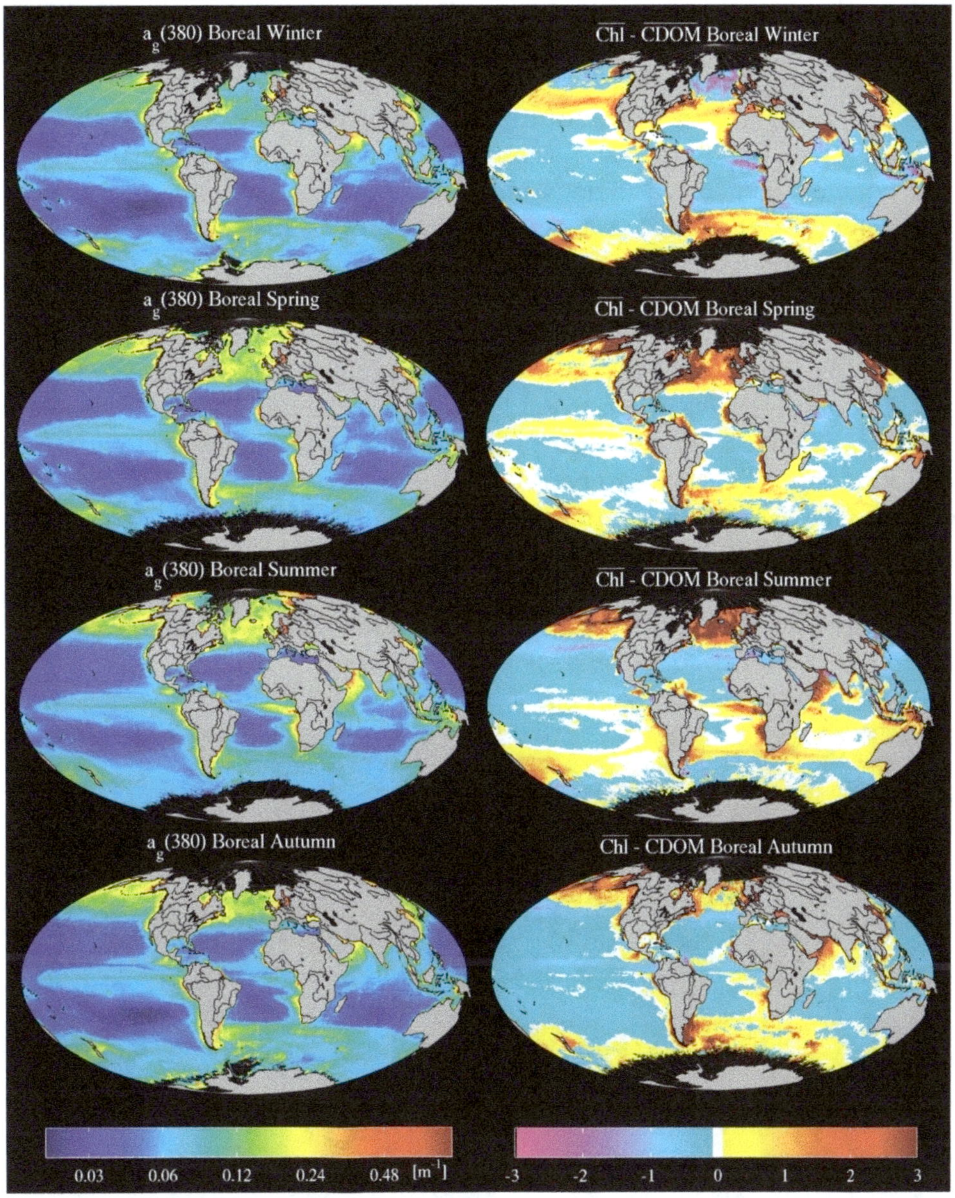

Figure 12. MLR retrievals of $a_g(380)$ by season over the entire MODIS Aqua era (left column), and residuals between Chl and CDOM (right column). Imagery was binned from 4 km resolution monthly composites between August 2002 and January 2014.

Longer time-scale variability in the Aqua-retrieved CDOM was also apparent from the roughly twelve years of monthly, 4 km satellite composites. An examination of the monthly CDOM anomaly ($\Delta a_g(\lambda)$; the monthly $a_g(\lambda)$ divided by the Aqua-era averages for each month) and slope anomaly, $\Delta S_g(\lambda)$, revealed several regions characterized by sharp declines in CDOM during certain years, and elevations in others, as well as the expected inverse proportionality between CDOM and slope in the UV. Figure 13 shows an example of this from seasonal Aqua composites of $\Delta a_g(380)$ and $\Delta S_g(275)$ averaged over periods of El Niño (2002–2005) when surface temperatures are higher, inhibiting vertical nutrient transport and leading to lower primary productivity, and periods of La Niña (2007, 2008, 2010, 2011), which exhibit roughly the opposite dynamics. A feature in the western equatorial Pacific stands out starkly as a crescent stretching from about 10° N to 15° S and spanning nearly the entire 100° longitude range from South America to the Solomon Islands. For brevity, we refer to this as the Western Pacific Crescent (WPC). To test the link between El Niño Southern Oscillation dynamics and the CDOM anomaly in the WPC, an average of monthly $\Delta a_g(380)$ within the WPC is compared with the multivariate ENSO index (MEI [72]) and $\Delta S_g(275\text{–}295)$ (Figure 14). The MEI provides a convenient index for tracking the dominant characteristics associated with ENSO, namely sea-surface pressure, temperature, wind stress, and cloudiness. Positive MEI represent the warmer El Niño cycle associated with lower wind stress, a flattening in the trans-Pacific thermocline, and inhibited productivity in the equatorial Pacific, and negative MEI represents La Niña, which is cooler, and more productive. The coupling between MEI and $\Delta a_g(380)$ and MEI and $\Delta S_g(275)$ is remarkably strong and well correlated (Figure 14; r = −0.77, r = 0.80, respectively, and $p << 0.01$ in each case). As expected, CDOM and UVB slope are also very well correlated (r = −0.90, $p << 0.01$). The tight correlation between CDOM anomaly, CDOM slope anomaly and MEI may help to predict broad changes in surface CDOM in a future in which warmer sea surface temperatures are expected, particularly in the western Pacific, as the long-term warming trend leads to oceanic conditions favorable to El Niño-like conditions [73]. In fact, sustained deficits in surface CDOM available for photooxidation and microbial remineralization across the WPC, as demonstrated here, is likely to result in a lower partial pressure of CO_2 derived from CDOM, and may therefore increase the flux of CO_2 into the ocean from the atmosphere, although this effect would largely be offset by the decrease in solubility associated with warmer temperatures in the surface ocean. Another consequence of lower CDOM across this region in a warming regime may be decreased surface heating through CDOM absorption, potentially providing some degree of negative feedback to the surface warming trend.

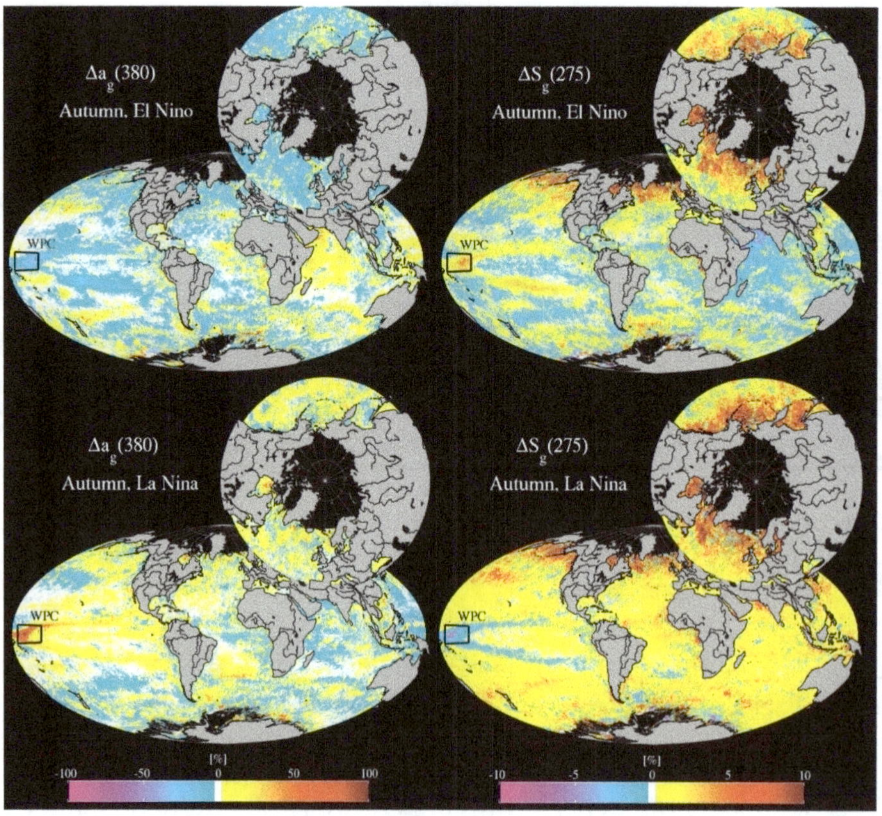

Figure 13. CDOM anomaly (left) and slope anomaly (right) from MLR applied to MODIS Aqua during Autumn in El Niño years (2002–2005; top panel) and La Niña years (2007, 2008, 2010, 2011; bottom panel). The Western Pacific Crescent (WPC) feature is defined here as the broad region exhibiting a notable decline in CDOM during El Niño years, and enhancement during La Niña. UV slope shows the opposite pattern, with lower slopes during La Niña, although the percentage change is roughly an order of magnitude lower. The box shows the portion of the WPC subsampled for comparison with MEI (See Figure 14).

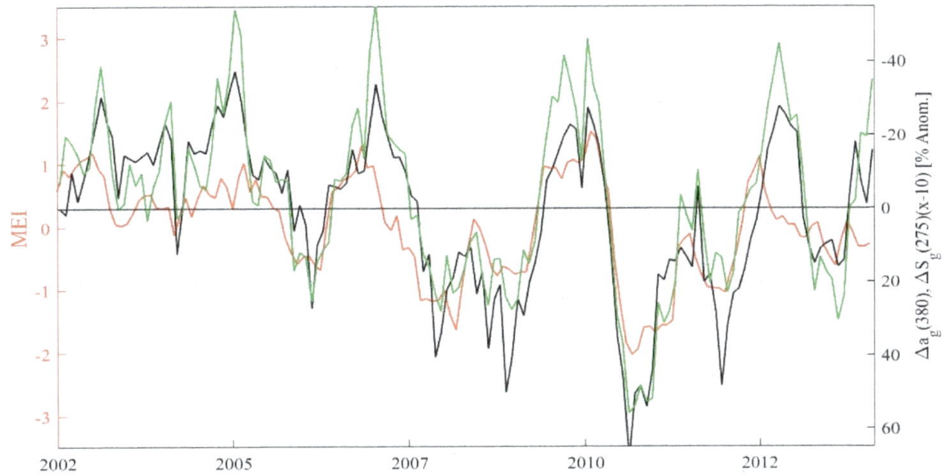

Figure 14. Multivariate ENSO Index (MEI, in red), CDOM anomaly at 380 nm (black), and UVB slope anomaly (green, scaled by a factor of −10 for clarity) over the entire MODIS Aqua era for the region of interest highlighted in Figure 13. Strong negative and positive correlations exist between MEI and CDOM and slope anomalies, respectively (see text).

4. Summary

The importance of characterizing and tracking change in global oceanic dissolved carbon over climatic evolutions is only possible synoptically using earth-observing technology. As methods to measure DOM sources and sinks continue to improve using laboratory and in situ optical techniques, algorithms and orbital sensor technology must keep pace. With technological and methodological improvements, however, come inevitable challenges. The nearly three decades of field data presented here are by necessity compromised in that, for example, no one standard method was employed for the measurement of spectral CDOM absorption. Similarly, there is obviously no standard algorithm for global ocean color retrievals of CDOM, as the algorithms must also continuously evolve with our knowledge of the parameters they retrieve. Many approaches have proven robust in retrieving CDOM absorption and its spectral slope over the years, though most are regionally optimized with little or no provision for what ties them together (e.g., proxies for optical water types). Global algorithms have been hampered by relatively small datasets of coincident radiometry with CDOM and CDOM slope extending into the UV, and DOC retrievals have been especially challenging due to the highly variable and often unpredictable fraction of chromophoric content.

In this study, we aggregate a global dataset approximately forty times the size of previous global, bio-optical databases. Naturally, despite our best efforts to ensure consistency in the data through quality control, the data within are subject to error and uncertainty, largely because methodologies and technology have evolved over thirty years. Quantification of the uncertainty in field estimates of the parameters retrieved here must necessarily precede uncertainty estimates in the algorithms used to derive them. Efforts are currently underway at NASA and elsewhere to do just that. New field data are always being collected and archived all over the world by various academic and public-sector agencies, but only a fraction is broadly distributed through invaluable archives like SeaBASS, in part because submission is only required of those principal investigators funded by NASA. Future algorithm development efforts should facilitate more cooperation and collaboration with other agencies collecting field data around the world encouraging sharing of data within a reasonable time after collection. To date, GOCAD and SeaBASS coverage in regions such as the Mediterranean and the oceans around Australia is astonishingly poor. Efforts must be sustained to continue bringing newly collected and historical data into GOCAD, NOMAD, and similar global, long-term bio-optical databases, and to expand them to include an even more comprehensive suite of inherent optical properties, which help support the development of more robust semi-analytical approaches. GOCAD was designed using nested and comprehensive Matlab structures conducive to expansion for both newly collected datasets as well as more complete suites of physical, optical, radiometric, and biogeochemical data. Based on our experience with SeaBASS, more rigorous quality control and documentation standards should be applied not only to recent and new submissions, but retroactively to historic data as well. While some algorithms performed better than others in this non-exhaustive comparison, it is important that algorithms continue to evolve as the data used to develop them improves, incorporating more than minor adjustments to empirical coefficients, and moving algorithms for oceanic carbon closer to theoretical, analytically-based approaches.

A representational suite of approaches, including empirical, semi-analytical, and machine-learning algorithms was evaluated against GOCAD field data for retrieving a_g in six wavebands between 275 and about 490 nm, S_g in eight wavebands in the UV and VIS, and DOC using a wide variety of metrics. Ultimately, the most versatile and best performing of those tested was a simple, empirical set of relationships based on multiple linear regression between four wavebands of remote sensing reflectance (440–555 nm), with the exception of DOC which also required sea surface salinity (e.g., from Aquarius) to act as a proxy to optical water type. Results varied, with CDOM retrievals showing regression coefficients to field data (r^2) generally over 0.80 for field radiometry to within 16%–34%, depending on the wavelength, and within 33%–54% for MODIS Aqua validation. CDOM slopes retrievals were best in the UVB (e.g., r^2 = 0.62, *MAPD* = 11% in satellite validation of S_g(275–295)), while DOC algorithms only optimized well after the inclusion of salinity (r^2 = 0.91, *MAPD* = 15%), and did not perform well in validation (e.g., RMSE = 27–29 μmol L^{-1}). Our analysis of the sensitivity of the DOC algorithm performance to factors such as salinity, DOC, water column depth, and geographic location ultimately proved inconclusive, exposing only a small anomaly involving overestimates of DOC in the southern oceans using MODIS Aqua imagery. Further validation—particularly in mid–ocean gyres where DOC varies very weakly or not at all with CDOM absorption, and salinity changes are very small—is clearly warranted prior to application of the DOC algorithm in those regions.

Application of CDOM algorithms to monthly and climatological Aqua imagery demonstrated that global retrievals of CDOM do not covary well with similar empirical retrievals of Chl, but rather appear to follow Chl on a seasonal lag depending on the region and source of dissolved material. This helps explain the lack of correlation between CDOM and Chl found in global GOCAD field data and described in previous studies, and further challenges the use of Case 1 assumptions in bio-optical remote sensing. Surface CDOM concentration varies in regions such the western equatorial Pacific by about 150% over the course of long-term climatological shifts associated with ENSO, fluctuating in tight correlation with the MEI and CDOM slope. Algorithms developed here may be applied to tracking ENSO behavior in the future, as well as observing changes in CDOM character and concentration associated with global warming.

Author Contributions: Conceptualization, D.A. and A.M.; Data Curation, D.A.; Formal Analysis, D.A. and D.L.; Funding Acquisition, A.M.; Investigation, D.A.; Methodology, D.A. and A.M.; Project Administration, A.M. and D.A.; Resources, A.M.; Software, D.A. and D.L.; Validation, D.A. and D.L.; Visualization, D.A.; Writing—original draft, D.A.; Writing—review & editing, D.A. and A.M.

Acknowledgments: We would like to acknowledge all investigators who have contributed to the SeaBASS database and the various Hansell/Carlson datasets (see Table 2). Thanks also to Jeremy Werdell, Chris Proctor, Norman Kuring, Chris Moellers, and Jason Leffler of the NASA Ocean Biology Processing Group for data acquisition and programming support.

References

1. Siegenthaler, U.; Sarmiento, J. Atmospheric carbon dioxide and the ocean. *Nature* **1993**, *365*, 119–125. [CrossRef]
2. Hedges, J.I. Why dissolved organics matter. *Biogeochem. Mar. Dissolved Org. Matter* **2002**, 1–33.

3. Belanger, S.; Babin, M.; Xie, H.; Krotkov, N.; Larouche, P.; Vincent, W.F. Cdom Photooxidation in the Arctic Coastal Waters: New Approach Using Satellite Information and Implications of Climate Change. In Proceedings of the Ocean Optics XVIII, Montreal, QC, Canada, 9–13 October 2006.
4. Carder, K.L.; Steward, R.G.; Harvey, G.R.; Ortner, P.B. Marine humic and fulvic acids: Their effects on remote sensing of chlorophyll. *Limnol. Oceanogr.* **1989**, *34*, 68–81. [CrossRef]
5. Lee, Z.; Arnone, R.; Hu, C.; Werdell, P.J.; Lubac, B. Uncertainties of optical parameters and their propagations in an analytical ocean color inversion algorithm. *Appl. Opt.* **2010**, *49*, 369–381. [CrossRef] [PubMed]
6. Sempéré, R.; Cauwet, G. Occurrence of organic colloids in the stratified estuary of the krka river (Croatia). *Estuar. Coast. Shelf Sci.* **1995**, *40*, 105–114. [CrossRef]
7. Blough, N.V.; Del Vecchio, R. Chromophoric dom in the coastal environment. In *Biogeochemistry of Marine Dissolved Organic Matter*; Hansell, D.A., Carlson, C.A., Eds.; Academic Press: San Diego, CA, USA, 2002; pp. 509–546.
8. Nelson, N.B.; Siegel, D.A. Chromophoric dom in the open ocean. *Biogeochem. Mar. Dissolved Org. Matter* **2002**, 547–578.
9. Twardowski, M.S.; Boss, E.; Sullivan, J.M.; Donaghay, P.L. Modeling the spectral shape of absorption by chromophoric dissolved organic matter. *Mar. Chem.* **2004**, *89*, 69–88. [CrossRef]
10. Aurin, D.A.; Dierssen, H.M.; Twardowski, M.S.; Roesler, C.S. Optical complexity in long island sound and implications for coastal ocean color remote sensing. *J. Geophys. Res. Oceans* **2010**, *115*. [CrossRef]
11. Del Vecchio, R.; Subramaniam, A. Influence of the amazon river on the surface optical properties of the western tropical north atlantic ocean. *J. Geophys. Res. Oceans* **2004**, *109*. [CrossRef]
12. Sauer, M.J.; Roesler, C.S. Seasonal and spatial cdom variability in the gulf of maine. In Proceedings of the Ocean Optics XVIII, Montreal, QC, Canada, 9–13 October 2006.
13. Siegel, D.A.; Maritorena, S.; Nelson, N.B.; Hansell, D.A.; Lorenzi-Kayser, M. Global distribution and dynamics of colored dissolved and detrital organic materials. *J. Geophys. Res. Oceans* **2002**, *107*. [CrossRef]
14. Pegau, W.S. Inherent optical properties of the central arctic surface waters. *J. Geophys. Res. Oceans* **2002**, *107*, 8035. [CrossRef]
15. Carder, K.L.; Chen, R.F.; Lee, Z.; Hawes, S.K.; Cannizzaro, J.P. *MODIS* Ocean Science Team Algorithm Theoretical Basis Document. *ATBD* **2003**, *19*, 7–18.
16. Morel, A.; Prieur, L. Analysis of variations in ocean color. *Limnol. Oceanogr.* **1977**, *22*, 709–722. [CrossRef]
17. Lee, Z. *Remote Sensing of Inherent Optical Properties: Fundamentals, Tests of Algorithms, and Applications*; International Ocean–Colour Coordinating Group: Dartmouth, NS, Canada, 2006.
18. Del Vecchio, R.; Blough, N.V. Photobleaching of chromophoric dissolved matter in natural waters: Kinetics and modeling. *Mar. Chem.* **2002**, *78*, 231–253. [CrossRef]
19. Helms, J.R.; Stubbins, A.; Ritchie, J.D.; Minor, E.C. Absorption spectral slopes and slope ratios as indicators of molecular weight, source, and photobleaching of chromophoric dissolved organic material. *Limnol. Oceanogr.* **2008**, *53*, 955–969. [CrossRef]
20. Stedmon, C.A.; Markager, S. Behavior of the optical properties of coloured dissolved organic matter under conservative mixing. *Estuar. Coast. Shelf Sci.* **2003**, *57*, 973–979. [CrossRef]
21. Del Castillo, C.E.; Miller, R.L. Horizontal and vertical distributions of colored dissolved organic matter during the southern ocean gas exchange experiment. *J. Geophys. Res. Oceans* **2011**, *116*. [CrossRef]
22. Stedmon, C.A.; Markager, S.; Kaas, H. Optical properties and signatures of chromophoric dissolved organic matter (CDOM) in danish coastal waters. *Estuar. Coast. Shelf Sci.* **2000**, *51*, 267–278. [CrossRef]
23. Cartisano, C.M.; Del Vecchio, R.; Bianca, M.R.; Blough, N.V. Investigating the sources and structure of chromophoric dissolved organic matter (CDOM) in the north pacific ocean (NPO) utilizing optical spectroscopy combined with solid phase extraction and borohydride reduction. *Mar. Chem.* **2018**, *204*, 20–35. [CrossRef]
24. Twardowski, M.S.; Donaghay, P.L. Photobleaching of aquatic dissolved materials: Absorption removal, spectral alteration, and their interrelationship. *J. Geophys. Res. Oceans* **2002**, *107*. [CrossRef]
25. D'Sa, E.J.; Hu, C.; Muller–Karger, F.E.; Carder, K.L. Estimation of colored dissolved organic matter and salinity fields in case 2 waters using seawifs: Examples from florida bay and florida shelf. *Proc. Indian Acad. Sci. Earth Planet. Sci.* **2002**, *111*, 197–207. [CrossRef]
26. Johannessen, S.C.; Miller, W.L.; Cullen, J.J. Calculation of uv attenuation and colored dissolved organic matter absorption spectra from measurements of ocean color. *J. Geophys. Res. Oceans* **2003**, *108*, 3301. [CrossRef]

27. Kahru, M.; Mitchell, B.G. Seasonal and nonseasonal variability of satellite–derived chlorophyll and colored dissolved organic matter concentration in the california current. *J. Geophys. Res. Oceans* **2001**, *106*, 2517–2529. [CrossRef]
28. Mannino, A.; Russ, M.E.; Hooker, S.B. Algorithm development and validation for satellite–derived distributions of doc and cdom in the u.S. Middle atlantic bight. *J. Geophys. Res.* **2008**, *113*, C07051. [CrossRef]
29. Matthews, M.W. A current review of empirical procedures of remote sensing in inland and near–coastal transitional waters. *Int. J. Remote Sens.* **2011**, *32*, 6855–6899. [CrossRef]
30. Tehrani, N.; Sa, E.; Osburn, C.; Bianchi, T.; Schaeffer, B. Chromophoric dissolved organic matter and dissolved organic carbon from sea–viewing wide field–of–view sensor (SEAWIFS), moderate resolution imaging spectroradiometer (MODIS) and meris sensors: Case study for the northern gulf of mexico. *Remote Sens.* **2013**, *5*, 1439–1464. [CrossRef]
31. Shanmugam, P. A new bio-optical algorithm for the remote sensing of algal blooms in complex ocean waters. *J. Geophys. Res. Oceans* **2011**, *116*. [CrossRef]
32. Tiwari, S.P.; Shanmugam, P. An optical model for the remote sensing of coloured dissolved organic matter in coastal/ocean waters. *Estuar. Coast. Shelf Sci.* **2011**, *93*, 396–402. [CrossRef]
33. Werdell, P.J.; Bailey, S.W. An improved in situ bio–optical data set for ocean color algorithm development and satellite data product validation. *Remote Sens. Environ.* **2005**, *98*, 122–140. [CrossRef]
34. Maritorena, S.; Lee, Z.; Du, K.P.; Loisel, H.; Doerffer, R.; Roesler, C.; Lyon, P.; Tanaka, A.; Babin, M.; Kopelevich, O.V. Chapter 2: Synthetic and in situ data sets for algorithm testing. In *Remote Sensing of Inherent Optical Properties: Fundamentals, Tests of Algorithms and Applications*; Lee, Z., Ed.; International Ocean Colour Coordinating Group: Dartmouth, NS, Cabada, 2006; pp. 13–18.
35. Gordon, H.R.; Brown, O.B.; Evans, R.H.; Brown, J.W.; Smith, R.C.; Baker, K.S.; Clark, D.K. A semi–analytic radiance model of ocean color. *J. Geophys. Res. Atmos.* **1988**, *93*, 10909–10924. [CrossRef]
36. Gordon, H.R.; Brown, O.B.; Jacobs, M.M. Computed relationships between the inherent and apparent optical properties of a flat homogenous ocean. *Appl. Opt.* **1975**, *14*, 417–427. [CrossRef] [PubMed]
37. Garver, S.A.; Siegel, D.A. Inherent optical property inversion of ocean color spectra and its biogeochemical interpretation. 1. Time series from the sargasso sea. *J. Geophys. Res. Oceans* **1997**, *102*, 18607–18625. [CrossRef]
38. Lee, Z.P.; Carder, K.L. Effect of spectral band numbers on the retrieval of water column and bottom properties from ocean color data. *Applied Optics* **2002**, *41*, 2191–2201. [CrossRef] [PubMed]
39. Roesler, C.S.; Perry, M.J. In–situ phytoplankton absorption, fluorescence emission, and particulate backscattering spectra determined from reflectance. *J. Geophys. Res. Oceans* **1995**, *100*, 13279–13294. [CrossRef]
40. Dong, Q.; Shang, S.; Lee, Z. An algorithm to retrieve absorption coefficient of chromophoric dissolved organic matter from ocean color. *Remote Sens. Environ.* **2013**, *128*, 259–267. [CrossRef]
41. Matsuoka, A.; Babin, M.; Doxaran, D.; Hooker, S.; Mitchell, B.; Bélanger, S.; Bricaud, A. A synthesis of light absorption properties of the pan–arctic ocean: Application to semi–analytical estimates of dissolved organic carbon concentrations from space. *Biogeosci. Discuss.* **2013**, *10*, 17071–17115. [CrossRef]
42. Odermatt, D.; Gitelson, A.; Brando, V.E.; Schaepman, M. Review of constituent retrieval in optically deep and complex waters from satellite imagery. *Remote Sens. Environ.* **2012**, *118*, 116–126. [CrossRef]
43. Swan, C.M.; Nelson, N.B.; Siegel, D.A.; Fields, E.A. A model for remote estimation of ultraviolet absorption by chromophoric dissolved organic matter based on the global distribution of spectral slope. *Remote Sens. Environ.* **2013**, *136*, 277–285. [CrossRef]
44. Tilstone, G.H.; Peters, S.W.; van der Woerd, H.J.; Eleveld, M.A.; Ruddick, K.; Schönfeld, W.; Krasemann, H.; Martinez–Vicente, V.; Blondeau–Patissier, D.; Röttgers, R. Variability in specific–absorption properties and their use in a semi–analytical ocean colour algorithm for meris in north sea and western english channel coastal waters. *Remote Sens. Environ.* **2012**, *118*, 320–338. [CrossRef]
45. Zhu, W.; Yu, Q.; Tian, Y.Q.; Chen, R.F.; Gardner, G.B. Estimation of chromophoric dissolved organic matter in the mississippi and atchafalaya river plume regions using above-surface hyperspectral remote sensing. *J. Geophys. Res. Oceans* **2011**, *116*. [CrossRef]
46. Stramski, D.; Woźniak, S.B. On the role of colloidal particles in light scattering in the ocean. *Limnol. Oceanogr.* **2005**, *50*, 1581–1591. [CrossRef]
47. Nelson, N.B.; Siegel, D.A. The global distribution and dynamics of chromophoric dissolved organic matter. *Annu. Rev. Mar. Sci.* **2013**, *5*, 447–476. [CrossRef] [PubMed]

48. Weishaar, J.L.; Aiken, G.R.; Bergamaschi, B.A.; Fram, M.S.; Fujii, R.; Mopper, K. Evaluation of specific ultraviolet absorbance as an indicator of the chemical composition and reactivity of dissolved organic carbon. *Environ. Sci. Technol.* **2003**, *37*, 4702–4708. [CrossRef] [PubMed]
49. Sullivan, J.M.; Twardowski, M.; Zanefeld, J.R.V.; Moore, C.M.; Barnard, A.H.; Donaghay, P.L.; Rhoades, B. Hyperspectral temperature and salt dependencies of absorption by water and heavy water in the 400–750 nm spectral range. *Appl. Opt.* **2006**, *45*, 5294–5309. [CrossRef] [PubMed]
50. Twardowski, M.S.; Sullivan, J.M.; Donaghay, P.L.; Zaneveld, J.R.V. Microscale quantification of the absorption by dissolved and particulate material in coastal waters with an ac–9. *J. Atmos. Ocean. Technol.* **1999**, *16*, 691–707. [CrossRef]
51. Zaneveld, J.R.V.; Kitchen, J.C.; Moore, C.M. Scattering error correction of reflecting–tube absorption meters. In *Ocean optics XII*; Jaffe, J.S., Ed.; International Society for Optics and Photonics: Bellingham, WA, USA, 1994; pp. 44–55.
52. Mitchell, B.G.; Stramska, M.; Wieland, J. Determination of spectral coefficients of particles, dissolved material and phytoplankton for discrete water samples. In *Ocean Optics Protocols for Satellite Ocean Color Sensor Validation*; National Aeronautics and Space Administration: Greenbelt, MA, USA, 2003.
53. Mueller, J.L. Inherent optical properties: Instruments, characterizations, field measurements and data analysis protocols. In *Ocean Optics Protocols for Satellite Ocean Color Sensor Validation*; National Aeronautics and Space Administration: Greenbelt, MA, USA, 2003.
54. Thuillier, G.; Hersé, M.; Foujols, T.; Peetermans, W.; Gillotay, D.; Simon, P.; Mandel, H. The solar spectral irradiance from 200 to 2400 nm as measured by the solspec spectrometer from the atlas and eureca missions. *Sol. Phys.* **2003**, *214*, 1–22. [CrossRef]
55. Bailey, S.W.; Werdell, P.J. A multi–sensor approach for the on–orbit validation of ocean color satellite data products. *Remote Sens. Environ.* **2006**, *102*, 12–23. [CrossRef]
56. Aurin, D.A.; Mannino, A. A database for developing global ocean color algorithms for colored dissolved organic material, cdom slope, and dissolved organic carbon. In Proceedings of the Ocean Optics XXI, Glasgow, Scotland, 8–12 October 2012.
57. Taylor, K.E. Summarizing multiple aspects of model performance in a single diagram. *J. Geophys. Res. Atmos.* **2001**, *106*, 7183–7192. [CrossRef]
58. Jolliff, J.K.; Kindle, J.C.; Shulman, I.; Penta, B.; Friedrichs, M.A.; Helber, R.; Arnone, R.A. Summary diagrams for coupled hydrodynamic–ecosystem model skill assessment. *J. Mar. Syst.* **2009**, *76*, 64–82. [CrossRef]
59. Carlson, C.A.; Hansell, D.A.; Tamburini, C. Doc persistence and its fate after export within the ocean interior. *Microb. Carbon Pump Ocean* **2011**, 57–59.
60. Hansell, D.A.; Carlson, C.A. Biogeochemistry of total organic carbon and nitrogen in the sargasso sea: Control by convective overturn. *Deep Sea Res. Part II* **2001**, *48*, 1649–1667. [CrossRef]
61. Breiman, L. Random forests. *Mach. Learn.* **2001**, *45*, 5–32. [CrossRef]
62. Lary, D.J.; Zewdie, G.K.; Liu, X.; Wu, D.; Levetin, E.; Allee, R.J.; Malakar, N.; Walker, A.; Mussa, H.; Mannino, A.; Aurin, D. Machine learning applications for earth observation. In *Earth Observation Open Science and Innovation*; Mathieu, P.P., Aubrecht, C., Eds.; Springer: Cham, Switzerland, 2018.
63. Lee, Z.P.; Carder, K.L.; Arnone, R. Chapter 10: The quasi–analytical algorithm. In *Remote Sensing of Inherent Optical Properties: Fundamentals, Tests of Algorithms, and Applications*; Stuart, V., Ed.; IOCCG: Dartmouth, NS, Canada, 2006; pp. 73–79.
64. Lee, Z.P.; Carder, K.L.; Arnone, R.A. Deriving inherent optical properties from water color: A multiband quasi–analytical algorithm for optically deep waters. *Appl. Opt.* **2002**, *41*, 5755–5772. [CrossRef] [PubMed]
65. Werdell, P.J.; Franz, B.A.; Bailey, S.W.; Feldman, G.C.; Boss, E.; Brando, V.E.; Dowell, M.; Hirata, T.; Lavender, S.J.; Lee, Z. Generalized ocean color inversion model for retrieving marine inherent optical properties. *Appl. Opt.* **2013**, *52*, 2019–2037. [CrossRef] [PubMed]
66. Morel, A.; Bricaud, A. Theoretical results concerning light absorption in a discrete medium and application to the specific absorption of phytoplankton. *Deep Sea Res.* **1981**, *28*, 1357–1393. [CrossRef]
67. Twardowski, M.S.; Boss, E.; Macdonald, J.B.; Pegau, W.S.; Barnard, A.H.; Zaneveld, J.R.V. A model for estimating bulk refractive index from the optical backscattering ratio and the implications for understanding particle composition in case i and case ii waters. *J. Geophys. Res. Oceans* **2001**, *106*, 14129–14142. [CrossRef]

68. Siegel, D.; Behrenfeld, M.; Maritorena, S.; McClain, C.; Antoine, D.; Bailey, S.; Bontempi, P.; Boss, E.; Dierssen, H.; Doney, S. Regional to global assessments of phytoplankton dynamics from the seawifs mission. *Remote Sens. Environ.* **2013**, *135*, 77–91. [CrossRef]
69. Dai, A.; Qian, T.; Trenberth, K.E.; Milliman, J.D. Changes in continental freshwater discharge from 1948 to 2004. *J. Clim.* **2009**, *22*, 2773–2792. [CrossRef]
70. Lentz, S.J. The amazon river plume during amasseds: Subtidal current variability and the importance of wind forcing. *J. Geophys. Res. Oceans* **1995**, *100*, 2377–2390. [CrossRef]
71. Maritorena, S.; Siegel, D.A.; Peterson, A.R. Optimization of a semianalytical ocean color model for global–scale applications. *Appl. Opt.* **2002**, *41*, 2705–2714. [CrossRef] [PubMed]
72. Wolter, K.; Timlin, M.S. Monitoring enso in coads with a seasonally adjusted principal component index. In Proceedings of the 17th Climate Diagnostics Workshop, Norman, OK, USA, January 1993.
73. Collins, M.; An, S.-I.; Cai, W.; Ganachaud, A.; Guilyardi, E.; Jin, F.-F.; Jochum, M.; Lengaigne, M.; Power, S.; Timmermann, A. The impact of global warming on the tropical pacific ocean and el niño. *Nat. Geosci.* **2010**, *3*, 391–397. [CrossRef]

6

Characterization of the Light Field and Apparent Optical Properties in the Ocean Euphotic Layer based on Hyperspectral Measurements of Irradiance Quartet

Linhai Li [1], Dariusz Stramski [1,*] and Mirosław Darecki [2]

[1] Scripps Institution of Oceanography, University of California San Diego, La Jolla, CA 92093-0238, USA; lil032@ucsd.edu

[2] Institute of Oceanology, Polish Academy of Sciences, 81-712 Sopot, Poland; darecki@iopan.gda.pl

* Correspondence: dstramski@ucsd.edu

Abstract: Although the light fields and apparent optical properties (AOPs) within the ocean euphotic layer have been studied for many decades through extensive measurements and theoretical modeling, there is virtually a lack of simultaneous high spectral resolution measurements of plane and scalar downwelling and upwelling irradiances (the so-called irradiance quartet). We describe a unique dataset of hyperspectral irradiance quartet, which was acquired under a broad range of environmental conditions within the water column from the near-surface depths to about 80 m in the Gulf of California. This dataset enabled the characterization of a comprehensive suite of AOPs for realistic non-uniform vertical distributions of seawater inherent optical properties (IOPs) and chlorophyll-*a* concentration (*Chl*) in the common presence of inelastic radiative processes within the water column, in particular Raman scattering by water molecules and chlorophyll-*a* fluorescence. In the blue and green spectral regions, the vertical patterns of AOPs are driven primarily by IOPs of seawater with weak or no discernible effects of inelastic processes. In the red, the light field and AOPs are strongly affected or totally dominated by inelastic processes of Raman scattering by water molecules, and additionally by chlorophyll-*a* fluorescence within the fluorescence emission band. The strongest effects occur in the chlorophyll-*a* fluorescence band within the chlorophyll-*a* maximum layer, where the average cosines of the light field approach the values of uniform light field, irradiance reflectance is exceptionally high approaching 1, and the diffuse attenuation coefficients for various irradiances are exceptionally low, including the negative values for the attenuation of upwelling plane and scalar irradiances. We established the empirical relationships describing the vertical patterns of some AOPs in the red spectral region as well as the relationships between some AOPs which can be useful in common experimental situations when only the downwelling plane irradiance measurements are available. We also demonstrated the applicability of irradiance quartet data in conjunction with Gershun's equation for estimating the absorption coefficient of seawater in the blue-green spectral region, in which the effects of inelastic processes are weak or negligible.

Keywords: oceanic light field; irradiance quartet; apparent optical properties; inelastic processes; Gershun equation; ocean euphotic zone

---

1. Introduction

The ocean epipelagic zone that extends from the surface to approximately 200 m depth is extremely important for ocean-atmosphere interactions with implications to climate and supporting life on Earth. Most solar radiation incident on the ocean is absorbed within this layer [1,2]. In the euphotic layer that overlaps with the upper portion or the entire epipelagic zone in very clear ocean waters, there is enough light to support the process of photosynthesis, which contributes nearly half of the world's total biological primary production [3]. The studies of light propagation and light field characteristics are crucial for understanding many physical and biological processes in the upper ocean, which are driven by or depend on solar radiation. For example, for studying the heating rate, the radiometric quantities of spectral downward plane irradiance, $E_d(z, \lambda)$, and the spectral upward plane irradiance, $E_u(z, \lambda)$, are essential [4]. The symbols z (units of m) and λ (units of nm) stand for depth in the ocean and light wavelength in vacuum, respectively. In photosynthesis studies the key quantity of photosynthetically available radiation (PAR) is best represented on the basis of radiometric quantity of spectral scalar irradiance, $E_o(z, \lambda)$, and its integration over the spectral range of 350–700 nm or 400–700 nm with conversion from energy units to quantum units [5–9]. This measure of PAR is hereafter referred to as the PAR quantum scalar irradiance, E_{oPAR}. In the past, however, the PAR estimates in the ocean have been often based on the measurements of downward plane irradiance [10], so this measure of PAR will be referred to as PAR quantum downward irradiance, E_{dPAR}. The PAR estimate in terms of E_{oPAR} is superior to E_{dPAR}, mainly because it does not ignore the contribution of upwelling light to photosynthetically available radiation and the measurement of E_o with a spherical collector gives equal weight to quanta arriving at a point from all possible directions, in contrast to the cosine weighting in the measurement of E_d with a plane collector. For simplicity, the explicit dependence of optical quantities on z and λ is omitted hereafter unless causing ambiguity (symbols and definitions are summarized in Table 1, see also [9] for terminology and definitions used in hydrologic optics).

Table 1. Symbols of basic variables used in this study.

Symbol	Description	Units
λ	Light wavelength in vacuum	nm
z	Depth in water	m
θ_s	Solar zenith angle	degree
a	Absorption coefficient	m^{-1}
b	Scattering coefficient	m^{-1}
c	Beam attenuation coefficient (sum of a and b)	m^{-1}
L_u	Spectral upwelling radiance at zenith direction	$W\ m^{-2}\ sr^{-1}\ nm^{-1}$
E_d, E_u	Spectral downwelling and upwelling plane irradiances	$W\ m^{-2}\ nm^{-1}$
E_o, E_{od}, E_{ou}	Spectral total, downwelling, and upwelling scalar irradiances	$W\ m^{-2}\ nm^{-1}$
K_x	Diffuse attenuation coefficients for irradiance or radiance x	m^{-1}
$\overline{\mu}$, $\overline{\mu}_d$, $\overline{\mu}_u$	Average cosines of total, down- and upwelling light fields	dimensionless
R	Irradiance reflectance	dimensionless
Chl	Chlorophyll-a concentration	$mg\ m^{-3}$
Subscripts		
w	Water	
p	Suspended particulate matter	
g	Colored dissolved organic matter (CDOM)	

Radiometric measurements in the ocean started in the early 1930s and a summary of pioneering work can be found in [11]. Example early studies include measurements of spectral irradiance, e.g., [12,13] and the angular distribution of radiance, e.g., [14–16], and determinations of some apparent optical properties (AOPs) of the ocean from light measurements, e.g., [12,17,18]. Early reports on radiometric and other optical measurements in the ocean can be also found in the Russian literature, e.g., [19].

More recent studies have considerably expanded these early measurements by using advanced instrumentation and various deployment approaches, e.g., [20–32]. Much research interest in recent decades has been concentrated on ocean reflectance in relation to applications of ocean color remote sensing. For these applications, many radiometric measurements have been collected for spectral upwelling (zenith) radiance, L_u, and E_d, and optionally also for E_u [32–34]. Note that the zenith radiance indicates that the measured light travels toward zenith. The analysis of radiometric measurements driven by the interest in ocean color has been typically focused on near-surface layer, approximately the top 10–20 m. With regard to AOPs, most experimental efforts have been placed on the determinations of (i) the spectral diffuse attenuation coefficients K_d, K_u, and K_{Lu} for E_d, E_u, and L_u, respectively; (ii) the spectral irradiance reflectance, $R = E_u/E_d$; and (iii) the spectral remote-sensing reflectance just above the sea surface, $R_{rs}(z = 0^+) = L_w(z = 0^+)/E_d(z = 0^+)$ where L_w is the spectral water-leaving radiance and $z = 0^+$ is just above the sea surface, e.g., [32,34–43]. Note also that the measurements of both E_d and E_u allow for determination of the net irradiance, $E = E_d - E_u$, and the diffuse attenuation coefficient for net irradiance, K_E.

In contrast to the plane irradiances E_d and E_u, the measurements of spectral scalar irradiance, E_o, including its downward, E_{od}, and upward, E_{ou}, components have seldom been conducted. In particular, simultaneous measurements of the irradiance quartet that includes E_d, E_u, E_{od}, and E_{ou} have been very rare [44] and we are unaware of explicit presentation of experimental data of the complete irradiance quartet in the literature. This greatly limits the availability of experimental data of the average cosines that provide a simple way of specifying the angular structure of the light field. The average cosines include: (i) the average cosine for the downwelling light field, $\overline{\mu}_d = E_d/E_{od}$; (ii) the average cosine for the upwelling light field, $\overline{\mu}_u = E_u/E_{ou}$; and (iii) the average cosine for the entire light field, $\overline{\mu} = (E_d - E_u)/(E_{od} + E_{ou}) \equiv E/E_o$ [8,9]. We note that according to the original concept and definition of AOPs introduced by Preisendorfer [45,46], the average cosines (and the reciprocal of the average cosines referred to as the distribution functions) can be considered AOPs. Preisendorfer [46] provides the following definition of AOPs: "The apparent optical properties of a natural hydrosol are those radiometrically determined scattering- and absorbing-induced quantities which generally depend on the geometrical structure of the light field (i.e., whether the light field is more or less collimated or diffuse) but which have enough regular features and enough stability to be entitled to the appellation optical property". In some studies, however, for example in the work of Kirk [10,47], the average cosines are considered just as simple parameters that specify the angular structure of the light field, and not AOPs. In this paper, we choose to refer to the average cosines as AOPs, which is consistent with the traditional view [9,46,48] and simplifies the narrative structure of the presentation of our results for the three classes of parameters (i.e., K-coefficients, reflectances, and average cosines) which are derived from radiometric quantities. The issue of whether or not we refer to the average cosines as AOPs has no impact on the presented results.

The scarcity of simultaneous measurements of E and E_o has also greatly limited the potential use of the so-called Gershun equation [49] for estimating the absorption coefficient of seawater, a, from radiometric measurements or AOPs. This equation can be derived directly from the 1-D scalar (i.e., depth dependence only and no consideration of polarization of light) radiative transfer equation (RTE) for radiance, and takes a simple form $a = K_E \cdot \overline{\mu}$ when the inelastic radiative processes and internal sources are ignored [9]. The inelastic processes in the ocean include Raman scattering by water molecules, fluorescence by phytoplankton pigments, especially chlorophyll-a, and fluorescence by colored dissolved organic matter (CDOM). Few experimental studies have incorporated measurements of both spectral plane and scalar irradiances [18,44] or derived the irradiance quartet from radiance measurements limited to a single or a few wavebands [20,29,31,50] to examine the application of Gershun's equation. Højerslev [51] demonstrated that the irradiances involved in Gershun's equation, E and E_o, can be measured with two irradiance sensors equipped with spherical collectors oriented in upward and downward looking directions and masked so that they each collect light over a hemispherical portion of the collector. This latter approach was also used by Spitzer and Wernand [52] to determine the absorption coefficient of seawater from Gershun's equation.

In principle, theoretical simulations of radiative transfer (RT) in the ocean enable the analysis of essentially all radiometric quantities of underwater light field and AOPs for various pre-defined scenarios of input data of inherent optical properties (IOPs) of seawater and boundary conditions at the sea surface and ocean bottom. There is a large body of literature in this area with a great majority of studies focusing on the so-called inverse problem or developing methods for estimating

IOPs from radiometric measurements of natural light field or AOPs in a water body (see [53] for a review of this topic). In general, the limitations of such RT models are set by the assumptions used in their development, which include the assumptions regarding the vertical structure of IOPs and inelastic processes. The RT simulations have usually assumed a homogeneous water column and ignored all or some inelastic processes [47,54–60]. If included, the vertical stratification normally requires some idealized parameterization of the depth profile of IOPs, e.g., [60–64]. The models based on such simplifying assumptions can be satisfactory within certain limitations for the purposes of their development, for example the retrieval of average IOPs in the near-surface ocean layer. Notwithstanding the simplifying assumptions about the ocean environment, the RT modeling has provided insights into the complexity of the effects of the vertical inhomogeneity of the water column [60,62,63,65,66] and inelastic processes [60,67–75] on in-water and water-leaving light fields as well as AOPs. However, a comprehensive theoretical analysis of all (or nearly all) radiometric quantities (including plane and scalar irradiances) as well as the three AOP classes (*K*-coefficients, reflectances, and average cosines) throughout the water column for a variety of complex environmental scenarios is not easily tractable, especially for various naturally-occurring vertical profiles of IOPs within the upper ocean. Such analysis has usually remained beyond the scope of RT modeling; rare examples are found in Morel and Gentili [60], which focused on the well-lit upper ocean layer, and Li et al. [64], which focused on the dimly-lit mesopelagic zone ($z > 200$ m). Morel and Gentili [60] demonstrated, for example, the intricate patterns of the influences of many factors, including the sun position, seawater IOPs (as driven primarily by chlorophyll-*a* concentration, *Chl*, and co-varying water constituents), and Raman scattering, on the various AOPs within a hypothetically homogeneous ocean.

In spite of decades of experimental and theoretical studies of optical radiometry of the ocean, there exist some gaps or interesting questions, which deserve further attention. This study has been motivated by a few such questions, in particular (i) how the spectral and vertical patterns of light field characteristics and AOPs are affected by the interplay of inelastic processes (i.e., Raman scattering and chlorophyll-*a* fluorescence) and the actual non-uniform vertical distributions of *Chl* (and hence the vertically varying IOPs) within the euphotic layer; (ii) what relationships can be established between different AOPs using the depth-resolved data within the euphotic layer; and (iii) what is the quality and applicability of estimation of seawater absorption coefficient from measured AOPs and Gershun's equation under real conditions of non-uniform vertical distributions of *Chl* and IOPs, and inelastic processes within the euphotic layer. To address these questions, we collected simultaneous hyperspectral measurements of the irradiance quartet (E_d, E_u, E_{od}, and E_{ou}) and the zenith radiance L_u from the surface to depths of 60–80 m in the Gulf of California. This dataset is unique because it allows conducting a case study of a comprehensive set of radiometric quantities and all essential AOPs for a broad range of specific optical and biogeochemical conditions within the water column as described by the actual vertical profiles of *Chl* and IOPs.

2. Methods

2.1. Study Area

The field data were collected in the Gulf of California onboard the R/V New Horizon in June 2010 and June–July 2011 (Figure 1). A total of 11 stations were investigated in 2010 and 14 stations in 2011. These stations were located within three regions of the Gulf of California, Guaymas Basin, Carmen Basin, and Farallon Basin, which differ significantly in terms of water optical properties. The Guaymas Basin is the central part of the Gulf of California, Farallon Basin is the southernmost part of the Gulf of California, and Carmen Basin is located between the Guaymas and Farallon Basins. The Gulf of California has unique oceanographic and ecosystem characteristics [76,77]. According to Jerlov's optical classification of natural water bodies based on near-surface values of $K_d(\lambda)$ [11], the waters in the Guaymas Basin range typically between the oceanic Type IB and Type II, the Carmen Basin waters between the oceanic Type I and Type IA, and the Farallon Basin waters between Type II and coastal Type 1.

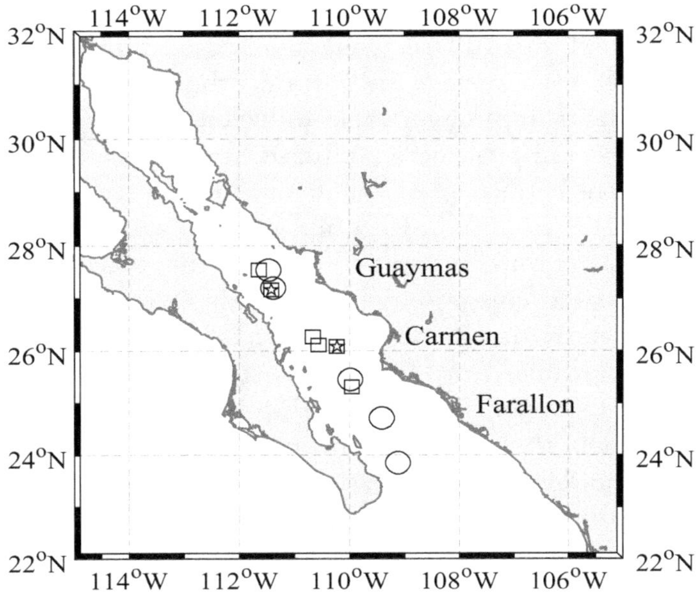

Figure 1. Location of stations in the Gulf of California where underwater radiometric and ancillary measurements were made. Circles indicate the station sites in 2010 and squares in 2011. Optical measurements were made at four, three, and four stations in the Guaymas, Carmen, and Farallon Basins, respectively. Stars indicate two stations where the absorption and beam attenuation measurements with an ac-9 instrument were made in 2011.

2.2. CTD Measurements

At each station, a conductivity-temperature-depth (CTD) package equipped with Niskin bottles was deployed to determine the depth profiles of seawater physical parameters (temperature T, salinity S, and density anomaly σ), chlorophyll-*a* fluorescence, and non-water optical beam attenuation coefficient at 660 nm, $c_{nw}(z, 660)$ (i.e., the total beam attenuation of seawater after subtraction of pure seawater contribution). Water samples were also collected at discrete depths with the CTD-Rosette for subsequent determinations of *Chl* in the laboratory.

The CTD package was deployed from water surface to ~500 m depth. The acquired raw data were processed using a manufacturer's software, SBEDataProcessing v7.23.2 (Sea-Bird Electronics Inc., Bellevue, WA, USA), and converted to values in physical units with calibration files. The processing software automatically aligned the measurements of each sensor to the common depth vector. The calibrated data were subsequently split into down- and up-casts. Preliminary inspection of the data showed that the up-casts of CTD measurements often exhibited less noise and thus were chosen for further processing. Depth profiles of each parameter were despiked using a running median method and smoothed using a Savitzky-Golay filter in Matlab R2015b (MathWorks, Natick, MA, USA). The profiles were then binned to 0.5 m depth resolution to yield the final data of T, S, σ, chlorophyll-*a* fluorescence, and $c_{nw}(660)$. The $c_{nw}(660)$ was considered to be equivalent to the particulate attenuation coefficient at 660 nm, $c_p(660)$, because the absorption of dissolved material at 660 nm was assumed negligible.

2.3. Chlorophyll-a Concentration

The chlorophyll-*a* concentration, *Chl*, was determined for discrete water samples obtained from the Niskin bottles deployed during the CTD-Rosette casts. The particulate material was collected by filtration onto glass-fiber filters (GF/F) and stored in liquid nitrogen for post-cruise analysis. The samples collected in 2010 were analyzed with High-Performance Liquid Chromatography (HPLC) using the method described in [78]. The total *Chl*, which represents the summed contributions of monovinyl chlorophyll-*a*, divinyl chlorophyll-*a*, chlorophyllide-*a*, and the allomeric and epimeric forms of chlorophyll-*a*, was determined. In 2010, *Chl* was usually determined for water samples collected at two depths, 1–3 m below the sea surface and around the depth of chlorophyll-*a* fluorescence maximum, z_{Chlmax}. In 2011, water samples were also collected below z_{Chlmax}, resulting in three water samples at each station. These samples were analyzed spectrophotometrically to determine *Chl*. The absorbance spectra of acetone extracts of pigments were measured inside an integrating sphere of a spectrophotometer (Perkin-Elmer Lambda 18 equipped with a 15-cm sphere) from 290 to 860 nm with 1-nm interval. The *Chl* concentrations were computed from absorbance values at 630, 647, 665, and 691 nm using the equation of Ritchie [79].

Based on *Chl* determinations at discrete depths, depth profiles of *Chl* were derived by calibrating the chlorophyll-*a* fluorescence profiles from the CTD cast. The values of fluorescence at depths of *Chl* determination were used to establish a relationship between the fluorescence signal and *Chl*. The squared correlation coefficient r^2 of this relationship was 0.942 and 0.938 for the data collected in 2010 and 2011, respectively. With this relationship, the depth profiles of *Chl* were constructed using the depth profiles of measured fluorescence at all stations (see Figure 2 for an example). We note that the analysis and discussion of our results do not, however, rely strongly on exact values of *Chl*, but rather on relative changes in *Chl*.

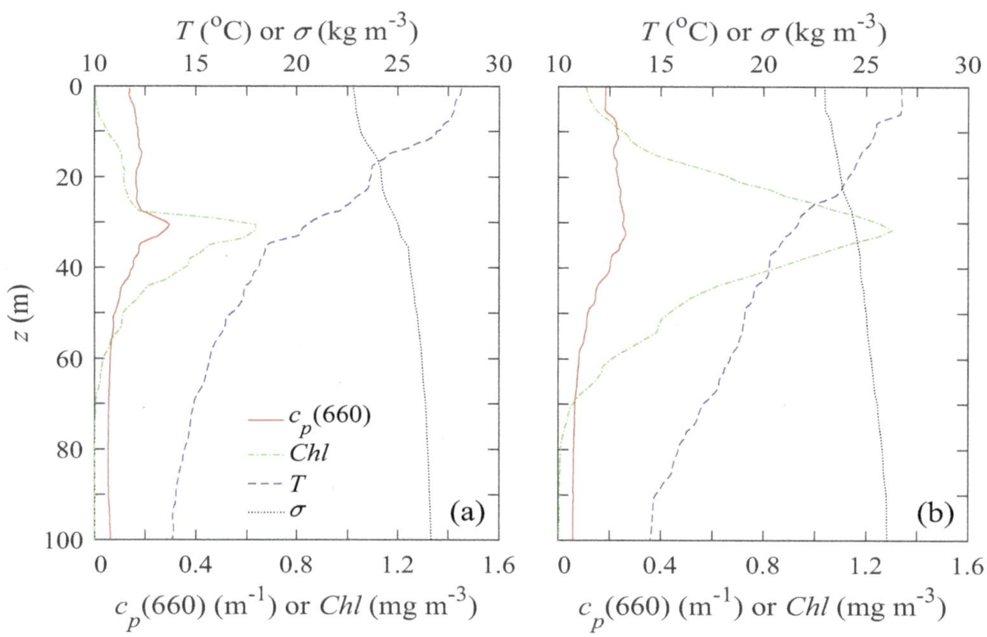

Figure 2. Vertical profiles of water temperature T, density anomaly σ, chlorophyll-*a* concentration *Chl*, and beam attenuation coefficient of particles at 660 nm $c_p(660)$ measured at two contrasting stations. (**a**) Station in the Guaymas Basin (27.54° N, 111.64° W) with lower *Chl* and c_p and smaller solar zenith angle of ~5° and (**b**) station in the Farallon Basin (25.28° N, 109.58° W) with higher *Chl* and c_p and larger solar zenith angle of ~57°.

2.4. Radiometric Quantities and Apparent Optical Properties

Underwater radiometric measurements, including the irradiance quartet, E_d, E_u, E_{od}, and E_{ou} (for all stations) and upwelling zenith radiance L_u (for a few stations in 2011), were collected using an instrument package equipped with five RAMSES sensors (TriOS Mess- und Datentechnik GmbH, Rastede, Germany). Concurrent in-air downwelling plane irradiance, $E_s(\lambda)$, was also measured with an additional sensor mounted on board the ship. These sensors typically acquired signal over the approximate spectral range 320–950 nm (with slight difference among sensors) with ~3.3 nm spectral interval across the spectrum. The radiance sensor had a 20° field of view in air. All five in-water sensors were mounted on a common frame and aligned to the exact same depth level (Figure 3). The in-water package was lowered into water from the sun-exposed side of the vessel at a distance of ~15 m from the vessel to minimize ship-induced light field perturbations [80,81]. Vertical profile measurements were taken at several discrete depths from a few meters below the surface to depths of 50 to 80 m depending on the station. The package was stopped and held every 3–10 m to acquire 5–10 spectral scans at each measurement depth. The integration time of each sensor was automatically adjusted to the light intensity at each measurement depth to ensure an adequate signal to noise ratio. The depth resolution of discrete measurements was highest near the surface and decreased to 10 m below the 20 m depth. Generally, at least one measurement was made near the depth of the *Chl* maximum as indicated by the chlorophyll-*a* fluorescence profile from the CTD cast.

Figure 3. The instrument package used to measure the irradiance quartet and upwelling radiance. As shown, the five radiometric sensors are aligned to the same depth level.

The raw radiometric data acquired in the field were converted to physical units of spectral irradiance (W m^{-2} nm^{-1}) and spectral radiance (W m^{-2} sr^{-1} nm^{-1}). For the underwater measurements, the correction for immersion effect was made for each sensor with the coefficients provided by the manufacturer, which were verified and revised in the laboratory experiment [82]. All measurements were matched to each other by their time stamp and assigned depth. Each spectrum was filtered using the Savitzky-Golay filter in Matlab R2015b to minimize noise, and the linear interpolation was then applied to obtain the final spectral data with a 1 nm spectral interval. The quality of radiometric measurements was best when the irradiance magnitude was higher than the detection limit of 10^{-6} W m^{-2} nm^{-1}. The measurements of L_u below ~30–40 m were often not possible because of insufficient sensitivity of the sensor. All measurements with a signal below the detection limit were excluded from the analysis. As a result, at some stations the greatest depth of measurements of upwelling irradiance in the red spectral region was somewhat smaller (by ~10–20 m) compared with measurements of upwelling irradiance in the blue-green spectral region or measurements of downwelling irradiances. The underwater irradiance and radiance data that passed quality criteria were adjusted to eliminate the effects of temporal changes in sky conditions during vertical profiling with the underwater system. This was accomplished by appropriate normalization of underwater data acquired at different depths using simultaneously acquired data of E_s.

The data of irradiance quartet were used to derive a comprehensive set of AOPs at each discrete depth, including, $\overline{\mu}_d$, $\overline{\mu}_u$, $\overline{\mu}$, R (all dimensionless), and various diffuse attenuation coefficients, collectively denoted with a symbol K (units of m^{-1}). The K coefficients for various irradiances were derived from

$$K_x(z_1 + \Delta z, \lambda) = -\frac{\ln E_x(z_2, \lambda) - \ln E_x(z_1, \lambda)}{z_2 - z_1}, \quad (1)$$

where x represents d, u, od, and ou, z_1 is a depth corresponding to the upper boundary of the layer over which K_x is determined, z_2 is the lower boundary of this layer, and $\Delta z = (z_2 - z_1)/2$. Similarly, other K_x coefficients were also calculated, specifically the diffuse attenuation coefficient of net irradiance, K_E, by replacing E_x with net irradiance, $E = E_d - E_u$, and the diffuse attenuation coefficient of upwelling zenith radiance, K_{Lu}, by replacing E_x with L_u. PAR quantum scalar irradiance, E_{oPAR} (in units of μmol quanta m^{-2} s^{-1}), was calculated by the conversion of spectral E_o from energy units to quantum units and integration between 400 and 700 nm. Similarly, PAR quantum downward irradiance, E_{dPAR}, was calculated from spectral E_d. The corresponding diffuse attenuation coefficients, K_{oPAR} and K_{dPAR}, were calculated by replacing E_x with E_{oPAR} and E_{dPAR}, respectively, in Equation (1).

The spectra of all AOPs were inspected to ensure the quality of data. Because at each depth multiple radiometric measurements were taken, AOPs were initially determined from each radiometric measurement. The unrealistic AOPs, for example $R > 1$ or average cosines > 1, or obvious outliers were then discarded. The abnormal data were often observed near the surface, most likely owing to the effects of wave-induced light fluctuations in E_d and E_{od} [83,84]. After this quality control, the accepted spectra of radiometric quantities were averaged to generate final data of irradiances and radiance for each depth of measurement. The average spectra of irradiances and radiance were then used to compute the final AOPs. Such final data from 11 stations (five in 2010 and six in 2011; see Figure 1) that included valid measurements at more than four depths were selected for further analysis in this study.

2.5. Absorption and Beam Attenuation Coefficients

The non-water absorption coefficient, $a_{nw}(z, \lambda) = a(z, \lambda) - a_w(\lambda)$, and non-water beam attenuation coefficient, $c_{nw}(z, \lambda) = c(z, \lambda) - c_w(\lambda)$, were measured at five stations in 2011 with an ac-9 instrument equipped with nine spectral bands; 412, 440, 488, 510, 532, 555, 650, 676, and 715 nm (WET Labs, Inc., Philomath, OR, USA). $a(z, \lambda)$ and $c(z, \lambda)$ are the total absorption and beam attenuation coefficients of seawater and $a_w(\lambda)$ and $c_w(\lambda)$ represent the pure seawater contributions to these coefficients, respectively. Note that $a_{nw}(z, \lambda)$ is often written as a sum of $a_p(z, \lambda)$ and $a_g(z, \lambda)$ where the subscripts p and g represent the contributions associated with suspended particulate matter and CDOM, respectively. The coefficient $c_{nw}(z, \lambda)$ can be written as a sum of $a_{nw}(z, \lambda)$ and $b_p(z, \lambda)$, where $b_p(z, \lambda)$ represents the particulate component of the total scattering coefficient of seawater, $b(z, \lambda)$.

Vertical profiles of ac-9 data were obtained from the surface to ~500 m depth. The first step of data processing included quality checking of downcast and upcast measurements and rejecting doubtful and excessively noisy data. Because data were often missing for the upcast profiles within the top ~30–50 m layer, only the downcast data were used for further processing, which included temperature and salinity correction [85], subtraction of a baseline determined with Milli-Q water before the cruise, and binning to 0.5 m depth resolution. The $a_{nw}(z, \lambda)$ values were additionally corrected for the scattering error by subtracting the values at 715 nm at all ac-9 wavelengths. After quality control, only two stations (indicated in Figure 1) provided ac-9 data with concurrent radiometric measurements described in Section 2.4.

2.6. Regression Analysis

The presented results of regression analysis are mostly based on the Model I regression method because our objective was to formulate a functional relationship for estimating the ordinate variable y from the abscissa variable x [86]. The ordinary least squares method was applied for all linear relationships and the nonlinear least squares method was applied to all non-linear relationships. The regression calculations were performed using standard fitting methods in Matlab R2015b.

The ordinary (untransformed) variables x and y were used as input to the fitting procedure. The fitting process in Matlab R2015b also outputs root mean squared error, *RSME*, that is calculated as:

$$RMSE = \sqrt{\frac{\sum_{i=1}^{N} (y_i - x_i)^2}{N - m}}, \quad (2)$$

where N is the total number of data points and m is the number of coefficients fitted via the regression analysis.

In one case of our data analysis, we applied the Model II regression method in which the relationship was estimated using the slope of major (principal) axis of the bivariate dataset (Section 3.3). In this specific analysis we compared the estimates of the absorption coefficient obtained with two different experimental methods based on irradiance quartet and ac-9 measurements.

3. Results and Discussion

3.1. Contrasting Examples of Light Field Characteristics and AOPs

Our measurements show that the bio-optical properties within the Gulf of California exhibited a large range of variability, which is summarized in terms of *Chl* and $c_p(660)$ data in Table 2. The examined water column extended from the surface to a maximum measurement depth of 80 m. The depth of euphotic layer, z_{eu}, was determined using a conventional criterion by estimating a depth at which the surface PAR (i.e., E_{oPAR}) was reduced to 1%. The depth z_{eu} ranged from about 30 to 60 m depending on time and location of measurement. Typically, our measurements (at least in the blue-green spectral region) extended ~10–30 m below z_{eu}, and these maximum measurement depths corresponded to ~0.1–0.5% of surface E_{oPAR}. Because phytoplankton have specific energetic demand rather than demand for percentage of light, it is reasonable to expect that the compensation depth of phytoplankton photosynthesis for specific latitude, season, and optical water types investigated in our study was greater than z_{eu}, probably closer to 0.1% of surface PAR [87]. Therefore, we refer to the entire investigated water column as the euphotic layer.

Table 2. Summary of data characterizing the shallowest and greatest depths of radiometric measurements, the depth of chlorophyll-*a* maximum, and the corresponding chlorophyll-*a* concentration, *Chl*, and particulate beam attenuation coefficient at 660 nm, $c_p(660)$, for these depths based on all stations during the 2010 and 2011 cruises in the Gulf of California. For each of these variables the average value ± standard deviation with the minimum and maximum values in parenthesis are given. In addition, the minimum and maximum values of the solar zenith angle θ_s and the data characterizing the euphotic depth z_{eu} (the average value ± standard deviation with the minimum and maximum values in parenthesis) are shown for all stations.

Depth of Measurement	Shallowest	Greatest	*Chl* Maximum
z (m)	5.4 ± 2.9 (1.2, 10.5)	64.9 ± 12.1 (47.6, 80.7)	33.0 ± 9.9 (20.4, 50.5)
Chl (mg m^{-3})	0.067 ± 0.067 (0.005, 0.24)	0.13 ± 0.15 (0.004, 0.42)	0.81 ± 0.54 (0.25, 2.17)
$c_p(660)$ (m^{-1})	0.173 ± 0.035 (0.103, 0.238)	0.089 ± 0.048 (0.043, 0.202)	0.239 ± 0.076 (0.125, 0.405)
Solar zenith angle θ_s (degrees)	0.5–57.1		
Euphotic depth z_{eu} (m)	40.5 ± 8.9 (29.3, 60.0)		

Within the examined water column, *Chl* varied from extremely low levels on the order of 10^{-3} to more than 2 mg m^{-3} with the highest values observed within the *Chl* maximum (Chl_{max}) layer and the lowest values either near the surface or at the deepest measurement depths (~60–80 m). The shape of the $Chl(z)$ vertical profile also changed significantly within the study area. The depth z_{Chlmax} of Chl_{max} ranged from about 20 m to 50 m. The full width half maximum (FWHM) of the Chl_{max} layer, when fitted to a Gaussian function, ranged from 5 m to 30 m. Significant variations spanning one order of magnitude from about 0.04 to 0.4 m^{-1} were also observed in $c_p(z, 660)$. In addition, our measurements were conducted under a broad range of the solar zenith angle, θ_s, varying between 0.5° (i.e., the sun nearly at zenith) to 57.1°. Such broad range of environmental conditions is evident in terms of diverse scenarios in the measured light field characteristics and AOPs.

In this section we discuss two contrasting scenarios of light fields and AOPs within the euphotic layer, one was located in the Guaymas Basin which was observed in 2011 (Figure 2a) and the other in the Farallon Basin which was observed in 2010 (Figure 2b). Figure 2 compares the vertical profiles of T, σ, *Chl*, and $c_p(660)$ at both stations. These two stations have highly contrasting *Chl* profiles and also exhibit noticeable differences in the $c_p(660)$ profiles. Although both stations had a similar depth of Chl_{max} (~30 m) the magnitude and FWHM of the Chl_{max} layer were very different. Specifically, the magnitude of Chl_{max} in the Farallon Basin (1.3 mg m^{-3}) was about twice as high as compared with that observed in the Guaymas Basin (0.64 mg m^{-3}). The FWHM was also considerably larger in the Farallon Basin, i.e., 25 m vs. 10 m in the Guaymas Basin. In general, the magnitudes of $c_p(660)$ for these two stations did not differ significantly but the vertical distributions were different. In particular, there was a noticeable local maximum of $c_p(660)$ near the depth of Chl_{max} in the Guaymas Basin, and no similar feature was observed in the Farallon Basin.

The observed optical differences between the stations could have been caused by differences in the characteristics of suspended particulate matter such as organic and mineral composition. In addition, whereas the measurements were made under clear skies, the solar zenith angle was dramatically different at these two stations, 5.5° in the Guaymas Basin (the measurement was taken around noon at 12:45 p.m. local time) and 57.1° in the Farallon Basin (the measurement taken late afternoon at 4:35 p.m.). Thus, the comparison of these two stations provides an illustration of differences in the light field and AOPs, which were affected not only by water optical properties but also solar zenith angle. We will first present the data of irradiance quartet (Figure 4) and AOPs (Figures 5 and 6) for the station in the Guaymas Basin, and then the AOPs for the station in the Farallon Basin (Figure 7).

The spectra of irradiances, E_d, E_u, E_{od}, and E_{ou}, at several discrete depths within the top 70 m of the water column are presented in Figure 4a–d. These results illustrate the well-known general pattern of the vertical change in the irradiance spectra in the upper ocean, including a decrease in the magnitude and narrowing of the spectral distribution with an increase in depth, e.g., [11]. In the investigated waters, all four irradiances exhibited well-pronounced maximum at a wavelength of about 495 nm at the deepest measurement depth of 67 m, but the spectral distributions of downwelling light, E_d and E_{od}, were clearly broader with relatively more light in the green compared with the distributions for the upwelling light, E_u and E_{ou}.

A remarkable feature associated with vertically non-uniform fluorescence of chlorophyll-*a* was also evident in the irradiance spectra. At a depth of 29 m that coincides with z_{Chlmax} the spectra of E_d and E_{od} showed a distinct maximum in the spectral band of maximum fluorescence centered around 683 nm. This maximum was even more pronounced in the spectra of E_u and E_{ou}. This feature weakened greatly for E_d and E_{od} at depths shallower or deeper than z_{Chlmax}, but was preserved in the spectra of E_u and E_{ou} throughout the water column, which was expected as long as *Chl* did not drop to a very low level.

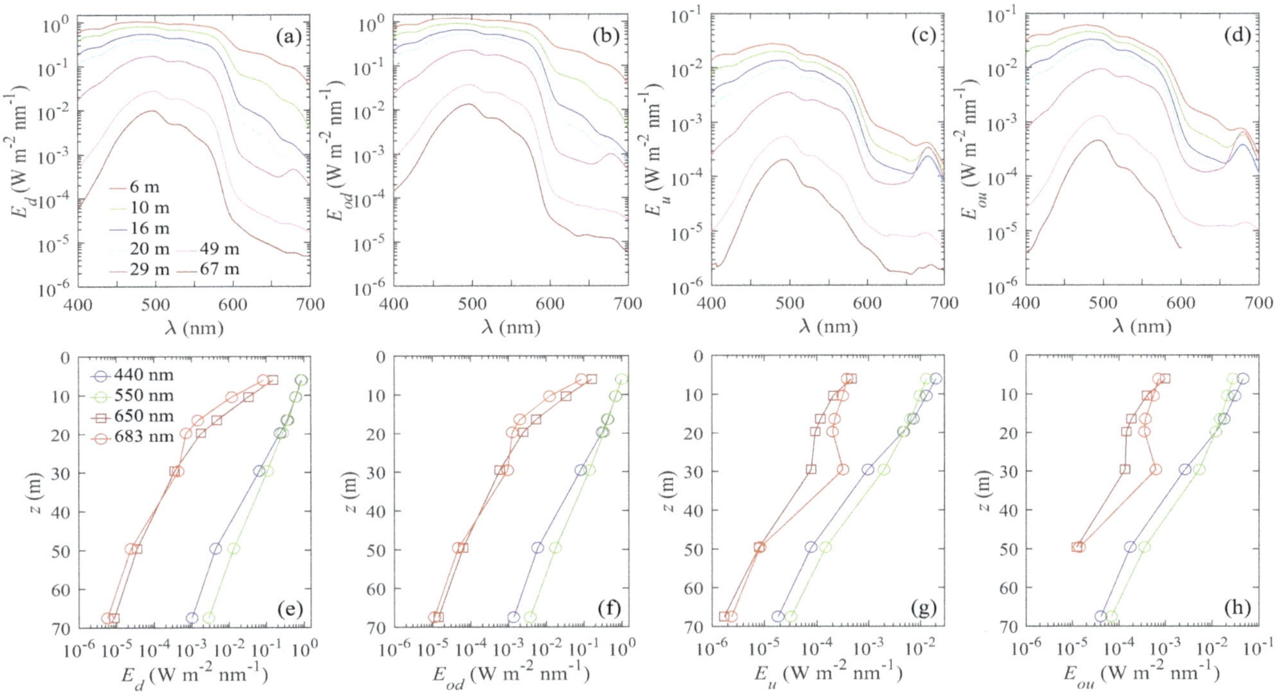

Figure 4. Spectra and vertical profiles of the irradiance quartet, E_d (**a**,**e**), E_{od} (**b**,**f**), E_u (**c**,**g**), and E_{ou} (**d**,**h**), measured at the station in the Guaymas Basin shown in Figure 2a. In panels (**a**–**d**) different colors represent the measurement depths as indicated in (**a**). In (**e**–**h**) different colors and symbols represent selected light wavelengths as indicated in (**e**).

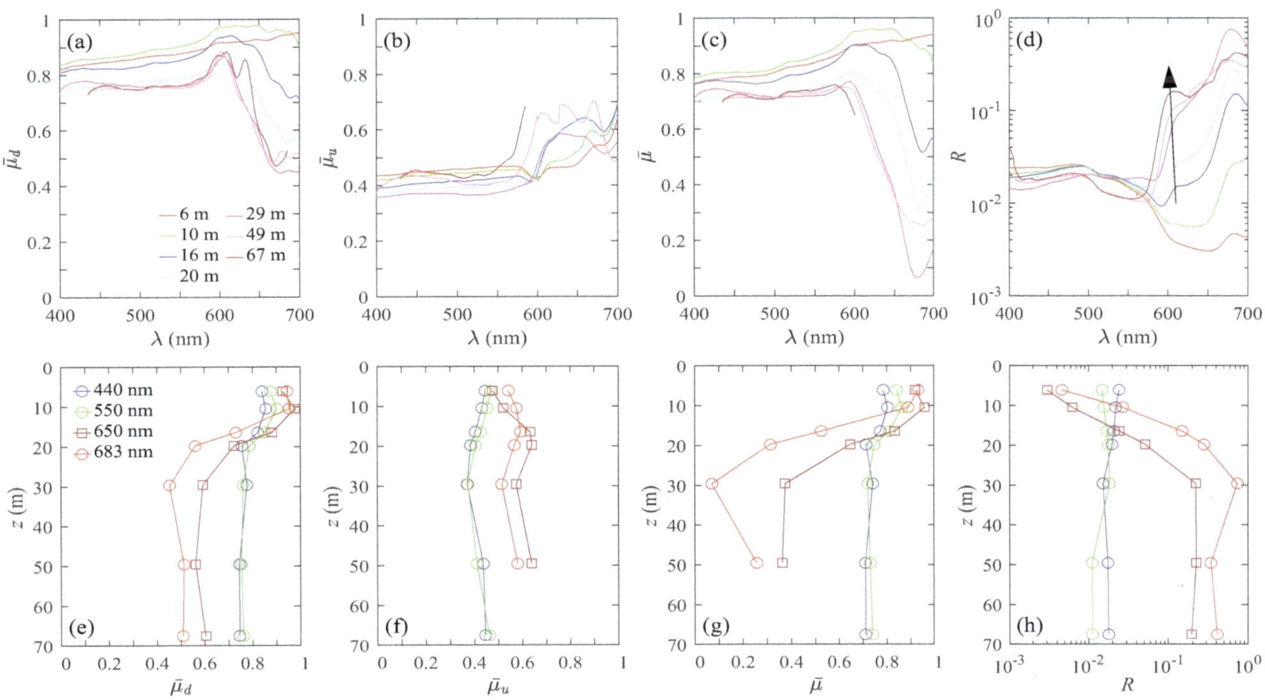

Figure 5. Spectra and vertical profiles of average cosines of underwater light field, $\overline{\mu}_d$ (**a**,**e**), $\overline{\mu}_u$ (**b**,**f**) and $\overline{\mu}$ (**c**,**g**) as well as irradiance reflectance, R (**d**,**h**), derived from the irradiance quartet for Guaymas Basin shown in Figure 4. In panels (**a**–**d**) different colors represent depths as indicated in (**a**). The black arrow in (**d**) indicates a spectral feature caused by Raman scattering of water molecules. In (**e**–**h**) different colors and symbols represent light wavelengths as indicated in (**e**).

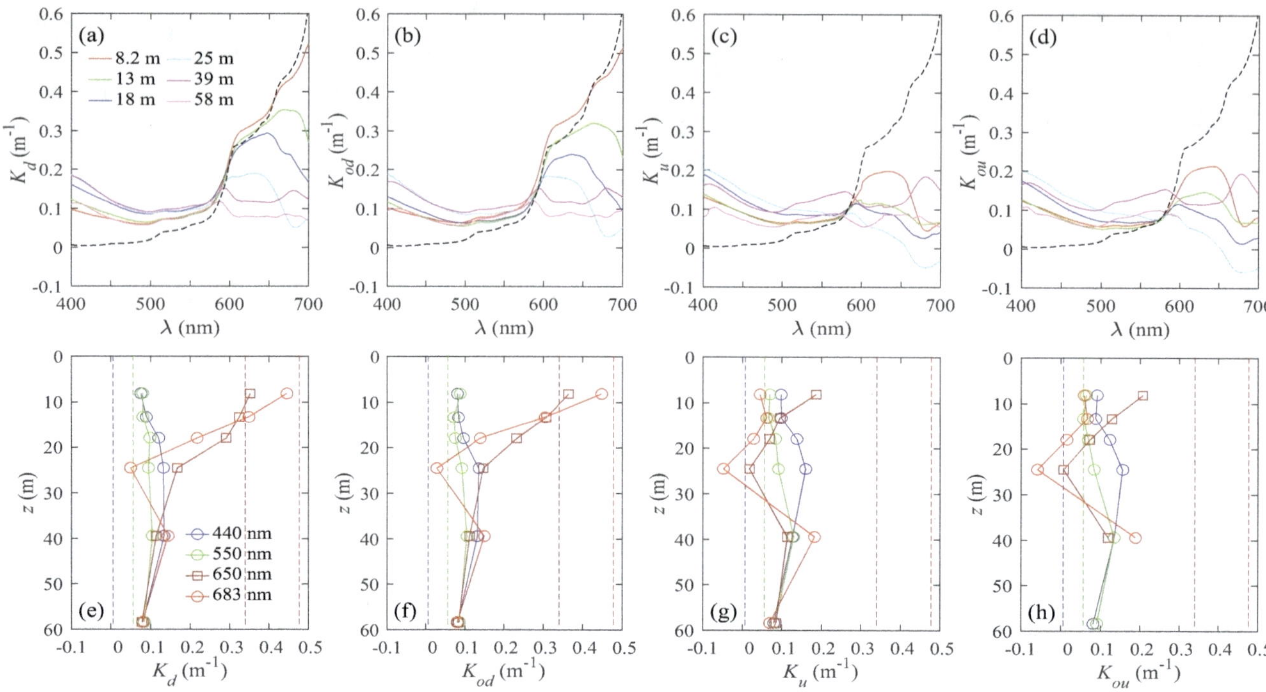

Figure 6. Spectra and vertical profiles of diffuse attenuation coefficients, K_d (**a**,**e**), K_{od} (**b**,**f**), K_u (**c**,**g**), and K_{ou} (**d**,**h**), derived from the irradiance quartet for Guaymas Basin in Figure 4. In panels (**a**–**d**) different colors represent the mid-point depths of the layers within which the K coefficients were determined as indicated in panel (**a**). Dashed lines represent the spectrum of pure water absorption coefficient, which is a theoretical minimum of K for a hypothetical case when no inelastic processes and true emission sources are present. In (**e**–**h**) different colors and symbols represent the light wavelengths as indicated in panel (**e**) and dashed lines represent pure water absorption at respective wavelengths.

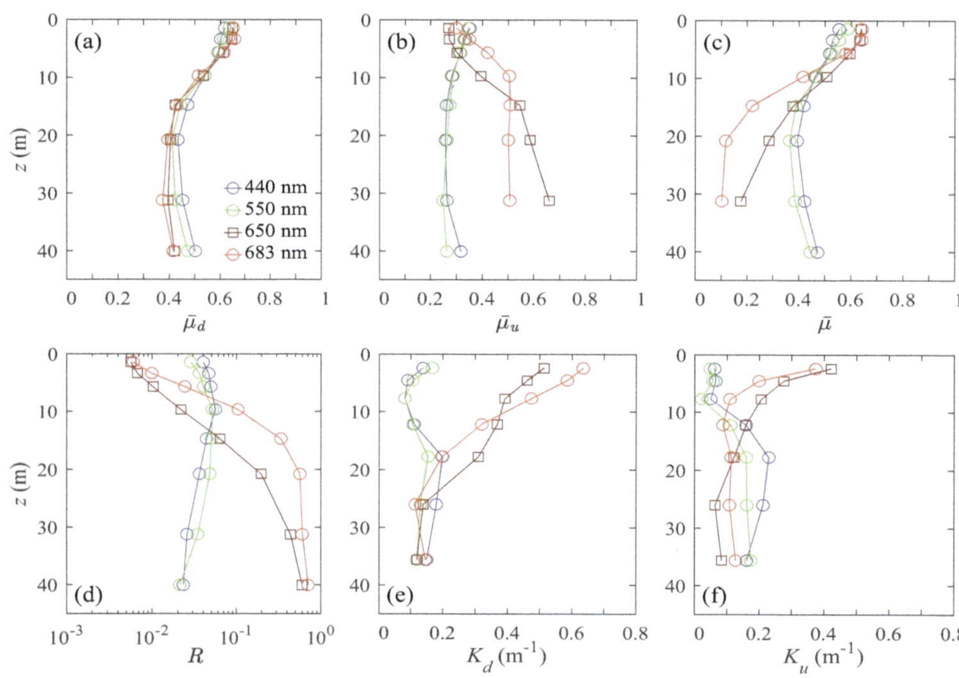

Figure 7. Vertical profiles of average cosines, irradiance reflectance, and diffuse attenuation coefficients for the station in the Farallon Basin shown in Figure 2b. Data for $\overline{\mu}_d$ panel (**a**); $\overline{\mu}_u$ (**b**); $\overline{\mu}$ (**c**); R (**d**); K_d (**e**); and K_u (**f**) are shown for selected light wavelengths as indicated in panel (**a**).

The effects of vertically non-uniform IOPs and chlorophyll-*a* fluorescence were also clearly revealed in the depth profiles of the irradiance quartet at four selected wavelengths (Figure 4e–h). Because the unscattered and elastically scattered light in the red spectral region was efficiently removed within the near-surface layer owing to strong absorption by water molecules, the vertical profiles in the red (650 and 683 nm) were affected by inelastic processes (Raman scattering and chlorophyll-*a* fluorescence) to much greater extent compared with shorter wavelengths, and hence differed significantly from the profiles in the blue and green (440 and 550 nm). The slope of the profiles in the blue and green showed relatively small variations with depth, which were driven primarily by vertical inhomogeneity in IOPs associated with the strength of absorption and elastic scattering processes. For example, at 440 nm all four irradiances showed steeper profiles (i.e., higher attenuation rate with depth) within the $Chl_{\max}$ layer, which can be attributed largely to enhanced absorption of blue light by phytoplankton. In contrast, the slope of irradiance profiles in the red varied within the water column to much greater extent compared with shorter wavelengths. For example, in the 650 nm band the profiles of E_d and E_{od} became gradually less steep with depth, which can be attributed largely to the increasing contribution of Raman scattering. In the 683 nm band where the contribution of chlorophyll-*a* fluorescence was strongest, E_d and E_{od} remained nearly constant with increasing depth within the upper portion of the $Chl_{\max}$ layer (note that no data is available to resolve the depth variations throughout the lower portion of the $Chl_{\max}$ layer as the next data point at 50 m was near the base of the $Chl_{\max}$ layer). The effects of inelastic processes were, however, most pronounced for E_u and E_{ou} in the red spectral region. At 683 nm the effect of chlorophyll-*a* fluorescence on the upwelling light field was so strong that both E_u and E_{ou} increased with depth within the upper portion of the $Chl_{\max}$ layer. Although Raman scattering is known to have very important contribution to the upwelling light in the red [9,60,64,71,88], our data show that the fluorescence effect in the 683 nm band can be dominant in the presence of $Chl_{\max}$ layer. At 650 nm E_u and E_{ou} remain nearly constant within the upper portion of the $Chl_{\max}$ layer. This suggests that the fluorescence fingerprint in the upwelling irradiances is still noticeable at this wavelength, which can be attributed to relatively broad emission band of chlorophyll-*a* fluorescence (FWHM of ~20–25 nm) [89,90].

The vertical variations in the spectral AOPs derived from the irradiance quartet measurements reveal features that were clearly associated with the effects of inelastic processes in the presence of non-uniform profiles of *Chl* and IOPs (Figures 5 and 6). The spectra and depth profiles of the average cosines, $\overline{\mu}_d$, $\overline{\mu}_u$, and $\overline{\mu}$, were relatively weakly variable within the blue-green spectral region where the effects of inelastic processes are much weaker compared with the long-wavelength portion of the spectrum (Figure 5a–c,e–g). In the spectral range of 400–580 nm, the values of $\overline{\mu}_d$ and $\overline{\mu}$ were in the range 0.7–0.9 with higher values observed at shallower depths and related to the high solar elevation in the clear sky during the measurement. Thus, the downwelling light field or the entire light field was characterized by the predominant downward direction of light propagation within 45° of the vertical. The values of $\overline{\mu}_u$ in the blue-green were around 0.4 suggesting more uniform angular distribution of upwelling light field. Our data of average cosines in the blue-green are in the range of theoretical and experimental values reported in other studies [9,10,29,52,60,73]. In the red spectral region, the average cosines exhibited large variations owing to the effects of inelastic processes within the non-uniform water column. We observed a large rapid decrease in $\overline{\mu}_d$ and $\overline{\mu}$ below the near-surface layer (below 10 m) to about 0.5 and 0.05, respectively, at depths ≥20 m. For $\overline{\mu}_u$ in the red we observed a variation between about 0.45 and 0.65. The observed values of average cosines in the red differ significantly from those which would have occurred if inelastic processes were absent [9,73]. In the presence of inelastic processes, the light field in the red approached a more uniform distribution with increasing depth quite rapidly, and could achieve a nearly isotropic distribution within moderately strong Chl_{max} layer at relatively shallow depths, such as ~30 m in the examined case (Figure 5e–g). Note also that, unlike Raman scattering that leads gradually to a more uniform light field in the red with increasing depth, the vertically non-uniform chlorophyll-*a* fluorescence caused additional depth variations in the angular distribution of light field within the 683 nm band. In contrast, in the blue-green region where the effects of inelastic processes are much weaker, the light field remained highly directional throughout the examined euphotic layer (Figure 5e–g).

Similar to the average cosines, the spectral irradiance reflectance R (Figure 5d,h) showed relatively little variation with depth in the blue-green spectral region as the R values ranged between 0.01 and 0.025. In contrast, R exhibited very large vertical variations at wavelengths longer than about 580 nm. In the red R varied from about 0.003 at 6 m to 0.8 at z_{Chlmax} = 29 m, and relatively high values (0.2–0.4) persisted below the Chl_{max} layer (Figure 5h). Similar observations were previously reported by Dirks and Spitzer [37]. This result was caused by inelastic processes, including a dominant effect of chlorophyll-*a* fluorescence that produces a maximum in the R spectra within the fluorescence emission band centered at 683 nm (Figure 5d). At shorter wavelengths within the approximate range 580–650 nm that was away from significant effect of chlorophyll-*a* fluorescence, Raman scattering appeared to exert a dominant influence on R throughout much of the examined water column. Owing to this inelastic process a shoulder-like feature was seen in the R spectra around 600 nm at intermediate measurement depths (16–49 m), which developed into a distinct maximum with further increase of depth to 67 m (Figure 5d). This pattern is consistent with an increasing role of Raman scattering with depth [37,69,70,73]. These results also suggest that the effects of Raman scattering were more easily detectable or distinguishable from the effects of chlorophyll-*a* fluorescence in the data of AOPs than in the irradiance data.

Figure 6 shows the attenuation coefficients for the four irradiances, specifically K_d, K_u, K_{od}, and K_{ou}, as derived from the measurements of irradiance quartet, E_d, E_u, E_{od}, and E_{ou}, respectively. Similar to other AOPs discussed above, the K-coefficients were less variable in the blue-green than in the red portion of the spectrum, both in terms of the spectral shape (Figure 6a–d) and vertical profiles (Figure 6e–h). One of the most remarkable features was that the K-coefficients in the red (more specifically for λ > 580–600 nm) were generally smaller than the pure water absorption coefficient, a_w. This feature resulted from inelastic processes and was more pronounced for K_u and K_{ou} than K_d and K_{od}. In the chlorophyll-*a* fluorescence band, K_d and K_{od} approached zero in the upper portion of the Chl_{max} layer (Figure 6e,f), which is consistent with an earlier observation that E_d and E_{od} are nearly constant within that layer (Figure 4e,g). A stronger effect within the 683 nm band was observed for the K_u and K_{ou} coefficients that assume negative values in the upper portion of the Chl_{max} layer (Figure 6g,h). This is again consistent with the measurements of E_u and E_{ou} which both increased with depth within that layer (Figure 4f,h). Similar increase of E_u in the red band (685 nm) with depth within the Chl_{max} layer was previously observed by Dirks and Spitzer [37]. The presence of negative values for the attenuation of upwelling irradiance was also demonstrated through RT simulations of light field within highly scattering near-surface layer of bubble clouds entrained by wave breaking [91]. Other interesting features of the long-wavelength portion of K-spectra include a maximum at 683 nm below z_{Chlmax} (for all K-coefficients, see Figure 6a–d) or a maximum that could peak at wavelengths longer than about 600 nm but shorter than 683 nm at depths above z_{Chlmax} (especially for K_d and K_{od}, see Figure 6a,b). These features arise from differential spectral effects of inelastic processes as a function of increasing wavelength. It is also interesting to note that a distinct feature at these intermediate depths was observed in the spectra of $\overline{\mu}_d$ and $\overline{\mu}$ (Figure 5b,d). This can be explained by higher attenuation coefficient leading to more light concentrated around the vertical direction.

In contrast to the red portion of the spectrum, in the blue-green spectral region where the effects of inelastic processes are much weaker, the K-coefficients were higher than $a_w(\lambda)$. At these wavelengths, especially at 440 nm, the vertical profiles of K-coefficients exhibited maximum values within the Chl_{max} layer, and hence generally resembled the shape of $Chl(z)$ profile. It is also noteworthy that the K-coefficients at all different wavelengths became nearly identical at the greatest depth of about 60 m used in these determinations (Figure 6e–h). This was primarily because the effect of Raman scattering generally led to a gradual decrease of K-coefficients in the red, for example at 650 nm, with increasing depth until approaching the level of K-coefficients in the blue and green. This pattern that eventually led to relatively flat K-coefficients has been demonstrated previously in experimental data and theoretical results from RT simulations [37,64,73]. It is, however, important to emphasize that we do not suggest that the light field in the examined case has reached an asymptotic regime within the deepest portion of the examined layer (i.e., ~60–70 m) where we observe such similarity of K-coefficients and relatively constant AOPs with depth. While such patterns can be associated with the asymptotic regime, previous deep-sea RT simulations demonstrated that a nearly asymptotic light field could generally be reached only at significantly greater depths within the ocean mesopelagic zone [64,73].

For comparison with the results discussed above in the Guaymas Basin, Figure 7 shows the vertical profiles of AOPs obtained from measurements at a station in the Farallon Basin where the Chl_{max} layer was significantly broader with a two-fold higher Chl_{max} and the solar elevation was significantly lower (see Figure 2). As a result of these differences in the environmental conditions, which could also include differences in the IOPs, some features of AOPs measured in the Farallon Basin were different compared to those in the Guaymas Basin. We note, however, that the data from the Farallon Basin do not extend throughout the entire Chl_{max} layer but end at 40 m which is about 10 m below z_{Chlmax}. The values of the average cosines in the blue-green spectral region in the Farallon Basin were much lower (Figure 7a–c) than those in the Guaymas Basin (Figure 5f–h). The lower values of $\overline{\mu}_d$ and $\overline{\mu}$ in the blue and green can be explained by the higher solar zenith angle and slightly higher IOPs as indicated by the $c_p(z, 660)$ profiles (Figure 2b). This result is consistent with previous experimental and modeling results [29,60,92–94]. The observation of very low values (0.25–0.35) of $\overline{\mu}_u$ in the blue and green (Figure 7b) will be discussed is some detail in Section 3.2.1, but it is interesting to note that the values less than 0.35 were previously reported on the basis of RT simulations of highly scattering environment within near-surface bubble clouds [91]. In the red where the effects of inelastic processes are strong, the vertical profiles of average cosines exhibited similar features to those described earlier for the station in the Guaymas Basin. Overall, our data of the average cosines support the notion that sun position in a clear sky sets the values of $\overline{\mu}_d$ and $\overline{\mu}$ near the ocean surface, which then change with depth in a manner dependent on the interplay of the vertical structure of IOPs and inelastic processes, and $\overline{\mu}_u$ may vary within a broader range than commonly assumed. It is thus important to emphasize that the average cosines are sensitive to surface illumination, especially for the downwelling and total light in the near-surface layer, so these average cosines may potentially serve as AOPs or useful descriptors of near-surface water bodies only under specified surface boundary conditions.

The vertical profiles of R in the Farallon Basin (Figure 7d) showed patterns that were similar to those observed in the Guaymas Basin (Figure 5e), but the R values were generally higher because of higher Chl and water turbidity at the station in the Farallon Basin. The properties of particulate assemblages, including the composition and size distribution, could have also been different at these stations. The K_d and K_u values within the Chl_{max} layer in the Farallon Basin (Figure 7e,f) were not as low as in the examined case from the Guaymas Basin (Figure 6e,g). This observation also applies to K_{od} and K_{ou} (not shown in Figure 7). In particular, unlike in the Guaymas Basin case, the chlorophyll-a fluorescence within the much stronger Chl_{max} layer in the Faralllon Basin did not produce a minimum of K_u with negative values around the depth of z_{Chlmax} (Figure 7f). This lack of distinct negative minimum of K_u can be attributed to the weakening of the effects of inelastic processes when the water column becomes more turbid, as previously demonstrated by RT simulations [60]. In addition, it is also likely that the influence of Chl_{max} is less profound when the FWHM of the Chl_{max} layer is large, which is the case for the station in the Farallon Basin.

3.2. Overall Variability and Relationships Involving the Apparent Optical Properties

In this section we examine the overall variability of AOPs and the relationships involving AOPs and some environmental factors (depth, solar zenith angle, Chl) using the entire dataset collected in the Gulf of California.

3.2.1. Average Cosines

As indicated earlier the average cosines serve as simple proxies of the angular structure of the light field and they have or have not been treated as AOPs by different investigators in the past [9,10,46,48]. Based on our measurements of irradiance quartet the average cosine of underwater light field, $\overline{\mu}$, ranged from about 0.38 to 0.88 in the blue and green spectral regions. In the red, the range was much larger, from 0.07 to 0.94, owing to strong effects of inelastic processes. The average cosine for downwelling light field, $\overline{\mu}_d$, varied from 0.41 to 0.91 in the blue and green and 0.37 to 0.98 in the red. The approximate range for the average cosine for the upwelling light field, $\overline{\mu}_u$, was 0.2–0.5 in the blue and green and 0.27–0.75 in the red. Our data are consistent but generally exhibit a broader range of values compared with relatively few experimental data of average cosines reported in literature. Højerslev [18] reported on measurements of E_o, E_d, and E_u in the Mediterranean Sea and Norwegian fjord. Using his data, we calculated that $\overline{\mu}$ was in the range of 0.47–0.95 in the blue-green spectral region (427, 477, and 532 nm) between the sea surface and 70 m depth, and 0.7–0.97 in the red (633 nm) within top 20 m layer. Spitzer and Wernand [52] reported on measurements of $\overline{\mu}$ in the range 0.7–0.9 at wavelengths shorter than 600 nm and depths between 24 and 55 m. Lewis et al. [29] measured the full angular distribution of radiance at a single wavelength of 555 nm in surface waters of the Pacific Ocean off Hawaii Islands, Santa Barbara Channel, and in Bedford Basin, and used these measurements to determine the ranges for $\overline{\mu}$, $\overline{\mu}_d$, and $\overline{\mu}_u$, which were approximately 0.41–0.82, 0.5–0.94, and 0.2–0.48, respectively. Our determinations in the blue and green are generally consistent with the data from literature. The comparison for the red spectral region is, however, more limited because only few fragmentary data in the red have been previously reported. Our measurements of the average cosines in the red exhibited a very wide range of variability within the water column, providing a more comprehensive characterization of the average cosines that are strongly affected by inelastic processes in this spectral region.

It has long been recognized that at near-surface depths in relatively clear ocean waters, the downwelling radiance distribution, and hence $\overline{\mu}_d$, are strongly dependent on the sun position in clear sky or more generally on sky conditions, for example clear skies vs. overcast skies [11]. Such sensitivity of the downwelling average cosine to surface illumination conditions cautions against its indiscriminate use as a general descriptor (in the sense of AOPs) of water bodies, especially in the surface layers, and emphasizes a need for specifying surface boundary conditions. Our data collected within the top 10 m layer in the Gulf of California support the general trend of a decrease in near-surface $\overline{\mu}_d$ with increasing solar zenith angle regardless of light wavelength (Figure 8a). For comparison, the best fit lines to experimental data of Aas and Højerslev [94] obtained at a 5 m depth (Mediterranean Sea) in the blue (465 and 474 nm) and experimental data of Lewis et al. [29] at a 5 m depth (open ocean waters in the Pacific) in the green (555 nm) are also shown in Figure 8a. In contrast, our data of $\overline{\mu}_u$ at near-surface depths did not exhibit a clear dependence on the solar zenith angle (Figure 8b). We note, however, that Aas and Højerslev [94] and Lewis et al. [29] suggested a very weak dependence based on their data, as indicted in Figure 8b.

The increased role of inelastic processes, especially Raman scattering, with increasing depth is illustrated for the average cosines in the red part of the spectrum in Figure 9. This effect was particularly well-pronounced for the downwelling light field in terms of a well-behaved trend of a decrease in $\overline{\mu}_d$ with increasing depth towards $\overline{\mu}_d$ = 0.5, which is clearly seen in our data at the two examined wavelengths in the red, 650 nm and 683 nm (Figure 9a). The best-fit exponential functions to the data are:

$$\overline{\mu}_d(650) = 0.5058 \exp(-0.0460\, z) + 0.5140\ (r^2 = 0.828, RMSE = 0.061, N = 65); \quad (3a)$$

$$\overline{\mu}_d(683) = 0.6512 \exp(-0.0715\, z) + 0.4659\ (r^2 = 0.855, RMSE = 0.067, N = 65), \quad (3b)$$

where r^2 is the determination coefficient and N the number of observations. A distinct group of outlying data marked in Figure 9a was collected at significantly larger solar zenith angle (θ_s = 57.1°)

than the remaining data. These outliers were not included in the regression analysis. Whereas at 650 nm Raman scattering is expected to have a dominant effect, at 683 nm chlorophyll-*a* fluorescence can have an additional significant or even dominant effect at certain depths, depending on chlorophyll-*a* concentration and its variation within the water column. The combined effects of Raman scattering and chlorophyll-*a* fluorescence at 683 nm result in faster decrease of $\overline{\mu}_d(683)$ with depth from the near surface through the lower portion of the Chl_{max} layer compared with $\overline{\mu}_d(650)$. In addition, in our dataset the interplay of vertically variable chlorophyll-*a* fluorescence and Raman scattering throughout the water column does not seem to have a noticeable deteriorating effect on the strength of the relationship $\overline{\mu}_d(683)$ vs. z as compared with $\overline{\mu}_d(650)$ vs. z.

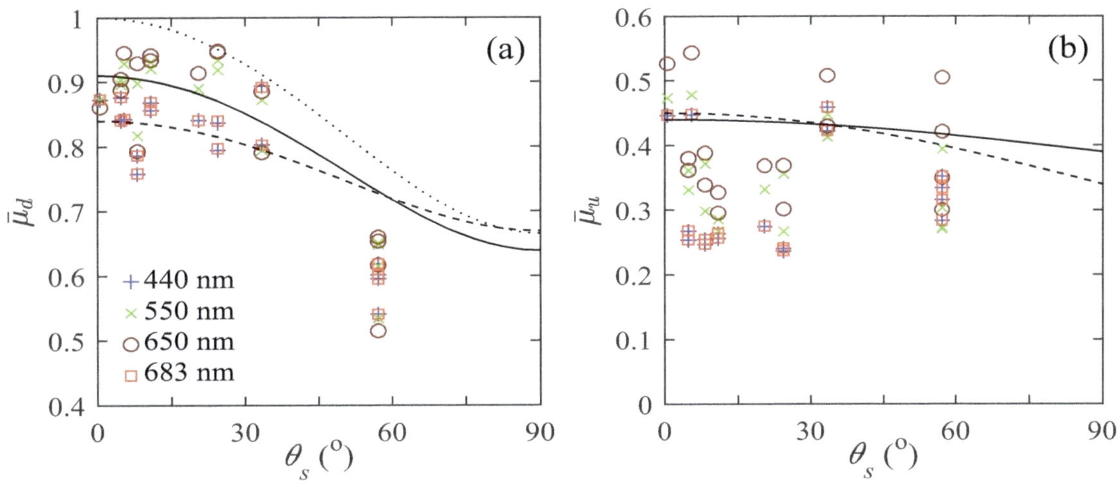

Figure 8. (**a**) Relationship between $\overline{\mu}_d$ at selected light wavelengths as indicated and solar zenith angle θ_s. (**b**) Same as (**a**) but for $\overline{\mu}_u$. The presented data were collected at all stations within the top 10 m layer. For comparison, solid lines in (**a**,**b**) represent the relationship reported by Lewis et al. [29] and dashed lines by Aas and Højerslev [94]. Dotted line in (**a**) represents $\overline{\mu}_d = \cos(\theta_{sw})$ where θ_{sw} is the zenith angle of the refracted solar beam just beneath the ocean surface obtained from Snell's law.

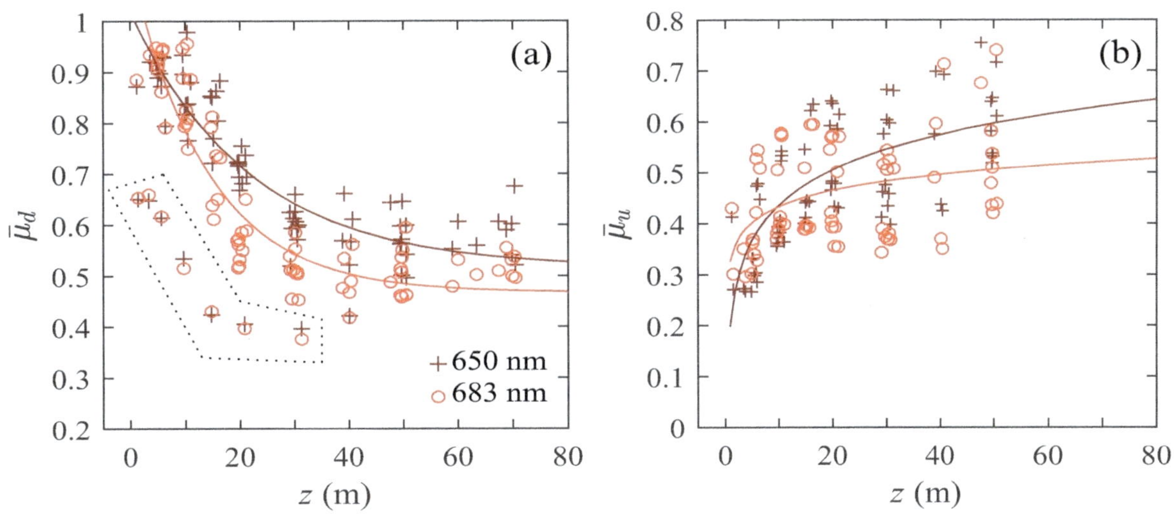

Figure 9. Average cosines in the red spectral bands (650 and 683 nm) plotted as a function of depth z using data from all stations and depths. (**a**) $\overline{\mu}_d(z, 650)$ (crosses) and $\overline{\mu}_d$ $(z, 683)$ (open circles) vs. z. (**b**) Same as (**a**) but for $\overline{\mu}_u$. Data points surrounded by dotted lines in (**a**) are for the station in the Farallon Basin with the solar zenith angle significantly larger (~57°) than at other stations. Solid lines show the best-fit regression functions for the average cosines vs. depth.

In contrast to $\overline{\mu}_d$, $\overline{\mu}_u$ in the red did not exhibit such strong dependence on depth although it is clear that $\overline{\mu}_u$ assumed generally lower values near the surface (Figure 9b). The r^2 values for the $\overline{\mu}_u$ vs. z data dropped to 0.488 at 650 nm and 0.147 at 683 nm. The differences in the patterns of depth dependencies for $\overline{\mu}_d$ and $\overline{\mu}_u$ can be attributed to relatively smaller role played by inelastic processes at near-surface depths for the downwelling light field compared with the upwelling light field in the red spectral region. It is also important to point out that in comparison with our dataset the dependence of $\overline{\mu}_d$ on depth is expected to be reduced for datasets collected in turbid waters, at large solar zenith angles, and/or heavily overcast skies. Under such conditions the near-surface values of $\overline{\mu}_d$ are expected to be lower (i.e., closer to 0.5) compared with most of our near-surface data, and hence the expected changes of $\overline{\mu}_d$ with depth would not be as well pronounced. Thus, the strong sensitivity of near surface $\overline{\mu}_d$ to illumination conditions at the sea surface suggests that no single relationship can describe a change of $\overline{\mu}_d$ with depth. This is illustrated in Figure 9a which includes an outlying case of data points collected at significantly larger θ_s than the remaining data, emphasizing a need to pay particular attention to the possible effect of solar zenith angle. In addition, it is important to emphasize that in the blue-green spectral region where the effects of inelastic processes are weaker, the average cosines exhibited much less distinct dependency on depth (see Figures 5e–g and 7a–c).

Our analysis also shows that regardless of light wavelength, none of the average cosines were related significantly to *Chl* in our dataset. The highest r^2 of 0.42 was found between $\overline{\mu}_d(z, 650)$ and $Chl(z)$ (not shown). For $\overline{\mu}_d$ in the blue-green where inelastic processes are less important than in the red, and for $\overline{\mu}_u$ at all examined spectral bands, the relation was much weaker or essentially indiscernible. This lack of relation with *Chl* is not surprising because the average cosine within the water column, if not dominated by inelastic processes, depends on the interplay of the inherent absorption and scattering properties of seawater in rather complex way [85], and *Chl* cannot adequately represent such effects.

Similar patterns observed in our data of $\overline{\mu}$ and $\overline{\mu}_d$ (e.g., Figure 5) indicate that these two average cosines were well correlated. The relationships between $\overline{\mu}$ and $\overline{\mu}_d$ obtained with our data are presented in Figure 10a. Specifically, the linear relationship for the blue-green spectral region as determined from the data collected at various depths and two wavelengths, 440 nm and 550 nm, is:

$$\overline{\mu} = 0.996\ \overline{\mu}_d - 0.0352 \quad \left(r^2 = 0.978, RMSE = 0.042, N = 158\right). \tag{4a}$$

On the basis of data at two wavelengths in the red, 650 and 683 nm, the relationship is:

$$\overline{\mu} = 1.51\ \overline{\mu}_d - 0.472 \quad \left(r^2 = 0.929, RMSE = 0.167, N = 134\right). \tag{4b}$$

These relationships indicate that reasonably good estimates of $\overline{\mu}$ can be obtained from downwelling irradiance measurements. As seen, $\overline{\mu}$ and $\overline{\mu}_d$ in the blue-green were typically close to one another within the examined euphotic layer. In the red, the correlation is marginally lower and still very strong but $\overline{\mu}$ tends to be somewhat smaller than $\overline{\mu}_d$ over a nearly complete data range with the exception of the highest values.

In contrast to $\overline{\mu}$ vs. $\overline{\mu}_d$ relationship, $\overline{\mu}_u$ and $\overline{\mu}_d$ were not correlated or very weakly correlated (Figure 10b). The r^2 values for the blue-green and red spectral regions are 0.026 and 0.109, respectively. Therefore, $\overline{\mu}_u$ cannot be estimated from $\overline{\mu}_d$ or vice versa, which means that a complete irradiance quartet or radiance angular distribution is required for investigating both $\overline{\mu}_u$ and $\overline{\mu}_d$. Figure 10b also reveals that the $\overline{\mu}_u$ data in the blue-green appeared to be grouped in at least two distinct clusters, one with $\overline{\mu}_u > \sim 0.3$ and the other with $\overline{\mu}_u < \sim 0.3$. A large number of $\overline{\mu}_u$ measurements fells in the range from 0.2 to ~0.3, suggesting that much of the upwelling light in these cases propagated relatively close (within ~10°–20°) to the horizontal direction. This is not consistent with a common assumption that the angular distribution of upwelling light field is more isotropic. Nonetheless, such low values of $\overline{\mu}_u$ were also observed by Lewis et al. [29] and reported on the basis of RT simulations within highly-scattering bubble clouds at near-surface depths [91] and for certain combinations of absorption coefficient, scattering coefficient, and scattering phase function in the water column (C. Mobley, personal communication). A closer inspection of our data indicates that the group with $\overline{\mu}_u$ < ~0.3 was generally characterized by relatively larger values of $c_p(z, 660)$, especially within the top 30–40 m, compared to the group with $\overline{\mu}_u$ > ~0.3. This is seen in the example data shown in Figures 3, 5b,f and 7b, supporting the notion that certain suites of IOPs can play an important role for producing low values of $\overline{\mu}_u$.

Although the behavior of average cosines in natural waters can be complex, these quantities are usually assumed to be constant or vertically uniform when used in the context of ocean color remote sensing applications, bio-optical modeling, and/or estimation of water constituents, e.g., [39,95–99]. The assumption that $\overline{\mu}$ obtained from simulations for a homogeneous water column is representative of a stratified water column was also used [100,101]. The present study demonstrates that these simplifying assumptions are not valid. Further studies to determine how such assumptions lead to errors in data products of interest are needed. For example, Sathyendranath and Platt [102] reported that oceanic primary productivity could be systematically underestimated by 5 to 13% when ignoring the angular distribution of the underwater light field, for which the average cosines serve as simple yet useful proxies.

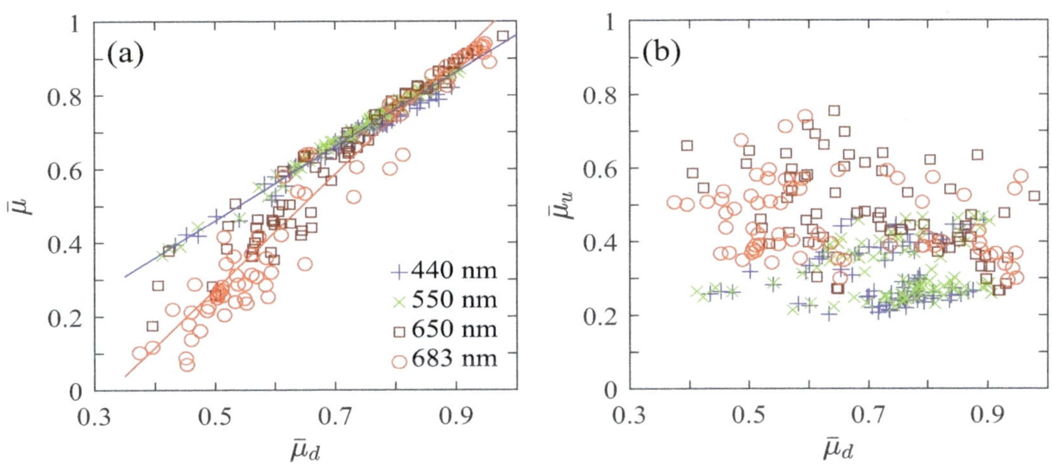

Figure 10. Relationships between average cosines based on measurements from all stations and depths. Data for (**a**) $\overline{\mu}$ vs. $\overline{\mu}_d$ and (**b**) $\overline{\mu}_u$ vs. $\overline{\mu}_d$ are shown for selected light wavelengths as indicated in panel (**a**). In (**a**) data points for the blue and green (440 nm and 550 nm) were combined to determine the best-fit regression function (blue line) and data points for the red (650 nm and 683 nm) were combined to determine the best-fit regression function (red line).

3.2.2. Irradiance Reflectance

Similar to the average cosines, the variations in the irradiance reflectance, R, throughout the euphotic layer are characterized by a much broader range in the red spectral region where the effects of inelastic processes are very strong compared with the blue-green part of the spectrum where these effects are much weaker. Our data of R in the blue-green ranged from 0.006 to 0.056, whereas in the red from 0.0015 near the surface to 0.75 around z_{chlmax}. The observed range in the blue-green is similar to that of 0.006–0.063 based on E_d and E_u measurements reported by Højerslev [18] for 427, 477, 532, and 572 nm between the surface and ~70 m. The R values at 633 nm in Højerslev's dataset range from 0.001 to 0.004, but these values were determined only at shallow depths, typically less than 5 m. Our data in the red extend to much higher values within the euphotic layer owing to strong inelastic effects in this spectral region, which is consistent with a few measurements from undisclosed locations reported by Dirks and Spitzer [37].

As emphasized before, the increasing role of inelastic processes with increasing depth is most significant in the long-wavelength part of the spectrum. This effect is clearly seen in our data of R

plotted as a function of depth for wavelengths of 650 and 683 nm (Figure 11). Both $R(650)$ and $R(683)$ showed gradual increase from the values of the order of 10^{-3} near the surface to more than 0.1 within the mid- to deep portions of the examined water column. These data suggest that depth can serve as a useful proxy for the vertical changes in R in the red part of the spectrum. The best-fit exponential functions to the data are:

$$R(650) = 10^{0.7633\ln(z)-3.6633} \quad (r^2 = 0.820, RMSE = 0.331, N = 74) \tag{5a}$$

$$R(683) = 10^{0.7690\ln(z)-3.2582} \quad (r^2 = 0.794, RMSE = 0.362, N = 74). \tag{5b}$$

Whereas these vertical trends in R reflect primarily the increasing effect of Raman scattering with depth, the effects of chlorophyll-*a* fluorescence were also evident within the 683 nm band. The values of $R(683)$ were comparable to $R(650)$ near the surface but increased faster with depth owing to increasing *Chl* and associated fluorescence until the Chl_{max} was reached. Consequently, $R(683)$ tended to have the highest values at depths close to z_{Chlmax}, which was about 30 m for most investigated stations in this study.

The dominant role of Raman scattering and additional complexities caused by chlorophyll-*a* fluorescence emission in the red spectral region indicate that this part of the spectrum is not well suited for establishing a relationship between R and *Chl* within the euphotic layer. The analysis of our entire dataset including all measurement depths confirmed the lack of relationship between R in the red and *Chl* (not shown). For example, a very low determination coefficient (r^2 in the range of ~0.091–0.155) was found between the log-transformed data of red (650 or 683 nm)-to-green (550 nm) reflectance ratios and *Chl*. In contrast, our data of the blue-to-green reflectance ratio, $R(440)/R(550)$, and *Chl* collected at all measurement depths exhibited a relationship with significantly higher determination coefficient of 0.608 (not shown). Although this may not seem surprising given the common use of the blue-to-green ratio of ocean surface reflectance (typically the remote-sensing reflectance rather than irradiance reflectance) for estimating surface *Chl* in the context of remote sensing applications, e.g., [103,104], it is important to emphasize a distinct difference which is that our data represent irradiance reflectance and *Chl* measured at different discrete depths within the entire euphotic layer down to depths as high as ~80 m.

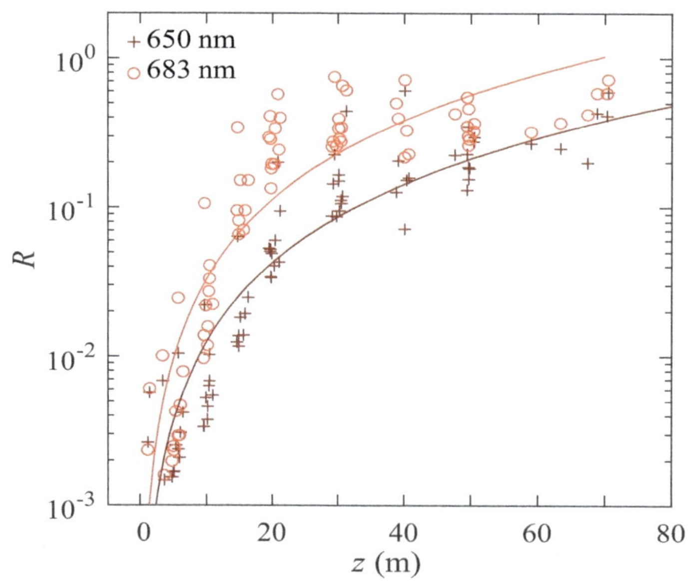

Figure 11. Irradiance reflectance R in the red spectral bands (650 and 683 nm as indicated) plotted as a function of depth z using data from all stations and depths. Solid lines represent the best-fit regression functions.

3.2.3. Diffuse Attenuation Coefficients

The diffuse attenuation coefficients K in our dataset span a relatively wide range of values. For example, the ranges for K_d at 440, 550, 650, and 683 nm are 0.031–0.246, 0.054–0.168, 0.072–0.514, and 0.044–0.635 m^{-1}, respectively. The ranges for K_u at the same wavelengths are 0.041–0.253, 0.021–0.192, 0.018–0.421, and −0.095–0.374 m^{-1}, respectively. The lowest values of K_u at longer wavelengths (green and red) were significantly smaller than those for K_d and can even include the negative values because the inelastic processes exerted a stronger reducing effect on K_u than K_d. The possibility of negative values of the K-coefficients at 685 nm within the Chl_{max} layer were also demonstrated through earlier observations and modeling [37].

Because E_d has been by far the most commonly measured radiometric quantity in the ocean, K_d in the surface layer is relatively well documented in the literature. An overview of early data of K_d, including the classification of optical water types based on the spectral K_d within the top 10 m layer of the ocean, is given in Jerlov [11]. The range of K_d values observed in our study in the blue spectral region indicates that the investigated waters in the Gulf of California fall between the oceanic water Types IB or II and the coastal water Type 1 according to Jerlov's classification. Following an increased interest in ocean color remote sensing in early 1970s and the introduction of the concept of bio-optical state of ocean waters representing a measure of the effect of biological processes on ocean optical properties [105], several studies examined experimental data of K_d in the context of the relationship with chlorophyll-*a* concentration over a broad range of these variables measured in various oceanic waters [35,38,39,106]. In those studies, K_d was typically estimated to represent the top layer from the ocean surface to one attenuation depth ($1/K_d$) or euphotic depth (1% of surface PAR). The lowest limiting values of spectral K_d for the clearest natural waters were also estimated, for example 0.017 m^{-1} or 0.00885 m^{-1} at 440 nm, 0.0648 m^{-1} or 0.05746 m^{-1} at 550 nm, and 0.35 m^{-1} or 0.34052 m^{-1} at 650 nm by Morel and Maritorena [39] and Smith and Baker [107], respectively. These estimates were generally consistent with the lowest measured values in the datasets presented in [11,38,39] although, to our knowledge, the lowest reported measurements of K_d in the blue are 0.012 m^{-1} at 420 nm [26]. The lowest K_d values in the blue in our dataset are thus 1.8 to 3.5 higher than these previous estimates for the clearest waters. In the red, our data of K_d collected within the top 10 m layer (Figures 6e and 7e) are similar or somewhat higher than the previous determinations for the clearest waters. At larger depths, K_d in the red was reduced owing to the increasing effect of inelastic processes with depth. This effect on the vertical changes in both K_d and K_u in the red spectral region is illustrated for our dataset in Figure 12. Both K coefficients decrease considerably with depth. This reduction is more pronounced at 683 nm than 650 nm. K_u decreased more rapidly than K_d and approached a relatively stable level of values at shallower depths (i.e., about 15–20 m) than K_d. The best-fit exponential functions to the data in Figure 12 are:

$$K_d(650) = 0.4792 \exp(-0.0451\ z) + 0.0387 \quad (r^2 = 0.925, RMSE = 0.033\ \mathrm{m}^{-1}, N = 66) \tag{6a}$$

$$K_d(683) = 0.7229 \exp(-0.0947\ z) + 0.0846 \quad (r^2 = 0.898, RMSE = 0.050\ \mathrm{m}^{-1}, N = 66) \tag{6b}$$

$$K_u(650) = 0.5614 \exp(-0.2258\ z) + 0.0730 \quad (r^2 = 0.806, RMSE = 0.032\ \mathrm{m}^{-1}, N = 64) \tag{6c}$$

$$K_u(683) = 1.0000 \exp(-0.4711\ z) + 0.0641 \quad (r^2 = 0.391, RMSE = 0.062\ \mathrm{m}^{-1}, N = 64). \tag{6d}$$

The relationship for $K_u(683)$ is greatly inferior because at this wavelength the effect of Raman scattering is reinforced by chlorophyll-*a* fluorescence emission, which resulted in a very steep decrease of $K_u(683)$ in the top 20 m (Figure 12b).

In contrast to K_d, other K coefficients are poorly documented in the literature. It is therefore useful to examine the relationships between K_d and other K coefficients using our dataset containing depth-resolved measurements within the euphotic layer. The results from this analysis are shown in Figure 13. Not shown in the figure are results that indicated a very good linear relationship with close agreement between the K_d and K_{od} values (r^2 = 0.985) and between K_u and K_{ou} (r^2 = 0.953).

This agreement was generally observed regardless of light wavelength and depth of measurement within our dataset.

The relationship between K_u and K_d is reasonably good in the blue and green spectral regions (Figure 13a). For example, at 440, 490, and 550 nm the best-fit linear relationships are:

$$K_u(z, 440) = 1.0115\, K_d(z, 440) + 9.47 \times 10^{-3} \quad (r^2 = 0.828, RMSE = 0.023\ \text{m}^{-1}, N = 66) \tag{7a}$$

$$K_u(z, 490) = 1.0220\, K_d(z, 490) + 1.71 \times 10^{-3} \quad (r^2 = 0.691, RMSE = 0.021\ \text{m}^{-1}, N = 66) \tag{7b}$$

$$K_u(z, 550) = 1.4699\, K_d(z, 550) - 4.83 \times 10^{-2} \quad (r^2 = 0.672, RMSE = 0.023\ \text{m}^{-1}, N = 66). \tag{7c}$$

As the wavelength increases into the red spectral region the relationship weakens considerably or nearly vanishes as suggested by the large scatter of data points for 650 nm and 683 nm in Figure 13a. For these wavelengths, the r^2 values drop to 0.444 and 0.079, respectively. This result can be attributed to increased effects of inelastic processes with increasing wavelength, which act differentially on the downwelling and upwelling light attenuation. It is noteworthy that one possible application of the relationships between K_u and K_d in the blue and green is to use the K_d measurements for extrapolating E_u measurements to deeper depths where upwelling light is no longer detectable with the E_u sensor but is still detectable with the E_d sensor. However, a special caution must be exercised for situations with large solar zenith angles, as evidenced by one outlying data point in our dataset which was obtained for relatively large θ_s of 57.1° compared with the remaining data. While it is known that at large θ_s K_d tends to increase significantly in the surface layer [11,17], the effect of solar angle is expected to be less pronounced for K_u. Such result is seen in our dataset in Figure 6e,f for the near-surface measurements taken at the example station in the Farallon Basin.

Our analysis shows that the linear relationship can be also used for estimating the diffuse attenuation coefficient of net irradiance, K_E, from K_d using the depth-resolved data in the euphotic layer (Figure 13b). For the blue-green spectral region the best-fit relationship is nearly a perfect 1:1 line with the correlation coefficient $r^2 > 0.998$. Specifically, the slope and offset parameters of the linear fit are: 1.0001 and -1.67×10^{-4} m^{-1} for 440 nm, 0.9999 and 1.66×10^{-5} m^{-1} for 490 nm, and 1.0015 and 1.78×10^{-5} m^{-1} for 550 nm ($N = 66$ for each wavelength). For the red spectral region, the best fit parameters are still very close to the 1:1 line, specifically 0.9746 and 1.26×10^{-2} m^{-1} for 650 nm and 0.9857 and 1.92×10^{-2} m^{-1} for 683 nm ($N = 62$ for each wavelength). The r^2 values at these wavelengths are only slightly lower than in the blue-green region and remained above 0.965. These results indicate that the magnitudes and behavior of K_E throughout the water column were consistent with those of K_d regardless of the spectral region including the long-wavelength portion of the spectrum where the effects of inelastic processes are strongest. Therefore, the use of a single relationship between K_E and K_d may be satisfactory regardless of wavelength. Such relationship determined from the combined data collected at wavelengths of 440, 550, and 650 nm is:

$$K_E(z) = 1.0093\, K_d(z) + 7.88 \times 10^{-4} \quad (r^2 = 0.996, RMSE = 0.006\ \text{m}^{-1}, N = 194). \tag{8}$$

This relationship is plotted in Figure 13b. It can be useful when K_E is required, for example for the analysis of Gershun's equation, when E_d measurements are available but are not accompanied by E_u measurements.

Similar to K_E vs. K_d, the linear relationship between the K_o and K_d values measured throughout the water column is also quite robust regardless of light wavelength (Figure 13c). The plotted regression line corresponds to the best fit obtained from the combined data collected at 440, 550, and 650 nm:

$$K_o(z) = 0.9480\, K_d(z) - 1.24 \times 10^{-3} \quad (r^2 = 0.984, RMSE = 0.015\ \text{m}^{-1}, N = 192). \tag{9}$$

This type of relationship can be useful in the absence of scalar irradiance measurements throughout the water column which is a common situation, but when the measurement or estimate of spectral E_o

just below the surface is available. In such case, E_o can be propagated throughout the water column on the basis of vertical measurements of spectral E_d using the relationship between K_o and K_d. The resulting depth profile of spectral E_o can, in turn, be subject to conversion to quantum units and spectral integration to obtain the depth-resolved PAR quantum scalar irradiance, E_{oPAR}.

Assuming the availability of E_{oPAR} just below the surface, e.g., [108], another relationship between the K coefficients can be useful for estimating E_{oPAR} at different depths from vertical measurements of spectral E_d. This relationship of K_{oPAR} vs. K_d on the basis of data collected throughout the water column is shown in Figure 13d. In this analysis we ignored, however, the data collected at near-surface depths (<5 m) because K_{oPAR} decreases rapidly with depth within the near-surface layer owing to high absorption by water molecules within the long-wavelength portion of the spectrum [10,109–111]. Figure 13d demonstrates that there is a reasonably good linear relationship between K_{oPAR} and K_d in the blue and green spectral regions, as illustrated for 440 nm, 490 nm, and 550 nm. The best-fit relationships are:

$$K_{oPAR}(z) = 0.5643\ K_d(z, 440) + 3.07 \times 10^{-2} \quad (r^2 = 0.911, RMSE = 0.029\ \text{m}^{-1}, N = 54) \tag{10a}$$

$$K_{oPAR}(z) = 0.8504\ K_d(z, 490) + 2.41 \times 10^{-2} \quad (r^2 = 0.923, RMSE = 0.015\ \text{m}^{-1}, N = 54) \tag{10b}$$

$$K_{oPAR}(z) = 1.2256\ K_d(z, 550) - 2.52 \times 10^{-2} \quad (r^2 = 0.856, RMSE = 0.012\ \text{m}^{-1}, N = 54). \tag{10c}$$

A closer examination also showed that $K_{PAR}(z) \approx [K_d(z, 490) + K_d(z, 550)]/2$:

$$K_{oPAR}(z) = 1.0182[K_d(z, 490) + K_d(z, 550)]/2 - 2.61 \times 10^{-3} \quad (r^2 = 0.908, RMSE = 0.010\ \text{m}^{-1}, N = 54). \tag{10d}$$

Being very close to the 1:1 line, this relationship provides a convenient approximation to $K_{oPAR}(z)$ based on $K_d(z)$ and is plotted in Figure 13d.

Because of unavailability or scarcity of scalar irradiance measurements, the measurements of downward plane irradiance have been typically used in the past for estimating photosynthetically available radiation in terms of PAR quantum downward irradiance, E_{dPAR} [11,105,109–111]. Kirk [10] provides a comprehensive overview of K_{dPAR} values for various oceanic, coastal, estuarine, and inland aquatic environments. Our dataset shows a strong linear relationship between K_{oPAR} and K_{dPAR} throughout the examined water column (Figure 13e):

$$K_{oPAR}(z) = 0.9516\ K_{dPAR}(z) - 3.12 \times 10^{-3} \quad (r^2 = 0.937, RMSE = 0.012\ \text{m}^{-1}, N = 54). \tag{11}$$

This relationship provides an alternative to Equation (10d) for the use in a situation when E_{oPAR} just below the surface and vertical measurements of spectral E_d are available.

It is noteworthy to mention that because of widespread interest in the determinations of remote-sensing reflectance the contemporary underwater radiometric systems are often configured to include only two sensors, one for the spectral measurements of E_d and another for the spectral measurements of upwelling zenith radiance, L_u. Because such systems lack the measurement of E_u, it is useful to examine the relationship between K_u and K_{Lu}. The analysis of our dataset indicates that a single linear relationship provides reasonably good estimates of K_u from K_{Lu} regardless of light wavelength and depth (Figure 13f):

$$K_u = 0.9838\ K_{Lu} - 0.0010 \quad (r^2 = 0.905, RMSE = 0.016\ \text{m}^{-1}, N = 52). \tag{12}$$

This relationship was obtained from data collected at three wavelengths, 440, 550, and 650 nm. Note that the slope coefficient of the regression is close to 1 but many data points of K_u in the blue and green spectral bands are higher than K_{Lu}. This tendency is not seen in the red band, however.

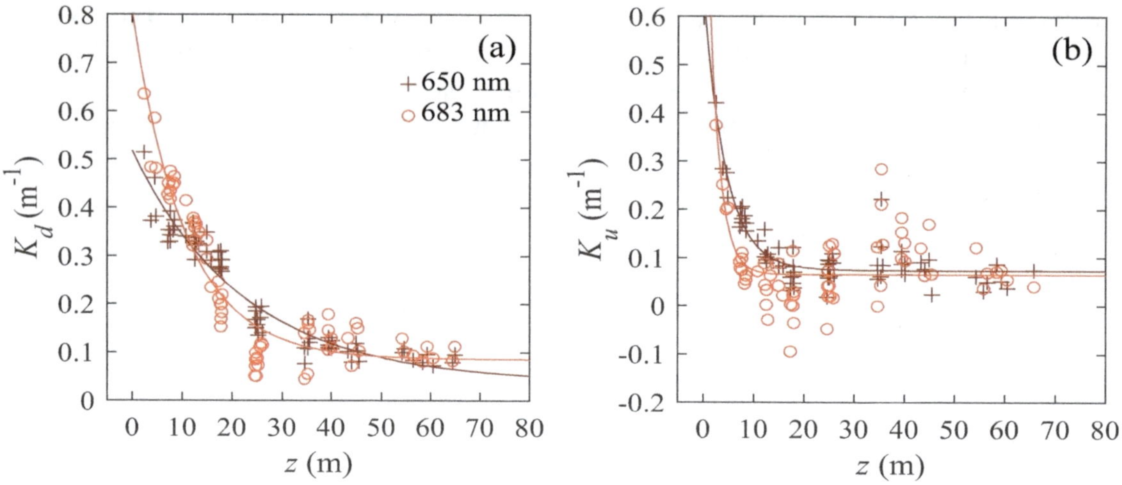

Figure 12. Diffuse attenuation coefficients in the red spectral bands (650 and 683 nm) plotted as a function of depth z using data from all stations and depths. (**a**) $K_d(z, 650)$ (crosses) and $K_d(z, 683)$ (open circles) vs. z. (**b**) Same as (**a**) but for K_u. Solid lines represent the best-fit regression functions.

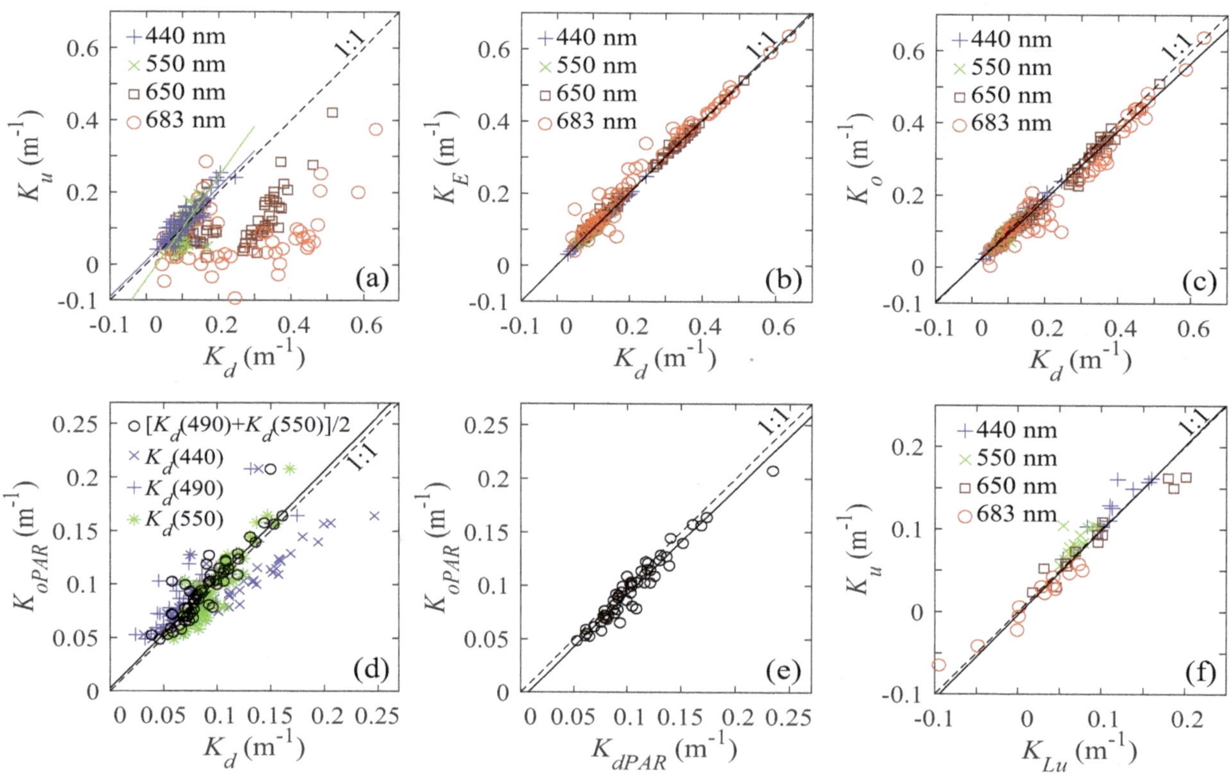

Figure 13. Relationships between diffuse attenuation coefficients based on measurements from all stations and depths. (**a**) K_u vs. K_d; (**b**) K_E vs. K_d; (**c**) K_o vs. K_d. (**d**) K_{dPAR} vs. K_d; (**e**) K_{oPAR} vs. K_{dPAR}; and (**f**) K_u vs. K_{Lu}. In (**a–c**,**f**), different symbols represent data at different light wavelengths as indicated. In (**d**) symbols indicate the relationships between $K_{oPAR}(z)$ and different K-coefficients as indicated, including the relationship between $K_{oPAR}(z)$ and $[K_d(z, 490) + K_d(z, 550)]/2$. The best fit regression functions (solid lines) are shown for (**a**) data at 440 nm and 550 nm analyzed separately, (**b**,**c**) combined data at 440 nm, 550 nm, and 650 nm, (**d**) data of $K_{oPAR}(z)$ vs. $[K_d(z, 490) + K_d(z, 550)]/2$, and (**e**,**f**) all data displayed in these panels. Dashed lines represent the 1:1 agreement between the variables.

3.3. Application of Gershun's Equation

The underwater radiometric measurements in conjunction with the use of Gershun's equation provide a method for determining the absorption coefficient of seawater, $a(\lambda, z)$, as a product of $K_E(\lambda, z)$ and $\overline{\mu}(\lambda, z)$. The main limitation of this radiometric method is that Gershun's equation was derived from a simplified radiative transfer equation that ignores inelastic processes of Raman scattering and fluorescence and internal sources (i.e., true light emission) such as bioluminescence, e.g., [9,11,49]. Therefore, in principle, this method can work satisfactorily only for cases in which the contributions of inelastic processes and internal sources are negligible. Because bioluminescence is typically intermittent and discrete under natural conditions in the ocean, the primary limitation of the applicability of Gershun's equation in the upper ocean or euphotic layer is associated with the presence of inelastic processes. In spite of these limitations, this radiometric method offers a unique benefit resulting from the estimation of the absorption coefficient that is representative of relatively large volume of seawater. This is because the radiometric measurements and AOPs are representative of much larger volumes of water compared with small volumes involved in direct IOP measurements including in situ absorption meters.

We examined the applicability of the radiometric method using our data collected at two stations, one in the Carmen Basin and another in the Farallon Basin (see Figure 1), where concurrent radiometric and ac-9 measurements were made. In this analysis, we compare the values of the absorption coefficient, $a_{ac9}(z, \lambda)$, measured with the ac-9 instrument with those derived from Gershun's equation, i.e., $a_{AOP}(z, \lambda) = K_E(z, \lambda) \cdot \overline{\mu}(z, \lambda)$. In order to enable such comparison, the data of $a_{ac9}(z, \lambda)$ and $\overline{\mu}(z, \lambda)$ were adjusted to common depths corresponding to the determinations of $K_E(z, \lambda)$. Specifically, $\overline{\mu}(z, \lambda)$ was linearly interpolated to obtain the values for depths at which $K_E(z, \lambda)$ was determined, i.e., the mid-points between the discrete depths (see Equation (1)) where the radiometric measurements, and hence $\overline{\mu}(z, \lambda)$, were made. The high depth resolution ac-9 data required only minor interpolation to these mid-point depths. In addition, the high spectral resolution data of $K_E(z, \lambda)$ and $\overline{\mu}(z, \lambda)$ were linearly interpolated to ac-9 wavelengths. Our comparative analysis of $a_{AOP}(z, \lambda)$ and $a_{ac9}(z, \lambda)$ is focused on five ac-9 wavelengths, 440, 488, 555, 650, and 676 nm.

In addition to the effects of inelastic processes and mismatch of spatial scales of AOP and ac-9 measurements, it is noteworthy to mention other sources of uncertainty that can affect the comparisons of $a_{ac9}(z, \lambda)$ and $a_{AOP}(z, \lambda)$. The accurate direct in situ measurements of $a(z, \lambda)$ with instruments such as ac-9 are difficult, mainly because of scattering error and stringent calibration requirements [112–114]. The determinations of $K_E(z, \lambda)$ and $\overline{\mu}(z, \lambda)$ require measurements with four (E_d, E_u, E_{od}, E_{ou}) or at least three (E_d, E_u, E_o) radiometric sensors, so these determinations are critically dependent on accurate calibrations of multiple sensors. The accuracy of determinations of $K_E(z, \lambda)$ for specific depths can be also affected by relatively low depth resolution of radiometric measurements that were taken at several discrete depths in our study.

In spite of multiple sources of uncertainty, we found reasonably good agreement between a_{AOP} and a_{ac9} throughout the euphotic layer for the blue and green spectral regions where the effects of inelastic processes were relatively weak (Figure 14a–c). As shown for the example data collected in the Farallon Basin, the applicability of Gershun's equation within the examined water column extends from the short-wavelength portion of the spectrum to about 580 nm, as indicated by the vertical dotted line in Figure 14a. At longer wavelengths this application failed owing to stronger effects of Raman scattering and chlorophyll-*a* fluorescence, even at the shallowest depth of 7.2 m examined in this case (Figure 14a,b). As the Raman scattering effects continue to accumulate with increasing depth the short-wavelength boundary of the affected spectral region is expected to shift towards shorter wavelengths [60,64,73]. Therefore, one can expect that at depths greater than those examined in our study the application of Gershun's equation may fail for wavelengths shorter than ~580 nm. Figure 14b also shows that somewhat larger discrepancy between a_{AOP} and a_{ac9} at 440 nm was observed within the Chl_{max} layer around the depth of z_{Chlmax}. This may be associated with the limitation that the radiometric measurements were taken at discrete depths with relatively low depth resolution (i.e., 10 m below z = 20 m), as opposed to nearly continuous depth profiles which would be required for better determinations of $K_E(z, \lambda)$ within the optically heterogeneous layer.

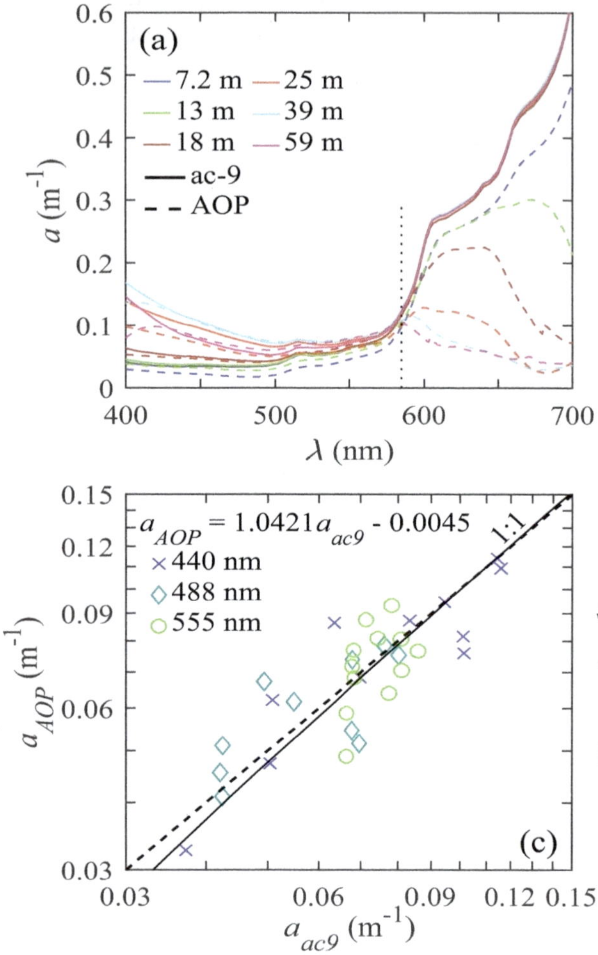

Figure 14. Comparison of the absorption coefficient of seawater measured with an ac-9 instrument (a_{ac9}) and estimated from AOPs using the irradiance quartet measurements and Gershun's equation (a_{AOP}). (**a**) Spectra of absorption coefficient at depths for which the determinations of K_E were made on the basis of radiometric measurements taken at one station in the Carmen Basin where ac-9 measurements were made during the 2011 cruise. The vertical dotted line indicates the transition region to long-wavelength portion of the spectrum where the use of Gershun's is inadequate because of the effects of inelastic processes. (**b**) Vertical profiles of absorption coefficient measured with the ac-9 instrument (solid lines) and estimated from AOPs (dashed lines) at selected light wavelengths as indicated. Data were collected at the same station as in panel (**a**). (**c**) Direct comparison between a_{AOP} and a_{ac9} at selected blue and green spectral bands as indicated. Data were collected at one station in the Guaymas Basin and one station in the Carmen Basin where ac-9 measurements were made during the 2011 cruise. (**d**) Relationship between $K_E \cdot \overline{\mu}$ and $K_d \cdot \overline{\mu}_d$ for the blue-green spectral regions based on data at the three selected light wavelengths as indicated. Solid lines in (**c**,**d**) represent the best-fit regression functions and dashed lines the 1:1 agreement between the variables.

Figure 14c shows that a_{AOP} and a_{ac9} compare favorably for the combined dataset from the two stations, which includes data points for the three wavelengths from the blue-green spectral region (440, 488, and 555 nm) and all discrete depths included in this analysis (i.e., where K_E was determined). The best linear fit to the data is very close to the 1:1 line and the correlation is strong (r^2 = 0.750). This was the only case in this study when Model II regression analysis was applied (see Section 2.6). If the ac-9-measured values are tentatively assumed to represent true values, then the statistical indicators suggest a very small bias of a_{AOP} estimates as the median value for the ratio of a_{AOP}/a_{ac9} is very close to 1 (0.989) and the mean difference between a_{AOP} and a_{ac9} is very small (-0.0016 m^{-1}). The median absolute percent difference between a_{AOP} and a_{ac9} is 12.4% and the root mean square deviation is 0.0116 m^{-1}.

Because of tight relationships $\overline{\mu}$ vs. $\overline{\mu}_d$ (Figure 10a) and K_E vs. K_d (Figure 13b), it is natural to expect a good relationship between the products $K_E \cdot \overline{\mu}$ and $K_d \cdot \overline{\mu}_d$. This is indeed the case as shown for our data combining the three wavelengths from the blue-green spectral region (440, 490, and 550 nm) and all examined discrete depths at all stations (Figure 14d). The best-fit linear fit to the data is:

$$K_E \cdot \overline{\mu} = 0.980\, K_d \cdot \overline{\mu}_d - 0.0017 \left(r^2 = 0.999, RMSE = 0.003\ m^{-1}, N = 36 \right). \tag{13}$$

By indicating that $K_E \cdot \overline{\mu}$ can be quite accurately estimated from $K_d \cdot \overline{\mu}_d$, this relationship has important practical ramifications. Specifically, the measurements of three (E_d, E_u, E_o) or four (E_d, E_u, E_{od}, E_{ou}) irradiances required for determining the absorption coefficient from Gershun's equation can be replaced with just two downwelling irradiance measurements (E_d and E_{od}), thus significantly simplifying the experimental requirements. We note that the use of two AOPs based on the downwelling light field, K_d and $\overline{\mu}_d$, has been previously introduced in the context of estimation of absorption by phytoplankton and CDOM, as well as chlorophyll-*a* concentration from measurements of E_d alone [96,115]. The modeling formalism based on E_d alone required assumptions about $\overline{\mu}_d$ and the backscattering coefficient of seawater.

4. Summary

Although quantitative radiometric measurements in the ocean have a long-recorded history with the beginning in the early 1930s, the measurements of scalar irradiance have been rare and simultaneous measurements of downward and upward plane and scalar irradiances (E_d, E_u, E_{od}, and E_{ou}), the so-called irradiance quartet, are virtually lacking. To our knowledge, this is the first-ever study to report data of high spectral resolution measurements of irradiance quartet within the ocean euphotic layer and the apparent optical properties (AOPs) derivable from this quartet, including a complete set of average cosines characterizing the angular structure of underwater light field, i.e., the downwelling ($\overline{\mu}_d$), upwelling ($\overline{\mu}_u$), and total ($\overline{\mu}$) light field. Because the experimental irradiance data collected in the past have been fragmentary in terms of incomplete set of irradiances or limited spectral coverage in rare experiments when E_d, E_u, and E_o were measured directly or derived from radiance measurements, currently available descriptions and understanding of underwater light field and AOPs associated with a complete irradiance quartet have relied mostly on theoretical simulations of radiative transfer in the ocean. Such simulations unavoidably involve some idealized assumptions about the real environment. Thus, the experimental data presented in this paper provide a unique and revealing resource on the patterns of irradiance quartet and associated suite of AOPs within a real water column that was characterized by the interplay of non-uniform vertical distribution of inherent optical properties (IOPs) of seawater and the effects of inelastic radiative processes such as Raman scattering by water molecules and chlorophyll-*a* fluorescence.

Our dataset was collected in the Gulf of California within the euphotic layer from the near-surface depths to ~80 m and covered a broad range of IOPs, sky conditions, and solar zenith angle. Our data show that in the blue and green spectral regions, the vertical patterns of AOPs were driven primarily by IOPs of seawater with weak or no discernible effects of inelastic radiative processes. In contrast, in the red spectral region the radiometric variables characterizing the light field and the AOPs were strongly affected or totally dominated by inelastic processes of Raman scattering, and additionally by chlorophyll-*a* fluorescence within the fluorescence emission band, the latter being particularly well-pronounced within the chlorophyll-*a* maximum layer. Some of the most notable features caused by inelastic processes in the red include the values of average cosines which approached those of the uniform light field, exceptionally high values of irradiance reflectance R approaching

1, and exceptionally low values (i.e., smaller than the seawater absorption coefficient) of the diffuse attenuation coefficients, K, for various irradiances, including the negative values for the attenuation of upwelling irradiances in the chlorophyll-*a* fluorescence emission band within the chlorophyll-*a* maximum layer.

Whereas details of the vertical patterns in the light field characteristics and AOPs depended on the interplay of inelastic processes and vertically non-uniform IOPs including the chlorophyll-*a* concentration, there were general vertical patterns associated with an increase in the contribution of inelastically produced light with increasing depth. We observed, for example, that the vertical patterns of some AOPs in the red, such as $\overline{\mu}_d$, $\overline{\mu}$, R, K_d, and K_u, can be approximated by relatively simple functions of depth. We also established empirical relationships between some AOPs, which can be useful in common experimental situations when only data on downwelling plane irradiance are available. For example, we proposed single relationships for estimating $K_E(z)$ and $K_o(z)$ from $K_d(z)$ at any depth z within the examined depth range regardless of light wavelength. Similarly, we also determined that a single relationship can be used for estimating $K_u(z)$ from $K_d(z)$ within the blue-green spectral region, in which the effects of inelastic processes are generally weak. Our results also demonstrated that below the near-surface layer (i.e., z approximately greater than 5 m in our dataset) the values of $K_{oPAR}(z)$ can be estimated from measurements of $K_d(z)$ in the blue-green bands using a single relationship. Typically, the empirical relationships are expected to be applicable within a range of environmental conditions consistent with the dataset used in the development of the relationships. This limitation also applies to the relationships established in this study, in which we covered a fairly broad range of conditions (Table 2).

We tested the use of irradiance quartet measurements in conjunction with Gershun's equation for estimating the absorption coefficient of seawater and found that in the blue-green spectral region such estimates agree reasonably well with direct determinations of the absorption coefficient from measurements with the ac-9 instrument. As the Gershun equation is inadequate in situations when inelastic processes are significant, no agreement was observed at light wavelengths longer than about 580 nm. Specifically, in the long-wavelength portion of the spectrum the absorption coefficients estimated from Gershun's equation were greatly underestimated due to the effects of inelastic processes. The analysis of our data from the entire examined depth range also showed that the product of two AOPs in the blue-green spectral region, $K_E \cdot \overline{\mu}$, which is used for estimating the absorption coefficient from Gershun's equation, can be adequately estimated from the product $K_d \cdot \overline{\mu}_d$. This is potentially useful because the determinations of $K_E \cdot \overline{\mu}$ require the measurements of irradiance quartet or at least three irradiances $E_d(z)$, $E_u(z)$, and $E_o(z)$, whereas the determinations of $K_d \cdot \overline{\mu}_d$ require a simpler experimental design with only two irradiances, E_d and E_{od}.

Author Contributions: L.L. and D.S. wrote the paper; M.D. collected the field data; L.L. and M.D. processed the data; L.L. analyzed the data; and all authors made intellectual contributions and reviewed the manuscript.

Acknowledgments: This study was supported by the U.S. Office of Naval Research (grant N00014-09-1-1053) as part of the Department of Defense Multidisciplinary University Research Initiative (MURI) Program. Partial support for Linhai Li was provided by the Scripps Institution of Oceanography. The scientists, crew members, and officers of R/V New Horizon are acknowledged for providing logistical support and help during the field work. In particular, we thank Alison Sweeney for providing data from CTD measurements and Yakir Gagnon for ac-9 data. We also thank Rick A. Reynolds and anonymous reviewers for providing comments on the manuscript.

References

1. Lewis, M.R.; Carr, M.E.; Feldman, G.C.; Esaias, W.; McClain, C. Influence of penetrating solar-radiation on the heat-budget of the equatorial Pacific Ocean. *Nature* **1990**, *347*, 543–545. [CrossRef]
2. Lee, Z.P.; Du, K.P.; Arnone, R.; Liew, S.C.; Penta, B. Penetration of solar radiation in the upper ocean: A numerical model for oceanic and coastal waters. *J. Geophys. Res.* **2005**, *110*, C09019. [CrossRef]
3. Field, C.B.; Behrenfeld, M.J.; Randerson, J.T.; Falkowski, P. Primary production of the biosphere: Integrating terrestrial and oceanic components. *Science* **1998**, *281*, 237–240. [CrossRef] [PubMed]

4. Mobley, C.D.; Chai, F.; Xiu, P.; Sundman, L.K. Impact of improved light calculations on predicted phytoplankton growth and heating in an idealized upwelling-downwelling channel geometry. *J. Geophys. Res. Oceans* **2015**, *120*, 875–892. [CrossRef]
5. Smith, R.C; Wilson, W.H., Jr. Photon scalar irradiance. *Appl. Opt.* **1972**, *11*, 934–938. [CrossRef] [PubMed]
6. Booth, C.R. The design and evaluation of a measurement system for photosynthertically active quantum scalar irradiance. *Limnol. Oceanogr.* **1975**, *21*, 326–336. [CrossRef]
7. Højerslev, N.K. Daylight measurements appropriate for photosynthetic studies in natural sea waters. *J. Cons. Int. Explor. Mer* **1978**, *38*, 131–146. [CrossRef]
8. Morel, A. Available, usable, and stored radiant energy in relation to marine photosynthesis. *Deep-Sea Res.* **1978**, *25*, 673–688. [CrossRef]
9. Mobley, C.D. *Light and Water: Radiative Transfer in Natural Waters*; Academic Press: San Diego, CA, USA, 1994.
10. Kirk, J.T. *Light and Photosynthesis in Aquatic Ecosystems*; Cambridge University Press: Cambridge, UK, 1994.
11. Jerlov, N.G. *Marine Optics*; Elsevier Scientific Publishing Company: Amsterdam, The Netherlands, 1976.
12. Jerlov, N.G. Optical studies of ocean water. *Rep. Swed. Deep-Sea Exped.* **1951**, *3*, 1–59.
13. Tyler, J.E.; Smith, R.C. *Measurements of Spectral Irradiance Underwater*; Gordon & Breach Publishing Group: Philadelphia, PA, USA, 1970.
14. Jerlov, N.G.; Liljequist, G.H. On the angular distribution of submarine daylight and on the total submarine illumination. *Svenska Hydrografisk-Biologiska Kommisionens Skrifter Ny Serie Hydrografi* **1938**, *14*, 1–15.
15. Whitney, L.V. The angular distribution of characteristic diffuse light in natural waters. *J. Mar. Res.* **1941**, *4*, 122–131.
16. Smith, R.C. Structure of solar radiation in the upper layers of the sea. In *Optical Aspects of Oceanography*; Jerlov, N.G., Steemann, E., Eds.; Academic Press: San Diego, CA, USA, 1974; pp. 95–117.
17. Lundgren, B.; Højerslev, N.K. *Daylight Measurements in the Sargasso Sea: Results from the "Dana" Expedition January–April 1966. Report 14*; University of Copenhagen, Institution of Physical Oceanography: Copenhagen, Denmark, 1971.
18. Højerslev, N.K. *Inherent and Apparent Optical Properties of the Western Mediterranean and the Hardangerfjord. Report 21*; University of Copenhagen, Institution of Physical Oceanography: Copenhagen, Denmark, 1973.
19. Ochakovskiy, Y.Y.; Kopelevich, O.V.; Voytov, V.I. *Light in the Sea*; (Edited translation from Russian) FTD-HT-24-105-72; Foreign Technology Division, Air Force System Command, US Air Force; National Technical Information Service, U.S. Department of Commerce: Springfieldm, VA, USA, 1972.
20. Voss, K.J. Electro-optic camera system for measurement of the underwater radiance distribution. *Opt. Eng.* **1989**, *28*, 384–387. [CrossRef]
21. Dickey, T.; Frye, D.; Jannasch, H.; Boyle, E.; Manov, D.; Sigurdson, D.; McNeil, J.; Stramska, M.; Michaels, A.; Nelson, N.; et al. Initial results from the Bermuda testbed mooring program. *Deep Sea Res. Part I* **1998**, *45*, 771–794. [CrossRef]
22. Hooker, S.B.; Maritorena, S. An evaluation of oceanographic radiometers and deployment methodologies. *J. Atmos. Ocean. Technol.* **2000**, *17*, 811–830. [CrossRef]
23. Clark, D.K.; Feinholz, M.; Yarbrough, M.; Johnson, B.C.; Brown, S.W.; Kim, Y.S.; Barnes, R.A. Overview of the radiometric calibration of MOBY. In *Earth Observing Systems VI*; Barres, W.L., Ed.; SPIE Proceedings; SPIE: Bellingham, WA, USA, 2002; Volume 4483, pp. 64–76. [CrossRef]
24. Bulgarelli, B.; Zibordi, G.; Berthon, J.F. Measured and modeled radiometric quantities in coastal waters: Toward a closure. *Appl. Opt.* **2003**, *42*, 5365–5381. [CrossRef] [PubMed]
25. Voss, K.J.; Chapin, A. Upwelling radiance distribution camera system, NURADS. *Opt. Express* **2005**, *13*, 4250–4262. [CrossRef] [PubMed]
26. Morel, A.; Gentili, B.; Claustre, H.; Babin, M.; Bricaud, A.; Ras, J.; Tieche, F. Optical properties of the "clearest" natural waters. *Limnol. Oceanogr.* **2007**, *52*, 217–229. [CrossRef]
27. Antoine, D.; d'Ortenzio, F.; Hooker, S.B.; Bécu, G.; Gentili, B.; Tailliez, D.; Scott, A.J. Assessment of uncertainty in the ocean reflectance determined by three satellite ocean color sensors (MERIS, SeaWiFS and MODIS-A) at an offshore site in the Mediterranean Sea (BOUSSOLE project). *J. Geophys. Res.* **2008**, *113*, C07013. [CrossRef]
28. Chang, G.C.; Dickey, T.D. Interdisciplinary sampling strategies for detection and characterization of harmful algal blooms. In *Real-Time Observation Systems for Ecosystem Dynamics and Harmful Algal Blooms*; Babin, M., Roesler, C.S., Cullen, J.J., Eds.; UNESCO: Paris, France, 2008; pp. 43–84.

29. Lewis, M.R.; Wei, J.; Van Dommelen, R.; Voss, K.J. Quantitative estimation of the underwater radiance distribution. *J. Geophys. Res.* **2011**, *116*, C00H06. [CrossRef]
30. Wei, J.; Dommelen, R.V.; Lewis, M.R.; McLean, S.; Voss, K.J. A new instrument for measuring the high dynamic range radiance distribution in near surface sea water. *Opt. Express* **2012**, *20*, 27024–27038. [CrossRef] [PubMed]
31. Antoine, D.; Morel, A.; Leymarie, E.; Houyou, A.; Gentili, B.; Victori, S.; Buis, J.-P.; Buis, N.; Meunier, S.; Canini, M.; et al. Underwater radiance distributions measured with miniaturized multispectral radiance cameras. *J. Atmos. Ocean. Technol.* **2013**, *30*, 74–95. [CrossRef]
32. Cunningham, A.; McKee, D. Measurement of hyperspectral underwater light fields. In *Subsea Optics and Imaging*; Watson, J.E., Zielinski, O., Eds.; Woodhead Publishing: Philadelphia, PA, USA, 2013; pp. 83–97.
33. Lewis, M.R. Measurement of apparent optical properties for diagonosis of harmful algal blooms. In *Real-Time Observation Systems for Ecosystem Dynamics and Harmful Algal Blooms*; Babin, M., Roesler, C.S., Cullen, J.J., Eds.; UNESCO: Paris, France, 2008; pp. 207–236.
34. Zibordi, G.; Voss, K.J. In situ optical radiometry in the visible and near infrared. In *Optical Radiometry for Ocean Climate Measurements*; Zibordi, G., Donlon, C.J., Parr, A.C., Eds.; Academic Press: San Diego, CA, USA, 2014; pp. 248–305.
35. Smith, R.C.; Baker, K.S. The bio-optical state of ocean waters and remote sensing. *Limnol. Oceanogr.* **1978**, *23*, 247–259. [CrossRef]
36. Siegel, D.A.; Dickey, T.D. Observations of the vertical structure of the diffuse attenuation coefficient spectrum. *Deep-Sea Res.* **1987**, *34*, 547–563. [CrossRef]
37. Dirks, R.W.J.; Spitzer, D. On the radiative transfer in the sea, including fluorescence and stratification effects. *Limnol. Oceanogr.* **1987**, *32*, 942–953. [CrossRef]
38. Morel, A. Optical modeling of the upper ocean in relation to its biogenous matter content (case I waters). *J. Geophys. Res.* **1988**, *93*, 10749–10768. [CrossRef]
39. Morel, A.; Maritorena, S. Bio-optical properties of oceanic waters: A reappraisal. *J. Geophys. Res.* **2001**, *106*, 7163–7180. [CrossRef]
40. Zibordi, G.; D'Alimonte, D.; Berthon, J.-F. An evaluation of depth resolution requirements for optical profiling in coastal waters. *J. Atmos. Ocean. Technol.* **2004**, *21*, 1059–1073. [CrossRef]
41. Yarbrough, M.A.; Houlihan, T.; Feinholz, M.; Flora, S.; Johnson, B.C.; Kim, Y.S.; Murphy, M.Y.; Ondrusek, M.; Clark, D. Results in coastal waters with high resolution in-situ spectral radiometry: The Marine Optical System ROV. In *Coastal Ocean Remote Sensing*; Frouin, R.J., Ed.; SPIE Proceedings; SPIE: Bellingham, WA, USA, 2007; Volume 6680, p. 66800I.
42. Antoine, D.; Hooker, S.B.; Bélanger, S.; Matsuoka, A.; Babin, M. Apparent optical properties of the Canadian Beaufort Sea—Part 1: Observational overview and water column relationships. *Biogeosciences* **2013**, *10*, 4493–4509. [CrossRef]
43. Lee, Z; Shaoling, S; Stavn, R.H. AOPs are not additive: On the biogeo-optical modeling of the diffuse attenuation coefficient. *Front. Mar. Sci.* **2018**, *5*, Article 8. [CrossRef]
44. Pegau, W.S.; Cleveland, J.S.; Doss, W.; Kennedy, C.D.; Maffione, R.A.; Mueller, J.L.; Stone, R.; Trees, C.C.; Weidemann, A.D.; Wells, W.H.; et al. A comparison of methods for the measurement of the absorption coefficient in natural waters. *J. Geophys. Res.* **1995**, *100*, 13201–13220. [CrossRef]
45. Preisendorfer, R.W. Application of radiative transfer theory to light measurements in the sea. *Int. Union Geod. Geophys. Monogr. Symp. Radiant Energy Sea* **1961**, *10*, 11–30.
46. Preisendorfer, R.W. *Hydrologic Optics. Volume I. Introduction*; US Department of Commerce, National Oceanic and Atmospheric Administration, Environmental Research Laboratories: Honolulu, HI, USA, 1976.
47. Kirk, J.T. Volume scattering function, average cosines, and the underwater light field. *Limnol. Oceanogr.* **1991**, *36*, 455–467. [CrossRef]
48. Gordon, H.R. Modeling and simulating radiative transfer in the ocean . In *Ocean Optics*; Spinrad, R.W., Carder, K.L., Perry, M.J., Eds.; Oxford University Press: New York, NY, USA, 1994; pp. 3–39.
49. Gershun, A. The light field. *J. Math. Phys.* **1939**, *18*, 51–151. [CrossRef]
50. Tyler, J.E.; Richardson, W.H.; Holmes, R.W. Method for obtaining the optical properties of large bodies of water. *J. Geophys. Res.* **1959**, *64*, 667–673. [CrossRef]
51. Højerslev, N.K. A spectral light absorption meter for measurements in the sea. *Limnol. Oceanogr.* **1975**, *20*, 1024–1034. [CrossRef]

52. Spitzer, D.; Wernand, M.R. In situ measurements of absorption spectra in the sea. *Deep-Sea Res.* **1981**, *28A*, 165–174. [CrossRef]
53. Gordon, H.R. Inverse methods in hydrologic optics. *Oceanologia* **2002**, *44*, 9–58.
54. Gordon, H.R.; Brown, O.B.; Jacobs, M.M. Computed relationships between the inherent and apparent optical properties of a flat homogeneous ocean. *Appl. Opt.* **1975**, *14*, 417–427. [CrossRef] [PubMed]
55. Kirk, J.T. Monte Carlo study of the nature of the underwater light field in, and of the relationship between optical properties of, turbid yellow waters. *Aust. J. Mar. Freshw. Res.* **1981**, *32*, 517–532. [CrossRef]
56. Gordon, H.R. Absorption and scattering estimates from irradiance measurements: Monte Carlo simulations. *Limnol. Oceanogr.* **1991**, *36*, 769–777. [CrossRef]
57. Sathyendranath, S.; Platt, T. Angular distribution of the submarine light field: Modification by multiple scattering. *Proc. R. Soc. Lond. A* **1991**, *433*, 287–297. [CrossRef]
58. Stramska, M.; Stramski, D.; Mitchell, B.G.; Mobley, C.D. Estimation of the absorption and backscattering coefficients from in-water radiometric measurements. *Limnol. Oceanogr.* **2000**, *45*, 628–641. [CrossRef]
59. Loisel, H.; Stramski, D. Estimation of inherent optical properties of natural waters from the irradiance attenuation coefficient and reflectance in the presence of Raman scattering. *Appl. Opt.* **2000**, *39*, 3001–3011. [CrossRef] [PubMed]
60. Morel, A.; Gentili, B. Radiation transport within oceanic (case 1) water. *J. Geophys. Res.* **2004**, *109*, C06008. [CrossRef]
61. Lewis, M.R.; Cullen, J.J.; Platt, T. Phytoplankton and thermal structure of the ocean: Consequences of nonuniformity in the chlorophyll profile. *J. Geophys. Res.* **1983**, *88*, 2565–2570. [CrossRef]
62. Gordon, H.R. Diffuse reflectance of the ocean: Influence of nonuniform phytoplankton pigment profile. *Appl. Opt.* **1992**, *31*, 2116–2129. [CrossRef] [PubMed]
63. Stramska, M.; Stramski, D. Effects of a nonuniform vertical profile of chlorophyll concentration on remote-sensing reflectance of the ocean. *Appl. Opt.* **2005**, *44*, 1735–1747. [CrossRef] [PubMed]
64. Li, L.; Stramski, D.; Reynolds, R.A. Characterization of the solar light field within the ocean mesopelagic zone based on radiative transfer simulations. *Deep Sea Res. Part I* **2014**, *87*, 53–69. [CrossRef]
65. Gordon, H. Remote sensing of optical properties in continuously stratified waters. *Appl. Opt.* **1978**, *17*, 1893–1897. [CrossRef] [PubMed]
66. Yang, Q.; Stramski, D.; He, M.-X. Modeling the effects of near-surface plumes of suspended particulate matter on remote-sensing reflectance of coastal waters. *Appl. Opt.* **2013**, *52*, 359–374. [CrossRef] [PubMed]
67. Stavn, R.H.; Weidemann, A.D. Optical modeling of clear ocean light fields: Raman scattering effects. *Appl. Opt.* **1988**, *27*, 4002–4011. [CrossRef] [PubMed]
68. Marshall, B.R.; Smith, R.C. Raman scattering and in-water ocean optical properties. *Appl. Opt.* **1990**, *29*, 71–84. [CrossRef] [PubMed]
69. Kattawar, G.W.; Xu, X. Filling in of Fraunhofer lines in the ocean by Raman scattering. *Appl. Opt.* **1992**, *31*, 6491–6500. [CrossRef] [PubMed]
70. Ge, Y.; Gordon, H.R.; Voss, K.J. Simulation of inelastic-scattering contributions to the irradiance field in the ocean: Variation in Fraunhofer line depths. *Appl. Opt.* **1993**, *32*, 4028–4036. [CrossRef] [PubMed]
71. Stavn, R.H. Effects of Raman scattering across the visible spectrum in clear ocean water: A Monte Carlo study. *Appl. Opt.* **1993**, *32*, 6853–6863. [CrossRef] [PubMed]
72. Hu, C.; Voss, K.J. In situ measurements of Raman scattering in clear ocean water. *Appl. Opt.* **1997**, *36*, 6962–6967. [CrossRef] [PubMed]
73. Berwald, J.; Stramski, D.; Mobley, C.D.; Kiefer, D.A. The effect of Raman scattering on the average cosine and the diffuse attenuation coefficient of irradiance in the ocean. *Limnol. Oceanogr.* **1998**, *43*, 564–576. [CrossRef]
74. Gordon, H.R. Contribution of Raman scattering to water-leaving radiance: A reexamination. *Appl. Opt.* **1999**, *38*, 3166–3174. [CrossRef] [PubMed]
75. Li, L.; Stramski, D.; Reynolds, R.A. Effects of inelastic radiative processes on the determination of water-leaving spectral radiance from extrapolation of underwater near-surface measurements. *Appl. Opt.* **2016**, *55*, 7050–7067. [CrossRef] [PubMed]
76. Lavín, M.F.; Marinone, S.G. An overview of the physical oceanography of the Gulf of California. In *Nonlinear Processes in Geophysical Fluid Dynamics*; Velasco Fuentes, O.U., Sheinbaum, J., Ochoa, J., Eds.; Springer: Dordrecht, The Netherlands, 2003; pp. 173–204.

77. Lluch-Cota, S.E.; Aragon-Noriega, E.A.; Arreguín-Sánchez, F.; Aurioles-Gamboa, D.; Bautista-Romero, J.J.; Brusca, R.C.; Cervantes-Duarte, R.; Cortés-Altamirano, R.; Del-Monte-Luna, P.; Esquivel-Herrera, A.; et al. The Gulf of California: Review of ecosystem status and sustainability challenges. *Prog. Oceanogr.* **2007**, *73*, 1–26. [CrossRef]
78. Ras, J.; Claustre, H.; Uitz, J. Spatial variability of phytoplankton pigment distributions in the Subtropical South Pacific Ocean: Comparison between in situ and predicted data. *Biogeosciences* **2008**, *5*, 353–369. [CrossRef]
79. Ritchie, R.J. Universal chlorophyll equations for estimating chlorophylls a, b, c, and d and total chlorophylls in natural assemblages of photosynthetic organisms using acetone, methanol, or ethanol solvents. *Photosynthetica* **2008**, *46*, 115–126. [CrossRef]
80. Gordon, H.R. Ship perturbation of irradiance measurements at sea 1: Monte Carlo simulations. *Appl. Opt.* **1985**, *24*, 4172–4182. [CrossRef] [PubMed]
81. Waters, K.J.; Smith, R.C.; Lewis, M.R. Avoiding ship-induced light-field perturbation in the determination of oceanic optical properties. *Oceanography* **1990**, *3*, 18–21. [CrossRef]
82. Zibordi, G.; Darecki, M. Immersion factors for the RAMSES series of hyper-spectral underwater radiometers. *J. Opt. A Pure Appl. Opt.* **2006**, *8*, 252–258. [CrossRef]
83. Stramski, D. Fluctuations of solar irradiance induced by surface waves in the Baltic. *Bull. Pol. Acad. Sci. Earth Sci.* **1986**, *34*, 333–344.
84. Darecki, M.; Stramski, D.; Sokólski, M. Measurements of high-frequency light fluctuations induced by sea surface waves with an Underwater Porcupine Radiometer System. *J. Geophys. Res.* **2011**, *116*, C00H09. [CrossRef]
85. Sullivan, J.M.; Twardowski, M.S.; Zaneveld, J.R.V.; Moore, C.M.; Barnard, A.H.; Donaghay, P.L.; Rhoades, B. Hyperspectral temperature and salt dependencies of absorption by water and heavy water in the 400–750 nm spectral range. *Appl. Opt.* **2006**, *45*, 5294–5309. [CrossRef] [PubMed]
86. Sokal, R.R.; Rohlf, F.J. *Biometry: The Principles and Practice of Statistics in Biological Research*, 3rd ed.; W.H. Freeman: New York, NY, USA, 1995.
87. Banse, K. Should we continue to use the 1% light depth convention for estimating the compensation depth of phytoplankton for another 70 years. *Limnol. Oceanogr. Bull.* **2004**, *13*, 49–52. [CrossRef]
88. Sugihara, S.; Kishino, M.; Okami, N. Contribution of Raman scattering to upward irradiance in the sea. *J. Oceanogr. Soc. Jpn.* **1984**, *40*, 397–404. [CrossRef]
89. Bristow, M.; Nielsen, D.; Bundy, D.; Furtek, R. Use of water Raman emission to correct airborne laser fluorosensor data for effects of water optical attenuation. *Appl. Opt.* **1981**, *20*, 2889–2906. [CrossRef] [PubMed]
90. Hoge, F.E.; Swift, R.N. Airborne simultaneous spectroscopic detection of laser-induced water Raman backscatter and fluorescence from chlorophyll a and other naturally occurring pigments. *Appl. Opt.* **1981**, *20*, 3197–3205. [CrossRef] [PubMed]
91. Stramski, D.; Tęgowski, J. Effects of intermittent entrainment of air bubbles by breaking wind waves on ocean reflectance and underwater light field. *J. Geophys. Res.* **2001**, *106*, 31345–31360. [CrossRef]
92. Bannister, T.T. Model of the mean cosine of underwater radiance and estimation of underwater scalar irradiance. *Limnol. Oceanogr.* **1992**, *37*, 773–780. [CrossRef]
93. Berwald, J.; Stramski, D.; Mobley, C.D.; Kiefer, D.A. Influences of absorption and scattering on vertical changes in the average cosine of the underwater light field. *Limnol. Oceanogr.* **1995**, *40*, 1347–1357. [CrossRef]
94. Aas, E.; Højerslev, N.K. Analysis of underwater radiance observations: Apparent optical properties and analytic functions describing the angular radiance distribution. *J. Geophys. Res.* **1999**, *104*, 8015–8024. [CrossRef]
95. Sathyendranath, S.; Platt, T. Ocean-color model incorporating transspectral processes. *Appl. Opt.* **1998**, *37*, 2216–2227. [CrossRef] [PubMed]
96. Nahorniak, J.S.; Abbott, M.R.; Letelier, R.M.; Pegau, W.S. Analysis of a method to estimate chlorophyll-a concentration from irradiance measurements at varying depths. *J. Atmos. Ocean. Technol.* **2001**, *18*, 2063–2073. [CrossRef]
97. Brown, C.A.; Huot, Y.; Purcell, M.J.; Cullen, J.J.; Lewis, M.R. Mapping coastal optical and biogeochemical variability using an autonomous underwater vehicle and a new bio-optical inversion algorithm. *Limnol. Oceanogr. Methods* **2004**, *2*, 262–281. [CrossRef]

98. Xing, X.; Morel, A.; Claustre, H.; d'Ortenzio, F.; Poteau, A. Combined processing and mutual interpretation of radiometry and fluorometry from autonomous profiling Bio-Argo floats: 2. Colored dissolved organic matter absorption retrieval. *J. Geophys. Res.* **2012**, *117*, C04022. [CrossRef]
99. Westberry, T.K.; Boss, E.; Lee, Z. Influence of Raman scattering on ocean color inversion models. *Appl. Opt.* **2013**, *52*, 5552–5561. [CrossRef] [PubMed]
100. Liu, C.C.; Carder, K.L.; Miller, R.L.; Ivey, J.E. Fast and accurate model of underwater scalar irradiance. *Appl. Opt.* **2002**, *41*, 4962–4974. [CrossRef] [PubMed]
101. Liu, C.C.; Miller, R.L.; Carder, K.L.; Lee, Z.; D'Sa, E.J.; Ivey, J.E. Estimating the underwater light field from remote sensing of ocean color. *J. Oceanogr.* **2006**, *62*, 235–248. [CrossRef]
102. Sathyendranath, S.; Platt, T. Computation of aquatic primary production: Extended formalism to include effect of angular and spectral distribution of light. *Limnol. Oceanogr.* **1989**, *34*, 188–198. [CrossRef]
103. O'Reilly, J.E.; Maritorena, S.; Mitchell, B.G.; Siegel, D.A.; Carder, K.L.; Garver, S.A.; Kahru, M.; McClain, C. Ocean color chlorophyll algorithms for SeaWiFS. *J. Geophys. Res.* **1998**, *103*, 24937–24953. [CrossRef]
104. O'Reilly, J.E.; Maritorena, S.; Siegel, D.A.; O'Brien, M.C.; Hooker, S.B.; Smith, R.; Menzies, D.; Mueller, J.L.; Kahru, M.; Toole, D.; et al. *SeaWiFS Postlaunch Calibration and Validation Analyses, Part. 3, NASA/TM-2000-206892*; Hooker, S.B., Firestone, E.R., Eds.; NASA Goddard Space Flight Center: Greenbelt, MD, USA, 2000; Volume 11.
105. Smith, R.C.; Baker, K.S. Optical classification of natural waters. *Limnol. Oceanogr.* **1978**, *23*, 260–267. [CrossRef]
106. Gordon, H.R.; Morel, A. *Remote Assessment of Ocean Color for Interpretation of Satellite Visible Imagery: A Review*; Springer: New York, NY, USA, 1983.
107. Smith, R.C.; Baker, K.S. Optical properties of the clearest natural waters (200–800 nm). *Appl. Opt.* **1981**, *20*, 177–184. [CrossRef] [PubMed]
108. Frouin, R.; Lingner, D.W.; Gautier, C.; Baker, K.S.; Smith, R.C. A simple analytical formula to compute clear sky total and photosynthetically available solar irradiance at the ocean surface. *J. Geophys. Res.* **1989**, *94*, 9731–9742. [CrossRef]
109. Morel, A. Optical properties and radiant energy in the waters of the Guinea dome and the Mauritanian upwelling area in relation to primary production. *J. Cons. Int. Explor. Mer* **1982**, *180*, 94–107.
110. Siegel, D.A.; Dickey, T.D. On the parameterization of irradiance for open ocean photoprocesses. *J. Geophys. Res.* **1987**, *92*, 14648–14662. [CrossRef]
111. Smith, R.C.; Marra, J.; Perry, M.J.; Baker, K.S.; Swift, E.; Buskey, E.; Kiefer, D.A. Estimation of a photon budget for the upper ocean in the Sargasso Sea. *Limnol. Oceanogr.* **1989**, *34*, 1673–1693. [CrossRef]
112. Moore, C.C.; Bruce, E.J.; Pegau, W.S.; Weidemann, A.D. WET Labs ac-9: Field calibration protocol, deployment techniques, data processing, and design improvements. In *Ocean Optics XIII*; SPIE Proceedings; SPIE: Bellingham, WA, USA, 1997; Volume 2963, pp. 725–731. [CrossRef]
113. Twardowski, M.S.; Sullivan, J.M.; Donaghay, P.L.; Zaneveld, J.R.V. Microscale quantification of the absorption by dissolved and particulate material in coastal waters with an ac-9. *J. Atmos. Ocean. Technol.* **1999**, *16*, 691–707. [CrossRef]
114. Röttgers, R.; McKee, D.; Woźniak, S.B. Evaluation of scatter corrections for ac-9 absorption measurements in coastal waters. *Methods Oceanogr.* **2013**, *7*, 21–39. [CrossRef]
115. Sathyendranath, S.; Platt, T. The spectral irradiance field at the surface and in the interior of the ocean: A model for applications in oceanography and remote sensing. *J. Geophys. Res.* **1988**, *93*, 9270–9280. [CrossRef]

7

Advantages and Limitations to the use of Optical Measurements to Study Sediment Properties

Emmanuel Boss [1,*], Christopher R. Sherwood [2], Paul Hill [3] and Tim Milligan [3]

1 University of Maine, Orono, ME 04469, USA
2 U.S. Geological Survey, Woods Hole, MA 02543, USA; csherwood@usgs.gov
3 Dalhousie University, 1355 Oxford Street, Halifax, NS B3H 4R2, Canada; paul.hill@dal.ca (P.H.); milligantg@gmail.com (T.M.)
* Correspondence: emmanuel.boss@maine.edu

Featured Application: Sediment characteristics and dynamics are studied using in situ optical measurements.

Abstract: Measurements of optical properties have been used for decades to study particle distributions in the ocean. They are useful for estimating suspended mass concentration as well as particle-related properties such as size, composition, packing (particle porosity or density), and settling velocity. Measurements of optical properties are, however, biased, as certain particles, because of their size, composition, shape, or packing, contribute to a specific property more than others. Here, we study this issue both theoretically and practically, and we examine different optical properties collected simultaneously in a bottom boundary layer to highlight the utility of such measurements. We show that the biases we are likely to encounter using different optical properties can aid our studies of suspended sediment. In particular, we investigate inferences of settling velocity from vertical profiles of optical measurements, finding that the effects of aggregation dynamics can seldom be ignored.

Keywords: particle dynamics; optical properties; suspended sediment

1. Introduction

Optical properties have long been used to study suspended particles and their dynamics (e.g., reviews by [1–3]). The most commonly measured optical properties are attenuation and scattering at different angles (both forward and back). Other optical devices, including ambient radiation sensors, cameras, and holographic instruments, also produce valuable data, but this paper will focus primarily on measurements of attenuation and scattering. Measurement volumes are typically small (from a few mL to tens of mL) and temporal averaging can increase the likelihood that rare large particles are sampled. Optical measurements can provide relatively direct estimates of mass or volume concentrations and particle size, and they also can be used to infer information about particle density, composition, and settling velocity. The primary advantages of using optical properties to study suspended particles are that they can be obtained at high frequency over long periods, and they are relatively non-invasive. Interpretation of optical measurements, however, are complicated by the fact that measurements are affected by all the particles in the suspension, but as we explain below, they do not respond to all particles equally. Other known disadvantages of optical instruments are that they saturate at high particle concentrations (e.g., [2]); they are intrusive and can produce turbulent wakes; they can be affected by ambient light; they can have large power demand; and they are susceptible to bio-fouling.

1.1. *Optical Proxies of Properties of Sediment Particles*

1.1.1. Volume and Mass Concentration

Most of the variation in optical signals measured in the field is due to changes in suspended particle mass concentrations (SPM; e.g., [4]). Optical estimates of SPM typically are made with measurements of attenuation or scattering at long visible wavelengths (e.g., 660 nm) or at infrared wavelengths (e.g., 850 nm), minimizing the impacts from varying dissolved materials and particulate absorption [5]. Transmitted light, which is reduced by scattering and absorption by particles, is measured in the near forward direction by so-called transmissometers, and scattered light can be measured at an angle near 90° by nephelometers, or at an angle greater than 90° by optical backscatter sensors (OBS). A multi-site comparison of the application of backscattering, side-scattering, and attenuation as proxies for SPM demonstrated their ability to predict SPM within 36%, 51%, and 54% respectively, for 95% of all cases [4]. The differences likely are due to variable sensitivity of each property to particle size, packing, and composition [1,6–8]. For example, the acceptance angle of transmissometers acts to filter out responses from larger particles [9]. SPM has also been estimated from space-based measurements of radiance (e.g., [10]). Remotely-sensed reflectance is most sensitive to the particulate backscattering coefficient in red and NIR wavelengths [11].

1.1.2. Size

Suspended sediment ranges in size from sub-μm-sized clay platelets to mm-sized sand and even larger flocs. It also encompasses plankton, non-algal organic particles, and aggregates that can be mixture of both organic and inorganic particles. We ignore in this paper particles capable of sinking at speeds >10 mm s^{-1}. These are rarely in suspension and, when they are in suspension, concentrations may saturate optical instruments. Optical proxies for size information include size distributions inverted from measurements of near forward scattering at several angles [12], the exponents of power-law fits of the particulate attenuation or backscattering spectrum [13–15] (but see [16]), and the fluctuation in optical signals, which can be used to obtain the average size of suspended particles [17]. Images of particles have also been used to derive size distributions, particularly of larger flocs and aggregates (e.g., [1,3]).

Theoretically, the maximal response of attenuation or scattering per volume (equivalent to mass if density is constant) occurs for single-grain sediment near $(D/\lambda)(n-1) \sim 1$ where λ is the wavelength in water ($= 0.75\lambda_{air}$), D is the particle diameter, and n is the index of refraction of the particle relative to water ([18], Figure 1). For solid inorganic particles with $n = 1.15$ at $\lambda_{air} = 660$ nm, the maximal attenuation per mass occurs for small particles with diameters between 0.8–3.2 μm. This dependency decreases as $1/D$ as D increases, and it increases for larger indices of refraction (Figure 1). Increases in the index of refraction are typically associated with increases in the inorganic fraction in the particle suspension [19].

The dependence of attenuation or scattering per unit of suspended mass on size should limit their use for estimation of SPM in suspensions with varying particle size, yet they are reasonably precise proxies for SPM across a range of environments, as discussed previously. This paradox is resolved if particles in suspension are not primarily single solid particles, but rather are agglomerations of particles separated by relatively large volumes of interstitial fluid and transparent organic material [1,20]. Terminology used to describe particle agglomerations in suspension can be ambiguous. Here we refer to a "floc" as an agglomeration of material that forms in suspension at relatively short times scales (e.g., tides) that is susceptible to breakup under increasing shear stress. We use the term "aggregate" to refer to an agglomeration that has undergone multiple cycles of resuspension and deposition, during which it has become more compacted and more strongly bound. While the component particles in flocs and aggregates may be similar, the density and settling velocity of aggregates is greater [21]. Flocs and aggregates typically sink faster than their component particles [22,23]. Flocs are broken by shear, including that generated by turbulence and by sinking, limiting their maximal size to about

1 cm [20,22,24–26]. Aggregates, because they are more compact and stronger, are not broken easily by shear. In terms of their optical properties, flocs, if sufficiently porous, can maintain the efficiency of scattering of the particles of which they are made [27], resulting in mass-specific optical properties that are independent of floc size. These theoretical results were validated in laboratory and field experiments that found that attenuation and backscattering to SPM ratios remained relatively constant despite large changes in particle size [1,28]. More densely packed aggregates theoretically should have mass-specific optical attenuation, scattering and backscattering coefficients that decrease with increasing aggregate size, but at a rate that is less than the $1/D$ dependence of solid particles. This theoretical relationship has not been demonstrated with measurements.

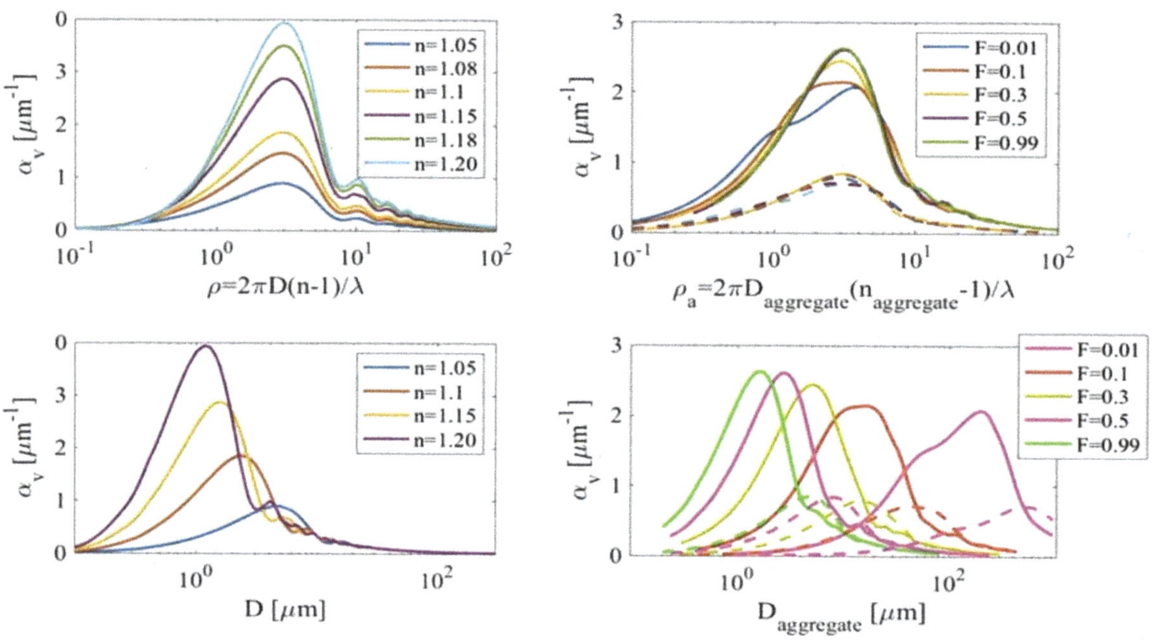

Figure 1. Top panels: volume-specific beam attenuation (α_v for solid particles (left) and aggregates (right) as function of $(2\pi D)/\lambda_{\text{water}}$ $(n-1)$ where D is diameter, λ_{water} the wavelength in water and n the index of refraction (which increases between organic and inorganic particles). For aggregates, $n_{aggregate} = 1 + F(n-1)$, where F is the solid fraction and n the index of refraction of the particles that comprise the aggregate. In all cases, the volume used to compute α_v is that of the solid fraction. Bottom panels: same as on top but plotted as function of particle diameter. In all the computations $\lambda_{air} = 660$ nm and $n' = 0.0001$, where n' is the imaginary part of the index of refraction, representing absorption. For aggregates we use $n = 1.15$ (solid) and $n = 1.05$ (dashed) typical of inorganic and organic materials, respectively.

1.1.3. Composition

Composition of suspended particles ranges from inorganic clays and silts of varied mineralogy to organic particles including both pigmented phytoplankton and non-algal particles. An optical proxy for composition (separating dominance by organic and inorganic particles) is the ratio of backscattering to total scattering [29,30]. Fluid-filled organic particles such as plankton have lower indexes of refraction compared to inorganic particles, resulting in a lower backscattering relative to total scattering. The ratio of chlorophyll-containing particles to total particles, estimated from the ratio of chlorophyll absorption to the particulate beam attenuation, has also been used as a compositional proxy in locations where the particle assembly includes a significant inorganic component [31]. In addition, the ratio of particulate organic carbon to SPM was found to correlate well with the ratio of particulate absorption at 675 nm to that at 570 nm [32]. It should be noted that optically derived knowledge of composition and/or size can lead to improved estimates of particulate concentration from optical measurements and associated composition (or size)-specific algorithms relative to those that do not account for variations in particle composition and/or size [1,16,32,33].

1.1.4. Packing

A proxy for bulk particle density has been derived from the ratio of beam attenuation (a mass concentration proxy) to the total particulate volume (obtained by inverting the forward scattering measurements with the Laser In Situ Scattering and Transmission (LISST) sensor [28,34,35]). If particle volume increases due to aggregation (with a significant fluid fraction), particle density will decrease. This will happen up to the point where the particle has a fractal dimension that is reduced below two, at which point most light passes through the particle and attenuation is caused only by component particles. Because the beam attenuation is also closely linked to the summed cross-sectional area of all the particles, it follows that the above proxy should behave similarly to the suspension's volume-to-cross-sectional-area ratio, or Sauter diameter [34].

1.2. Particle Dynamics

In this section, we focus on the bottom boundary layer (BBL; for a recent review of the fluid dynamics of the BBL see [36]). The particle assembly C is a sum of many types of particles $C = \sum_i C_i$, each with a specific settling speed, composition, size and any other property that is relevant to either hydrodynamic or optical properties.

The general conservation equation for particles can be written as:

$$\frac{\partial C_i}{\partial t} + \nabla(C_i(U - w_i)) = \nabla(K\nabla C_i) + \sum_j f(C_{i-j}, C_j) + S_i \tag{1}$$

where C_i is the mass concentration of particles of type i, t is time, U is the 3-D velocity field, w_i is the settling velocity of these particles (we use the convention that it is positive downward), K is a diffusion coefficient, $f(C_{i-j}, C_j)$ represents aggregation and disaggregation dynamics creating and destroying C_i-type particles, and S_i represents other sources and sinks (e.g., resuspension/deposition or biological production/consumption of particles).

To solve Equation (1) for a given flow field, one needs boundary conditions (e.g., flux or concentration of particles at the boundaries of the domain) and initial conditions (e.g., the state of the particles at time $t = 0$). Even then, Equation (1) cannot be solved analytically except for very simple cases such as we will address next.

Equation (1) can be simplified to represent only the vertical dimension (z, positive upwards), assuming that horizontal advection of horizontal gradients is negligible relative to vertical processes. After time averaging (or assuming steady-state), the steady balance between downward settling, upward diffusion, exchange among particle types (aggregation/disaggregation), and sources/sink is:

$$-w_i\frac{dC_i}{dz} = \frac{d(K_{eddy}dC_i)}{dz^2} + \sum_j f(C_{i-j}, C_i) + S_i \tag{2}$$

where C_i is the time- (and horizontally) averaged concentration of particles of type i, K_{eddy} is an eddy diffusion coefficient (representing the mixing by the small-scale turbulent field), and w_i is a constant settling velocity for each particle type i.

Large aggregation rates are associated with large particle concentrations, e.g., following a major resuspension event or during an algal bloom. Large disaggregation rates are associated with large fluid shears, e.g., in the wave boundary layer, which is a layer a few centimeters thick next to the bottom. If particle concentrations and fluid shears do not vary greatly throughout a boundary layer, then an equilibrium size distribution develops, and the aggregation and disaggregation terms cancel one another. An assumed equilibrium under certain circumstances is unlikely to emerge in bottom boundary layers, for example when near-bed wave-generated shears are much greater than current-generated shears higher in the boundary layer.

The law-of-the-wall is often invoked in the BBL, which is consistent with a linear eddy diffusivity profile where $K_{eddy} = \kappa u_* z$, and κ is von Kármán's constant (~0.4), and u_* is the friction velocity (a function of BBL turbulence, e.g., due to wave and current shear). Neglecting aggregation dynamics and sources/sinks, and integrating Equation (2) in the vertical and solving the resulting differential equation results in the Rouse equation [37,38] for a homogenous population of particles of type i:

$$C_i(z) = C_i(z_a)\left(\frac{z}{z_a}\right)^{-\frac{w_i}{\kappa u_*}} \tag{3}$$

where z_a is a reference elevation where a particle concentration is assigned. This is one of several different analytical solutions for the balance of settling and turbulent mixing, which vary depending on assumptions about the eddy diffusivity profile [39,40]. An alternative derivation of Equation (3) using probability theory has also been proposed [41]. Equation (3) predicts a profile that is linear in $\log(C_i)$ versus $\log(z)$, and all of the other forms predict similar decreases in concentration with elevation above the bottom. Because mass concentrations can be summed, the bulk concentration profile is:

$$C_b(z) = \sum_i C_i(z) = \sum_i \left(C_i(z_a)\left(\frac{z}{z_a}\right)^{-\frac{w_i}{\kappa u_*}} \right) \tag{4}$$

It follows that the bulk particle concentration will also decrease with height above the bottom. Note, however, that if a suspension comprises sub-populations of particle types with differing settling velocities, each will have a different vertical profile (Equation (3)), and the profile of the summed concentration will follow Equation (4), which does not, in general, follow the linear shape in log-log space. In fact, large deviations in the bulk profile from Equation (3), under steady conditions in a turbulent BBL, are likely indications that a suspension contains particles with diverse settling velocities.

1.3. An Equation for the Vertical Distribution of an Optical Property

The mass concentration of a population of particle type i is $C_i = N_i V_i \rho_I$, where ρ_i is individual particle density, N_i is the number concentration of particles of type i, and V_i is individual particle volume. By the Beer–Lambert law the bulk optical response (e.g., backscattering or beam attenuation) of a sub-population $b_{x,i}$, is the simple sum of the individual contributions so that:

$$b_{x,i} = \alpha_{v,i} N_i V_i \tag{5}$$

and the bulk response to the combined sub-populations is:

$$b_x = \sum_i b_{x,i} \tag{6}$$

where $\alpha_{v,i}$, is a volume-specific optical property. Volume- or mass-specific optical properties typically are calculated by using Mie theory (which assumes homogeneous spherical particles) for solid particles, or by using other models for flocs (e.g., [31]). The calculations of $\alpha_{v,i}$ require as input particle diameter, index of refraction (a function of composition), wavelength of light, and, for flocs, the fractal dimension that relates volume concentration to mass concentration. The results are resonance-like functions of size (Figure 1). Although the application of Mie solutions to backscattering has been challenged based on observations (e.g., [42]), it is reasonable to relate constant values of $\alpha_{v,i}$ to specific particle types. Both Equations (5) and (6) rely on the Beer–Lambert law, with underlying assumptions that the light is monochromatic with parallel rays and, most importantly, that the particles do not scatter the light multiple times. This is clearly not the case as particle concentrations rise but, at low concentrations, we may be able to assume that Equations (5) and (6) are valid. Substituting optical response for concentration in Equation (3) yields an equation for the optical-response profile for a homogenous sub-population i with identical $\alpha_{v,i}$, V_i, and w_i:

$$b_{x,i}(z) = b_{x,i}(z_a)\left(\frac{z}{z_a}\right)^{-\frac{w_i}{\kappa u_*}} \quad (7)$$

Equations (6) and (7) can be combined to yield an equation for the bulk response of a heterogeneous population:

$$b_x(z) = \sum_i b_{x,i}(z_a)\left(\frac{z}{z_a}\right)^{-\frac{w_i}{\kappa u_*}} = \sum_i \frac{\alpha_{v,i}C_i(z_a)}{\rho_i}\left(\frac{z}{z_a}\right)^{-\frac{w_i}{\kappa u_*}} \quad (8)$$

As with particle concentrations, we find that the information provided by a profile of optical properties will depend on how heterogeneous the suspension is. Unlike the concentration profile, the averaging done by the optical properties depends on how the optics respond to different particles (Equation (8)). Hence, the more biased an optical property is towards a specific particle type, the better it can provide information on its specific settling speed.

2. Observations

Field data were obtained from a profiling instrument platform deployed at the Martha's Vineyard Coastal Observatory (MVCO) south of Martha's Vineyard, MA, USA, at the 12-m isobath in the summer of 2011, as part of the Office of Naval Research (ONR)-funded Optics and Acoustics and Stress In Situ (OASIS) experiment [43]. The platform was mounted on a pivoting arm that profiled every 20 min from 10 cm to 2 m above bottom. The platform was equipped with variety of optical and acoustical sensors. Here we discuss data from a Sequoia LISST 100-X (Sequoia Scientific, Bellevue, WA, USA), a EcoBB2f (WETLabs, Philomath, OR, USA) triplet measuring dissolved organic fluorescence and backscattering at two wavelengths (532 and 650 nm), and a WETLabs AC-9 spectral absorption and beam attenuation meter. Water was pumped from an intake at the tip of the arm into the 10-cm pathlength sampling volume of the AC-9. An automatic valve periodically routed the water sample through a 0.2-μm filter to remove particulates, leaving the dissolved fraction, to obtain calibration-independent particulate properties [44]. Shear velocity u_{*c} [cm/s] associated with mean flow in the bottom boundary layer (BBL) was inferred from a pair of acoustic Doppler velocimeter (Sontek, San-Diego, CA, USA) measurements (cf. [45]), and model estimates of the wave–current combined maximum shear velocity in the wave boundary layer (relevant to sediment resuspension) were determined using a version of the Grant–Madsen model [46,47]. We have plotted sign(u) u_{*c} in Figure 2, where u is the east–west component of the current velocity, to differentiate flood (east, positive) from ebb (west, negative). The location is dominated by the east–west semi-diurnal tidal currents, northward swell, and periodic storms (Figure 2). The measurements discussed here were taken in the bottom boundary layer, which is mostly well mixed.

Waves and currents varied during the experiment, and we have selected four periods with distinctly different forcing and optical responses (Figure 2). Two periods (Maria and Ophelia) were associated with offshore passage of hurricanes and arrival of swell waves at the study site. Another period (Spring tides) was associated with moderate wave conditions and strong spring tidal currents. The fourth period (Calm) was characterized by low waves and weaker neap tidal currents. For each of the four periods identified in Figure 2, we show the distribution of the following optical properties (or properties inferred from optical measurements):

1. Beam attenuation $c_p(650)$ [m^{-1}] measured by the AC-9 and particulate backscattering coefficient ($b_{bp}(650)$ [m^{-1}] measured by the EcoBB2F provide proxies of particulate concentration (e.g., [5], where higher values associated with higher particle concentrations.
2. Exponent of the power-law fits of the particulate beam attenuation γ_{cp} [dimensionless] and backscattering γ_{bbp} [dimensionless] ($c_p = c_p(\lambda_0)(\lambda/\lambda_0)^{-\gamma_{cp}}$, with an analogous formula for γ_{bbp})

provide proxies for size distribution in the finer sizes (e.g., [13]). Lower values are associated with larger size averaged particles. γ_{cp} is biased towards the smaller (0.5 to 10 μm) particles in the population [9], and γ_{bbp} may be more sensitive to larger particles [15].

3. Sauter diameter D_s [μm] is determined from the ratio of LISST measurements of volume and area concentrations, summed over size classes i as $D_s = 1.5\sum V_i/\sum A_i$ and reciprocal of particle density $\rho_a{}^{-1} = \sum V_i/c_p$ [m ppm^{-1} = μm], using the LISST-based c_p. Both are proxies for packing: larger values of D_s indicate larger, less-dense particle populations, and larger values of $\rho_a{}^{-1}$ also indicate less-dense particle populations.
4. Particulate backscattering ratio $b_{bp}(532)/b_p(532)$ measured by the EcoBB2F ($b_{bp}(532)$) and by differencing of particulate attenuation and particulate absorption from the AC-9 ($b_p(532)$) was a proxy of composition. Increasing values of this ratio are associated with inorganic particles [29,30]. For very small particles, this ratio is also sensitive to size, increasing for smaller particles.
5. Chlorophyll to attenuation ratio $Chl/c_p(650)$ is another proxy of composition where higher values are associated with higher phytoplankton-based organic content [31].
6. LISST-based size distribution spanning from 2–250 μm at 32 size bins and using a spherical kernel.

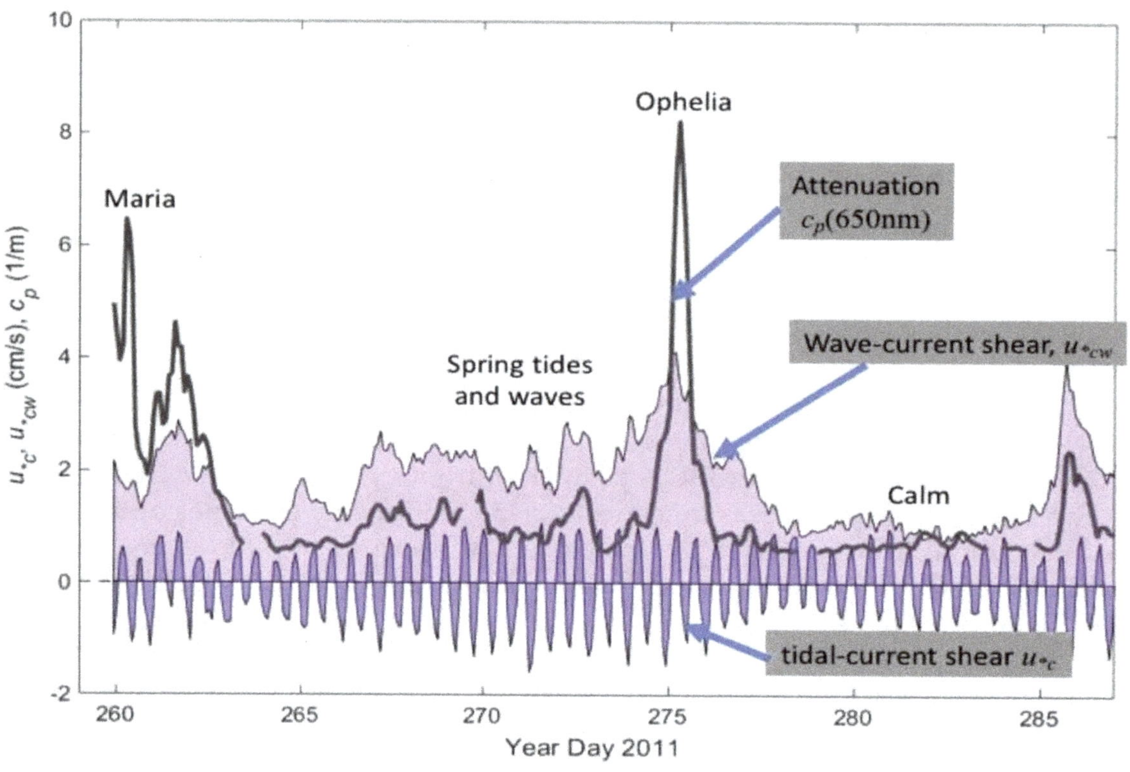

Figure 2. Time series of conditions at the 12-m Martha's Vineyard Coastal Observatory (MVCO) site during the Optics and Acoustics and Stress In Situ (OASIS) deployment in 2011: beam attenuation at 650 nm measured at 1 m above the bottom (gray), tidal current shear velocity sign(u)u_{*c} (blue), and combined wave-current shear velocity u_{*cw} (purple). Notice labels describing specific periods.

We selected 1-h intervals during each of the different periods identified in Figure 2 when the concentration profiles estimated from attenuation decreased monotonically with elevation above the seabed, and the profiles were relatively well approximated by a linear profile in log-log coordinates (Equation (5)). These profiles were consistent with the steady-state Rouse balance discussed above, suggesting that we might be able to neglect the effects of horizontal gradients and temporal transients. The data represent the average of three consecutive profiles, each of which took 20 min to complete. All properties are displayed as a function of elevation above the bottom in Figures 3–6. Trends in vertical profiles and representative values (mean and standard deviation) for each time period are summarized in Tables 1 and 2.

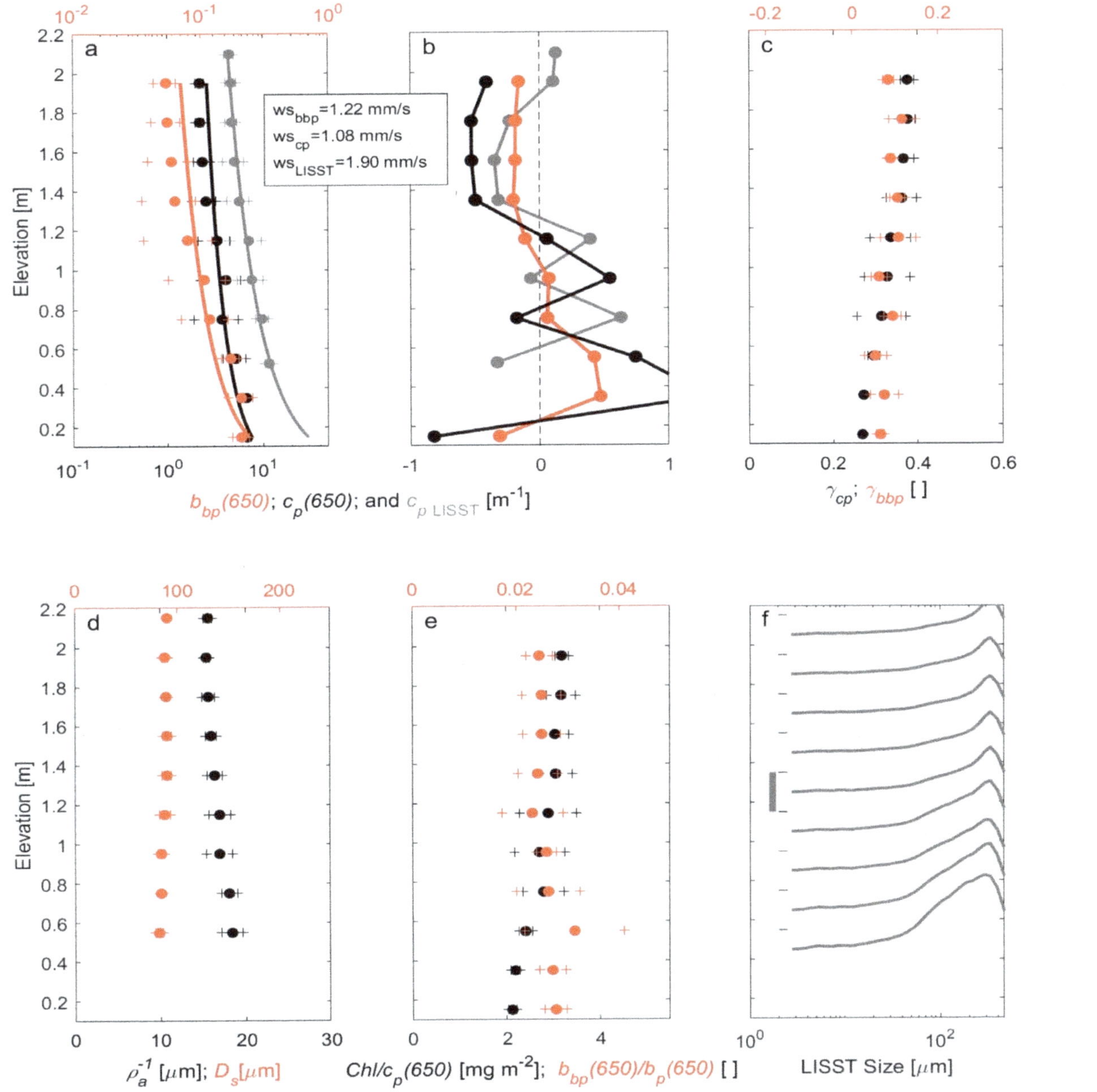

Figure 3. Profiles of optical parameters at the MVCO site on day 261.1 when waves from Hurricane Maria were coming to shore (see Figure 2). (**a**) Beam attenuation ($c_p(650)$, black), particulate backscattering ($b_{bp}(650)$, red), and LISST attenuation (gray). (**b**) Deviation from the Rouse-profile fits for the same parameters (i.e., observed minus fit). (**c**) Power-law exponent of $c_p(650)$ (black) and $b_{bp}(650)$ (red). (**d**) Sauter diameter (red) and inverse particle density (black). (**e**) Ratio of chlorophyll divided by $c_p(650)$ (black) and backscattering ratio (red). (**f**) Spectra of LISST volume concentration as a function of size at nine elevations. In panels (**a**, **c**, **d**, and **e**), standard deviation about the mean values for three consecutive profiles (60 min) are shown with crosses. Dashed lines in panel (**a**) are log-log (Rouse) fits to the data. In panel (**f**), the elevations for each spectrum are indicated by black lines, and the gray vertical scale indicates 10 μL/L. Numbers in the box denote settling velocities based on Rouse fits to backscattering, AC-9 particulate attenuation at 660 nm, and LISST attenuation at 670 nm.

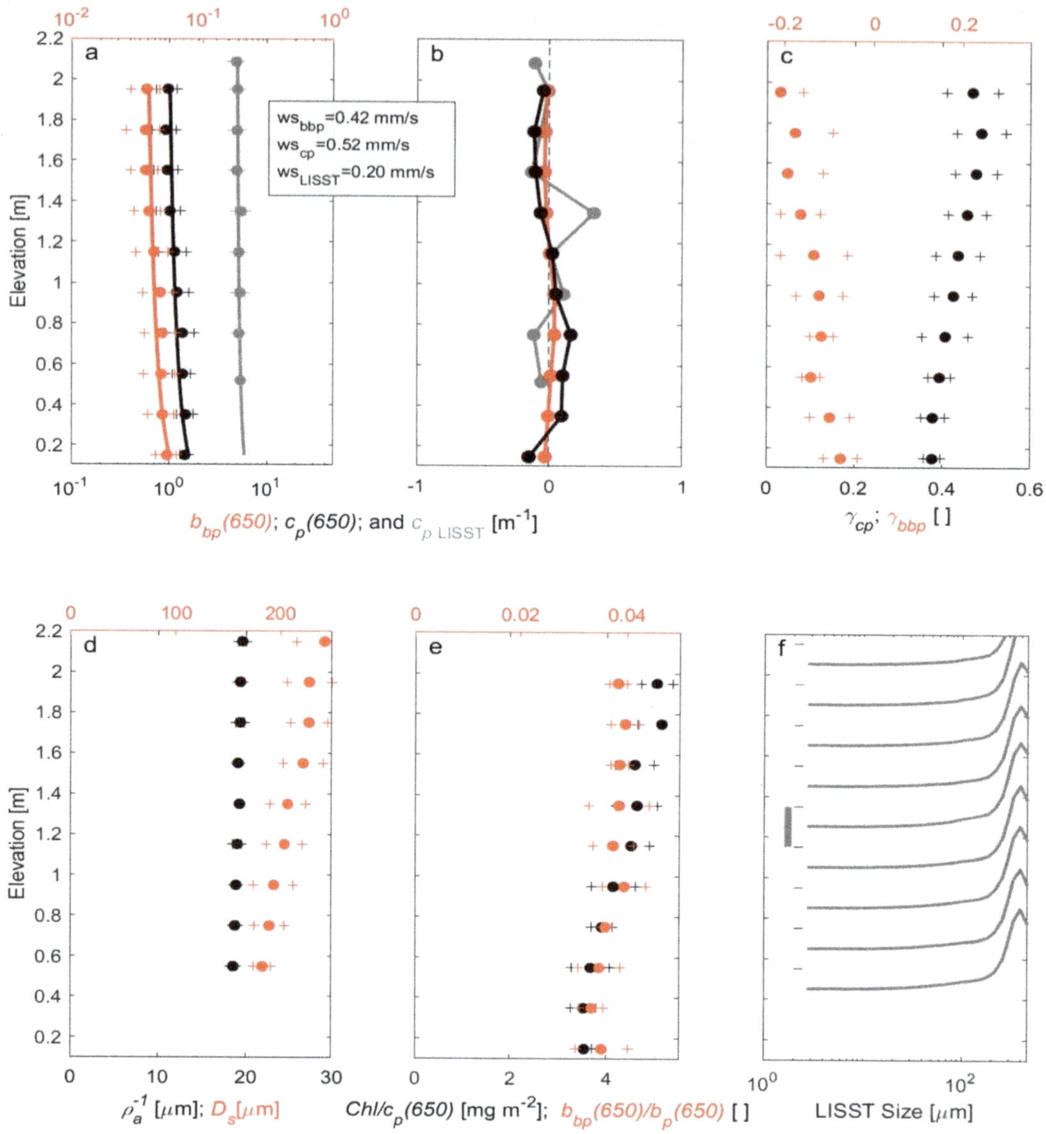

Figure 4. Profiles of optical parameters at MVCO on day 268.62 during spring tides and moderate waves. Panels are as described in Figure 3.

Table 1. Trends in particle parameters as function of depth based on the average of three profiles in the BBL in each of the four periods denoted in Figure 2. ↘ and ↗ denote profiles that are decreasing or increasing (respectively) with elevation above the bottom, ∼ denotes that the trend is weak, and | denotes that there is no trend with elevation.

Parameter	Maria	Spring Tide	Ophelia	Calm
$b_{bp}(650)$	|	↘	↘	∼↘
$c_p(650)$	↘	↘	↘	↘
γ_{bbp}	↗	↘	↗	∼↗
γ_{cp}	↗	↗	↗	|
D_s	|	↗	∼↗	∼↘
$\rho_a{}^{-1}$	∼↘	↗	∼↗	∼↘
$b_{bp}(650)/b_p(650)$	∼↗	↗	|	|
$Chl/c_p(650)$	↗	↗	∼↗	↗

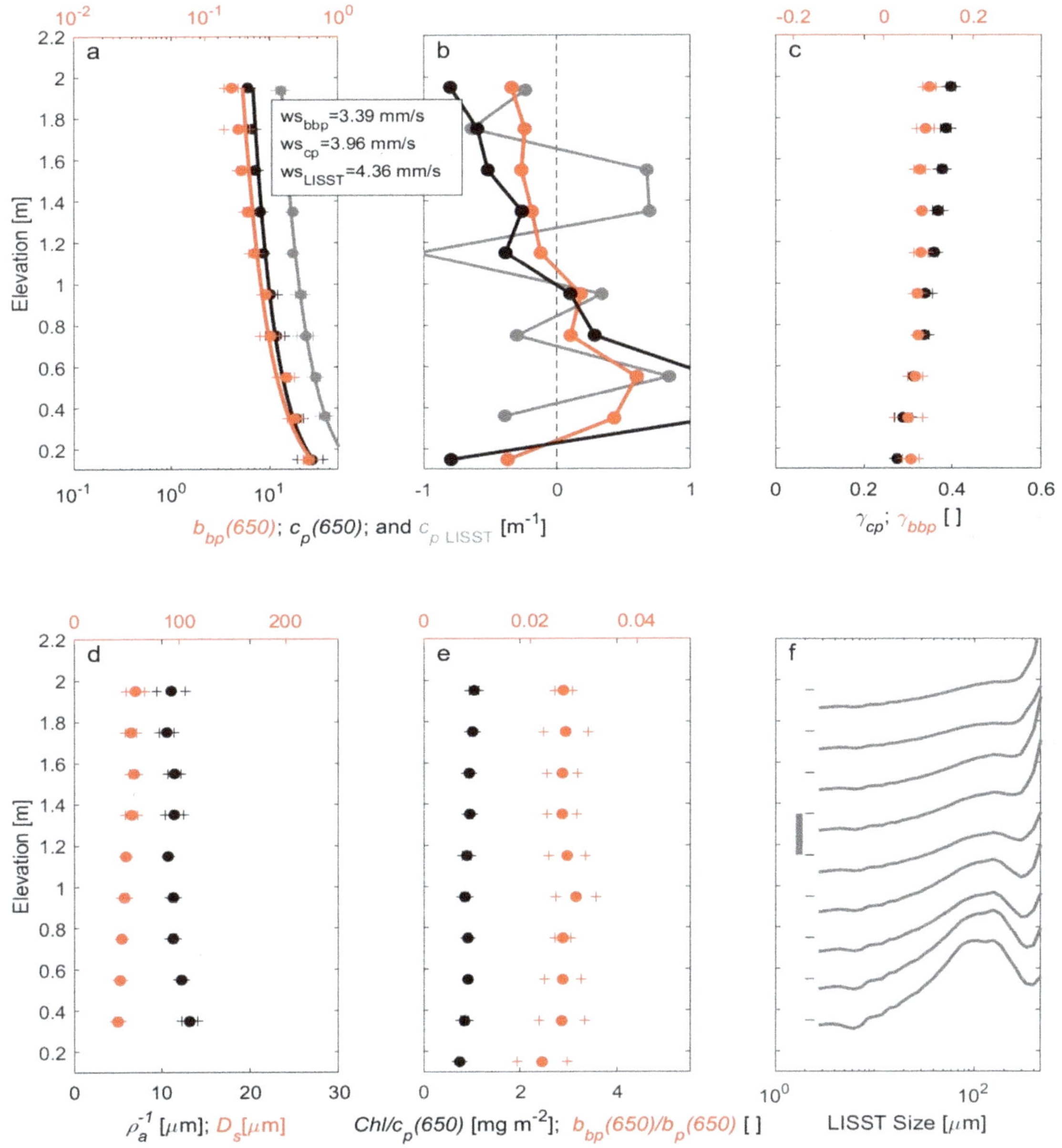

Figure 5. Profiles of optical parameters at MVCO on day 275.20 during the passage of Hurricane Ophelia. Panels are as described in Figure 3.

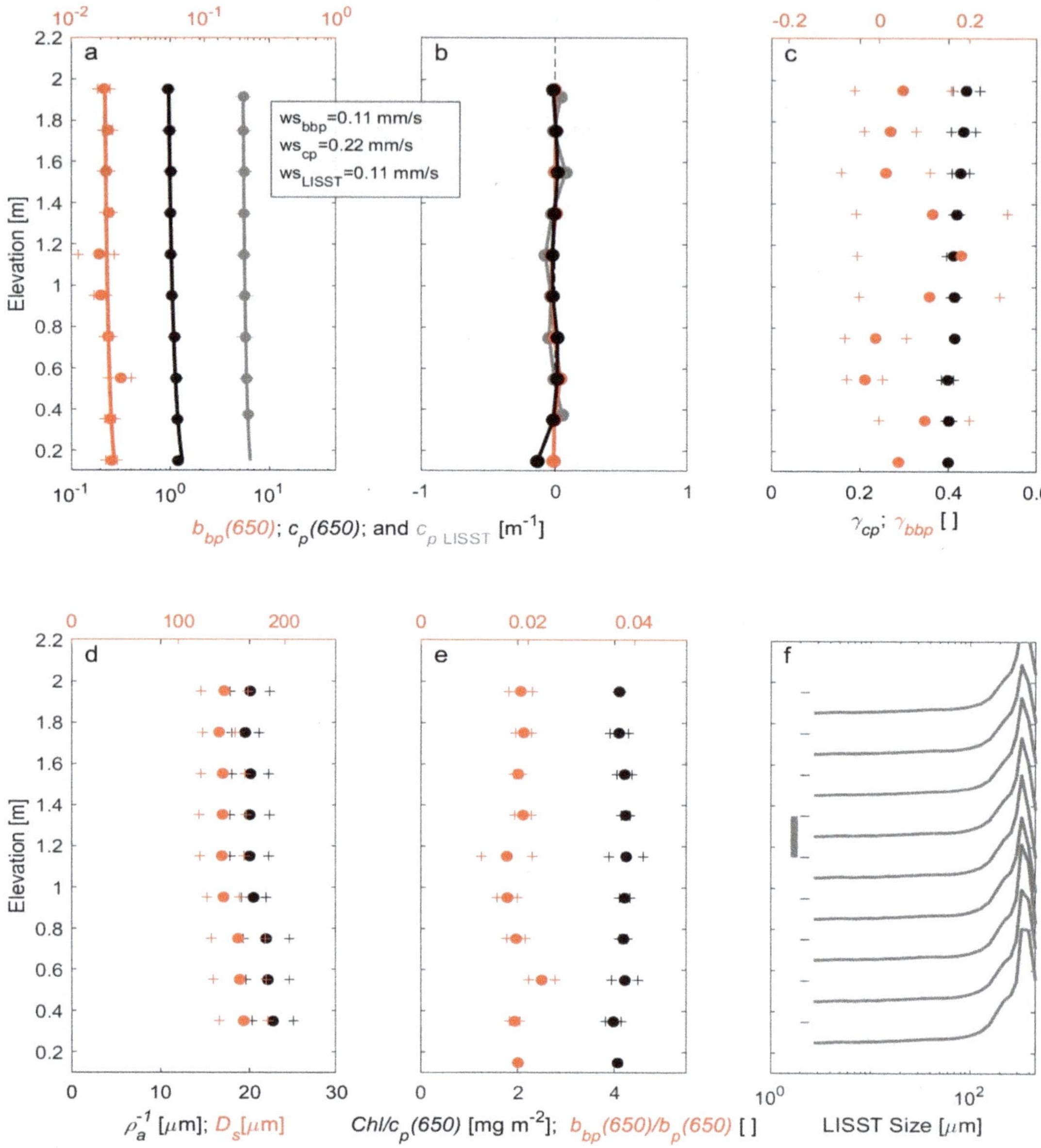

Figure 6. Profiles of optical parameters at MVCO on day 281.93 during calm conditions. Panels are as described in Figure 3.

3. Results

3.1. Suspended Particulate Material (SPM)

The overall concentration of SPM fluctuated through the experiment (Figure 2), as indicated by variations $c_p(650)$ (Figure 2) and $b_{bp}(650)$ (not shown). The time series of $c_p(650)$ at 1 m above the bottom indicates increased SPM during Maria and Ophelia, somewhat reduced SPM during the spring tides, and low SPM during the calm period. Mean values of $c_p(650)$ in the 60-min profiles ranged from about 1 m^{-1} during the calm period to 12 m^{-1} during Ophelia, when wave-induced resuspension increased SPM in the BBL. Mean values of $b_{bp}(650)$ generally covaried with $c_p(650)$, ranging from 0.02 m^{-1} to 0.30 m^{-1} for the same periods. The $c_p(650)$ and $b_{bp}(650)$ concentrations decreased with elevation, and the profiles examined here were nearly linear in log-log space (Rouse-like). However, there was typically more scatter in the profiles of $b_{bp}(650)$, as evidenced by their lower r^2 (Table 2), especially during periods of lower concentrations.

Table 2. Mean and standard deviation (in brackets) of optical particle parameters computed based on the average of three profiles in the BBL in each of the four periods denoted in Figure 2. The wave-current and current shear velocities and the settling velocities inferred from fitting the Rouse profile to c_p and b_{bp} are also listed, with regression coefficient r^2 in brackets.

Parameter	Maria	Spring Tide	Ophelia	Calm
$b_{bp}(650)$ [m^{-1}]	0.11(0.06)	0.04(0.01)	**0.30**(0.14)	0.02(0.002)
$c_p(650)$ [m^{-1}]	3.87(1.77)	1.19(0.22)	**11.81**(6.51)	1.06(0.09)
γ_{bbp}	**0.09**(0.02)	−0.14(0.06)	0.0(0.01)	0.06(0.07)
γ_{cp}	0.33(0.04)	**0.43**(0.04)	0.34(0.04)	0.42(0.01)
D_s [μm]	87.3(3.0)	**210.7**(20.5)	50.4(6.0)	147.0(8.7)
${\rho_a}^{-1}$ [m ppm^{-1}]	16.6(1.1)	19.1(0.3)	11.5(0.8)	**20.8**(1.1)
$b_{bp}(650)/b_p(650)$	0.03(0.002)	**0.04**(0.002)	0.03(0.002)	0.02(0.002)
$Chl/c_p(650)$ [μgm^{-2}]	2.8(0.4)	**4.3**(0.6)	0.9(0.1)	4.2(0.1)
u_{*wc} [cm/s]	2.3	2.4	**4.1**	0.8
u_{*c} [cm/s]	0.6	0.7	**1.8**	0.4
ws_{bbp} [cm/s]	1.22(0.72)	0.42(0.72)	**3.39**(0.94)	0.11(~0.0) *
ws_{cp} [cm/s]	1.08(0.80)	0.52(0.66)	**3.96**(0.99)	0.22(0.95)

* fit was not significant. Bold is used to highlight the largest values in each row.

3.2. Settling Velocities Assuming Rouse Profiles

Mean apparent settling velocities inferred from $c_p(650)$, $c_{p,\text{LISST}}$ and $b_{bp}(650)$ profiles ranged from about 0.1 to 4 mm/s. Although the settling velocities estimated from each instrument during a given period were similar, and showed similar trends over the four periods, $w_{s,cp(650)}$ was usually larger than $w_{s,bbp(650)}$. Particulate beam attenuation measured by the AC-9 is less sensitive to large, low-density particles, so settling-velocity estimates from $c_p(650)$ were likely to favor smaller, denser, faster-settling aggregates and single-grained particles. Inferred settling speeds were largest during passage of Maria and Ophelia. They were intermediate during spring tides, and they were lowest during the calm period. During passage of the two storms the residuals of the Rouse fits showed vertical structure, with negative residuals higher in the boundary layer and positive residuals lower in the boundary layer. This pattern indicates that, nearer the bed, the profiles during storms were steeper than the Rouse balance.

3.3. Size

The spectral exponents of attenuation and backscatter increased slightly with height above the bed during the calm periods and during the passage of Maria and Ophelia. During the spring tides, the spectral exponent of attenuation increased with height above the bed, but the spectral exponent of backscatter decreased with height above the bed. Values of γ_{cp} were smaller during passage of the two storms than during the calm period and during spring tides. Values of γ_{bbp} were smallest during spring tides, even taking on negative values. Values were higher during the storms, and they were scattered during the calm period.

Sauter diameter and the reciprocal of floc density generally were well correlated because they are based on similar measurements. Sauter diameter was smaller during passage of Ophelia, when it was ~50 μm, and during passage of Maria, when it was ~100 μm. During the spring tides, Sauter diameter was largest, with diameters ~200 μm. During the calm period, Sauter diameter was ~150 μm. During Maria, the spring tides, and the calm period, inverse densities were similar and equal to ~20 μm. Values were ~10 μm during the passage of Ophelia. During the passage of the two storms, inverse density and Sauter diameter values diverged near the bed, and during the spring tides, they diverged higher in the boundary layer.

Size distributions from the LISST provide a more complete understanding of the vertical and temporal changes in size distribution. During the calm period and during spring tides, the size distributions had well-defined modes at the upper end of the LISST range. The widths of the modes were relatively constant with height above the bed during the calm period, but they were wider nearer the bed during the spring tide period. During passage of Maria, distributions were unimodal, but they were more skewed to smaller sizes than during the calmer periods. During passage of Ophelia, curious bi-modal distributions emerged. One mode was centered on diameters of ~100 μm, and the other was located in the largest diameter class. This coarser mode was dominant farther away from the bed, and the finer mode grew larger nearer the bed.

3.4. Composition

The backscattering ratio was relatively high (>0.02) during the full deployment, indicating that the suspensions were dominated by inorganic particles. The ratio was smaller during the passage of the two storms than it was during the spring tides. The backscatter ratio was not correlated to the Chl/c_p ratio, which increased away from the bed in each period, consistent with chlorophyll-containing particles being at lower concentration relative to other particles closer to the bed. Chlorophyll profiles typically showed no trend with elevation (not shown), consistent with the hypothesis that the increase in the Chl/c_p ratio away from the bed was due to enrichment in faster-settling non-algal particles near the bottom.

4. Discussion

4.1. Inferences from Optical Properties

The vertical profiles of the attenuation and backscattering coefficients were fit reasonably well by Rouse profiles, and the estimated settling velocities increased with increasing shear stresses, as expected. The magnitudes of the estimated settling velocities are within ranges expected for flocs and for the fine sands typical of the seabed at the MVCO site. The ~0.1 mm s^{-1} estimated settling velocities during the calm period were small, and they indicate material that behaved as slowly sinking washload that is evenly distributed throughout the boundary layer. The ~0.4 mm s^{-1} estimated settling velocities during the spring tides are in the range of typical floc-settling velocities, and the ~1 mm s^{-1} estimated settling velocities during the passage of Maria are typical of flocs (e.g., [48]). The ~4 mm s^{-1} observed during the passage of Ophelia are larger than typical floc-settling velocities, but they are representative of very fine sand (D = ~80 μm; [49]) or densely packed aggregates [21].

Interestingly, boundary shear stresses during passage of Maria were similar to boundary shear stresses during the spring-tide period, yet the estimated settling velocities were more than two times larger during Maria. Residual plots of the Rouse fits indicate that the profiles had smaller gradients (lower settling velocities) during spring tides than the profiles near the bed during passage of Maria. In addition, Sauter diameters were smaller during passage of Maria, as were the spectral exponents of attenuation. Particle size was inverted during the spring tides, with larger particles farther above bottom. The spectral exponent of backscattering was lower during the spring tides than during passage of Maria, and it decreased with height above bottom.

Despite the similar stresses during these two periods, these observations suggest different particle dynamics. The larger Sauter diameters, inverted particle size, and smaller values of γ_{bbp} are consistent with the hypothesis that flocs dominated the suspension during the spring tides. The flocs were fragile, and they were disrupted near the bed, but they were able to reform higher in the boundary layer. The breakage of flocs near the bed decreased bulk estimated settling velocities from the profiles by transferring mass near the bed from faster-sinking large flocs (~1 mm s^{-1}) into slower-sinking smaller flocs or single grains. Overall, however, the extensive packaging of small particles into large flocs during spring tides was associated with small values of γ_{bbp} [15]. During the passage of Maria, we hypothesize that higher bottom shear stresses associated with the larger swell caused resuspension of fine sand typical of the site (D = ~125 μm; [50]) and caused a greater degree of floc breakup. Resuspension of sand caused larger gradients in optical properties near the bed. Resuspended sand also can account for the broadened peak in the particle size distributions. Greater floc breakup can account for smaller values of γ_{cp} during passage of Maria, because particle mass was transferred out of large flocs that were not sensed by the AC-9 into smaller flocs or single grains that were [15]. It can also account for large values of γ_{bbp} [15].

During passage of Ophelia, when stress was highest, particle sizes were smallest. This result suggests that floc breakup was more extensive during Ophelia (e.g., [26]). The finer mode in the LISST size distributions was similar to the fine sands at the site (D = ~125 μm), and the presence of this sand, with typical settling velocities of 8 mm s^{-1} [51], steepened the near-bed profiles of the optical properties. The appearance of the coarser mode at the upper limit of the LISST size range, however, is not consistent with greater floc breakup. We hypothesize that these particles were resuspended bed aggregates that were tougher than flocs and not prone to breakup (cf. [21]). The fact that they were relatively more abundant higher above the bottom suggests that they had settling velocities smaller than that of the fine sands (<8 mm s^{-1}), consistent with the estimated settling velocities in the outer part of the profile of 4 mm s^{-1}.

The hypothesis of sand resuspension during passage of the two storms can be reconciled with observed variations in the backscatter ratio, which was lower during passage of the two storms than it was during the period of spring tides. Permeable sand beds can store organic matter in the interstitial pore spaces [52]. We propose that during resuspension events, this organic material was resuspended, which caused backscatter ratios to decrease. The observation that the Chl/c_p ratio did not change indicates that the organic matter in the bed interstices was degraded and did not contain chlorophyllous material.

The relative magnitudes of Sauter diameter and inverse density in the different periods may also argue for different particle dynamics during passage of the storms. During these periods, the ratios of Sauter diameter to inverse density were smaller than during the periods of spring tides and calm. This ratio reduces to the ratio of the beam attenuation coefficient measured by the LISST to the particle area estimated by the LISST. The inversion of near-forward scattered light measured by the LISST to particle area assumes that a single refractive index applies to all particles. We hypothesize that small organic particles liberated from the bed during resuspension had attenuation to particle area ratios that were smaller than the ratios for particles during the spring tides and during the calm period.

4.2. Broader Advantages and Disadvantages of Optical Measurements

Measurements of optical properties often behave in ways consistent with our expectations. When they do, their advantages are that they provide robust, relatively non-intrusive, well-understood first-order information about the suspended particle populations. Their main disadvantages are their sensitivity to ambient light and limitations in turbid conditions, tendency to foul, and, in some cases, their large size, power requirements, or data storage limits. The more subtle disadvantages are the biases that each instrument has, often based on wavelength, angles of illumination, or acceptance angles. These disadvantages can be turned to advantage by using combinations of different measurements, each with a different bias, as we have shown here. Combinations of optical measurements (attenuation and backscatter at various wavelengths, measurements of chlorophyll, size data from the LISST) provide more information about the size and composition of the particle population.

Simple, time-tested optical measurements such as attenuation and backscatter remain valuable, even in complicated settings like MVCO. Their main advantage is they yield simple interpretations in terms of SPM that capture the first-order variations that often can be linked to wave- and current-induced bottom stress. The main disadvantages are that they can be biased by size and composition, and the conversion from attenuation or backscatter to SPM can vary, depending on instruments, particle size, composition [1,2].

The advantages of direct measurements of size from the LISST are clear: they provide information that cannot be measured directly with simpler optical sensors, and the size information is very useful for interpreting the dynamic behavior of the particles. One important disadvantage is that LISST instruments are large and intrusive. In addition, they have upper limits on sizes and, when larger particles are present, they can bias results by elevating reported volumes in smaller size classes [53]. Finally, when flocs are present, the LISST may be sensitive to smaller component particles, leading to over-representation of microflocs when macroflocs are present [54].

Near-bottom profiles of optical properties allow the observations of particle populations to be interpreted in terms of sediment dynamics. In particular, near-bottom profiles of size from the LISST have not been previously reported, and offer the advantage of direct observations of size changes in a region with strong gradients in concentration and turbulent shear.

The nearly-ubiquitous appearance of Rouse-like profiles in optical measurements provides estimates of settling velocity that can be linked to particle size. The disadvantage of these interpretations is that the bulk profile can be confounded by changes in particle density (and therefore settling velocity) or skewed by overlapping profiles of diverse particle populations with varying vertical distributions. In all the cases illustrated here, relatively good Rouse fits provided settling-velocity estimates, but they could vary by an order of magnitude between profiles. Additional information was required to evaluate the underlying particle dynamics, and both floc dynamics and resuspension were found to be important.

Addition of acoustical measurements [55] could further constrain the particle dynamics and distribution as their sensitivity to composition, size, and packing differs from optical measurements. For example, for acoustic backscattering in the MHz frequency range, aggregation decreases the signal [56] while being sensitive to large single-grain particles (whereas the optical response is not).

Models of resuspension and settling have been used for years to examine the dynamics giving rise to Rouse-like vertical profiles in the BBL [35,36,57–60]. However, these models have not been coupled to an optical model so that their output could be validated with optical (and/or acoustical) observations. Models that now include resuspension as well as aggregation/disaggregation dynamics (e.g., [61]) can be coupled with optical models (for single grains as well as flocs), making it possible to express modeled variables (e.g., spatial and temporal concentrations of different types of particles) as optical measurements and allow for direct comparison with field observations. This also opens the possibility for assimilating such observational data into the models. Thus, the development of optical models of sediment dynamics and their evaluation on real-world datasets of is critically important.

5. Conclusions

This paper has described a suite of optical measurements designed to provide information about suspended particle populations. We have discussed the biases of these measurements and demonstrated their value in analyses of field observations. In conclusion:

- Near-bottom profiles of optical properties are valuable because they sample particle populations in a region with strong gradients in turbulence and concentrations.
- Profiles with combinations of instruments can be used to make inferences about sediment dynamics in the bottom boundary layer. Resuspension of bottom material and dynamics of aggregation and disaggregation are especially important at the MVCO study site.
- Aggregation/disaggregation dynamics cannot be neglected when interpreting profiles of properties sensitive to the small particles (e.g., beam attenuation) as the flocs are both a sink and source for fine particles.
- Combinations of optical instruments provide information about suspended particle population that individual instruments cannot, because of their individual design and biases. Many of the disadvantages associated with individual optical sensors can be turned to advantages when multiple sensors are used.

Author Contributions: P.H., T.M., E.B., and C.S. conceived, designed, and performed the experiments. C.S. and E.B. processed the data; E.B. conceived the paper and wrote the first draft; C.S., P.H. and T.M. contributed to the writing.

Acknowledgments: This work was supported by the Office of Naval Research and the United States Geological Survey Coastal and Marine Geology Program. The unique instrument platform and data acquisition system was designed and built by technical staff lead by Marinna Martini at the United States Geological Survey Woods Hole Coastal and Marine Science Center. This team was also responsible for deployment and recovery of the instrumentation. We thank the Woods Hole Oceanographic Institution (WHOI) MVCO staff for support during this experiment, and we thank the captains and crews of the R/V Connecticut and the R/V Tioga. Any use of trade, product, or firm names is for descriptive purposes only and does not imply endorsement by the United States Government. This paper has benefited significantly from insightful comments from D. Stramski, A. Aretxabaleta and two anonymous reviewers.

Abbreviations

The following abbreviations are used in this manuscript:

BBL	bottom boundary layer
LISST	Laser In Situ Scattering and Transmission
MVCO	Martha's Vineyard Coastal Observatory
OASIS	Optics and Acoustics and Stress In Situ
ONR	Office of Naval Research
SPM	suspended particulate mass
USGS	USA Geological Survey
WHOI	Woods Hole Oceanographic Institution

References

1. Hill, P.S.; Bos, E.; Newgard, J.P.; Law, B.A.; Milligan, T.G. Observations of the sensitivity of beam attenuation to particle size in a coastal bottom boundary layer. *J. Geophys. Res.* **2011**, *116*, C02023. [CrossRef]
2. Downing, J. Twenty-five years with OBS sensors: The good, the bad, and the ugly. *Cont. Shelf Res.* **2006**, *26*, 2299–2318. [CrossRef]
3. Mikkelsen, O.A.; Hill, P.S.; Milligan, T.G.; Chant, R.G. In situ particle size distributions and volume concentrations from a LISST100 laser particle sizer and a digital floc camera. *Cont. Shelf Res.* **2005**, *25*, 1959–1978. [CrossRef]
4. Boss, E.; Taylor, L.; Gilbert, S.; Gundersen, K.; Hawley, N.; Janzen, C.; Johengen, T.; Purcell, H.; Robertson, C.; Schar, D.W.; et al. Comparison of inherent optical properties as a surrogate for particulate matter concentration in coastal waters. *Limnol. Ocean. Meth.* **2009**, *7*, 803–810. [CrossRef]
5. Stramski, D.; Babin, M.; Wozniak, S. Variations in the optical properties of terrigenous mineral-rich particulate matter suspended in seawater. *Limnol. Oceanogr.* **2007**, *52*, 2418–2433. [CrossRef]
6. Stemmann, L.; Boss, E. Plankton and particle size and packaging: From determining optical properties to driving the biological pump. *Annu. Rev. Mar. Sci.* **2012**, *4*, 263–290. [CrossRef] [PubMed]
7. Stramski, D.; Boss, E.; Bogucki, D.; Voss, K.J. The role of seawater constituents in light backscattering in the ocean. *Prog. Ocean.* **2004**, *61*, 27–55. [CrossRef]
8. Hill, P.S.; Bowers, D.G.; Braithwaite, K.M. The effect of suspended particle composition on particle 389 area-to-mass ratios in coastal waters. *Meth.Oceanogr.* **2013**, *7*, 95–109. [CrossRef]
9. Boss, E.; Slade, W.H.; Behrenfeld, M.; Dall'Olmo, G. Acceptance angle effects on the beam attenuation in the ocean. *Opt. Exp.* **2009**, *17*, 1535–1550. [CrossRef]
10. Stumpf, R.P. Sediment transport in chesapeake bay during floods: Analysis using satellite and surface observations. *J. Coast. Res.* **1988**, *4*, 1–15.
11. Nechad, B.; Ruddick, K.G.; Park, Y. Calibration and validation of a generic multisensor algorithm for mapping of total suspended matter in turbid waters. *Remote. Sens. Envir.* **2010**, *114*, 854–866. [CrossRef]
12. Agrawal, Y.; Pottsmith, H. Instruments for particle size and settling velocity observations in sediment transport. *Mar. Geol.* **2000**, *168*, 89–114. [CrossRef]
13. Slade, W.H.; Boss, E. Spectral attenuation and backscattering as indicators of average particle size. *Appl. Opt.* **2015**, *54*, 7264–7277. [CrossRef] [PubMed]
14. Boss, E.; Twardowski, M.S.; Herring, S. Shape of the particulate beam attenuation spectrum and its relation to the size distribution of oceanic particles. *Appl. Opt.* **2001**, *40*, 4885–4893. [CrossRef] [PubMed]

15. Tao, J.; Hill, P.S.; Boss, E.S.; Milligan, T.G. Variability of suspended particle properties using optical measurements within the Columbia River Estuary. *J. Geophys. Res. Oceans* **2018**, *123*. [CrossRef]
16. Reynolds, R.A.; Stramski, D.; Neukermans, G. Optical backscattering of particles in Arctic seawater and relationships to particle mass concentration, size distribution, and bulk composition. *Limnol. Oceanogr.* **2016**, *61*, 1869–1890. [CrossRef]
17. Briggs, N.T.; Slade, W.H.; Boss, E.; Perry, M.J. Method for estimating mean particle size from high-frequency fluctuations in beam attenuation or scattering measurements. *Appl. Opt.* **2013**, *52*, 6710–6725. [CrossRef]
18. van de Hulst, H.C. *Light Scattering by Small Particles*; John Wiley and Sons: Dover, UK, 1981.
19. Carder, K.L.; Betzer, P.R.; Eggimann, D.W. Physical, chemical and optical measures of suspended-particle concentrations: Their intercomparison and application to the West African Shelf. In *Suspended Solids in Water*; Gibbs, J., Ed.; Plenum: New York, NY, USA, 1974; pp. 173–193.
20. Winterwerp, J.C.; Van Kesteren., W.G.M. *Introduction to the Physics of Cohesive Sediment Dynamics in the Marine Environment*; Elsevier: Amsterdam, The Netherlands, 2004.
21. Milligan, T.G.; Kineke, G.C.; Blake, A.C.; Alexander, C.R.; Hill, P.S. Flocculation and sedimentation in the ACE Basin, South Carolina. *Estuaries* **2001**, *24*, 734–744. [CrossRef]
22. Hill, P.S.; Syvitiski, J.P.; Cowan, E.A.; Powell, R.D. In situ observations of floc settling velocities in Glacier Bay, Alaska. *Mar. Geol.* **1998**, *145*, 85–94. [CrossRef]
23. Johnson, C.; Li, X.; Logan, B. Settling velocities of fractal aggregates. *Environ. Sci. Technol.* **1996**, *30*, 1911–1918. [CrossRef]
24. Winterwerp, J.C. A simple model for turbulence induced flocculation of cohesive sediment. *J. Hydraulic Res.* **1998**, *36*, 309–326. [CrossRef]
25. Winterwerp, J.C.; Manning, A.J.; Martens, C.; de Mulder, T.; Vanlede, J. A heuristic formula for turbulence-induced flocculation of cohesive sediment. *Estuar. Coast. Shelf Sci.* **2006**, *68*, 195–207. [CrossRef]
26. Hill, P.S.; Voulgaris, G.; Trowbridge, J.H. Controls on floc size in a continental shelf bottom boundary layer. *J. Geophys. Res.* **2001**, *106*, 9543–9549. [CrossRef]
27. Boss, E.; Slade, W.H.; Hill, P. Effect of particulate aggregation in aquatic environments on the beam attenuation and its utility as a proxy for particulate mass. *Opt. Exp.* **2009**, *17*, 9408–9420. [CrossRef]
28. Slade, W.H.; Boss, E.; Russo, C. Effects of particle aggregation and disaggregation on their inherent optical properties. *Opt. Exp.* **2011**, *19*, 7945–7959. [CrossRef] [PubMed]
29. Twardowski, M.; Boss, E.; MacDonald, J.B.; Pegau, W.S.; Barnard, A.H.; Zaneveld, J.R.V. A model for estimating bulk refractive index from the optical backscattering ratio and the implications for understanding particle composition in case I and case II waters. *J. Geophys. Res.* **2001**, *106*, 129–142. [CrossRef]
30. Loisel, H.; Meriaux, X.; Berthon, J.-F.; Poteau, A. Investigation of the optical backscattering to scattering ratio of marine particles in relation to their biogeochemical composition in the eastern English Channel and southern North Sea. *Limnol. Oceanogr.* **2007**, *52*, 739–752. [CrossRef]
31. Boss, E.; Pegau, W.S.; Lee, M.; Twardowski, M.; Shybanov, E.; Korotaev, G.; Baratange, F. Particulate backscattering ratio at LEO 15 and its use to study particle composition and distribution. *J. Geophys. Res.* **2004**, *109*. [CrossRef]
32. Wozniak, S.B.; Stramski, D.; Stramska, M.; Reynolds, R.A.; Wright, V.M.; Miksic, E.Y.; Cichocka, M.; Cieplak, A.M. Optical variability of seawater in relation to particle concentration, composition, and size distribution in the nearshore marine environment at Imperial Beach, California. *J. Geophys. Res.* **2010**, *115*, C08027. [CrossRef]
33. Neukermans, G.; Reynolds, R.A.; Stramski, D. Optical classification and characterization of marine particle assemblages within the western Arctic Ocean. *Limnol. Oceanogr.* **2015**, *61*, 1472–1494. [CrossRef]
34. Neukermans, G.; Loisel, H.; Meriaux, X.; Astoreca, R.; McKee, D. In situ variability of mass-specific beam attenuation and backscattering of marine particles with respect to particle size, density, and composition. *Limnol. Oceanog.* **2012**, *57*, 124–144. [CrossRef]
35. Hurley, A.J.; Hill, P.S.; Milligan, T.G.; Law, B.A. Optical methods for estimating apparent density of sediment in suspension. *Meth. Oceanog.* **2016**, *17*, 153–168. [CrossRef]
36. Trowbridge, J.H.; Lentz, S.J. The bottom boundary layer. *Ann. Rev. Mar. Sci.* **2018**, *10*, 397–420. [CrossRef] [PubMed]
37. Rouse, H. Modern concepts of the mechanics of turbulence. *ASCE Trans.* **1937**, *102*, 463–543.

38. Rouse, H. *An Analysis of Sediment Transportation in the Light of Fluid Turbulence*; United States Department of Agriculture: Washington, DC, USA, 1939.
39. Dyer, K.R. *Coastal and Estuarine Sediment Dynamics*; John Wiley: Chichester, UK, 1986.
40. Orton, P.M.; Kineke, G.C. Comparing calculated and observed vertical suspended-sediment distributions from a Hudson River Estuary turbidity maximum. *Estuarine Coastal Shelf Sci.* **2001**, *52*, 401–410. [CrossRef]
41. Kumbhakar, M.; Ghoshal, K.; Singh, V.P. Derivation of Rouse equation for sediment concentration using Shannon entropy. *Phys. A* **2017**, *465*, 494–499. [CrossRef]
42. Dall'Olmo, G.; Westberry, T.K.; Behrenfeld, M.J.; Boss, E.; Slade, W.H. Significant contribution of large particles to optical backscattering in the open ocean. *Biogeosciences* **2009**, *6*, 947–967. [CrossRef]
43. Sherwood, C.R.; Dickhudt, P.J.; Martini, M.A.; Montgomery, E.T.; Boss, E.S. *Profile Measurements and Data from the 2011 Optics, Acoustics, and Stress In Situ (OASIS) Project at the Martha's Vineyard Coastal Observatory*; United States Geological Survey: Reston, VA, USA, 2012.
44. Slade, W.H.; Boss, E.; Dall'Olmo, G.; Langner, M.R.; Loftin, J.; Behrenfeld, M.J.; Roesler, C.; Westberry, T.K. Underway and moored methods for improving accuracy in measurement of spectral particulate absorption and attenuation. *J. Atmos. Ocean. Tech.* **2010**, *27*, 1733–1746. [CrossRef]
45. Trowbridge, J.H. On a technique for measurement of turbulent shear stress in the presence of surface waves. *J. Atmos. Oceanic Technol.* **1998**, *15*, 290–298. [CrossRef]
46. Grant, W.D.; Madsen, O.S. Combined wave and current interaction with a rough bottom. *J. Geophys. Res. Oceans* **1979**, *84*, 1797–1808. [CrossRef]
47. Madsen, O.S. Spectral Wave-Current Bottom Boundary Layer Flows. In Proceedings of the 24th International Conference Coastal Engineering Research Council, Kobe, Japan, 23–28 October 1994; ACSE: Reston, VA, USA, 1995.
48. Fox, J.M.; Hill, P.S.; Milligan, T.G.; Ogston, A.S.; Boldrin, A. Floc fraction in the waters of the Po River prodelta. *Cont. Shelf Res.* **2004**, *24*, 1699–1715. [CrossRef]
49. Wiberg, P.L.; Smith, J.D. Calculations of the critical shear stress for motion of uniform and heterogeneous sediments. *Water Resour. Res.* **1987**, *23*, 1471–1480.
50. Traykovski, P.; Richardson, M.D.; Mayer, L.A.; Irish, J.D. Mine burial experiments at the Martha's Vineyard Coastal Observatory. *IEEE J. Oceanic Eng.* **2007**, *32*, 150–166. [CrossRef]
51. Dietrich, W.E. Settling velocity of natural particles. *Water Resour. Res.* **1982**, *18*, 1615–1626. [CrossRef]
52. Law, B.A.; Hill, P.S.; Milligan, T.G.; Zions, V. Erodibility of aquaculture waste from different bottom substrates. *Aquacult. Environ. Interact.* **2016**, *8*, 575–584. [CrossRef]
53. Davies, E.J.; Nimmo-Smith, W.A.M.; Agrawal, Y.C.; Souza, A.J. LISST-100 response to large particles. *Mar. Geol.* **2012**, *307*, 117–122. [CrossRef]
54. Graham, G.W.; Davies, E.J.; Nimmo-Smith, W.A.M.; Bowers, D.G.; Braithwaite, K.M. Interpreting LISST-100X measurements of particles with complex shape using digital in-line holography. *J. Geophys. Res. Oceans* **2012**, *117*. [CrossRef]
55. Lynch, J.F.; Irish, J.D.; Sherwood, C.R.; Agrawal, Y.C. Determining suspended sediment particle size information from acoustical and optical backscatter measurements. *Cont. Shelf Res.* **1994**, *14*, 1139–1165. [CrossRef]
56. Russo, C. An Acoustical Approach to the Study of Marine Particles Dynamics Near the Bottom Boundary Layer. Ph.D. Thesis, University of Maine, Orono, ME, USA, 2011.
57. Dyer, K.R.; Soulsby, R.L. Sand transport on the continental shelf. *Annu. Rev. Fluid Mech.* **1988**, *20*, 295–324. [CrossRef]
58. McLean, S.R. On the calculation of suspended load for noncohesive sediments. *J. Geophys. Res. Oceans* **1992**, *97*, 5759–5770. [CrossRef]
59. Gelfenbaum, G.; Smith, J.D. *Experimental Evaluation of a Generalized Suspended-Sediment Transport Theory*; AAPG: Tulsa, OK, USA, 1986; pp. 133–144.
60. Pal, D.; Ghoshal, K. Vertical distribution of fluid velocity and suspended sediment in open channel turbulent flow. *Fluid Dyn. Res.* **2016**, *48*, 035501. [CrossRef]
61. Sherwood, C.R.; Aretxabaleta, A.L.; Harris, C.K.; Rinehimer, J.P.; Verney, R.; Ferré, B. Cohesive and mixed sediment in the Regional Ocean Modeling System (ROMS v3.6) implemented in the Coupled Ocean-Atmosphere-Wave-Sediment Transport Modeling System (COAWST r1234). *Geosci. Model Dev.* **2018**, *11*, 1849–1871. [CrossRef]

8

Ocean Color Analytical Model Explicitly Dependent on the Volume Scattering Function

Michael Twardowski [1,2,*] and Alberto Tonizzo [2,*]

1 Harbor Branch Oceanographic Institute, Florida Atlantic University, Ft. Pierce, FL 34946, USA
2 Sunstone Scientific LLC, Vero Beach, FL 32963, USA
* Correspondence: mtwardowski@fau.edu (M.T.); alberto.tonizzo@gmail.com (A.T.)

Featured Application: Ocean color remote sensing.

Abstract: An analytical radiative transfer (RT) model for remote sensing reflectance that includes the bidirectional reflectance distribution function (BRDF) is described. The model, called ZTT (Zaneveld-Twardowski-Tonizzo), is based on the restatement of the RT equation by Zaneveld (1995) in terms of light field shape factors. Besides remote sensing geometry considerations (solar zenith angle, viewing angle, and relative azimuth), the inputs are Inherent Optical Properties (IOPs) absorption a and backscattering b_b coefficients, the shape of the particulate volume scattering function (VSF) in the backward direction, and the particulate backscattering ratio. Model performance (absolute error) is equivalent to full RT simulations for available high quality validation data sets, indicating almost all residual errors are inherent to the data sets themselves, i.e., from the measurements of IOPs and radiometry used as model input and in match up assessments, respectively. Best performance was observed when a constant backward phase function shape based on the findings of Sullivan and Twardowski (2009) was assumed in the model. Critically, using a constant phase function in the backward direction eliminates a key unknown, providing a path toward inversion to solve for a and b_b. Performance degraded when using other phase function shapes. With available data sets, the model shows stronger performance than current state-of-the-art look-up table (LUT) based BRDF models used to normalize reflectance data, formulated on simpler first order RT approximations between r_{rs} and b_b/a or $b_b/(a + b_b)$ (Morel et al., 2002; Lee et al., 2011). Stronger performance of ZTT relative to LUT-based models is attributed to using a more representative phase function shape, as well as the additional degrees of freedom achieved with several physically meaningful terms in the model. Since the model is fully described with analytical expressions, errors for terms can be individually assessed, and refinements can be readily made without carrying out the gamut of full RT computations required for LUT-based models. The ZTT model is invertible to solve for a and b_b from remote sensing reflectance, and inversion approaches are being pursued in ongoing work. The focus here is with development and testing of the in-water forward model, but current ocean color remote sensing approaches to cope with an air-sea interface and atmospheric effects would appear to be transferable. In summary, this new analytical model shows good potential for future ocean color inversion with low bias, well-constrained uncertainties (including the VSF), and explicit terms that can be readily tuned. Emphasis is put on application to the future NASA Plankton, Aerosol, Cloud, and ocean Ecosystem (PACE) mission.

Keywords: ocean optics; ocean color; remote sensing; radiative transfer approximation; volume scattering function; NASA PACE mission

1. Introduction

Radiative transfer (RT) approximations linking inherent optical properties (IOPs), such as spectral absorption $a(\lambda)$ (m^{-1}) and spectral backscattering $b_b(\lambda)$ (m^{-1}) to ocean color remote sensing reflectance $R_{rs}(\lambda)$ are vital to interpreting R_{rs} because it is not possible to analytically invert the full RT equation [1,2]. Once forward RT approximations are developed, inversions can then be explored to solve for IOPs from R_{rs} and subsequently ocean biogeochemical properties [3]. Ocean color $R_{rs}(\lambda)$ (sr^{-1}) here is defined as $L_w(\lambda)/E_d(\lambda)$, or water leaving radiance (W m^{-2} sr^{-1} nm^{-1}) normalized to above water downwelling irradiance (W m^{-2} nm^{-1}) (see Reference [4] for complete definitions of all optical parameters and Appendix A, Table A1 for notation).

Ocean color expressions to date have almost exclusively relied on first order approximations of RT relating R_{rs} to b_b/a through a proportionality represented as f/Q [5–8], or to $b_b/(a+b_b)$ with multi-term polynomial expressions based originally on Gordon et al. [9] with coefficients represented as l or G ([9–12]; see review by Werdell et al. [13]). The coefficients describing the relationship between R_{rs} and IOPs are detailed in look-up tables (LUTs), or a neural network in the case of [8], with dependencies on geometry (i.e., solar zenith, viewing angle, relative azimuth) and in some cases wavelength, wind speed, atmospheric conditions, and/or chlorophyll concentration [Chl]. The LUTs are generated from full RT computations with so-called synthetic data sets where IOPs and their interrelationships are assumed and referenced to [Chl]. LUT coefficients are thus also implicitly dependent on IOPs.

These ocean color relationships have been tremendously useful to the ocean color community for decades. Morel [14] discusses the first order analytical relationship and the empiricism necessary to invert. Coefficients describing the relationship between R_{rs} and IOPs are dependent on the bidirectional reflectance distribution function (BRDF), which describes the transformation of downwelling irradiance for different solar zenith angles into the distribution of upwelling radiance. Morel et al. [7] describe the current state-of-the-art BRDF model (herein referred to as M02), currently implemented operationally by the ocean color community (NASA Ocean Biology Processing Group (OBPG); [15]) to convert L_w measured at any viewing geometry to a conceptual "exact normalized" L_w, $[L_w]_N^{ex}$, with Sun at zenith and nadir viewing in a non-attenuating atmosphere. This allows any measurement at any Sun-viewing geometry to be directly intercompared. Inversion algorithms to derive IOPs are then typically applied to $[L_w]_N^{ex}$ and are often based on the same type of first order R_{rs} to IOP approximation [13,16]. The current semi-analytical algorithm (SAA) application to water-leaving radiances is thus a two-step process (after applying calibration coefficients, georeferencing, and atmospheric correction): (1) A BRDF correction to carry out the conversion to $[L_w]_N^{ex}$, followed by (2) application of an inversion algorithm to derive IOPs (see [13]).

A key potential source of uncertainty in current SAA approaches is associated with the volume scattering function (VSF; $\beta(\psi)$ m^{-1} sr^{-1}, where ψ is scattering angle), and this uncertainty is ambiguously dispersed in both steps. The VSF dependency is of paramount importance; as Morel and Gentili [5] note, the BRDF "... is essentially controlled by the shape of the VSF ... ." Historically, very few measurements of the VSF have been available to develop and test SAAs. The M02 BRDF tables were developed from extensive RT modeling using VSF shapes (also known as phase functions $P(\psi)$, defined as the VSF normalized to total scattering, $P(\psi) = \beta(\psi)/b$, with units sr^{-1}) tied to estimated chlorophyll concentrations [Chl]. To obtain $P(\psi)$ for modeling, Morel et al. [7] first mixed two phase functions for populations dominated by large and small particles representing high and low [Chl] extremes, respectively, and then mixed that particulate phase function with the phase function for molecular seawater to a degree that was also linked to estimated [Chl]. The [Chl] estimate was initially derived from R_{rs} with an empirical band ratio algorithm. Uncertainty is thus associated with how close the assumed $P(\psi)$ used in the BRDF modeling matches the actual $P(\psi)$ associated with any given L_w measurement. While the M02 model has been validated in Case 1 waters with atmospheric conditions and IOPs that presumably agree with the underlying atmospheric and bio-optical models [17,18], there have been limited attempts to assess any embedded phase function uncertainties with actual VSF measurements in diverse water types [19,20]. No subsequent BRDF approach has demonstrated enhanced performance for Case 1 waters relative to the M02 approach while being feasible to implement for ocean color. However, Talone et al. [21] have recently shown the Lee et al. [11] LUT-based approach was more accurate for Case 2 waters sampled in the Adriatic, Baltic, and Black Seas.

Uncertainty also arises from exclusively using the b_b term in the inversion algorithm, i.e., in step two of the SAA approach described above. Effort has been devoted to trying to correct this uncertainty by layering additional VSF dependence in the f/Q proportionality [22]. Interestingly, after the current M02 BRDF normalization is applied (i.e., measurement geometry transferred to Sun at zenith and nadir viewing), the scattering parameter that should be most closely linked to $[L_w]_N^{ex}$, and thus presumably should provide the lowest associated uncertainties in IOP retrievals, is $\beta(\pi)$ if single scattering is assumed, which is typically a good approximation [23]. We are not aware however of any algorithm using $\beta(\pi)$ in lieu of b_b. The practical but still arbitrary choice of converting measured L_w to $[L_w]_N^{ex}$ with Sun at zenith and nadir viewing may thus not optimize uncertainties in inversion algorithms. The parameter $\beta(\pi)$ is almost completely unknown in the ocean, as there are virtually no direct measurements of $\beta(\pi)$ in the literature and typical models of particle scattering with simplified particle shapes do not account for possible particle-particle coherent scattering that may cause significant, but poorly understood, enhancement near $\beta(\pi)$ [24]. For example, a recent estimate of $b_b/\beta(\pi)$ made with a combination of airborne lidar and in situ b_b measurements [25] was 50% the value expected from extrapolation of available β measurements [26].

Although the M02 BRDF correction as currently applied can be considered "conceptual" at $L_w(0,\pi)$ without creating a problem from a geometry point of view (i.e., L_w at any viewing geometry can be corrected to a conceptual standard L_w at another geometry as long as it is consistent), it will be problematic if we try to develop algorithms based on $\beta(\pi)$ to reduce uncertainties because the necessary data for algorithm development and validation are lacking. This would also suggest a possible benefit in using the BRDF correction to obtain $[L_w]_N^{ex}$ at another unique geometry (solar zenith, viewing angle, and azimuth), one that was representative of single scattering at an angle that we can measure directly with available instrumentation. To minimize the magnitude of BRDF correction, this angle could be chosen as the centroid angle of the maximum in the frequency distribution of in-water single scattering angles observed for polar orbiting satellites, which is about 150° (Figure 1). Since we can accurately measure β at or near 150° with commercially available instrumentation (e.g., WET Labs ECO, In-situ Marine Optics IMO-SC6, and legacy HOBI Labs Hydroscat sensors), there would appear to be potential to reduce uncertainties with $\beta(150°)$-based algorithms matched to a $\beta(150°)$-based BRDF correction using M02 relative to the current b_b-based algorithms applied to a $\beta(\pi)$-based BRDF correction. In the former approach, uncertainties associated with the phase function are thus mostly restricted to the BRDF correction step.

Herein we explore a different approach, working with a RT expression from Zaneveld [27,28] that explicitly incorporates a dependency on VSF shape, as well as specific viewing geometry. A path to inversion to IOPs is also presented where SAA steps 1 and 2 mentioned above are combined in a single relationship and uncertainties related to VSF shape and other parameters can be directly assessed. This approach has several potential benefits, including (1) a single, fully analytical and invertible expression describing the RT process for all remote sensing geometries, (2) optimal retention of native RT relationships with more degrees of freedom than the first order approximation, directly linked to physically meaningful terms, (3) all parameters including the VSF are explicit in the model with readily characterized uncertainties, (4) model can be readily enhanced by tuning one or more terms rather than developing new LUTs from complete recomputations of full RT, and (5) native viewing geometries produce scattering at angles that can be resolved with available instrumentation. A challenge in an explicitly VSF-dependent approach is amenability to inversion, as some information about the phase function is ostensibly required [1,20]. However, it should be pointed out this is also the case with any model, including the M02 BRDF correction, and subsequent inversion algorithms and their inherent assumptions about VSF shape. In fact, as is assessed herein, the same assumptions of M02 in linking changes in VSF shape with an independent estimate of chlorophyll can be directly applied to this RT model, with the benefit of being able to directly quantify associated errors.

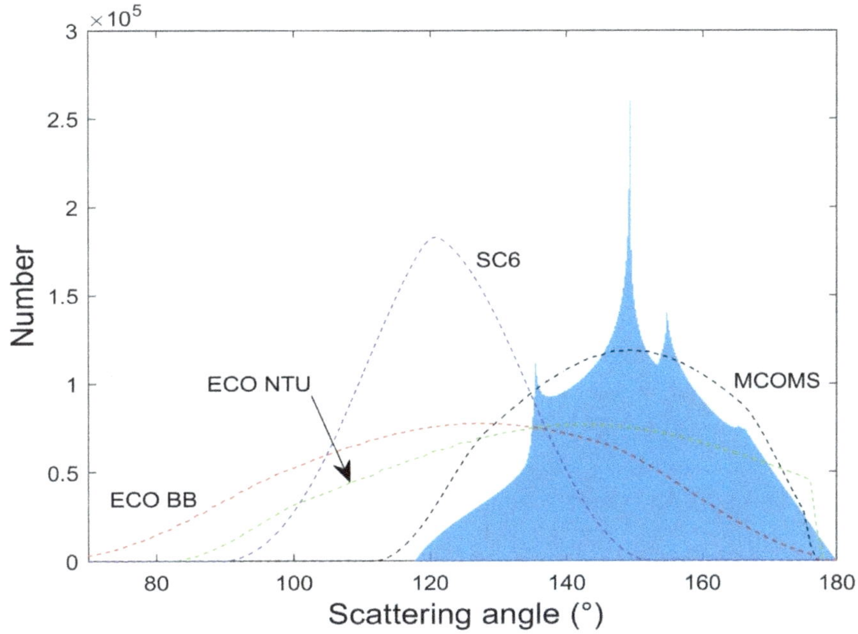

Figure 1. Refracted, in-water scattering angles made between the solar zenith and viewing angle, simulated for the upcoming NASA PACE satellite imager through a complete polar orbit (solid blue). Scattering angle distributions for SeaWiFS were similar. Data courtesy Bryan Franz (NASA GSFC). Angular weighting functions of commercial backscattering sensors WET Labs ECO-BB, ECO-NTU, and MCOMS and IMO-SC6 (see text) are overlaid after scaling by 5×10^6.

In this paper we focus on the performance of a forward implementation of the Zaneveld expression using a modified formulation in terms of IOPs that is now amenable to inversion. Performance of the inversion to IOPs is being fully assessed in ongoing work. A key advance promoting inversion is the finding by Sullivan and Twardowski [26] that the shape of the particulate VSF in the backward direction is relatively constant for a wide range of water types, and so may be represented by a constant function in an algorithm without introducing significant error. Importantly, we have also now overcome a limitation in practically assessing this approach by collecting a database of measured VSFs resolved over a large dynamic range concurrently with other high quality IOPs and radiometry under ideal cloud-free environmental conditions [29].

2. Forward Model Development

Zaneveld [27,28] derived an exact restatement of the RT equation assuming a plane-parallel, optically deep water column in terms of upwelling radiance L_u in viewing direction θ_v:

$$\frac{L_u(\theta_s,\theta_v,\phi)}{E_{od}} = \frac{f_b(\theta_s,\theta_v,\phi)\frac{b_b}{2\pi}}{-\cos(\theta_v)K_{L_u}(\theta_s,\theta_v,\phi)+c-f_L(\theta_s,\theta_v,\phi)b_f}, \tag{1}$$

where E_{od} is the scalar downwelling irradiance, K_{Lu} is the attenuation coefficient for upwelling radiance, b_f is forward scattering, and c is the attenuation coefficient. Parameters f_b and f_L are light field shape factors representing the path radiance term of the RT equation [27,28]. Factor f_b describes the redirection of downwelling radiance into the upwelling viewing angle for a given VSF and is normalized to the redirection that would be observed if the VSF was isotropic. Factor f_b is thus directly linked to the shape of the VSF in the backward direction. The factor f_L is defined similarly but describes the redirection of upwelling radiance into the upwelling viewing angle and is thus directly linked to the shape of the near forward VSF. These shape factors are expected to vary within a relatively narrow range near unity. All terms are a function of depth z and wavelength λ. The Zaneveld expression does not account for any inelastic processes, such as molecular Raman scattering or fluorescence [30–33]. Using physically reasoned approximations for f_b including the assumption of single scattering, the following was obtained by Zaneveld ([28]; his Equation (14)):

$$\frac{L_u(\theta_s, \theta_v, \phi)}{E_{od}} = \frac{\beta(\psi)}{-\cos(\theta_v) K_{Lu}(\theta_s, \theta_v, \phi) + c - f_L(\theta_s, \theta_v, \phi) b_f}, \tag{2}$$

where β is the VSF (including water) and ψ is the in-water scattering angle formed between peak incident sunlight at solar zenith angle θ_s and scattered light traveling in viewing direction (θ_v,Φ), where $\cos(\psi) = \cos(\theta_s)\cos(\theta_v) - \sin(\theta_s)\sin(\theta_v)\cos(\phi)$. Azimuth ϕ is relative to the Sun's direction. Zenith angles θ are in-water, refracted through the air-sea interface and determined from a vertically downward direction. For nadir viewing, $\psi = \pi - \theta_s$ and $-\cos(\theta_v) = 1$ in the denominator. Note approximations of Equation (2) provided in Zaneveld [28] assumed only nadir viewing while we retain the full BRDF functionality here.

In theoretical analyses, Weidemann et al. [34] showed b_b retrievals based on the Zaneveld expression had errors as large as -20% and +40%. However, an extraordinarily wide range of VSF shapes was applied in their simulated data, including VSF shapes for specific particulate subcomponents, such as bacteria, minerals, and phytoplankton. These populations were considered separately as quasi-monodispersions with phase function shapes computed from Lorenz-Mie theory, i.e., assuming homogeneous spheres. These phase function shapes thus had large oscillations with respect to angle, structure that we now know is extremely unrealistic for VSFs representative of bulk in situ particle populations (e.g., [26,35–37]). Weidemann et al. also showed with these phase functions that the Zaneveld approximation of the shape factor at nadir viewing, i.e., $f_b \approx 2\pi\beta(\pi - \theta_s)/b_b$, had an average error of only 5% (their Figure 11) and this included overcast conditions that would not occur in remote sensing. To our knowledge, the only study attempting to test the Zaneveld model with directly measured data was He et al. [20] with a subset of the NOMAD data set, which included no VSF measurements, where performance for the BRDF component of the model was comparable to the current state-of-the-art [7,11].

Equation (2) demonstrates the direct link between $L_u(\theta_s, \theta_v, \phi)$ and $\beta(\psi)$. However, for the Zaneveld expression to be a practical tool for ocean color, the terms K_{Lu} and f_L must be expressed in terms of IOPs. Furthermore, the IOPs in the model should ideally be coefficients that are closely linked to reflectance and directly measurable with good accuracy using existing sensor technology. For example, the term in the denominator $[c - f_L(\theta_s, \theta_v, \phi)b_f]$ has two IOPs that are difficult to determine directly because of acceptance angle issues with standard transmissometer designs [38,39]. However, since f_L is close to unity, this term is also closely related to $a + b_b$, immediately recognizable from commonly used first order approximations. Goals of the next several sections are: 1) To represent Equation (2) entirely in terms of such IOPs, 2) to rework in terms of the commonly used remote sensing reflectance L_u/E_d, since these are the measurements currently available in validation data sets, and 3) to include the inelastic effects associated with water Raman scattering.

2.1. *Diffuse Attenuation of Upwelling Radiance* K_{Lu}

For the term K_{Lu}, Zaneveld [28] suggested an assumption of equivalency to K_∞, the diffuse attenuation coefficient in the asymptotic regime. Asymptotic theory is based on the principle that the shape of the light field with depth gradually transforms from being dependent on the incident surface light field to being constant, azimuthally symmetric (so L is only a function of θ), and dependent only on IOPs. Attenuation coefficients for all aspects of the light field, i.e., for all radiances and therefore all irradiances, are equivalent in the asymptotic regime and are also IOPs. This is a critical assumption for the purposes of model development since K can then be described only in terms of IOPs. Zaneveld justified the assumption, even though the omnidirectional light field in surface waters is far from asymptotic, due to the decoupling of upwelling radiances to downwelling radiance distributions. Measurements of upwelling radiance fields from the 1960s and 1970s showed a near constant shape and attenuation rate with depth. Recently, Twardowski and Tonizzo [40] confirmed this assumption in RT simulations with no more than 3% error when the sun was at solar zenith and the following relationship held for *above water* solar zeniths $\theta_s{}'$ up to 75°:

$$\frac{K_{Lu} - K_\infty}{K_\infty} = F(\theta_s{}') = f_1\theta_s{}'^4 + f_2\theta_s{}'^3 + f_3\theta_s{}'^2 + f_4\theta_s{}' + f_5, \tag{3}$$

which included a full range of possible natural water types. Coefficients f are provided in Appendix A, Table A2. Angle $\theta_s{}'$ is related to θ_s by Snell's Law, i.e., $\theta_s{}' = \sin^{-1}(1.34 \sin(\theta_s))$. Only nadir viewing was considered in the simulations for K_{Lu} in Equation (3), and we make the assumption that K_{Lu} for other viewing angles that define a specific scattering angle ψ can be approximated by Equation (3) with nadir viewing geometry after assigning a $\theta_s{}'$ that provides an equivalent in-water ψ. For example, for $\theta_s{}' = 60°$, θ_s will be 40.3° and ψ will be 139.7° for nadir viewing; we assume the resulting $F(60°)$ will also be applicable to any off-nadir viewing angle with $\psi = 139.7°$. This assumption has been verified using Hydrolight RT simulations (methodology addressed in Section 3.2) to no worse than 2% in the solar plane and no worse than 5% within the upwelling hemisphere for in-water scattering angles consistent with remote sensing (i.e., Figure 1). This assumption implies a rotational reference frame, where the first order determinant of radiance field shape in the model, i.e., ψ, is preserved. Two potential drawbacks of this assumption are (1) the influence of skylight may be skewed in the rotated reference frame, and (2) the range of viewing angles is restricted since the smallest ψ is ~134° for underwater nadir viewing. The range ψ > 134°, however, comprises >95% of the expected scattering angles that will be measured by the PACE imager (Figure 1). Further work with field measurements is needed to verify this assumption. Reformulating Equation (3) in terms of dependency on the in-water scattering angle, we obtain:

$$F(\psi) = f_{A1}\psi^4 + f_{A2}\psi^3 + f_{A3}\psi^2 + f_{A4}\psi + f_{A5}. \tag{4}$$

Coefficients f_A are provided in Appendix A, Table A2.

From Gershun's Law we can set $K_\infty = a/\overline{\mu}_\infty$, where $\overline{\mu}_\infty$ is the average cosine of the asymptotic light field. After inserting Equation (4) into Equation (2) and allowing for $c = a + b$, the following is obtained:

$$\frac{L_u(\theta_s, \theta_v, \phi)}{E_{od}} \cong \frac{\beta(\psi)}{a\left(1 - \cos(\theta_v)\Psi_{K_{Lu}}(\psi)\,\overline{\mu}_\infty{}^{-1}\right) + b - f_L(\theta_s, \theta_v, \phi) b_f}, \tag{5}$$

$$\text{where } \Psi_{K_{Lu}}(\psi) = K_{Lu}/K_\infty = [1 + F(\psi)]. \tag{6}$$

The parameter $\Psi_{K_{Lu}}$ is assumed spectrally independent with errors over the full range of possible water types estimated at <2% [40]. Dividing numerator and denominator of Equation (5) by b_b, we obtain:

$$\frac{L_u(\theta_s, \theta_v, \phi)}{E_{od}} \cong \frac{\beta(\psi)}{b_b} \Big/ \left[\frac{a}{b_b}\left(1 - \cos(\theta_v)\Psi_{K_{Lu}}(\psi)\overline{\mu}_\infty{}^{-1}\right) + f_L(\theta_s, \theta_v, \phi)\left(1 - \widetilde{b}_b{}^{-1}\right) + \widetilde{b}_b{}^{-1}\right], \tag{7}$$

where $\widetilde{b}_b = b_b/b$ is the backscattering ratio. For IOPs we now have the backward phase function in the numerator; in the denominator we have the recognizable b_b/a, as well as $\widetilde{b}_b$, an IOP that incorporates information on bulk particle composition (see Section 2.4; [41]), and is not typically associated with ocean color remote sensing vis-à-vis the common first order approximation of r_{rs} assumed proportional to b_b/a.

2.2. Average Cosine of the Asymptotic Light Field $\overline{\mu}_\infty$

The $\overline{\mu}_\infty$ term in Equation (7) must be expressed in terms of IOPs to invert. Using a fit to theoretical calculations of radiance fields by Prieur and Morel [42], Zaneveld [28] recommended $1/\overline{\mu}_\infty$ be modeled empirically with respect to the single scattering albedo ω (=b/c) using a quadratic fit. Additionally, through more detailed RT computations, Berwald et al. [43] found a 4th order dependency of $\overline{\mu}_\infty$ on the albedo.

In the study by Twardowski and Tonizzo [40], $\overline{\mu}_\infty$ was parameterized in terms of b_b/a instead of ω, including an assessment across a full range of environmentally representative phase function shapes using the Fournier-Forand analytical model [44,45]. The data set used in the assessment was not a representative synthetic data set (i.e., [46]), as it included a full range of possible b_b/a values, possible phase functions, and permutations thereof. Representing $\overline{\mu}_\infty$ in terms of b_b/a has two distinct advantages. First, b_b and a can be measured with commercially available in situ instrumentation with accuracies of a few percent [47–49] to enable performance assessment for algorithms. Parameters c and b cannot be measured without significantly larger errors, typically >25–50%, because of the acceptance of near forward scattered light in conventional transmissometer designs [38,39]. Secondly, b and c are not parameters that are closely linked to r_{rs} without additional information, whereas b_b and a are (e.g., Gordon et al. [50]), and a goal here is to represent the entire model in terms of b_b and a to enable inversion.

The Twardowski and Tonizzo [40] parameterization also explicitly depended on η_{bb}, the fraction of b_b attributable to molecular scattering, $\eta_{bb} = b_{bw}/(b_{bp} + b_{bw})$ [6]. The natural range for η_{bb} is from ~0 to ~0.98 [40,51], and the range used for b_b/a was 10^{-4} to 10^{-1}. Since the water components of η_{bb} may be assumed known [52], the effective unknown here is b_{bp}. After extending the analysis from Reference [40] to include near zero b_b/a and increased resolution in η_{bb}, the resulting fit was obtained for $\overline{\mu}_\infty$:

$$\begin{aligned}\overline{\mu}_\infty\left(\tfrac{b_b}{a},\eta_{bb}\right) \approx\ & \left[m_1(\log\eta_{bb})^3 + m_2(\log\eta_{bb})^2 + m_3\log\eta_{bb} + m_4\right]\left(\log\tfrac{b_b}{a}\right)^3 \\ & + \left[m_5(\log\eta_{bb})^3 + m_6(\log\eta_{bb})^2 + m_7\log\eta_{bb} + m_8\right]\left(\log\tfrac{b_b}{a}\right)^2 \\ & + \left[m_9(\log\eta_{bb})^3 + m_{10}(\log\eta_{bb})^2 + m_{11}\log\eta_{bb} + m_{12}\right]\log\tfrac{b_b}{a} \\ & + m_{13}(\log\eta_{bb})^3 + m_{14}(\log\eta_{bb})^2 + m_{15}\log\eta_{bb} + m_{16}.\end{aligned} \tag{8}$$

Coefficients m are provided in Appendix A, Table A2. Fits to simulated data were again made with the polyfit function from MATLAB. Absolute errors $\%\delta_{abs}$ for this fit vary from 0.19 to 3.5 for η_{bb} ranging from 0.98 to 0.0098, respectively. Since both b_b/a and η_{bb} are spectrally dependent, $\overline{\mu}_\infty$ will be as well (not shown for clarity).

2.3. Backward Phase Function $\beta(\psi)/b_b$

The scattering parameters in Equation (7) must be expanded into water and particulate components. Expanding $\beta(\psi)/b_b$ gives:

$$\frac{\beta(\psi)}{b_b} = \frac{\beta_p(\psi) + \beta_w(\psi)}{b_{bp} + b_{bw}}, \tag{9}$$

where the p and w subscripts represent particles and molecular seawater, respectively. In Equation (9), the pure seawater terms, which are temperature and salinity specific, can be directly computed with an estimated error of no more than 2% [52]. Introducing a term for the particulate phase function in the backward direction, $P_{bb}(\psi) = \beta_p(\psi)/b_{bp}$, Equation (9) then becomes:

$$\frac{\beta(\psi)}{b_b} = \frac{P_{bb}(\psi)b_{bp} + \beta_w(\psi)}{b_{bp} + b_{bw}}. \tag{10}$$

Note b_{bp} and b_{bw} both have spectral dependencies as does β/b_b, and the unknowns are $b_{bp}(\lambda)$ and $P_{bb}(\psi)$. Equation (10) can be easily rewritten in terms of η_{bb}, with the same unknowns. In coastal waters where η_{bb} is near zero, $\beta_p >> \beta_w$, $b_{bp} >> b_{bw}$, and $\beta(\psi)/b_b$ will be approximated by $P_{bb}(\psi)$. For clear ocean waters, phase functions are represented by a mixture of both particles and pure seawater with VSF shapes dependent on η_{bb}.

2.4. Backscattering Ratio $\widetilde{b}_b$

Expanding the backscattering ratio $\widetilde{b}_b = b_b/b$ in Equation (7) we obtain:

$$\frac{b_b}{b} = \frac{b_{bp} + b_{bw}}{b_p + b_w} = \frac{b_{bp} + b_{bw}}{b_{bp}/\widetilde{b}_{bp} + b_w}, \tag{11}$$

where $\widetilde{b}_{bp}$ is the particulate backscattering ratio. This $\widetilde{b}_{bp}$ is the "true" $\widetilde{b}_{bp}$, distinct from the $\widetilde{b}_{bp}$ typically derived from measurements that include c data from transmissometers with significant acceptance angle errors (e.g., [38,39]) (we note models linking particle biogeochemical properties and measured $\widetilde{b}_{bp}$ should account for the acceptance angle of c measurements [41,53]). Bootstrapping exercises using Equation (7) readily show an impact on reflectance of up to several percent when $\widetilde{b}_{bp}$ is varied over the full ~0.003 to ~0.03 dynamic range observed in the oceanic environment [41,53–56]. All terms have spectral dependencies.

2.5. Shape Factor f_L

Zaneveld [28] recommended the term f_L, i.e., the dimensionless upwelling radiance shape factor, could be set to 1.05 with small error. The natural range was estimated at 1 to 1.12 [27]. In the Results, we develop a new model for this term.

2.6. Remote Sensing Reflectance Formulation

Nearly all ocean RT algorithm work over the last several decades has used reflectance defined as L_u/E_d instead of L_u/E_{od}. Zaneveld [28], however, pointed out, as is evident from Equation (1), L_u/E_{od} is most closely aligned with RT theory. The irradiance parameter E_{od} is also less dependent on solar zenith angle than E_d. Furthermore, sensor technology has been available to measure E_{od}, historically from companies Biospherical (www.biospherical.com), Satlantic (www.satlantic.com), and Trios (www.trio.de). Nonetheless, since nearly all field data over the last several decades has focused on E_d, any testing and validation of the RT model described here requires modification in terms of E_d. Substituting $E_d/E_{od} = \overline{\mu}_d$ (the average cosine of downwelling radiance, just below the air-water interface) into Equation (7) gives:

$$r_{rs} \cong \frac{1}{\overline{\mu}_d}\frac{\beta(\psi)}{b_b} \Big/ \left[\frac{a}{b_b}\left(1 - \cos(\theta_v)\Psi_{K_{Lu}}\overline{\mu}_\infty{}^{-1}\right) + f_L\left(1 - \widetilde{b}_b{}^{-1}\right) + \widetilde{b}_b{}^{-1}\right], \tag{12}$$

where r_{rs} is the classically known remote sensing reflectance just below the water surface. Full dependencies of variables not shown for clarity.

2.7. Average Cosine of the Downwelling Light Field $\overline{\mu}_d$

A model for $\overline{\mu}_d$ is now required in the expression from Equation (12). If the sky is ignored and we assume a negligible fraction of the incident solar beam is scattered in the near-surface, a reasonable first-order approximation of $\overline{\mu}_d$ should be $\mu_w \equiv \cos(\theta_s)$. Adding in a cardioidal radiance distribution for skylight, Morel and Prieur [57] obtained:

$$\frac{1}{\overline{\mu}_d} \approx \frac{0.6}{\mu_w} + \frac{0.4}{0.859}, \tag{13}$$

and noted that for sun angles between 8° and 62°, $\overline{\mu}_d$ varied only from 0.79 to 0.94. Thus, even without knowledge of solar zenith angle, a median value could be used with an accuracy better than 10%.

Numerator values of 0.6 and 0.4 in Equation (13) represent the fractions of direct (E_{dd}/E_d) and diffuse ($H = E_{ds}/E_d$) downwelling light, respectively, which together equal unity. These values primarily depend on θ_s' and horizontal visibility V, the latter of which depends on aerosol optical thickness (AOT). For $20° \leq \theta_s' \leq 60°$ and for H between 0.2 and 0.5, the variability of $\overline{\mu}_d$ is ~7% for a fixed θ_s'.

The term $\overline{\mu}_d$ can be factorized in two parts, one part dependent on the IOPs, the other dependent on the atmospheric conditions and geometry:

$$\overline{\mu}_d\left(\theta_s', V, \frac{b_b}{a}, \eta_{bb}\right) \approx M_d^+(\theta_s', V) \times M_d^*\left(\frac{b_b}{a}, \eta_{bb}\right). \tag{14}$$

The atmospheric component can be represented as:

$$M_d^+(\theta_s', V) = \frac{\left[\frac{1-H(\theta_s',V)}{\mu_w} + \frac{H(\theta_s',V)}{0.859}\right]^{-1}}{P^3[\cos(\theta_s')]}. \tag{15}$$

As mentioned, Morel and Prieur [57] assumed $H = 0.4$. Gregg and Carder [58] later provided a relationship for H as a function of θ_s' and V that included skylight. Specifically, the results of ([58]; their Figure 4) can be fit as follows:

$$\begin{aligned} H(\theta_s', V) = \; & [e_1V^2 + e_2V + e_3]\theta_s'^5 + [e_4V^2 + e_5V + e_6]\theta_s'^4 \\ & + [e_7V^2 + e_8V + e_9]\theta_s'^3 + [e_{10}V^2 + e_{11}V + e_{12}]\theta_s'^2 \\ & + [e_{13}V^2 + e_{14}V + e_{15}]\theta_s' + e_{16}V^2 + e_{17}V + e_{18}\,. \end{aligned} \tag{16}$$

Coefficients e are provided in Appendix A, Table A2. Note the typical default value used in Hydrolight is $V = 15$ km.

The P^3 term in the M_d^+ relationship is a 3rd order polynomial in $\cos(\theta_s')$ to correct where Morel and Prieur's original approximation deviates from the Gregg and Carder relationship at large θ_s'. The P^3 term is:

$$P^3[\cos(\theta_s')] = 0.7792\cos^3(\theta_s') - 1.7366\cos^2(\theta_s') + 1.1551\cos(\theta_s') + 0.7842.$$

The IOP-dependent component M_d^* was modeled using the approach in [40] with the extended b_b/a range and η_{bb} resolution discussed in Section 2.2. The final fitted relationship is:

$$\begin{aligned} M_d^*\left(\frac{b_b}{a}, \eta_{bb}\right) = \; & \left[m_{d,1}^*\log\eta_{bb} + m_{d,2}^*\right]\left(\log\frac{b_b}{a}\right)^3 + \left[m_{d,3}^*\log\eta_{bb} + m_{d,4}^*\right]\left(\log\frac{b_b}{a}\right)^2 \\ & + \left[m_{d,5}^*\log\eta_{bb} + m_{d,6}^*\right]\log\frac{b_b}{a} + m_{d,7}^*\log\eta_{bb} + m_{d,8}^*. \end{aligned} \tag{17}$$

Coefficients m_d^* are provided in Appendix A, Table A2. Absolute percent error with this relationship relative to computations using full radiative transfer (Hydrolight, see Section 3.2) varies from 0.06% to 0.4% across all θ_s'. Error for the full $\overline{\mu}_d$ expression is <1%. Spectral dependency enters the expression through the V term, which is dependent on spectral AOT.

2.8. Including Inelastic Water Raman Effects

The molecular water Raman scattering contribution to r_{rs} can be included in Equation (12) as an additive term, resulting in the expression:

$$\begin{aligned} r_{rs}(\theta_s, \theta_v, \phi, V, a, b_b) \cong \; & r_{rs,Raman}(\theta_s', a, b_b) + \frac{1}{\overline{\mu}_d\left(\theta_s', V, \frac{b_b}{a}, \eta_{bb}\right)}\frac{\beta(\psi)}{b_b} \Big/ \\ & \left[\frac{a}{b_b}\left(1 - \cos(\theta_v)\Psi_{K_{Lu}}(\psi)\overline{\mu}_\infty\left(\frac{b_b}{a}, \eta_{bb}\right)^{-1}\right) + f_L(\theta_s, \theta_v, \phi)\left(1 - \widetilde{b}_b^{-1}\right) + \widetilde{b}_b^{-1}\right]. \end{aligned} \tag{18}$$

Full dependencies for all parameters in the model are shown except for λ; all parameters exhibit dependence on λ in the model except for $\Psi_{K_{Lu}}(\psi)$. We note a similar approach has been taken in adding water Raman effects in other reflectance models (e.g., [59]; reviewed in [13]). The term $r_{rs,Raman}$ can be derived according to Westberry et al. ([60]; see their Equation (7)) with inputs of above water ($z = 0^+$) downwelling irradiance $E_d(0^+,\theta_s')$, a, and b_b. The NASA Generalized IOP (GIOP) inversion model implementation [16] also currently uses this Raman formulation. Terms for other inelastic effects, such as fluorescence from dissolved organic matter and pigments may also be added if representative models are available.

2.9. ZTT Model Summary

Equation (18) is the final model for ocean color reflectance defined as L_u/E_d, called the ZTT model hereafter. In Equation (18) the $\overline{\mu}_d$ term is approximated by Equation (14), the $\beta(\psi)/b_b$ term approximated by Equation (10), the $\Psi_{K_{Lu}}$ term described by Equations (4) and (6), the $\overline{\mu}_\infty$ term described by Equation (8), and $\widetilde{b}_b$ represented by Equation (11). The ultimate assignment of the f_L term is addressed in the Results. The geometry variables, $E_d(0^+,\theta_s')$, V, and molecular water scattering parameters in the above can be considered knowns. The model is fully spectral. Key unknowns are b_b and a (or b_b/a and η_{bb}), and since pure seawater absorption a_w in the visible is considered known with good accuracy [61] the effective unknowns are b_{bp} and absorption by non-water constituents a_{pg}. The two additional unknowns are $P_{bb}(\psi)$ and $\widetilde{b}_{bp}$. In the forward implementation here, these four parameters must be provided from direct measurements or through some assumptions. In the inversion implementation, these are the parameters that may be solved through techniques to minimize errors in the expression if there are enough spectral bands (i.e., degrees of freedom) in r_{rs}, although *a priori* assumptions may be required for $P_{bb}(\psi)$ and $\widetilde{b}_{bp}$.

For ocean color reflectance defined as L_u/E_{od}, the simpler relationship in Equation (7) can be used. Water Raman effects can still be considered by applying an algorithm, such as in Reference [60].

3. Methods

3.1. Synthetic Dataset

A synthetic data set referenced to [Chl] was developed to test the ZTT model and develop an expression for the f_L term (see Section 4.1). Twenty values of [Chl] were assumed, logarithmically spaced between 0.01 and 30 mg m^{-3}. Total absorption coefficient (a) was represented by a sum of four components:

$$a(\lambda) = a_w(\lambda) + a_{ph}(\lambda) + a_d(\lambda) + a_g(\lambda). \tag{19}$$

Pure water absorption a_w was taken from Reference [61]. Phytoplankton absorption a_{ph} was calculated from chlorophyll concentration [Chl] and from spectrally averaged absorption coefficients of micro- and pico-plankton ($\overline{a}_{micro}(\lambda)$ and $\overline{a}_{pico}(\lambda)$, respectively) [62]:

$$a_{ph}(\lambda) = [Chl] * \left(\left[S_f\overline{a}_{pico}(\lambda)\right] + \left[\left(1 - S_f\right)\overline{a}_{micro}(\lambda)\right]\right), \tag{20}$$

where S_f = [0.25, 0.5, 0.75] is the shape mixing factor. Non-algal particulate absorption a_d was given by [63]:

$$a_d(\lambda) = a_d(440) * e^{-0.011(\lambda-440)}, \tag{21}$$

where $a_d(440)$ is equal to $R_d\ a_{ph}(440)$, with R_d = [0.05, 0.1, 0.5, 1, 2]. Similarly, we used an exponentially decaying expression for the absorption of chromophoric dissolved organic matter, a_g [64]:

$$a_g(\lambda) = a_g(440) * e^{-0.014(\lambda-440)}, \tag{22}$$

where $a_g(440)$ is equal to $R_g\ a_{ph}(440)$, with R_g = [0.1, 0.3, 0.5, 1, 2].

Total backscattering b_b was represented by the sum of water and particulate components:

$$b_b(\lambda) = b_{bp}(\lambda) + b_{bw}(\lambda). \tag{23}$$

To derive particulate backscattering b_{bp}, total particulate scattering b_p was first empirically estimated at 550 nm [65]:

$$b_p(550) = 0.416[Chl]^{0.766}, \tag{24}$$

and then extrapolated spectrally [66]:

$$b_p(\lambda) = b_p(550) * \left(\frac{\lambda}{550}\right)^{v([Chl])}, \tag{25}$$

where v([Chl]) = 0.5(log[Chl] − 0.3) when $0.01 \leq$ [Chl] ≤ 2 mg m^{-3} and v([Chl]) = 0 when [Chl] > 2 mg m^{-3}. The empirical relationship from [41] between the particulate backscattering ratio $\tilde{b}_{bp}$ and [Chl] was then used to derive b_{bp}:

$$b_{bp} = b_p \tilde{b}_{bp} = b_p * 0.0096[Chl]^{-0.253}. \tag{26}$$

Pure seawater backscattering b_{bw} was calculated according to Zhang et al. [52].

3.2. *Radiative Transfer Simulations*

RT simulations of r_{rs} were performed with Hydrolight (Sequoia Scientific, Bellevue, WA), following the procedure in Tonizzo et al. [29]. Fournier-Forand analytical phase functions were derived for each [Chl] iteration following the method of Mobley et al. [67].

Inelastic water Raman scattering was not included in the simulations for the synthetic data set, as this is separately addressed in the model (Equation (18)); fluorescence from any seawater constituents was also not considered. Note water Raman effects were included in full RT simulations for all field data (see [29]). Output wavelengths ranged from 350 to 800 nm at 5 nm resolution. Hydrolight default atmospheric parameters were used for all runs. Computations were run for $\theta_s{}'$ of 10, 30 and 60°. Altogether, 1500 different IOP permutations based on [Chl] were simulated for each $\theta_s{}'$.

3.3. *Field Data Sets*

Model performance was assessed using two aggregate data sets. The first is a high quality data set of 23 stations from the Ligurian Sea, waters around the Marine Optics BuoY (MOBY) west of Lanai, the southern California coast, and the New York Bight, collected in 2008 and 2009 as part of NASA Spectral Ocean Radiance Transfer Investigation and Experiment (SORTIE) and NASA Ocean Color Validation (OCVAL) exercises. Radiometric closure was assessed in detail for these data by Tonizzo et al. [29]. [Chl] ranged from 0.24 to 23.85 mg m^{-3}. The VSF was directly measured using the custom Multi-Angle SCattering Optical Tool (MASCOT) [26,36] and radiometric uncertainties were rigorously assessed by Voss et al. [68]. Here we use WET Labs ac-9 absorption measurements corrected for scattering using independent VSF measurements (i.e., the VSF98P correction [29]; also see Stockley et al. [49] for detailed evaluation of this correction). The second data set is the NASA NOMAD database, where 80 data records were identified containing the parameters needed for performance assessments here, i.e., a, b_b, $\tilde{b}_{bp}$, [Chl], and r_{rs}. [Chl] ranged from 0.23 to 10.68 mg m^{-3} in this data set.

3.4. *Depth Weighting IOPs*

Depth-weighted contributions of IOPs to water-leaving radiance were derived after Zaneveld et al. [69] using a 2-stream, first derivative approximation. The approach finds the IOP for a conceptual homogeneous ocean that reproduces the water-leaving radiance observed in a specific stratified ocean case. The remote sensing depth weighting for generic IOP X is:

$$\langle X\rangle = \int_{z=0}^{z=\infty} Xf(z)dz, \text{ where } f(z) = \frac{d}{dz}exp\left[-2\int_{z=0}^{z=\infty} K(z)dz\right]. \tag{27}$$

The exponential term can also be found in Gordon and Clark [70], where it is mentioned the $2K(z)$ term should actually be $K_u + K_d$, but approximating with K_d alone was not expected to lead to appreciable errors. Several approximations of K_d in terms of IOPs are available from the literature (e.g., [71,72]). The approximation by Lee et al. [73] for averaged K_d within the euphotic zone is used here and conveniently represented in terms of a, b_b and above water solar zenith $\theta_s{}'$:

$$\overline{K}_d = (1+0.005\theta_s{}')a + 4.18\left(1-0.52e^{-10.8a}\right)b_b. \tag{28}$$

For evaluating RT approximations, all IOP terms should ideally be weighted together. For example, Zaneveld et al. [69] considered the simple $R \propto b_b/a$ approximation and demonstrated that $<b_b>/<a>$ was not equivalent to $<b_b/a>$ for an IOP stratified ocean. However, b_b and a are considered independently in the model here and an ultimate objective is to invert the approximation to derive b_b and a independently, so each IOP was individually depth weighted. For all stations sampled in the validation data set, the difference between surface b_b and depth weighted b_b was <1%; with a, this was also the case for most stations, but reached a difference of 5% for one station sampled.

3.5. Metrics for Error Assessment

Mean absolute percent error (MAPE), $\%\delta_{abs}$, is a commonly used metric in assessing performance in r_{rs} match ups:

$$\%\delta_{abs} = 100 * \frac{\delta_{abs}}{\overline{y}}, \quad \delta_{abs} = \frac{\sum_{i=1}^{n}|y_i - \hat{y}_i|}{n}. \tag{29}$$

The MAPE metric takes into account the absolute magnitude of the residuals, giving them equal weight. Other commonly used metrics include root mean square error (RMSE), a measure of accuracy and potential forecasting errors in simulating r_{rs} when the errors may be assumed to be unbiased and normally distributed [74]. RMSE gives greater weight to larger errors than δ_{abs}. Since bias errors are often expected to be more significant than random, normally distributed errors in simulations, $\%\delta_{abs}$ is expected to be the most appropriate metric to assess match ups [29]. Other common metrics (e.g., [75]) are shown in the following plots as appropriate. Also see Werdell et al. [13] for a detailed discussion of performance metrics.

4. Results

4.1. Developing an Expression for the f_L Term

Dynamics of f_L were explored in the synthetic data set. As found by Hoge et al. [76], dependencies on wavelength and $\theta_s{}'$ were most important, and an average spectral shape $f_{L,ave}(\lambda)$ scaled to solar zenith according to $\sin(\theta_s{}')$ was ultimately found to be most representative (Figure 2):

$$f_L(\theta_s', \lambda) = f_{L,ave}(\lambda)\left[0.05959\sin(\theta_s{}') + 0.9728\right]. \tag{30}$$

Function $f_{L,ave}(\lambda)$ is provided in Appendix A, Table A3. Equation (30) was developed by minimizing residual errors between results from full radiative transfer and the ZTT model for each wavelength and each solar zenith angle for the full synthetic data set. Spectral shapes for f_L were relatively consistent in the results, so the term $f_{L,ave}(\lambda)$ was derived from an average for these results after spectral normalization. The scaling factor in brackets was a suitable function to minimize errors with respect to solar zenith angle. Only nadir viewing was again considered in the simulations for

f_L, and we again make the assumption from Section 2.1 that f_L for other viewing angles that define a specific in-water scattering angle ψ can be approximated by f_L observed at the nadir viewing geometry with equivalent ψ. The potential caveats mentioned in Section 2.1 apply here as well. Restating Equation (30) in terms of ψ, the following is obtained:

$$f_L(\psi, \lambda) = f_{L,ave}(\lambda)[0.07762 \sin(\psi) + 1.0405]. \tag{31}$$

Similarities between $f_{L,ave}(\lambda)$ and a typical ocean absorption spectrum are noted (Figure 2), resulting from the effects of multiple scattering, where higher relative scattering (lower relative absorption) promotes a radiance field closer to that which would result from isotropic scattering, i.e., where f_L approaches unity.

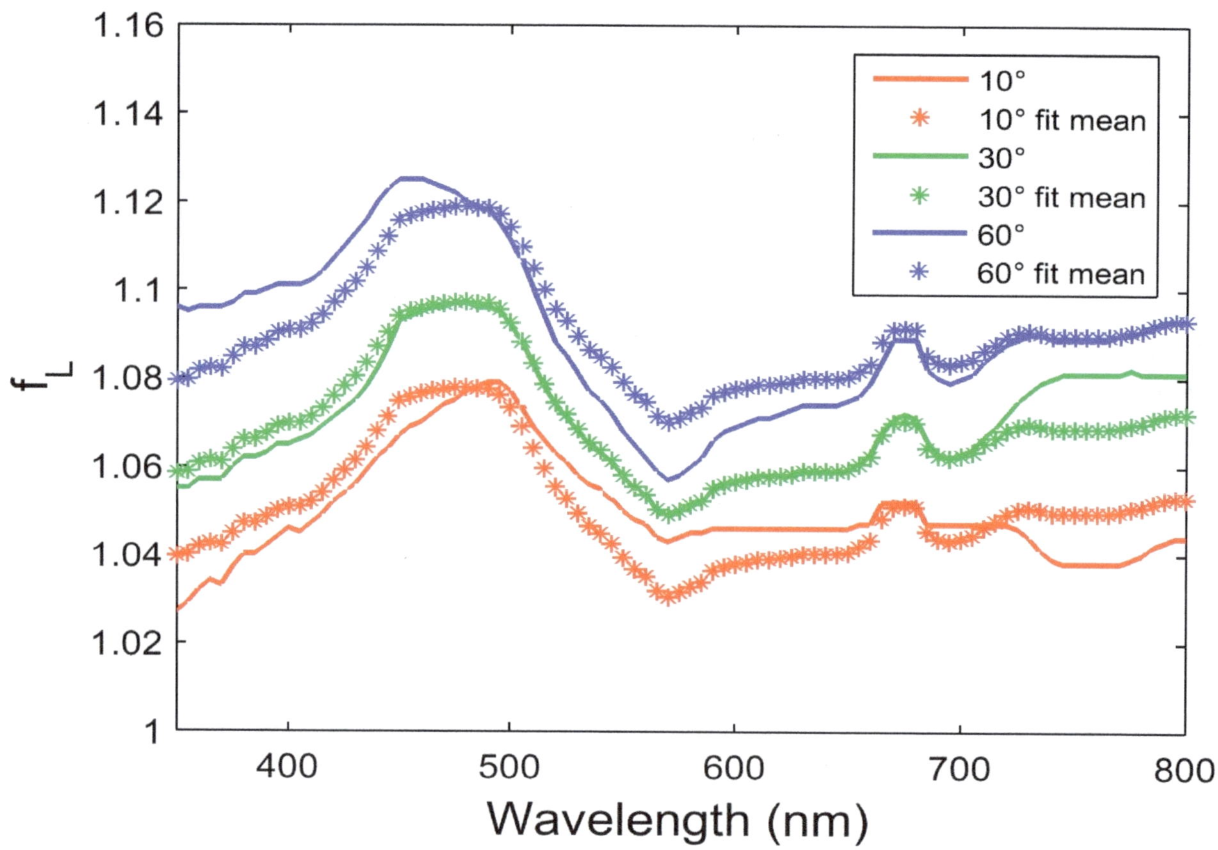

Figure 2. Spectral dependency of the f_L shape function at different $\theta_s{}'$. Thin lines are solved f_L functions for each $\theta_s{}'$ for the synthetic data set and stars are from the model approximation with constant shape described by $f_{L,ave}(\lambda)$ in Equation (30).

Figure 3A shows results deriving r_{rs} using the ZTT model for the synthetic data set, applying the approximation from Equation (31) for f_L and using all synthetic data IOPs as input. This match up is, thus, effectively assessing combined errors in the ZTT model from the approximations of f_L, f_b, K_{Lu}, and $\overline{\mu}_d$. We note f_L was specifically optimized to this data set, so in this respect Figure 3A shows what may be considered a best-case matchup. MAPE $\%\delta_{abs}$ for the full data set was relatively small at 2.68%. The largest errors were observed in three r_{rs} spectra, with the model showing overestimation bias centered around 570 nm. These spectra were associated with the highest [Chl] modeled in the data set and lowest $\widetilde{b}_{bp}$.

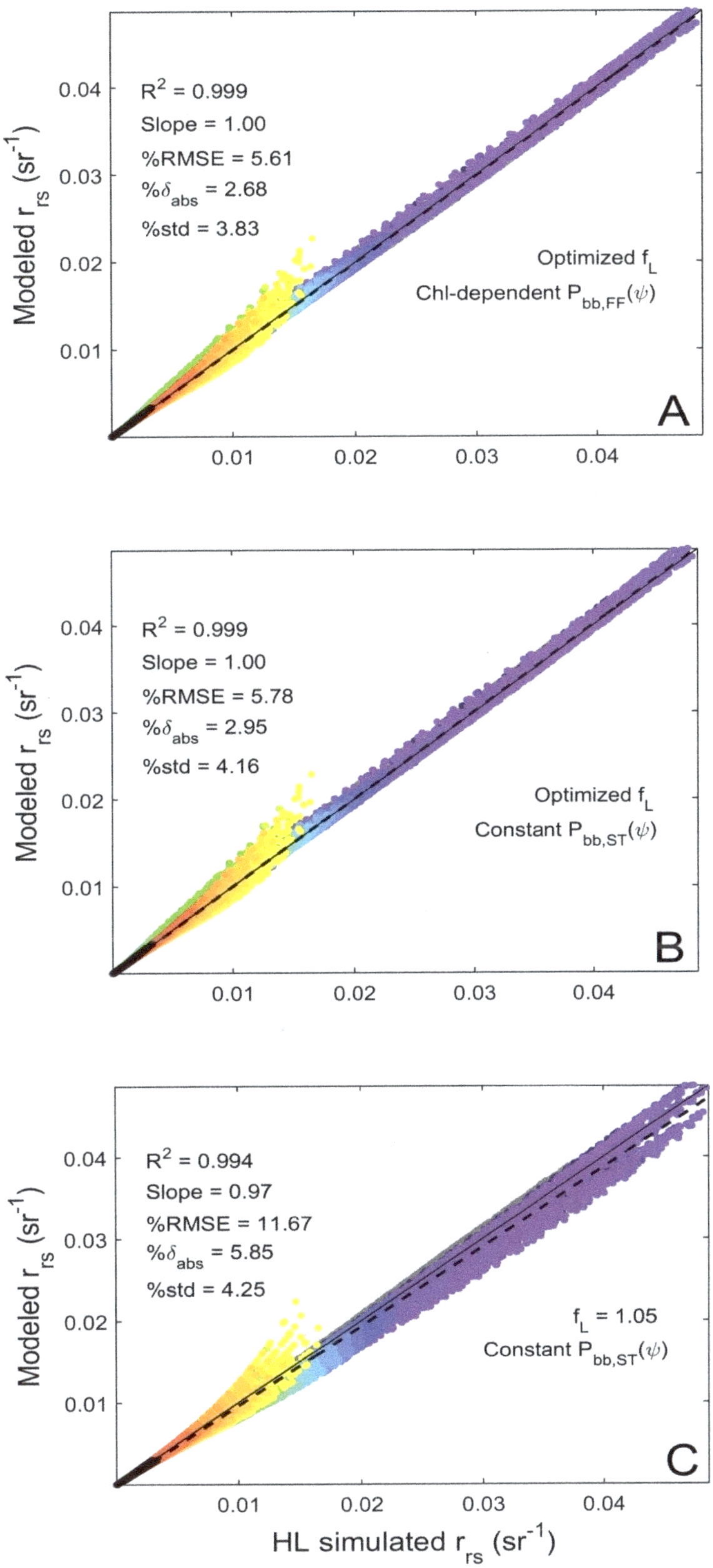

Figure 3. r_{rs} derived from the ZTT model with the synthetic data set compared to r_{rs} simulated using Hydrolight (HL): (**A**) f_L optimized spectrally (i.e., Equation (31) for the synthetic data set with Fournier-Forand phase functions individually determined for each [Chl]; (**B**) f_L optimized spectrally for the synthetic data set with constant $P_{bb,ST}(\psi)$ (see text); and (**C**) f_L set to 1.05 with constant $P_{bb,ST}(\psi)$. Colors represent wavelengths, from 400 to 700 nm with gray used for $350 < \lambda < 400$ nm and $700 < \lambda < 800$ nm.

4.2. Assessing Assumption of Constant $\beta_p(\psi)/b_{bp}$

Sullivan and Twardowski [26] found a high degree of consistency in the shape of the particulate VSF in the backward direction, i.e., $\beta_p(\psi)/b_{bp}$, for a large data set covering a wide dynamic range in bulk particle composition. Deviations from this relationship were no more than 5% through the entire backward angular range. While it is well known the phase function shape over the full angular range varies substantially (e.g., [37,41,45,77]), only the shape in the backward direction that is important for remote sensing and the ZTT model is considered here.

If we assume $\beta_p(\psi)/b_{bp}$ is a constant shape $P_{bb,ST}(\psi)$ after Sullivan and Twardowski [26] in the ZTT model, replacing $P_{bb,FF}(\psi,\tilde{b}_{bp})$ from the synthetic data set, errors increase by only ~0.3% (Figure 3B). We assume $P_{bb,ST}(\psi)$ is constant spectrally after [41] and others. This result is significant, as it demonstrates the potential for eliminating one of the key unknowns in the ZTT model. We note that an averaged Fournier-Forand phase function in the backward direction could be used in the model, which would enhance the analytical character of the model, but the Sullivan and Twardowski [26] averaged phase function, derived from extensive measurements in a wide range of water types, may be more representative of the natural environment. This is assessed further in the next section. Figure 3C shows results of setting f_L to a constant 1.05 after [28] and using a constant $P_{bb,ST}(\psi)$. MAPE increases significantly to 5.85%, showing the importance of using the f_L model in the ZTT. Each of the three distinct data groupings along the 1:1 relationship are associated with the individual solar zeniths that were used, i.e., 10, 30, and 60°, highlighting the influence of the f_L model in removing the effects of solar zenith. All ZTT model runs hereafter use the f_L model from Equation (31). The constant backward phase function $P_{bb,ST}(\psi)$ is used unless another phase function is specified.

4.3. Assessment with High Quality Validation Data

Figure 4A shows results for the ZTT model, with inputs of constant $P_{bb,ST}(\psi)$ and direct measurements of a_{pg}, b_{bp}, and $\tilde{b}_{bp}$, compared to measured r_{rs} for the validation data set [29]. MAPE $\%\delta_{abs}$ was 16% for all λ (summarized in Table 1; spectral $\%\delta_{abs}$ provided in Table 2). Interestingly, this result is slightly more favorable than r_{rs} computed with the full radiative transfer (i.e., Hydrolight) using measured phase functions, where $\%\delta_{abs}$ was 17% for the same data (as reported in Tonizzo et al. [29]). Based on closure analyses for these data [29], this 17% error represented the aggregate inherent error from all sources within the data and computations, i.e., from the IOP and radiometric measurements, as well as any errors from the assumptions within the Hydrolight RT code. For the ZTT model evaluated using measured phase functions instead of the constant $P_{bb,ST}(\psi)$, i.e., the same approach followed with the full RT computations, $\%\delta_{abs}$ was also 17% (Table 1). The different backward phase functions applied in the ZTT model are shown in Figure 5. Measured backward phase functions showed more variability at individual angles than the average $P_{bb,ST}(\psi)$ derived from a large data set of diverse Case 1 and Case 2 water types, likely the result of small scale hydrosol patchiness along the different light paths for individual measurements at each angle. This observation is addressed further in Section 5. Moreover, the strong agreement in absolute error between full radiative transfer simulations and the ZTT analytical approximation for this diverse data set is encouraging.

Excluding the effects of water Raman from the ZTT model increased absolute error by 3%, demonstrating Raman had a significant effect, especially >580 nm (results not shown).

Figure 4B shows ZTT model performance with a further constraint, setting $\tilde{b}_{bp}$ constant at 0.006, so the only IOP inputs were measured a_{pg} and b_{bp}. With respect to inversion, this approach has the same unknowns, i.e., a_{pg} and b_{bp}, as current algorithms based on the simple first order approximation [16]. Resulting MAPE $\%\delta_{abs}$ was 17%, comparable to full RT computations. ZTT runs with other fixed $\tilde{b}_{bp}$ are shown in Table 1. Some influence from $\tilde{b}_{bp}$ is apparent, which has not been fully appreciated in previous models based on the first order approximation of r_{rs} to b_b/a.

ZTT model runs with Fournier-Forand phase functions for the particulate component, computed from measured $\tilde{b}_{bp}$ following the method of [67], resulted in $\%\delta_{abs}$ of 19%, 3% higher than results using the constant $P_{bb,ST}(\psi)$ (Table 1).

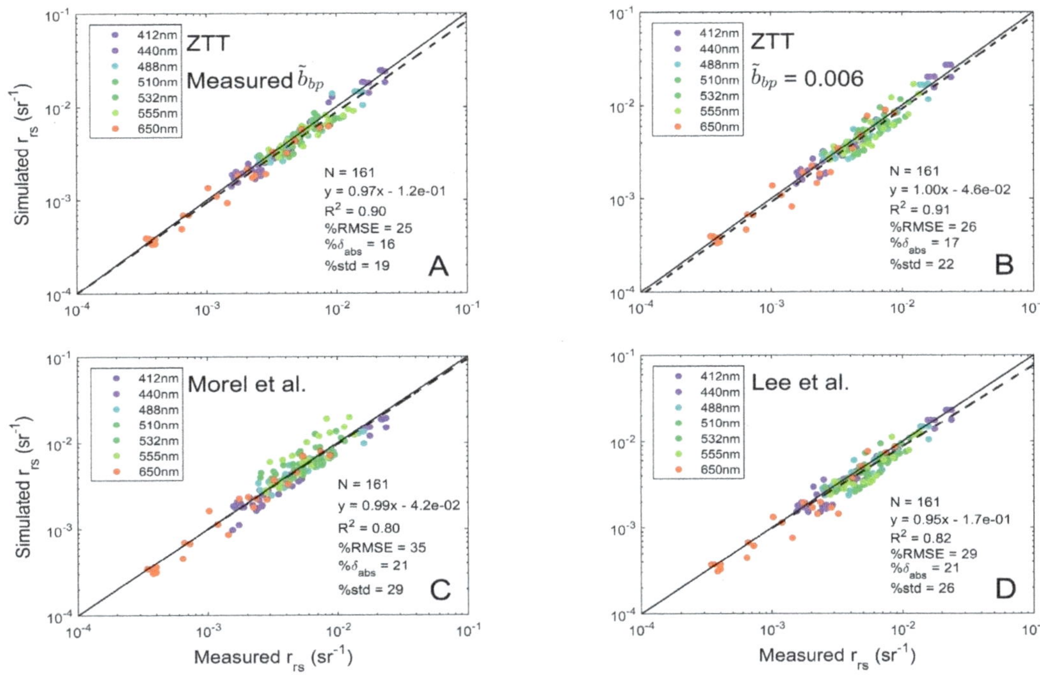

Figure 4. r_{rs} computed from (**A**) ZTT, (**B**) ZTT with $\tilde{b}_{bp}$ fixed at 0.006, (**C**) M02, and (**D**) L11 models, compared to measured r_{rs} in the high quality validation data set from Tonizzo et al. [29]. Note L11 does not account for Raman scattering. Chlorophyll input for M02 was derived from spectral absorption using the Nardelli and Twardowski [78] line height algorithm. MAPE $\%\delta_{abs}$ was 16%, 17%, 21%, and 21%, in A through D, respectively.

Table 1. Summary MAPE $\%\delta_{abs}$ results for simulated or calculated r_{rs} vs measured r_{rs} for all λ.

Approach	$\beta_p(\psi)/b_{bp}$ **Input**	$\tilde{b}_{bp}$ **Input**	$\%\delta_{abs}$	
			SORTIE and OCVAL (23 Stations) [1]	**NOMAD (80 Stations)**
Full RT [2]	directly measured	N/A	17	nd [4]
Full RT [2]	Fournier-Forand [3]	measured	20	nd [4]
ZTT	directly measured	measured	17	nd [4]
ZTT	Fournier-Forand [3]	measured	19	25
ZTT	$P_{bb,ST}(\psi)$	measured	16	20
ZTT	$P_{bb,ST}(\psi)$	0.005	18	22
ZTT	$P_{bb,ST}(\psi)$	0.006	17	23
ZTT	$P_{bb,ST}(\psi)$	0.008	18	25
ZTT	$P_{bb,ST}(\psi)$	0.010	19	27
ZTT	$P_{bb,ST}(\psi)$	0.015	22	29
ZTT	Large and small population phase functions with $\tilde{b}_{bp}$ of 0.19% and 1.4%, blended according to [Chl] [5]	measured	23	26
Morel et al. [7] (M02)	Large and small population phase functions with $\tilde{b}_{bp}$ of 0.19% and 1.4%, blended according to [Chl] [5]	N/A	21	25
Lee et al. [11] (L11)	Blend of Petzold[6] average and 1% $\tilde{b}_{bp}$ Fournier-Forand [3]	N/A	21	26

[1] see [29]; [2] Hydrolight; [3] derived from $\tilde{b}_{bp}$ according to [67]; [4] not determined; [5] algorithm of [7]; [6] [79].

Match ups in r_{rs} were also assessed applying the M02 model, the current standard in BRDF corrections for ocean color remote imagery (Figure 4C). The proportionality between r_{rs} and b_b/a is defined as f/Q, which is a function of viewing geometry, wind speed, and [Chl]. [Chl] can be estimated from an empirical r_{rs} band ratio algorithm. For the assessment here, [Chl] input was either quantified directly from discrete samples or derived from spectral particulate absorption measurements using the line height method with specific absorption a_p*(676) of 0.0108 m^2 mg^{-1} after [78]. Note that bootstrapping exercises have shown relatively large errors in this estimated [Chl], i.e., up to +/−50%, affect $\%\delta_{abs}$ in match ups by typically no more than 1%. The other inputs were measured a_{pg} and b_{bp}. Match up $\%\delta_{abs}$ for M02 was 21% for the data set from [29] (Table 1). Notably, M02 performed well in the blue spectral region (Table 2).

To derive phase functions, Morel et al. [7] used randomly-oriented spheroidal particles, computing the scattering phase function with the T-matrix method. Because the computations for spheroids were lengthy, a single refractive index n_p of 1.06 was considered with sizes ranging from 0.02 to 14 µm. A Junge-type particle size distribution was assumed, i.e., a power law model, with exponents of 3.1 and 4.2 representing two end-member populations. Particulate scattering phase functions were then calculated from (see Figure 5):

$$P_p(\psi) = \alpha_s([Chl])P_{ps}(\psi) + \alpha_l([Chl])P_{pl}(\psi),$$

$$\text{with } \alpha_s + \alpha_l = 1, \text{ and } \alpha_s([Chl]) = 0.855\left[0.5 - 0.25\log_{10}([Chl])\right].$$

If the Morel et al. [7] approach for deriving the phase function is used as input into the ZTT model, replacing the constant shape $P_{bb,ST}(\psi)$, MAPE $\%\delta_{abs}$ increases significantly from 16% to 23% (Table 1).

Finally, match ups in r_{rs} were assessed in a recently published BRDF model from Lee et al. ([11]; called L11 hereafter) that, like M02, developed a LUT based on RT simulations with a [Chl]-referenced synthetic IOP data set (Figure 4D; Tables 1 and 2). L11 uses a quadratic form of $b_b/(a + b_b)$ split into particulate and molecular seawater components with each term scaled by a G coefficient, with solutions provided in LUTs. In the synthetic data set, Lee et al. [11] used an averaged Petzold phase function and Fournier-Forand phase function assuming $\widetilde{b}_{bp}$ = 1% as particulate phase function endmembers. L11 however does not include water Raman effects. Like M02, match up MAPE $\%\delta_{abs}$ was also 21% for all λ.

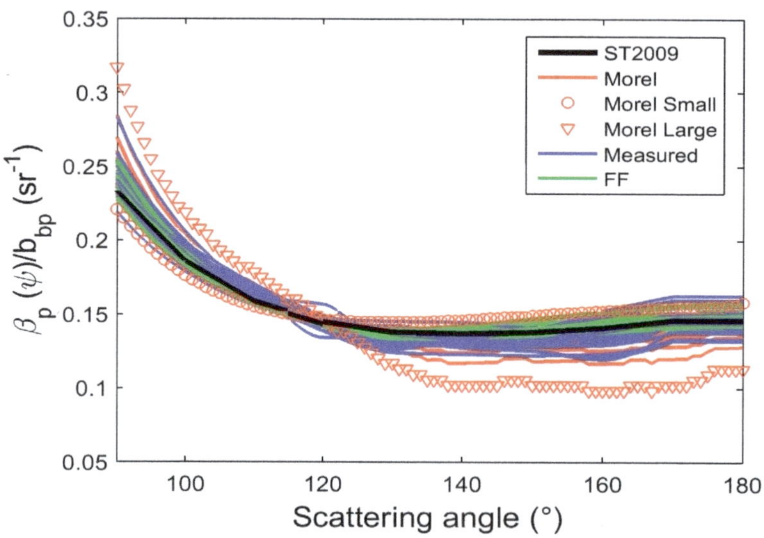

Figure 5. Phase functions in the backward direction used to assess the ZTT model. "ST2009" is $P_{bb,ST}(\psi)$ from [26] with the assumption $P_{bb,ST}(\psi > 170°) = P_{bb,ST}(170°)$; "Morel" is from Morel et al. [7] with endmember populations dominated by small and large particles also shown; "Measured" were directly measured; and "FF" are Fournier-Forand phase functions $P_{bb,FF}(\psi,\widetilde{b}_{bp})$ derived from measured $\widetilde{b}_{bp}$ following [67].

4.4. Assessment with Global NOMAD Data Set

Figure 6A shows results for the ZTT model applied to a subset of 80 measurements from the NASA NOMAD global data set. IOPs were averaged over the first optical depth. Match up MAPE $\%\delta_{abs}$ of 20% was observed for this global data set, comparing favorably to the error of 16% observed with the high quality data set from Tonizzo et al. [29] (Table 1; see Table 2 for spectral breakdown). This is especially true considering the different IOP and r_{rs} processing approaches (e.g., there are several scattering error correction options for in-water absorption measurements) used for these data sets. If $\tilde{b}_{bp}$ is constrained to 0.006, MAPE $\%\delta_{abs}$ increases to 23% (Figure 6B). Results from ZTT runs for other fixed $\tilde{b}_{bp}$ are shown in Table 1, where the dependency of performance on $\tilde{b}_{bp}$ is again apparent.

If the Morel et al. [7] approach for deriving the phase function from [Chl] is used as input into the ZTT model, $\%\delta_{abs}$ increased from 20% to 26% for the NOMAD data (Table 1). If measured $\tilde{b}_{bp}$ are used to derive $P_{bb,FF}(\psi,\tilde{b}_{bp})$ following [67], $\%\delta_{abs}$ increased to 25% for the ZTT model.

Table 2. Summary MAPE $\%\delta_{abs}$ results for calculated r_{rs} vs measured r_{rs} for each λ, for SORTIE and OCVAL data sets (see Figure 4) and NOMAD data (see Figure 6).

Approach	$\beta_p(\psi)/b_{bp}$ Input	$\tilde{b}_{bp}$ Input	λ (nm)	$\%\delta_{abs}$ SORTIE and OCVAL (23 Stations)	$\%\delta_{abs}$ NOMAD (80 Stations)
ZTT	$P_{bb,ST}(\psi)$	measured	412 (410)	20	17
			440	16	20
			488 (490)	16	20
			510	19	19
			532	13	-
			555	14	23
			650	13	-
			665	-	63
ZTT	$P_{bb,ST}(\psi)$	0.006	412 (410)	20	20
			440	16	23
			488 (490)	19	23
			510	13	25
			532	14	-
			555	13	26
			650	22	-
			665	-	63
Morel et al. [7] (M02)	Large and small population phase functions with $\tilde{b}_{bp}$ of 0.19% and 1.4%, blended according to [Chl]	N/A	412 (410)	17	22
			440	14	25
			488 (490)	16	25
			510	18	26
			532	22	-
			555	38	21
			650	17	-
			665	-	71
Lee et al. [11] (L11)	Blend of Petzold average and 1% $\tilde{b}_{bp}$ Fournier-Forand	N/A	412 (410)	22	22
			440	21	26
			488 (490)	24	27
			510	21	27
			532	20	-
			555	19	24
			650	26	-
			665	-	66

Match ups are also shown for implementation of the full M02 and L11 models with the NOMAD data set (Figure 6C,D). MAPE statistics are summarized in Tables 1 and 2. There are several possibilities for the systematically lower measured r_{rs} in the red spectral region for these data: (1) Suboptimal surface extrapolation of radiances due to water Raman effects [80,81], (2) residual scattering errors in reported in-water absorption measurements that may promote a bias to higher values of modeled r_{rs} (e.g., [29,49]), and/or (3) suboptimal corrections for sensor self-shading, which can be strong in clear water at longer wavelengths due to water absorption [82,83].

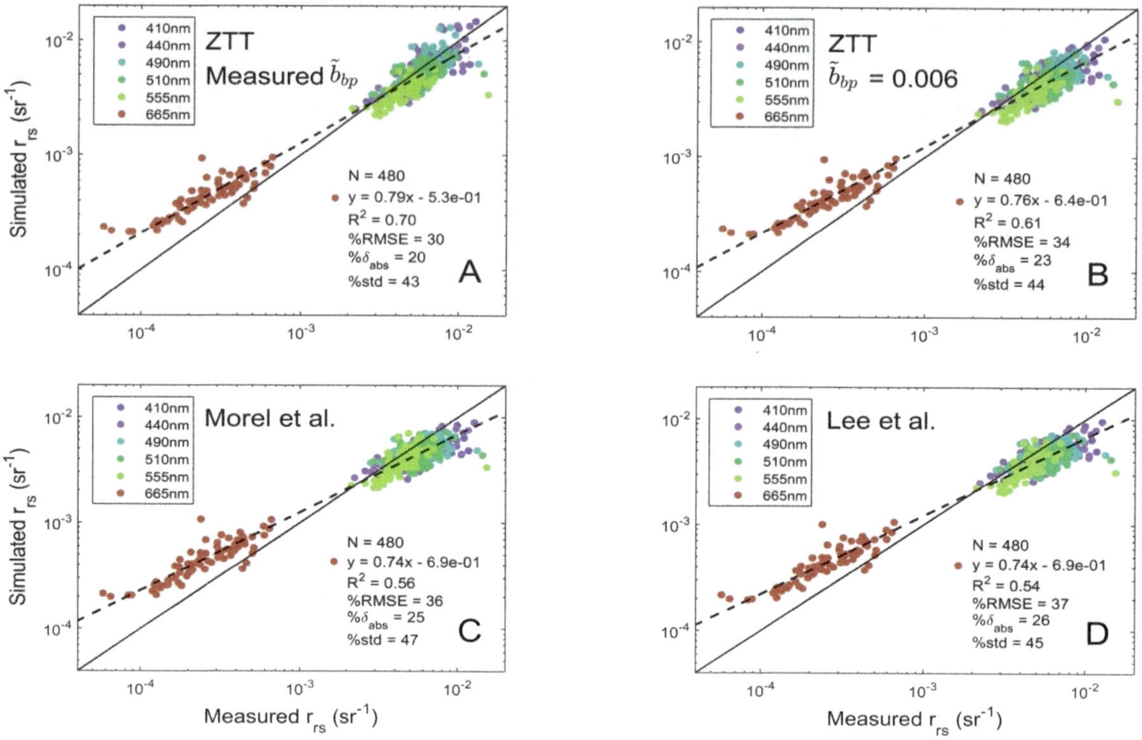

Figure 6. As in Figure 4, but for the NASA NOMAD data set containing a, b_b, r_{rs}, and [Chl]. MAPE $\%\delta_{abs}$ were 20%, 23%, 25%, and 26% in (**A–D**), respectively.

5. Discussion

The ZTT analytical model for remote sensing reflectance is based on the restatement of the RT equation by Zaneveld [27,28] with the following refinements: (1) Full bidirectionality was retained in the final model (no assumption of nadir viewing), (2) the assumption of asymptotic diffuse attenuation for upwelling radiances at the surface was analytically described, (3) the entire model was formulated in terms of b_b and a to enable inversion, (4) the VSF was explicitly included with $\beta_p(\psi)/b_{bp}$ set to a constant shape defined by $P_{bb,ST}(\psi)$ [26], which showed the strongest performance of any phase function, including measured phase functions in disparate, diverse data sets, (5) water Raman effects were added, and (6) for r_{rs}, an analytical model for $\overline{\mu}_d$ was developed that included dependency on atmospheric visibility. IOP inputs to the ZTT model are b_{bp}, a_{pg}, $\beta_p(\psi)/b_{bp}$, and $\tilde{b}_{bp}$. Moreover, the ZTT model enables systematic assessment of the effect of IOPs on the BRDF for the first time.

ZTT showed strong performance in match ups with field data sets spanning a wide range of water types. In terms of MAPE, the model with constant phase function shape $P_{bb,ST}(\psi)$ performed as well as full RT computations. Stronger performance was observed for these available data sets relative to current state-of-the-art LUT-configured models based on the first order approximation between r_{rs} and b_b/a or $b_b/(a + b_b)$, i.e., M02 and L11.

MAPE for the ZTT model was slightly better using $P_{bb,ST}(\psi)$ ($\%\delta_{abs}$ of 16%) when compared to using directly measured backward phase functions for the validation data set (where $\%\delta_{abs}$ was 17%). It may be argued, under certain conditions, the broad average $P_{bb,ST}(\psi)$ may be more accurate than an isolated measurement because specific bias errors in any single VSF measurement at an individual angle may be avoided. This of course assumes such biases may be larger than the natural variability in the shape of the backward phase function. Another explanation is error assessments for the validation data set were made over a relatively few number of stations, 23. To roughly estimate uncertainty in this error metric, single stations were sequentially left out of the error determination for ZTT using $P_{bb,ST}(\psi)$, which gave a total range in $\%\delta_{abs}$ of 14.8 to 16.8% with standard deviation of 0.6%. A ~1% difference in error may thus not be particularly meaningful.

The ZTT model is fully analytical in the sense it is described entirely by equations and "almost fully analytical" with respect to the inclusion of empiricism. Empiricism comes from optimizing the f_L parameter to our synthetic data set based on [Chl]-based bio-optical models, with MAPE improving by a significant ~3% compared to setting f_L equal to a spectrally constant 1.05. Using the constant $P_{bb,ST}(\psi)$ backward phase function shape also introduced empiricism. If sample-specific Fournier-Forand analytical phase functions, based on $\widetilde{b}_{bp}$ were used, errors in the validation data set increased by 3–5% (Table 1).

Particulate phase function shapes from Morel et al. [7] were modeled using the T-matrix approach, assuming homogeneous spheroidal shapes with constant n_p of 1.06. Although the endmember phase function shapes were not consistent with $P_{bb,ST}(\psi)$ and $P_{bb,FF}(\psi,\widetilde{b}_{bp})$ shapes, many of the blended phase functions were. If $P_{bb,ST}(\psi)$ may be considered widely representative, increasing errors may be expected in applying the M02 model to the lowest and highest Chl conditions where the endmember phase functions would fully manifest. Phase functions in the Chl range 1 to 10 mg m^{-3} were in closest agreement to $P_{bb,ST}(\psi)$. Applying phase functions derived using the Morel et al. [7] approach in the ZTT model increased errors relative to using a constant $P_{bb,ST}(\psi)$ by a significant 7% for the validation data set of [29] (from 16% to 23%) (Table 1). This error was also larger than the 21% error observed for M02 (addressed further below).

Since L11 used a Fournier-Forand phase function where $\widetilde{b}_{bp} = 1\%$ and the average Petzold phase function as endmembers [shown in 26, their Figure 3], many blended shapes from L11 should be relatively similar in the backward to the $P_{bb,ST}(\psi)$ used in ZTT (although the Petzold shape does exhibit systematic discrepancies). This particular $P_{bb,FF}$ with $\widetilde{b}_{bp}$ = 1% is actually a very close match to $P_{bb,ST}$. L11 was developed through least-squares fitting with respect to Hydrolight results, so all the blended phase functions are weighted in the results.

A "typical marine atmosphere" and associated incident sky radiance distribution were used in the Hydrolight RT computations for the synthetic data set, the effects of which enter the ZTT model through spectral optimization of the upwelling radiance shape factor f_L (Section 4.1). The parameters $\Psi_{K_{Lu}}$ and $\overline{\mu}_\infty$ were also solved with Hydrolight RT simulations assuming the same atmosphere. The M02 and L11 models also assumed a single "typical marine atmosphere" in their simulations using Hydrolight based on Reference [58], so this aspect of the models should be consistent. The $\overline{\mu}_d$ term in the ZTT model has added flexibility in accommodating varying atmospheric visibility (closely related to AOT; [58]) inasmuch as V is an explicit term for $\overline{\mu}_d$.

One question arising from this work is, if the phase function shape $P_{bb,ST}$ may indeed be considered highly representative, how would a LUT-based model following the approach of M02 but using $P_{bb,ST}$ perform? To test this, we developed a LUT based on the first order r_{rs} proportionality to b_b/a, using the synthetic data set described in Section 3.1. The particle phase function was constant, consisting of $P_{bb,ST}$ in the backward and a Fournier-Forand phase function with $\widetilde{b}_{bp}$ = 0.006 for the forward direction (this had the best performance for the ZTT model in the validation data set [29], with MAPE = 17%). The LUT approach resulted in a MAPE of 22% for the validation data set and 26% for the NOMAD data. This performance is very close to M02 (21% and 25%, respectively) and L11 (21% and 26%, respectively). An insight we can take from this analysis is that a component, such as phase function shape in the different models cannot be considered in a vacuum, i.e., each model has a large set of assumptions that ultimately influence final performance. Furthermore, although not always explicitly stated in the literature, field validation efforts are likely to have had a role in updating assumptions through the course of development and testing of some models to ensure the model was well aligned with available data. For example, in our case, the f_L expression (Section 4.1) was developed from an assumed synthetic data set and the $\widetilde{b}_{bp}$ value of 0.006 was recommended as it ultimately provided the best results with the field validation data sets available.

Recent work by Zhang et al. [84] has suggested the possibility of substantial variability in phase function shape in the backward direction, including up to a 40% increase for $\beta_p(\psi)/b_{bp}$ in the near backward relative to $\beta_p(120°)/b_{bp}$. These results are in stark contrast to the comprehensive study of [26] and other published phase functions in the literature (e.g., [35,85]). Other recent works showing increases in the VSF at $\psi > 150°$ [86–88] were made in enclosed cuvettes and did not include any correction for internal reflections (e.g., [89]), which are typically a difficult problem in laboratory VSF devices, since the scattered signal in the backward direction is orders of magnitude smaller than the forward. The Zhang et al. [84] results are also inconsistent with the phase functions used in previous BRDF models, such as M02, which have been shown to have satisfactory performance in validation efforts [17,18], even in many Case II coastal waters [19,90]. The recent theoretical r_{rs} modeling study of Xiong et al. [91] based on the phase function shapes presented in [84], where up to 50% disagreement with L11 was observed, consequently appear unrealistic.

Calibration and possible bias errors in the VSF device MASCOT used in [26] to derive $P_{bb,ST}(\psi)$ have been evaluated in detail [36,48]. The MASCOT is an open path, in situ device with 17 independently calibrated detectors resolving the VSF from 10° to 170° in 10° increments. There are two calibration coefficients for each channel, a dark offset (obtained with the source occluded) and a scaling factor to convert digital counts to VSF units m^{-1} sr^{-1}. Moreover, the observation of a remarkably consistent shape in the backward phase function for VSFs spanning more than four orders of magnitude in dynamic range cannot be explained by any known bias error in these coefficients. In fact this observation argues against any systematic bias, as consistency in phase function shapes during serial particle suspension experiments is a useful method to assess the accuracy of calibrations.

Recently, He et al. [20] developed a BRDF model also based on Zaneveld [28] but it differs from our implementation is several ways: (1) only the BRDF aspect was considered; (2) one wavelength was considered; (3) the formulation was in terms of a LUT (similar to M02 and L11) for aggregated terms; (4) widely varying backward phase function shapes were assumed after [84] described above; (5) the $\overline{\mu}_d$ model depended only on "typical average" sky conditions; and (6) water Raman effects were excluded. Direct comparisons were thus not possible here.

5.1. Assessing Residual Bias in the Model

The assumption of single scattering in the simplification of f_b (Section 2, Equation (2)) is not expected to invoke significant errors at turbidity levels found for most oceanic and coastal waters [5,13]. As shown by Gordon [23], secondary scattering events for light redirected toward the backward direction, i.e., downwelling radiance scattered into the upwelling radiance field, has a negligible effect on the shape of the upwelling radiance distribution because the vast majority of these scattering events are in the near forward direction. The full effects of multiple scattering were included in relationships for other terms in the model as they were determined using full RT modeling.

MASCOT VSF measurements used to determine $P_{bb,ST}(\psi)$ only extended to 170°, so the application of this function for $\psi > 170°$ in the model is uncertain. There were no $\psi > 170°$ in the data sets tested here, although such scattering angles may be expected for some viewing geometries for the PACE imager (Figure 1). There is a need for a better understanding of phase function shapes for $\psi > 170°$ and this extends to other applications, such as lidar [92].

Only nadir viewing was considered for the analytical relationships for K_{Lu} and f_L and we suggest these relationships may be transferable to other viewing geometries as long as ψ is preserved. This assumption has been verified in Hydrolight simulations to have small errors for representative cases, but cannot be assessed with the field data sets available here, which all have nadir viewing. This is being considered in future work, and, moreover, emphasizes the need for routine measurements of full upwelling radiance distributions in ocean color validation work as validation only with nadir radiance $L_u(\theta_s,\theta_v = \pi)$ neglects the BRDF effects that are essential to any algorithm.

The optimization of f_L was based on the [Chl]-based synthetic data set. A significant deviation from the bio-optical model assumptions of that synthetic data set could lead to a discrepancy in match ups, however it worth pointing out (1) f_L has a relatively weak dependency on IOPs except for the VSF shape in the backward direction (which is effectively accounted for in the solar zenith dependency in Equation (30)), and (2) other state-of-the-art models (i.e., M02 and L11) are constructed entirely around such synthetic data sets. A benefit of the ZTT model is that an analytical relationship for a specific term of the model may be replaced later if a more optimal approach is demonstrated.

As shown in Zaneveld [27,28], L_u normalized to E_{od} is more closely linked to the RT Equation than L_u/E_d, where an additional $\overline{\mu}_d$ term must be added to the model (Section 2.6). Adding this term is expected to increase uncertainty in the RT model and inversion. Scalar irradiance has the additional advantage of being less dependent on solar zenith angle and sky radiance distributions. As discussed in Section 2.6, sensor technology is available to measure E_{od} and inclusion of these measurements in field efforts focused on algorithm development and validation should be encouraged. When the $\overline{\mu}_d$ term must be included in association with E_d measurements, AOT is also a useful validation measurement that may be made in the field [93] and used to derive visibility V [58] for input in Equation (16) or full RT computations.

Lastly, potential uncertainties in the full RT computations must be considered, as this was used to parameterize several ZTT terms. Polarization is not considered in the ZTT model or in Hydrolight, which is likely important [29,33]. There is a need in the research community for commercially available RT code with a focus on the ocean that includes full polarization, along with concomitant measurements of polarized radiance in the field. Neither the Hydrolight simulations or ZTT model included inelastic effects from DOM fluorescence, which could lead to bias errors as high as about 10% in the mid-visible [33]. Another potential source of uncertainty is the water Raman estimation [60], where bias errors as high as 10% can be typical.

5.2. Suitability for Inversion

The ZTT model can be used to directly solve for b_b/a using least-squares minimization of a r_{rs} measurement at any wavelength. This is shown in Figure 7 for the validation data set from Tonizzo et al. [29], assuming a constant $\widetilde{b}_{bp}$ of 0.006. The typical value for V of 15 km was assumed. MAPE $\%\delta_{abs}$ in the inversion was 17%, which matches results from full RT simulations of closure in the forward direction. Residual errors thus appear almost entirely comprised of the inherent errors of the data set arising from IOP and radiometric measurements [29]. With respect to the use of $P_{bb,ST}(\psi)$, another potential approach to estimating VSF shape as it relates to the upcoming PACE mission is using planned multi- and hyper-angular polarimetry data from the Hyper-Angular Rainbow Polarimeter-2 (HARP2; contributed by University of Maryland, Baltimore County) and the Spectrometer for Planetary Exploration—one (SPEXone; contributed by the Netherlands Institute for Space Research) [94]. If information on the backward shape of the VSF could be gleaned with sufficient accuracy from angular polarimetry data (only a few angles would be necessary), this could be used directly in the ZTT model.

The ZTT model described here only addresses in-water RT, although the current approach to normalize water-leaving radiances measured by a satellite imager is expected to directly apply (see [15], Sections 3.1 and 6]). The BRDF is built-in to the Zaneveld model so an additional step to conceptually shift the geometry to a normalized geometry standard would not be required for IOP inversion. A normalized geometry (to nadir viewing or other geometry) could still be applied to intercompare reflectances in images. Translation to water-leaving radiance by accounting for all the reflection and refraction effects through the air-sea interface i.e., $\mathcal{R}(\theta_s,\theta_v)$ [6,15], can still be applied. Similarly, the effects of varying Earth-Sun distance and atmospheric attenuation can be removed following the current approach. Moreover, we would expect a practical implementation of this analytical model to have similar computational requirements as the current M02 LUT-based approach.

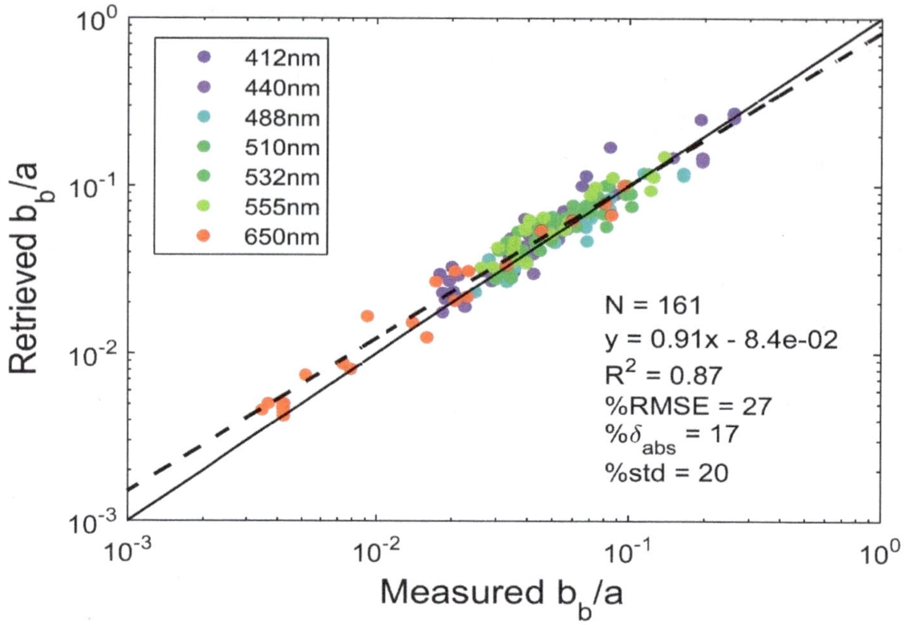

Figure 7. Comparing measured b_b/a to results of inverting ZTT using least-squares minimization individually at each wavelength. Water Raman effects were removed before applying the ZTT model.

The ZTT model shows $\widetilde{b}_{bp}$ exerts some influence on r_{rs} (Table 1) and the potential to invert r_{rs} for $\widetilde{b}_{bp}$ is also being investigated. The parameter $\widetilde{b}_{bp}$ has not historically been associated with r_{rs} and could provide new insights into particle composition in natural waters (e.g., [41]). Even a retrieval of $\widetilde{b}_{bp}$ with relatively large errors could be useful for many research applications.

The ZTT model is fully spectral and can be applied over the full anticipated ocean color hyperspectral wavelength range for the future PACE mission (350–800 nm). A primary goal in model development was parameterization entirely in terms of a and b_b. Inversion approaches, such as the General IOP (GIOP) inversion using a least-squares minimization to solve for IOP subcomponents [16] may consequently be readily applied. Spectral b_{bp} and a_{pg} can be solved using assumptions for subcomponent IOP spectra through least squares minimization of Equation (18). This is the step where empiricism prominently enters the problem as spectral shapes for IOP subcomponents must be assumed. This is being examined in ongoing work. Further investigation is needed to determine IOP subcomponent spectra with a high spectral resolution for error minimization over this full spectral range in inversion, which is currently being assessed by the PACE Science Team [13,95]. Hyperspectral (i.e., 5 nm) resolution for PACE is expected to improve fidelity in inverting for subcomponents (and potentially $\widetilde{b}_{bp}$) as there will be higher degrees of freedom than for multi-spectral imagers [96,97].

Code for the ZTT model in MATLAB is available at ioccg.org.

Author Contributions: Conceptualization, M.T.; Methodology, M.T. and A.T.; Software, A.T.; Validation, A.T. and M.T.; Formal Analysis, M.T. and A.T.; Investigation, A.T. and M.T.; Resources, M.T.; Data Curation, M.T.; Writing—Original Draft Preparation, M.T. and A.T.; Writing—Review and Editing, M.T. and A.T.; Visualization, A.T. and M.T.; Supervision, M.T.; Project Administration, M.T.; Funding Acquisition, M.T.

Acknowledgments: Thanks to Tim Moore for kindly providing the data subset from NOMAD, Bryan Franz for providing simulated geometry parameters for the expected orbit of the NASA PACE mission, David Antoine for providing the Morel et al. (2002) endmember phase functions, and to Zhongping Lee for access to the current PACE Science Team synthetic data sets. The authors thank Ron Zaneveld, Tim Moore, Curt Mobley, and Georges Fournier for helpful discussions. We also appreciate very thorough and constructive comments by Dariusz Stramski, Zhongping Lee, and two other anonymous reviewers.

Appendix A

Table A1. Notation.

Symbol	Description	Units
a	total absorption coefficient	m^{-1}
a_x	absorption coefficient, where subscript $x = w, p, d, ph, g, pg, pico$, and $micro$ specifies water, particulate, non-phytoplankton particulate, phytoplankton, dissolved, particulate plus dissolved, picoplankton, and microplankton	m^{-1}
a^*	specific absorption coefficient	$m^2\ mg^{-1}$
b	total scattering coefficient	m^{-1}
b_p	particulate scattering coefficient	m^{-1}
b_f	forward scattering coefficient	m^{-1}
b_{bx}	backscattering coefficient, where subscript $x = w$ and p specifies water and particulate	m^{-1}
$\widetilde{b}_b$	backscattering ratio	-
$\widetilde{b}_{bp}$	particulate backscattering ratio	-
β	total volume scattering function (VSF)	$m^{-1}\ sr^{-1}$
β_x	volume scattering function (VSF), where subscript $x = w$ and p specifies seawater and particulate	$m^{-1}\ sr^{-1}$
c	total attenuation coefficient	m^{-1}
[Chl]	chlorophyll concentration	$mg\ m^{-3}$
δ_x	error, where subscript $x = abs$ and rel specifies absolute and relative	-
E_x	planar irradiance, where subscript $x = d, ds$, and dd specifies downwelling, diffuse downwelling, and direct downwelling	$W\ m^{-2}\ nm^{-1}$
E_{od}	downwelling scalar irradiance	$W\ m^{-2}\ nm^{-1}$
f	model coefficient for relating irradiance reflectance to b_b/a	-
f_b, f_L	radiance shape factors	-
Φ	azimuth angle relative to solar plane	°, rad
G	model coefficient for above-surface remote sensing reflectance	sr^{-1}
ψ	scattering angle	°, rad
H	fraction of diffuse downwelling light (E_{ds}/E_d)	-
η_{bb}	fraction of total backscattering contributed by b_{bw}	-
K_x	diffuse attenuation coefficient, where subscript $x = L_u, u, d$, and ∞ specifies upwelling radiance, upwelling irradiance, downwelling irradiance, and asymptotic	m^{-1}
L_u	upwelling radiance	$W\ m^{-2}\ nm^{-1}\ sr^{-1}$
L_w	water-leaving radiance	$W\ m^{-2}\ nm^{-1}\ sr^{-1}$
λ	Wavelength	nm
M_d^+	Atmospheric component of $\overline{\mu}_d$	-
M_d^*	IOP component of $\overline{\mu}_d$	-
$\overline{\mu}_\infty$	average cosine of the asymptotic light field	-
$\overline{\mu}_d$	average cosine of the downwelling light field	-
μ_w	cosine of the in-water solar zenith	-
n_p	particulate refractive index, relative to water	-
P	phase function (β/b)	sr^{-1}
P_x	particulate phase function (β_p/b_p), where subscript $x = p, ps$, and pl specifies particulate, particulate small-dominant, and particulate large-dominant	sr^{-1}
$P_{bb,x}$	backward particulate phase function (β_p/b_{bp}), where subscript $x = ST$ and FF specifies functions from References [26,44]	sr^{-1}
Q	ratio of upwelling irradiance to nadir radiance	sr
r_{rs}	remote sensing reflectance, the ratio of upwelling subsurface radiance to downwelling irradiance	sr^{-1}
R_{rs}	remote sensing reflectance, the ratio of water-leaving radiance to downwelling irradiance	sr^{-1}
R_d	scaling factor for a_d	-
R_g	scaling factor for a_g	-
S_f	mixing factor for a_{ph}	-
θ_x	zenith angle, where $x = s$ and v specifies solar and viewing	°, rad
θ_s'	above water solar zenith angle	°, rad
v	exponent of empirical spectral b_p function	-
V	atmospheric horizontal visibility	km
ω	albedo (b/c)	-
$\Psi_{K_{Lu}}$	ratio of diffuse upwelling attenuation coefficient to asymptotic attenuation coefficient	-
z	depth	m

Table A2. Equation Coefficients. Full 15 decimal scaled fixed point precision in MATLAB is provided to ensure accuracy in calculations.

Equation	Symbol	Value
Equation (3)	f_1	$-5.98948784303628 \times 10^{-8}$
	f_2	$5.95904039870752 \times 10^{-6}$
	f_3	$-6.975283717755 \times 10^{-4}$
	f_4	$2.07111856771792 \times 10^{-3}$
	f_5	$2.69046922858858 \times 10^{-2}$
Equation (4)	f_{A1}	$-3.79435531537314 \times 10^{-7}$
	f_{A2}	$2.42117623125973 \times 10^{-4}$
	f_{A3}	$-5.76056692150838 \times 10^{-2}$
	f_{A4}	6.04944577004764
	f_{A5}	−236.166389774491
Equation (16)	e_1	$-3.37021020153209 \times 10^{-12}$
	e_2	$2.25040435584125 \times 10^{-10}$
	e_3	$-2.25897880448836 \times 10^{-9}$
	e_4	$4.98402568695743 \times 10^{-10}$
	e_5	$-3.67440351688922 \times 10^{-8}$
	e_6	$4.02677827509591 \times 10^{-7}$
	e_7	$-2.52448256032736 \times 10^{-8}$
	e_8	$2.09631870150827 \times 10^{-6}$
	e_9	$-2.43068373614361 \times 10^{-5}$
	e_{10}	$5.98295717192273 \times 10^{-7}$
	e_{11}	$-5.36922068813161 \times 10^{-5}$
	e_{12}	$6.84105803724285 \times 10^{-4}$
	e_{13}	$-5.34168078899319 \times 10^{-6}$
	e_{14}	$4.95201118318049 \times 10^{-4}$
	e_{15}	$-6.09578731164684 \times 10^{-3}$
	e_{16}	$5.32097604773773 \times 10^{-4}$
	e_{17}	$-2.91276619216202 \times 10^{-2}$
	e_{18}	0.589340234481004
Equation (17)	$m^*_{d,1}$	0.00611094400155735
	$m^*_{d,2}$	−0.00104841847722295
	$m^*_{d,3}$	0.0498255758922950
	$m^*_{d,4}$	−0.0117672820980625
	$m^*_{d,5}$	0.128019358635212
	$m^*_{d,6}$	−0.0429896134897322
	$m^*_{d,7}$	0.103528931695373
	$m^*_{d,8}$	0.950921179229178

Table A3. $f_{L,ave}(\lambda)$ function, see Equations (30) and (31).

Wavelength	Value	Wavelength	Value	Wavelength	Value
350	0.990	505	1.018	655	0.992
355	0.990	510	1.013	660	0.993
360	0.992	515	1.009	665	0.998
365	0.992	520	1.005	670	1.000
370	0.992	525	1.002	675	1.001
375	0.995	530	0.999	680	1.000
380	0.997	535	0.996	685	0.995
385	0.997	540	0.995	690	0.994
390	0.998	545	0.992	695	0.993
395	1.000	550	0.989	700	0.994
400	1.000	555	0.987	705	0.994
405	1.000	560	0.985	710	0.996
410	1.002	565	0.982	715	0.997

Table A3. *Cont.*

Wavelength	Value	Wavelength	Value	Wavelength	Value
415	1.003	570	0.981	720	0.999
420	1.006	575	0.982	725	1.000
425	1.008	580	0.983	730	1.000
430	1.010	585	0.984	735	1.000
435	1.013	590	0.986	740	0.999
440	1.016	595	0.987	745	0.999
445	1.020	600	0.988	750	0.999
450	1.023	605	0.988	755	0.999
455	1.024	610	0.989	760	0.999
460	1.025	615	0.989	765	0.999
465	1.025	620	0.989	770	0.999
470	1.026	625	0.990	775	1.000
475	1.026	630	0.990	780	1.000
480	1.026	635	0.990	785	1.001
485	1.026	640	0.990	790	1.002
490	1.026	645	0.990	795	1.002
495	1.024	650	0.990	800	1.002
500	1.022				

References

1. Gordon, H.R. Inverse methods in hydrologic optics. *Oceanologia* **2002**, *44*, 9–58.
2. Zaneveld, J.R.V.; Twardowski, M.S.; Lewis, M.; Barnard, A. Radiative transfer and remote sensing. In *Remote Sensing of Coastal Aquatic Waters*; Miller, R., Del-Castillo, C., McKee, B., Eds.; Springer: Dordrecht, The Netherlands, 2005; pp. 1–20.
3. Twardowski, M.; Lewis, M.; Barnard, A.; Zaneveld, J.R.V. In-water instrumentation and platforms for ocean color remote sensing applications. In *Remote Sensing of Coastal Aquatic Waters*; Miller, R., Del-Castillo, C., McKee, B., Eds.; Springer: Dordrecht, The Netherlands, 2005; pp. 69–100.
4. Mobley, C.D. *Light and Water: Radiative Transfer in Natural Waters*; Academic Press: San Diego, CA, USA, 1994; 592p.
5. Morel, A.; Gentili, B. Diffuse reflectance of oceanic waters. II. Bidirectional aspects. *Appl. Opt.* **1993**, *32*, 6864–6879. [CrossRef] [PubMed]
6. Morel, A.; Gentili, B. Diffuse reflectance of oceanic waters. III. Implication of bidirectionality for the remote-sensing problem. *Appl. Opt.* **1996**, *35*, 4850. [CrossRef] [PubMed]
7. Morel, A.; Antoine, D.; Gentili, B. Bidirectional reflectance of oceanic waters: Accounting for Raman emission and varying particle phase function. *Appl. Opt.* **2002**, *41*, 6289–6306. [CrossRef] [PubMed]
8. Fan, Y.; Li, W.; Gatebe, C.K.; Stamnes, K. Neural network method to correct bidirectional effects in water leaving radiance. *Appl. Opt.* **2016**, *55*, 10–21. [CrossRef] [PubMed]
9. Gordon, H.; Brown, O.B.; Evans, R.H.; Brown, J.W.; Smith, R.C.; Baker, K.S.; Clark, D.K. A semianalytic radiance model of ocean color. *J. Geophys. Res.* **1988**, *93*, 10909–10924. [CrossRef]
10. Park, Y.-J.; Ruddick, K. Model of remote sensing reflectance including bidirectional effects for case 1 and case 2 waters. *Appl. Opt.* **2005**, *44*, 1236–1249. [CrossRef]
11. Lee, Z.-P.; Du, K.-P.; Voss, K.J.; Zibordi, G.; Lubac, B.; Arnone, R.; Weidemann, A. An inherent optical property centered approach to correct the angular effects in water-leaving radiance. *Appl. Opt.* **2011**, *50*, 3155–3167. [CrossRef]
12. Hlaing, S.; Gilerson, A.; Harmel, T.; Tonizzo, A.; Weidemann, A.; Arnone, R.; Ahmed, S. Assessment of a bidirectional reflectance distribution correction of above-water and satellite water-leaving radiance in coastal waters. *Appl. Opt.* **2012**, *51*, 220–237. [CrossRef]
13. Werdell, J.; McKinna, L.I.W.; Boss, E.; Ackleson, S.G.; Craig, S.E.; Gregg, W.W.; Lee, Z.-P.; Maritorena, S.; Roesler, C.S.; Rousseaux, C.S.; et al. An overview of approaches and challenges for retrieving marine inherent optical properties from ocean color remote sensing. *Prog. Oceanogr.* **2018**, *160*, 186–212. [CrossRef]

14. Morel, A. In-water and remote measurements of ocean color. *Bound.-Lay. Meteorl.* **1980**, *18*, 177–201. [CrossRef]
15. Mobley, C.D.; Werdell, J.; Franz, B.; Ahmad, Z.; Bailey, S. *Atmospheric Correction for Satellite Ocean Color Radiometry*; NASA Report NASA/TM–2016-217551; NASA: Washington, DC, USA, 2016; 73p.
16. Werdell, P.J.; Franz, B.A.; Bailey, S.W.; Feldman, G.C.; Boss, E.; Brando, V.E.; Dowell, M.; Hirata, T.; Lavender, S.J.; Lee, Z.-P.; et al. Generalized ocean color inversion model for retrieving marine inherent optical properties. *Appl. Opt.* **2013**, *52*, 2019–2037. [CrossRef] [PubMed]
17. Voss, K.; Morel, A. Bidirectional reflectance function for oceanic waters with varying chlorophyll concentrations: Measurements versus predictions. *Limnol. Oceanogr.* **2005**, *50*, 698–705. [CrossRef]
18. Voss, K.J.; Morel, A.; Antoine, D. Detailed validation of the bidirectional effect in various Case 1 waters for application to ocean color imagery. *Biogeosciences* **2007**, *4*, 781–789. [CrossRef]
19. Gleason, A.; Voss, K.; Gordon, H.R.; Twardowski, M.; Sullivan, J.; Trees, C.; Weidemann, A.; Berthon, J.-F.; Clark, D.; Lee, Z.-P. Detailed validation of ocean color bidirectional effects in various Case I and Case II waters. *Opt. Express* **2012**, *20*, 7630–7645. [CrossRef]
20. He, S.; Zhang, X.; Xiong, Y.; Gray, D. A bidirectional subsurface remote sensing reflectance model explicitly accounting for particle backscattering shapes. *J. Geophys. Res. Oceans* **2017**, *122*, 8614–8626. [CrossRef]
21. Talone, M.; Zibordi, G.; Lee, Z.-P. Correction for the non-nadir viewing geometry of AERONET-OC above water radiometry data: An estimate of uncertainties. *Opt. Express* **2018**, *26*, A541–A561. [CrossRef] [PubMed]
22. Hirata, T.; Hardman-Mountford, N.; Aiken, J.; Fishwick, J. Relationship between the distribution function of ocean nadir radiance and inherent optical properties for oceanic waters. *Appl. Opt.* **2009**, *48*, 3129–3138. [CrossRef] [PubMed]
23. Gordon, H.R. Sensitivity of radiative transfer to small-angle scattering in the ocean: Quantitative assessment. *Appl. Opt.* **1993**, *32*, 7505–7511. [CrossRef] [PubMed]
24. Mishchenko, M.I.; Travis, L.D.; Lacis, A.A. *Scattering, Absorption, and Emission of Light by Small Particles*; Cambridge University Press: Cambridge, UK, 2002; pp. 74–82.
25. Hair, J.; Hostetler, C.; Hu, Y.; Behrenfeld, M.; Butler, C.F.; Harper, D.B.; Hare, R.; Berkoff, R.; Cook, A.; Collins, J.; et al. Combined Atmospheric and Ocean Profiling from an Airborne High Spectral Resolution Lidar. *Eur. Phys. J. Conf.* **2016**, *119*, 22001. [CrossRef]
26. Sullivan, J.M.; Twardowski, M.S. Angular shape of the volume scattering function in the backward direction. *Appl. Opt.* **2009**, *48*, 6811–6819. [CrossRef] [PubMed]
27. Zaneveld, J.R.V. Remotely sensed reflectance and its dependence on vertical structure: A theoretical derivation. *Appl. Opt.* **1982**, *21*, 4146–4150. [PubMed]
28. Zaneveld, J.R.V. A theoretical derivation of the dependence of the remotely sensed reflectance on the inherent optical properties. *J. Geophys. Res.* **1995**, *100*, 13135–13142. [CrossRef]
29. Tonizzo, A.; Twardowski, M.; McLean, S.; Voss, K.; Lewis, M.; Trees, C. Closure and uncertainty assessment for ocean color reflectance using measured volume scattering functions and reflective tube absorption coefficients with novel correction for scattering. *Appl. Opt.* **2017**, *56*, 130–146. [CrossRef]
30. Wolanin, A.; Rozanov, V.; Dinter, T.; Bracher, A. Detecting CDOM Fluorescence Using High Spectrally Resolved Satellite Data: A Model Study. In *Towards an Interdisciplinary Approach in Earth System Science: Advances of a Helmholtz Graduate Research School*; Springer International Publishing: Cham, Switzerland, 2015; pp. 109–121.
31. Zhai, P.W.; Hu, Y.; Winker, D.M.; Franz, B.A.; Boss, E. Contribution of Raman scattering to polarized radiation field in ocean waters. *Opt. Express* **2015**, *23*, 23582–23596. [CrossRef]
32. Rozanov, V.; Dinter, T.; Rozanov, A.; Wolanin, A.; Bracher, A.; Burrows, J. Radiative transfer modeling through terrestrial atmosphere and ocean accounting for inelastic processes: Software package SCIATRAN. *J. Quant. Spectrosc. Radiat. Transf.* **2017**, *194*, 65–85. [CrossRef]
33. Zhai, P.W.; Hu, Y.; Winker, D.M.; Franz, B.A.; Werdell, J.; Boss, E. Vector radiative transfer model for coupled atmosphere and ocean systems including inelastic sources in ocean waters. *Opt. Express* **2017**, *25*, A223–A239. [CrossRef]
34. Weidemann, A.; Stavn, R.; Zaneveld, J.R.V.; Wilcox, M.R. Error in predicting hydrosol backscattering from remotely sensed reflectance. *J. Geophys. Res.* **1995**, *100*, 163–177. [CrossRef]
35. Berthon, J.-F.; Shybanov, E.; Lee, M.; Zibordi, G. Measurements and modeling of the volume scattering function in the coastal northern Adriatic Sea. *Appl. Opt.* **2007**, *46*, 5189–5203. [CrossRef]

36. Twardowski, M.; Zhang, X.; Vagle, S.; Sullivan, J.; Freeman, S.; Czerski, H.; You, Y.; Bi, L.; Kattawar, G. The optical volume scattering function in a surf zone inverted to derive sediment and bubble particle subpopulations. *J. Geophys. Res.* **2012**, *117*, C00H17. [CrossRef]
37. Lee, M.; Korchemkina, E.N. Chapter 4: Volume scattering function of seawater. In *Springer Series in Light Scattering*; Kokhanovsky, A., Ed.; Springer International Publishing: New York, NY, USA, 2017; pp. 151–195.
38. Voss, K.; Austin, R. Beam attenuation measurement error due to small angle scattering acceptance. *J. Atmos. Ocean. Technol.* **1993**, *10*, 113–121. [CrossRef]
39. Boss, E.; Slade, W.H.; Behrenfeld, M.; Dall'Olmo, G. Acceptance angle effects on the beam attenuation in the ocean. *Opt. Express* **2009**, *17*, 1535–1550. [CrossRef] [PubMed]
40. Twardowski, M.; Tonizzo, A. Scattering and absorption effects on asymptotic light fields in seawater. *Opt. Express* **2017**, *25*, 18122–18130. [CrossRef] [PubMed]
41. Twardowski, M.S.; Boss, E.; Macdonald, J.B.; Pegau, W.S.; Barnard, A.H.; Zaneveld, J.R.V. A model for estimating bulk refractive index from the optical backscattering ratio and the implications for understanding particle composition in Case I and Case II waters. *J. Geophys. Res.* **2001**, *106*, 14129–14142. [CrossRef]
42. Prieur, L.; Morel, A. Etude theorique du regime asymptotique: Relations entre characteristiques optiques et coefficient d'extinction relatif a la penetration de la lumibr, e du jour. *Cah. Oceanogr.* **1971**, *23*, 35–48.
43. Berwald, J.; Stramski, D.; Mobley, C.D.; Kiefer, D.A. Influences of absorption and scattering on vertical changes in the average cosine of the underwater light field. *Limnol. Oceanogr.* **1995**, *40*, 1347–1357. [CrossRef]
44. Fournier, G.; Forand, F. Analytic phase function for ocean water. *Ocean Opt. XII* **1994**, *2558*, 194–201.
45. Jonasz, M.; Fournier, G. *Light Scattering by Particles in Water*; Academic Press: Amsterdam, The Netherlands, 2007; 704p.
46. IOCCG. Remote Sensing of Inherent Optical Properties: Fundamentals, Tests of Algorithms, and Applications. In *Reports of the International Ocean-Colour Coordinating Group, No. 5*; Lee, Z.-P., Ed.; IOCCG: Dartmouth, NS, Canada, 2006; 123p.
47. Twardowski, M.S.; Sullivan, J.M.; Donaghay, P.L.; Zaneveld, J.R.V. Microscale quantification of the absorption by dissolved and particulate material in coastal waters with an ac-9. *J. Atmos. Ocean. Technol.* **1999**, *16*, 691–707. [CrossRef]
48. Sullivan, J.; Twardowski, M.; Zaneveld, J.R.V.; Moore, C. Measuring optical backscattering in water. In *Light Scattering Reviews 7: Radiative Transfer and Optical Properties of Atmosphere and Underlying Surface*; Kokhanovsky, A., Ed.; Springer Praxis Books: New York, NY, USA; pp. 189–224. [CrossRef]
49. Stockley, N.D.; Rottgers, R.; McKee, D.; Lefering, I.; Sullivan, J.M.; Twardowski, M.S. Assessing uncertainties in scattering correction algorithms for reflective tube absorption measurements made with a WET Labs ac-9. *Opt. Express* **2017**, *25*, A1139–A1153. [CrossRef]
50. Gordon, H.R.; Brown, O.B.; Jacobs, M.M. Computed relationships between the inherent and apparent optical properties of a flat homogeneous ocean. *Appl. Opt.* **1975**, *14*, 417–427. [CrossRef] [PubMed]
51. Twardowski, M.S.; Claustre, H.; Freeman, S.A.; Stramski, D.; Huot, Y. Optical backscattering properties of the "clearest" natural waters. *Biogeosciences* **2007**, *4*, 1041–1058. [CrossRef]
52. Zhang, X.D.; Hu, L.B.; He, M.-X. Scattering by pure seawater: Effect of salinity. *Opt. Express* **2009**, *17*, 5698–5710. [CrossRef] [PubMed]
53. Twardowski, M.; Jamet, C.; Loisel, H. Analytical Model to Derive Suspended Particulate Matter Concentration in Natural Waters by Inversion of Optical Attenuation and Backscattering. *Proc. SPIE Ocean Sens. Monit. X* **2018**. [CrossRef]
54. Boss, E.; Pegau, W.S.; Lee, M.; Twardowski, M.S.; Shybanov, E.; Korotaev, G.; Baratange, F. The particulate backscattering ratio at LEO 15 and its use to study particles composition and distribution. *J. Geophys. Res.* **2004**, *109*, C01014. [CrossRef]
55. Sullivan, J.M.; Twardowski, M.S.; Donaghay, P.L.; Freeman, S.A. Using scattering characteristics to discriminate particle types in US coastal waters. *Appl. Opt.* **2005**, *44*, 1667–1680. [CrossRef] [PubMed]
56. Stramski, D.; Reynolds, R.A.; Babin, M.; Kaczmarek, S.; Lewis, M.R.; Röttgers, R.; Sciandra, A.; Stramska, M.; Twardowski, M.S.; Franz, B.A.; et al. Relationships between the surface concentration of particulate organic carbon and optical properties in the eastern South Pacific and eastern Atlantic Oceans. *Biogeosciences* **2008**, *5*, 171–201. [CrossRef]
57. Morel, A.; Prieur, L. Analyse spectrale des coefficients d'attenuation diffuse, de retrodiffusion pour diverses regions marines. *Cent. Rech. Oceanogr.* **1975**, *17*, 1–157.

58. Gregg, W.; Carder, K. A simple spectral solar irradiance model for cloudless maritime atmospheres. *Limnol. Oceanogr.* **1990**, *35*, 1657–1675. [CrossRef]
59. Loisel, H.; Stramski, D. Estimation of the inherent optical properties of natural waters from the irradiance attenuation coefficient and reflectance in the presence of Raman scattering. *Appl. Opt.* **2000**, *39*, 3001–3011. [CrossRef]
60. Westberry, T.K.; Boss, W.; Lee, Z.-P. Influence of Raman scattering on ocean color inversion models. *Appl. Opt.* **2013**, *52*, 5552–5561. [CrossRef]
61. Pope, R.M.; Fry, E. Absorption spectrum (380–700 nm) of pure water. II. Integrating cavity measurements. *Appl. Opt.* **1997**, *36*, 8710–8723. [CrossRef]
62. Ciotti, A.M.; Lewis, M.R.; Cullen, J.J. Assessment of the relationships between dominant cell size in natural phytoplankton communities and the spectral shape of the absorption coefficient. *Limnol. Oceanogr.* **2002**, *47*, 404–417. [CrossRef]
63. Roesler, C.S.; Perry, M.J. In situ phytoplankton absorption, fluorescence emission, and particulate backscattering spectra determined from reflectance. *J. Geophys. Res.* **1995**, *100*, 13279–13294. [CrossRef]
64. Twardowski, M.S.; Boss, E.; Sullivan, J.M.; Donaghay, P.L. Modeling the spectral shape of absorbing chromophoric dissolved organic matter. *Mar. Chem.* **2004**, *89*, 69–88. [CrossRef]
65. Loisel, H.; Morel, A. Light scattering and chlorophyll concentration in case 1 waters: A reexamination. *Limnol. Oceanogr.* **1998**, *43*, 847–858. [CrossRef]
66. Morel, A. Are the empirical relationships describing the bio-optical properties of case 1 waters consistent and internally compatible? *J. Geophys. Res.* **2009**, *114*, C01016. [CrossRef]
67. Mobley, C.D.; Sundman, L.K.; Boss, E. Phase function effects on oceanic light fields. *Appl. Opt.* **2002**, *41*, 1035–1050. [CrossRef] [PubMed]
68. Voss, K.J.; McLean, S.; Lewis, M.; Johnson, C.; Flora, S.; Feinholz, M.; Yarbrough, M.; Trees, C.; Twardowski, M.; Clark, D. An example crossover experiment for testing new vicarious calibration techniques for satellite ocean color radiometry. *J. Atmos. Ocean. Technol.* **2010**, *27*, 1747–1759. [CrossRef]
69. Zaneveld, J.R.V.; Barnard, A.H.; Boss, E. A theoretical derivation of the depth average of remotely sensed optical parameters. *Opt. Express* **2005**, *13*, 9052–9061. [CrossRef]
70. Gordon, H.R.; Clark, D.K. Remote sensing optical properties of a stratified ocean: An improved interpretation. *Appl. Opt.* **1980**, *19*, 3428–3430. [CrossRef]
71. Gordon, H.R. Dependence of the diffuse reflectance of natural waters on the sun angle. *Limnol. Oceanogr.* **1989**, *34*, 1484. [CrossRef]
72. Kirk, J.T.O. Estimation of the absorption and scattering coefficients of natural waters by the use of underwater irradiance measurements. *Appl. Opt.* **1994**, *33*, 3276–3278. [CrossRef] [PubMed]
73. Lee, Z.-P.; Du, K.-P.; Arnone, R. A model for the diffuse attenuation coefficient of downwelling irradiance. *J. Geophys. Res.* **2005**, *110*, C02016. [CrossRef]
74. Chai, T.; Draxler, R.R. Root mean square error (RMSE) or mean absolute error (MAE)?—Arguments against avoiding RMSE in the literature. *Geosci. Model. Dev.* **2014**, *7*, 1247–1250. [CrossRef]
75. Brewin, R.J.W. The Ocean Colour Climate Change Initiative: III. A round-robin comparison on in-water bio-optical algorithms. *Remote Sens. Environ.* **2013**. [CrossRef]
76. Hoge, F.; Lyon, P.E.; Mobley, C.D.; Sundman, L.K. Radiative transfer equation inversion: Theory and shape factor models for retrieval of oceanic inherent optical properties. *J. Geophys. Res.* **2003**, *108*, 3386. [CrossRef]
77. Agrawal, Y.; Mikkleson, O. Empirical forward scattering phase functions from 0.08 to 16 deg. for randomly shaped terrigenous 1-21 μm sediment grains. *Opt. Express* **2009**, *17*, 8805–14. [CrossRef] [PubMed]
78. Nardelli, S.; Twardowski, M.S. Improving assessments of chlorophyll concentration from in situ optical measurements. *Opt. Express* **2016**, *24*, A1374–A1389. [CrossRef] [PubMed]
79. Petzold, T.J. *Volume Scattering Functions for Selected Ocean Waters, Report 72–78*; Scripps Institution of Oceanography: La Jolla, CA, USA, 1972.
80. Li, L.; Stramski, D.; Reynolds, R.A. Effects of inelastic radiative processes on the determination of water-leaving spectral radiance from extrapolation of underwater near-surface measurements. *Appl. Opt.* **2016**, *55*, 7050–7067. [CrossRef]
81. Voss, K.; Gordon, H.; Flora, S.; Johnson, B.C.; Yarbrough, M.; Feinholz, M.; Houlihan, T. A method to extrapolate the diffuse upwelling radiance attenuation coefficient to the surface as applied to the Marine Optical Buoy (MOBY). *J. Atmos. Ocean. Technol.* **2017**, *34*, 1423–1432. [CrossRef]

82. Gordon, H.; Ding, K. Self-shading of in-water optical instruments. *Limnol. Oceanogr.* **1992**, *37*, 491–500. [CrossRef]
83. Leathers, R.A.; Downes, T.V.; Mobley, C.D. Self-shading correction for oceanographic upwelling radiometers. *Opt. Express* **2004**, *12*, 4709–4718. [CrossRef] [PubMed]
84. Zhang, X.; Fournier, G.R.; Gray, D.J. Interpretation of scattering by oceanic particles around 120 degrees and its implication in ocean color studies. *Opt. Express* **2017**, *25*, A191–A199. [CrossRef] [PubMed]
85. Lotsberg, J.K.; Marken, E.; Stamnes, J.; Erga, S.R.; Aursland, K.; Olseng, C. Laboratory measurements of light scattering from marine particles. *Limnol. Oceanogr. Methods* **2007**, *5*, 34–40. [CrossRef]
86. Tan, H.; Doerffer, R.; Oishi, T.; Tanaka, A. A new approach to measure the volume scattering function. *Opt. Express* **2013**, *21*, 18697–18711. [CrossRef] [PubMed]
87. Harmel, T.; Hieronymi, M.; Slade, W.; Röttgers, R.; Roullier, F.; Chami, M. Laboratory experiments for intercomparison of three volume scattering meters to measure angular scattering properties of hydrosols. *Opt. Express* **2016**, *24*, A234–A256. [CrossRef] [PubMed]
88. Tan, H.; Oishi, T.; Tanaka, A.; Doerffer, R.; Tan, Y. Chlorophyll-a specific volume scattering function of phytoplankton. *Opt. Express* **2017**, *25*, A564–A573. [CrossRef] [PubMed]
89. Volten, H.; de Haan, J.F.; Hovenier, J.W.; Schreurs, R.; Vassen, W.; Dekker, A.G.; Hoogenboom, H.J.; Charlton, F.; Wouts, R. Laboratory measurements of angular distributions of light scattered by phytoplankton and silt. *Limnol. Oceanogr.* **1998**, *43*, 1180–1197. [CrossRef]
90. Loisel, H.; Morel, A. Non-isotropy of the upward radiance field in typical coastal (Case 2) waters. *Int. J. Remote Sens.* **2001**, *22*, 275–295. [CrossRef]
91. Xiong, Y.; Zhang, X.; He, S.; Gray, D.J. Re-examining the effect of particle phase functions on the remote-sensing reflectance. *Appl. Opt.* **2017**, *56*, 6881–6888. [CrossRef]
92. Hostetler, C.; Behrenfeld, M.J.; Hu, Y.; Hair, J.W.; Schulien, J.A. Spaceborne lidar in the study of marine systems. *Annu. Rev. Mar. Sci.* **2018**, *10*, 121–147. [CrossRef]
93. Dayou, J.; Chang, J.H.W.; Sentian, J. *Ground-Based Aerosol Optical Depth Measurement Using Sunphotometers*; Springer: Singapore, 2014; pp. 9–30.
94. The NASA PACE Mission. Available online: https://pace.oceansciences.org/mission.htm (accessed on 16 June 2018).
95. Boss, E.; Remer, L.A. A novel approach to a satellite mission's science team. *Eos* **2018**, *99*. [CrossRef]
96. Bracher, A.; Bouman, H.A.; Brewin, R.J.W.; Bricaud, A.; Brotas, V.; Ciotti, A.M.; Clementson, L.; Devred, E.; Di Cicco, A.; Dutkiewicz, S.; et al. Obtaining phytoplankton diversity from ocean color: A scientific roadmap for future development. *Front. Mar. Sci.* **2017**, *4*. [CrossRef]
97. Vandermeulen, R.A.; Mannino, A.; Neeley, A.; Werdell, J.; Arnone, R. Determining the optical spectral sampling frequency and uncertainty thresholds for hyperspectral remote sensing of ocean color. *Opt. Express* **2017**, *25*, A785–A797. [CrossRef] [PubMed]

The Fundamental Contribution of Phytoplankton Spectral Scattering to Ocean Colour: Implications for Satellite Detection of Phytoplankton Community Structure

Lisl Robertson Lain [1,*] and Stewart Bernard [1,2]

[1] Department of Oceanography, University of Cape Town, Cape Town 7700, South Africa; sbernard@csir.co.za
[2] NRE, Centre for Scientific and Industrial Research (CSIR), Cape Town 7700, South Africa
* Correspondence: lislrobertson@gmail.com

Abstract: There is increasing interdisciplinary interest in phytoplankton community dynamics as the growing environmental problems of water quality (particularly eutrophication) and climate change demand attention. This has led to a pressing need for improved biophysical and causal understanding of Phytoplankton Functional Type (PFT) optical signals, in order for satellite radiometry to be used to detect ecologically relevant phytoplankton assemblage changes. Biophysically and biogeochemically consistent phytoplankton Inherent Optical Property (IOP) models play an important role in achieving this understanding, as the optical effects of phytoplankton assemblage changes can be examined systematically in relation to the bulk optical water-leaving signal. The Equivalent Algal Populations (EAP) model is used here to investigate the source and magnitude of size- and pigment-driven PFT signals in the water-leaving reflectance, as well as the potential to detect these using satellite radiometry. This model places emphasis on the determination of biophysically consistent phytoplankton IOPs, with both absorption and scattering determined by mathematically cogent relationships to the particle complex refractive indices. All IOPs are integrated over an entire size distribution. A distinctive attribute is the model's comprehensive handling of the spectral and angular character of phytoplankton scattering. Selected case studies and sensitivity analyses reveal that phytoplankton spectral scattering is most useful and the least ambiguous driver of the PFT signal. Key findings are that there is the most sensitivity in phytoplankton backscatter ($b_{b\phi}$) in the 1–6 μm size range; the backscattering-driven signal in the 520 to 570 nm region is the critical PFT identifier at marginal biomass, and that, while PFT information does appear at blue wavelengths, absorption-driven signals are compromised by ambiguity due to biomass and non-algal absorption. Low signal in the red, due primarily to absorption by water, inhibits PFT detection here. The study highlights the need to quantitatively understand the constraints imposed by phytoplankton biomass and the IOP budget on the assemblage-related signal. A proportional phytoplankton contribution of approximately 40% to the total b_b appears to a reasonable minimum threshold in terms of yielding a detectable optical change in R_{rs}. We hope these findings will provide considerable insight into the next generation of PFT algorithms.

Keywords: phytoplankton; PFT; ocean colour; satellite radiometry; radiative transfer; optical modelling

---

1. Introduction

Phytoplankton across the world's oceans represent about half of all primary production on our planet [1,2]. Their growth and function are fundamental to sustaining life: they constitute the foundation of the aquatic food web, and serve critical roles in the recycling of essential elements such as carbon and nitrogen, as well as in remineralisation [3–5]. Being so responsive to nutrient availability and water temperature, these tiny organisms are key indicators of ecosystem change, and understanding their community dynamics is key to answering some of the most challenging earth science questions of our time about the impacts of climate change on local, regional and global scale aquatic systems and the carbon cycle. The widespread distribution and integral role of phytoplankton in global marine ecosystems means that these fields of study depend heavily on modelling together with satellite data for any large scale analysis. In situ data collection is indispensable for local scale investigations and for ground truthing of satellite and model data, but simultaneous large scale direct measurements are logistically impossible. Optical measurements in natural waters are challenging: they are expensive and logistically difficult, technically complex due to large dynamic ranges of the signal, and overall require delicate, rigorously calibrated instrumentation with precise knowledge of sources of error. Remote sensing and moored in situ instrumentation are the only feasible ways to acquire continuous data series, but these largely involve bulk measurements of the total optical signal. Isolating the respective optical components for laboratory assessment is a significant further undertaking. In situ and laboratory measurements are consequently extremely valuable, and appropriate bio-optical models provide essential tools for the analysis and understanding of these bulk measurements, whether above- or sub-surface.

It has long been appreciated that phytoplankton have a direct effect on the observable colour of the ocean, and broad scale biomass estimates based on Chl *a* concentrations derived from satellite radiometry are widely relied upon despite persistent uncertainty in the accuracy of information derived from satellite imagery [6,7]. Recently, there has been considerable interest in more detailed information on phytoplankton assemblage characteristics [8–11], but it has not been widely ascertained to what degree Phytoplankton Functional Type (PFT) information can be gleaned from satellite data, and at what level of confidence. Furthermore, descriptions of PFTs differ with context—and the potential for identifying relationships between the ecological roles of phytoplankton and their optical properties must also be considered. Understanding the causal effect of biophysical phytoplankton characteristics on the optical water-leaving signal is at the heart of addressing these questions, and this is undoubtedly an outstanding topic in ocean optics.

Any useable radiometric PFT-related signal results directly from the interaction of phytoplankton with their light environment, but the physical basis of this interaction is not well understood in terms of observed variability across the wide diversity of aquatic environments and phytoplankton assemblages [12,13]. Generally, in oceanic waters, it is the strong absorption by phytoplankton which dominates the phytoplankton contribution to the ocean colour signature, and has therefore been identified as a promising signal in terms of PFT identification. However, distinguishing the effects of variable phytoplankton absorption due to biomass changes from the effects due to functional type changes (and further from changes induced by photoacclimation and photoprotective pigments) is not straightforward. This ambiguity in the phytoplankton community signal is at the core of the PFT problem. It is then overlaid with further complexity, given that a potential PFT signal from the phytoplankton component of a water body's optical constituents must be considered in the context of the other components in the water, recognising the contributions from the non-algal sources of optical variability: absorption due to CDOM (Coloured Dissolved Organic Matter) and detrital particles, i.e., $a_{gd}(\lambda)$, and non-algal backscatter i.e., $b_{bnap}(\lambda)$ [13]. The blue spectral region of maximum phytoplankton absorption is also the region most affected by CDOM and detrital absorption. (It should also be noted that, in the context of satellite radiometry, blue spectral bands display the largest absolute measurement uncertainties [7,14]. While the blue water-leaving signal may be large in oceanic regions, resulting in a small relative uncertainty, this is when the signal is overwhelmingly dominated by the backscattering of water, decreasing confidence in the lesser contributions of $a_{gd}(\lambda)$ and a_{ϕ}. Generally, a_{gd} product retrievals from the satellite tend to be less robust than those of other IOPs [15,16].)

A comprehensive guide to PFT approaches is given by Mouw et al. [17], dividing them into four categories: abundance-based e.g., Hirata et al. [18] and Brewin et al. [19,20]; radiance based e.g., the PHYSAT method: Alvain et al. [21,22]; absorption based e.g., Devred et al. [23], Ciotti and Bricaud [24], and PhytoDOAS: Bracher et al. [25]; and scattering based e.g., Kostadinov et al. [10,26] (all references in [17]). Existing scattering-based approaches [10,11] assume a Jungian (exponential) particle size distribution and rely on Mie modelling, which does not adequately represent phytoplankton angular scattering [27], and there are consequently high uncertainties in PSD retrieval where the particle size distribution slope is low, i.e., highly productive and coastal areas dominated by relatively large cells [10]. Low biomass (Chl $a < 1$ mg·m^3) oceanic conditions with an absorption-dominated phytoplankton component of the water-leaving signal can exhibit good relationships with differential pigment absorption e.g., the diagnostic pigment approach used on satellite R_{rs} in Uitz et al. [28]. It follows that, in the context of additional non-algal absorption, differentiated spectra show better similarity than non-differentiated [29], and also that high spectral resolution measurements show better potential for retrieving phytoplankton assemblage information than multi-spectral R_{rs} [29] when retrieving diagnostic features of the first derivative of R_{rs}. However, other methods using the fourth derivative of pigment absorption and R_{rs} to identify fine-scale phytoplankton absorption features have found that objective discrimination of pigment groups from hyperspectral R_{rs} may not be feasible at low biomass < 1 mg·m^3 due to the high similarities in the derivative spectra [30]. This suggests that the primary phytoplankton signal in R_{rs} is due to biomass (Chl a) rather than the accessory pigments, and that with the exception of uniquely diagnostic pigment absorption outside the spectral regions of that of Chl a, phytoplankton information cannot be retrieved without assumptions about PFT relationships with biomass. The PhytoDOAS method also employs a fourth derivative analysis [25,31] but is performed on hyperspectral top-of-atmosphere satellite measurements, avoiding the uncertainties associated with poor atmospheric correction, but the sensitivity of this high spectral resolution approach to biomass and both algal and non-algal scattering contributions to the IOP budget has yet to be determined.

Generally, phytoplankton absorption- and abundance-based methods rely on empirical relationships between biomass, functional type and CDOM. Where these quantities co-vary predictably or are exactly known, empirical PFT algorithms may be successful. However, Brewin et al. [6] acknowledges that both the abundance-based approaches as well as approaches relying on differential pigment absorption break down in environments that do not conform to the generalised relationships between community structure and biomass upon which these approaches are based, usually in elevated biomass comprised of small cells. The relative contributions of phytoplankton absorption and scatter to light emerging from seawater change with biomass, size and other functional type traits, and as the $a_{gd}(\lambda)$ and $b_{bnap}(\lambda)$ components vary (see Stramski et al. [32] and references therein). The total water-leaving signal is a delicate balance of the frequently opposing optical effects of biomass and phytoplankton assemblage variability such as size, pigments and ultrastructure, together with the optical effects of the non-algal in-water constituents. An interactive webpage demonstrating the first order effect of variability in these parameters on Rrs is available in the Supplementary Material. It was observed by Brown et al. [13] that backscatter anomaly maps (i.e., backscatter independent of variability due to biomass) correlate approximately with PFT distribution maps calculated from optical anomalies which were initially attributed to differences in phytoplankton accessory pigments [21]. This leads to the suggestion that radiance-based methods, e.g., the Alvain (PHYSAT) criteria used to distinguish PFTs, are in fact primarily due to backscattering characteristics [9,13], indicating that phytoplankton groups either directly determine, or perhaps are simply associated with, backscattering variability around the mean.

Brown et al. [13] conclude that these relationships can only be fully explored if a method is applied where the phytoplankton groups are causally linked to the optical conditions. The Equivalent Algal Populations (EAP) model provides exactly such a method, and is used here to investigate the impact of size- and pigment-based PFT variability on the optical signal, and to confirm the assertion that biomass drives the largest part of observed variability in the water-leaving signal, and that the radiometric signal in the blue is ambiguous due to the effects of $a_{gd}(\lambda)$, and the additional effects of $b_{bnap}(\lambda)$ [13,33].

The EAP model is a fully physics-based two-layered spherical model, which calculates, from first principles, biophysically linked phytoplankton absorption and scattering characteristics from particle refractive indices reflecting the primary light-harvesting pigments of various phytoplankton groups. IOPs are calculated at high spectral resolution between 400 and 900 nm and are integrated over an entire equivalent size distribution [34,35], simulating the dominant optical characteristics of natural phytoplankton assemblages. The EAP is used here only as a forward model: the intention of this study is to isolate the biophysical driver(s) of PFT optical signals and determine the associated implications for detecting PFT changes from satellite radiometry.

In this study, the term "Phytoplankton Functional Type" is used in a broad sense of the dominant characteristics of a phytoplankton assemblage, with respect to both cell size and accessory pigments, from an optical perspective.

Study Objectives and Outline

The aim of this work is to investigate the magnitude and spectral location of optical water-leaving signals resulting from phytoplankton assemblage changes; to determine how these signals respond to changes in biomass and functional type; to evaluate their optical ambiguity in the context of the optical effects of other in-water constituents; and to assess their robustness against measurement uncertainties in satellite radiometry. This work does not present a PFT detection method, but instead aims to identify the reasonable limits of PFT detection from satellite, inferred from appropriate illuminative case studies and sensitivity analyses demonstrating the source and magnitude of PFT signals in terms of both cell size (assemblage D_{eff}) and accessory pigments.

To give context to the discussion on the case studies, an analysis is first made of the contribution of the phytoplankton-driven signal to the bulk R_{rs}, and how this relates to the proportional contribution of phytoplankton to the IOP budget. A Southern Ocean based case study then demonstrates the optical impact on the R_{rs} of transitioning assemblages in terms of both biomass and D_{eff} changes. This discussion is developed further with a Benguela-like example more representative of productive upwelling systems investigating the relative magnitude of pigment-driven PFT changes. A sensitivity analysis then shows the spectral position and magnitude of the accessible phytoplankton optical signal in R_{rs} as biomass and D_{eff} vary. The source of these signals is traced back to phytoplankton backscatter and its relationship with biomass, and ambiguity associated with non algal variability is evaluated.

It is clear that the case studies reflect simplified representative examples of much wider pigment- and size-related variability in nature, but the described dependence of absorption-driven pigment signals versus scattering-driven cell size signals on biomass holds across assemblage types. Optical PFT effects are most easily identified in relatively high biomass environments (Chl $a > 1$ mg·m^3) [36–38], and where the IOP budget is dominated by phytoplankton [37,39], and so the case studies deal with these water types. However, as the sensitivity analysis shows, together with the contextual discussion around ambiguity and uncertainty in satellite R_{rs}, the conclusions of this study have implications for the identification of PFT changes from satellite R_{rs} across all water types.

2. Methods: Modelling Approach

2.1. The Requirement for a Biophysically Consistent PFT Optical Model

The EAP model was developed to understand the causality-driven impact of different phytoplankton assemblages on the water-leaving optical signal. Optical variability in phytoplankton is known to be driven by particle size (effective diameter D_{eff}) [32,40,41], pigment quantity and type, cellular material, shape and internal structure, fine-scale morphology, and aggregation [42–45].

The model focuses primarily on particle size as given by the D_{eff} parameter, which is of fundamental importance both optically and ecologically [10,46].

Due to immense species diversity and variability in distribution, the Phytoplankton Functional Type (PFT) approach (e.g., Sathyendranath et al. [8], Alvain et al. [21], Ciotti and Bricaud [24], Bouman [47]) groups phytoplankton species according to their biogeochemical function and attempts to relate this to their biophysical characteristics, with size as a major consideration [10,46,48]. This approach is important for oceanic waters, characterised by widespread but low biomass, which contribute the largest proportion of global oceanic primary production [1]. Cell size governs many biological traits [49]; smaller phytoplankton are ubiquitous and play an important role in nutrient recycling, while larger phytoplankton often display the highest growth rates [49]. The dynamics of phytoplankton ecology have profound and intricate influence not only on oceanic biogeochemistry (e.g., acidification, and its effects on both CO_2 uptake and on marine life) but also at higher trophic levels e.g., on fish ecology, as certain phytoplankton environments promote the development of different fish populations [48]. A size-based PFT approach is particularly meaningful in the context of carbon sequestration [46], as particle size determines sinking rates for a large part.

However, phytoplankton ecology is complex, and modelling PFTs with adequate parameterisation in a biogeochemical context is consequently extremely challenging [12]. Following the EAP's conceptual intent to understand the impact of D_{eff} as the primary optical determinant once the effect of biomass has been accounted for, other sources of bio-optical variability are intentionally constrained. PFTs can therefore, to the first order, be approached from a size-based perspective, and the EAP model consequently lends itself extremely well to PFT sensitivity studies in terms of its ability to isolate small differences in reflectance resulting only from variability in assemblage size distribution [37]. The model does additionally provide scope for varying other biophysical attributes within a population (such as the pigment-determined spectral refractive indices, the shape of the size distribution itself, the ratio of core to shell sphere volumes, and the cellular Chlorophyll *a* density of the cells in the distribution), as required. It should be noted, however, that the model is not intended as a full representation of phytoplankton optical complexities, and there is certainly ecologically significant natural variability in phytoplankton IOPs e.g., dependent on their growth state [50], in response to growth irradiance, nutrient availability and water temperature [51–53] and diel cycles [54,55]. These effects can be a large (e.g., 80% increase in the phytoplankton scattering cross section between sunrise and sunset [54]), and while they are not explicitly addressed here, they serve to add further uncertainty to PFT retrievals from the optical water-leaving signal.

Empirically based phytoplankton abundance-type approaches, following observed relationships between phytoplankton assemblage taxonomic information (e.g., pigments) and biomass, show good results in low biomass conditions (i.e., where phytoplankton absorption dominates the phytoplankton IOP contribution), and where the covariability of the phytoplankton optical contribution with that of other in-water constituents generally holds [56], but do not address the sources of second order variability or optical causality [13], or the likelihood that these empirical relationships will not withstand the ecological shifts resulting from changing climatic conditions [6]. A biophysical approach to PFTs not only allows improved analysis of sensitivity and causality but is likely to have greater validity in a future ocean (see also [57]).

The optical impact of a phytoplankton assemblage interacting with its aquatic environment is by no means straightforward, and a rigorous IOP model such as the EAP can systematically vary phytoplankton biogeophysical attributes in the context of likely additional non-algal absorption and scatter, and can examine the resulting effects on the light field when used in combination with a Radiative Transfer (RT) model. The value of this reductionist approach has been demonstrated [58,59] (and furthermore by Stramski et al. [32]) for separating and understanding the effects of various phytoplankton groups and accompanying in-water constituents on the oceanic light field and emergent R_{rs}. There is a bulk effect attributable simply to biomass, for which Chlorophyll *a* (Chl *a*) is used as a proxy, and which for the most part dominates the phytoplankton-related signal in Case 1 waters [60]

(It is acknowledged that Chl *a* concentration and biomass are not equivalent, as biomass includes non-pigmented biological matter in quantities which may not be proportional to pigmented matter. However, for the purposes of this study, biomass and Chl *a* concentration are used interchangeably, as this work is approached from a purely optical perspective and ignores non-pigmented biological matter.). PFT characteristics generally result in optical effects secondary to those of Chl *a*: accessory pigments dominate assemblage absorption characteristics [61], and particle size is usually the primary determinant of phytoplankton scattering characteristics [62] (excepting the influence of ultrastructure in certain species, e.g., highly scattering liths or vacuoles). Natural waters are also subject to non-algal absorption, the dissolved part of which is frequently referred to as Coloured Dissolved Organic Matter (CDOM) or gelbstoff, but which may also have a particulate component in addition to non-algal scatter that can include scatter by detrital matter, sediment, bacteria, and/or bubbles. These quantities absorb and scatter light with spectral signatures distinct from those of phytoplankton, and their subsequent optical interactions and resulting effect on the total water-leaving signal are highly complex. Understanding the interaction between cells' biophysical characteristics and the light field in the presence of these additional optically active constituents is central to determining which parts of the optical signal are useable for PFT diagnostics, and, likewise, where signal ambiguity is prohibitive.

2.2. *Equivalent Algal Populations Model: Principal Attributes*

The EAP model has been used for a variety of applications [35,37,63,64]. It can be assumed that a model demonstrated as successful in phytoplankton-dominated waters [65] addresses the phytoplankton component accurately. Models designed for low biomass, with simplistic and absorption-decoupled phytoplankton scattering models, tend to underperform in higher biomass conditions when phytoplankton IOPs dominate [65]. It follows that the phytoplankton component of the combined optical properties is not generally well represented in these models. Following a reductionist approach, good information on the phytoplankton component is a prerequisite for any quantitative comment on the optical contribution of respective PFTs, or identifying changes in the bulk optical properties of seawater as dominant PFTs change. Only when representing the detailed nature of phytoplankton optics, with absorption and scattering biophysically consistent—as they are in nature—is a causal understanding of their interactive effect on the optical signal possible [32,66].

The EAP model exhibits a two-layered sphere particle and equivalent size-based community structure [27], which enables the calculation of phytoplankton IOPs from first principles, presenting a valuable opportunity for furthering the understanding of causal relationships between phytoplankton physiology and their optical characteristics based on quantified community structure. It is emphasised that this is not an empirical model and its use here is not to provide optical closure, but rather to identify and understand the biophysical drivers of phytoplankton optics and their contribution to an observable signal in the context of different water types.

At the core of the model are the phytoplankton particle refractive indices, with the imaginary part of the refractive index approximately representing that portion of light that is absorbed by the cell, and the real part of the refractive index representing that portion of light which is scattered. The imaginary and real parts of the refractive index spectra are numerically linked through the Kramers–Kronig relations [67], whereby the real part of the refractive index $n(\lambda)$ is calculated as the imaginary part of a Hilbert transform of the imaginary refractive index, originally derived from cellular absorption measurements. It should be noted that the imaginary refractive index characterises the absorption of the intracellular material and has no dependency on cell size. This has implications for the applicability of the model to a wide range of cell sizes, and is discussed further in Appendix A.1.

With a real refractive index of 1.12 for the 'chloroplast' sphere, and as 1.02 for the 'cytoplasm' sphere, this yields an overall particle spectral real refractive index of between 1.03 and 1.04 for phytoplankton cells (see also Stramski et al. [32], Aas [68]). Full details of the refractive index calculations can be found in Bernard et al. [69].

In this model, the imaginary part of the refractive index is also numerically linked to the specified intracellular Chl *a* concentration [27,52,54,55,70]. For eukaryotic particles, a core sphere represents the cytoplasm (which contains approximately 80% water, and is almost colourless), while an outer sphere represents the more refractive chloroplast, where the pigmented material (generally Chl *a* in the largest part) is also strongly absorbing.

A critical feature of the model is that Chl *a*-specific absorption ($a^*\phi$) is constrained at 675 nm to reflect the theoretical maximum absorption by unpackaged phytoplankton of 0.027 mg/m^2 as per Johnsen et al. [71]. This is incorporated into the calculation of the imaginary refractive index of the chloroplast layer n'_{chlor} (outer sphere), based on the assumption that the cytoplasm layer (inner sphere) has no signficant absorption at 675 nm:

$$n'_{chlor}(675) = \frac{675}{n_{media}} \frac{\pi c_i a^*_{sol}(675)}{4Vv}, \tag{1}$$

where n_{media} = 1.334 and Vv is the relative chloroplast volume, c_i is the intracellular Chl a, and $a^*_{sol}(675)$ is the Chl *a*-specific absorption at 675 nm of that pigment in solution, i.e., unpackaged [27].

The effect of constraining the unpackaged absorption in this way is to establish a quantitative relationship between the intracellular Chl *a* and the cell volume; a relationship that is biophysically consistent as the cell size varies [27]. This results in an effectively decreasing Chl *a*-specific absorption with increasing size, observable in the resulting optics as the "package effect" [40,72].

When coupled with a radiative transfer model—here, Hydrolight-Ecolight (Numerical Optics, Ltd., Devon, UK) is used—the interactions of phytoplankton IOPs (in combination with those of other in-water constituents) with the surrounding light field can be examined systematically. A full physics-based model such as this has the additional advantage of providing not only biophysically interrelated particle absorption, scattering and backscattering, but IOPs for assemblages that are integrated over the entire assemblage size distribution, and which are fully angularly resolved. This presents the unique opportunity of closely examining simuluated phytoplankton phase functions, which are notoriously difficult to measure, and whose behaviour in terms of variability in particle size and wavelength is poorly understood. With no decoupling of absorption and backscattering, and IOPs integrated over the entire size distribution, the model provides an unprecedented opportunity to examine the drivers of variability in phytoplankton optical signals systematically.

2.3. Case Study Methods

The complex optical interactions of D_{eff} and biomass, and the question of whether they can be separated into a useable PFT signal from a background environment of further non-algal optical complexity, is best addressed by investigating specific ecological events of interest to the remote sensing community.

The case studies outlined in the *Introduction* consider phytoplankton from two groups—a Chl *a*-carotenoid phytoplankton group, representing phytoplankton dominated by Chl *a*, and fucoxanthin and/or peridinin; and a Chl *a*- and phycoerythrin-containing group [27]. The former group is chosen as representative of a wide range of phytoplankton across size classes, and the latter for the unique absorption characteristics of phycoerythrin-associated phytoplankton species. This selection is intentionally kept limited in order to assess the relative magnitude of particle size- vs. accessory pigment-related optical signals in likely ecological scenarios, in the context of changing biomass.

Refractive indices for the chloroplast spheres are derived from measurements of cells from blooms in the Benguela—dinoflagellate and diatoms, dominated by Chl *a* and the carotenoid pigments fucoxanthin and peridinin—as well as for a phycoerythrin-associated cryptophyte group (based on a *Mesodinium rubrum*/*Myrionecta rubra*—dominated assemblage [27]). A justification for using these derived refractive indices across wide size ranges of modelled phytoplankton assemblages is included in Appendix A.1. Phytoplankton assemblages are modelled using a Standard Normal size distribution with a nominal effective variance of 0.6, recognising that while Jungian (exponential) distributions are frequently used for bulk particulate in oceanic conditions, the former is more appropriate for representing the increased species monospecifity associated with elevated biomass ([34]), and the case studies refer mainly to biomass > 1 mg·m^3. Assemblages are modelled with appropriate effective diameters to represent the effective diameters of measured size distributions in the case studies. The resulting IOPs are presented, with explanatory notes, in Appendix A.2. (A cyanobacterial group with substantially altered geometry to represent vacuolated cells has also been developed [63]).

Phytoplankton IOPs are combined, in various proportions as indicated, with appropriate non-algal optical constituents as detailed for each experiment. The phytoplankton-related optical signal is assessed against variability in the non-algal contributions (detailed in Appendices A.3 and A.4), so their absolute magnitude is not critical. Both $a_{gd}(\lambda)$ and $b_{bnap}(\lambda)$ do, however, assume a smooth spectral shape with predictable spectral structure. The potential for additional spectral features in these contributions is not addressed here, and would add further complexity (and hence ambiguity) to resolving phytoplankton scattering characteristics. Water types are considered homogenous with depth (i.e., IOPs constant with depth), generic atmospheric and geographic conditions, and the full radiative transfer solution is calculated by Hydrolight at a spectral resolution of 5 nm. Given the technical challenges with using EAP phase functions for modelling high resolution spectra [35], a Fournier Forand phase function chosen for the backscatter fraction of the combined particulate IOPs is used at each wavelength throughout these experiments. A basic fluorescence efficiency model is included for completeness (detailed in Appendix B.2), but modelling this spectral region accurately is challenging and outside of the scope of this work, so the features of this spectral region are not discussed in terms of PFT sensitivity.

2.3.1. Southern Ocean Case Study: Separating the Effects of Biomass From the Effects of D_{eff} Change

As shown in Lain et al. [65] and Lain et al. [35], where the water-leaving signal is phytoplankton-dominated (e.g., in the Benguela system), it is quite reasonable to expect that some PFT information may be derived from the bulk radiometric signal. However, the challenge for the ocean colour community is determining the PFT signal in low biomass oceanic conditions, for example in the Southern Ocean.

Phytoplankton dynamics in the Southern Ocean are particularly important for their role in uptake of anthropogenic CO_2 (around half of all oceanic uptake), and hence carbon sequestration [3,4]. Variability in phytoplankton ecology is directly linked to mineral and nutrient cycles: assemblages of large diatoms drive primary productivity and carbon export, while assemblages of small phytoplankton play a significant role in nutrient recycling although the net productivity is very low [73].

The third Southern Ocean Seasonal Cycle Experiment (SOSCEx III) undertaken on the SANAE 55 cruise (austral winter 2015) provides the phytoplankton size distribution and Chl *a* data for this experiment [74]. Assemblage D_{eff} were calculated from Coulter Counter measurements, and Chl *a* determined by fluorometric analysis [5]. The additional $a_{gd}(\lambda)$ and $b_{bnap}(\lambda)$ components were estimated guided by observations in [75,76] respectively, noting that these are simply used to approximate the bulk R_{rs} and do not influence any of the other results, as they are discussed in terms of likely variability rather than absolute magnitudes. EAP phytoplankton IOPs with generalised Chl *a*-carotenoid eukaryotic refractive indices were calculated according to the measured D_{eff} and Chl *a* concentrations, and were combined with these estimates and run through Hydrolight to produce the modelled R_{rs}.

Given that the refractive indices used to model the EAP IOPs for this example are from the generalised Chl *a*-carotenoid group suitable for diatom and dinoflagellate species, the likelihood of encountering *Phaeocystis* sp. in the Southern Ocean must be addressed. Given the oceanographic context, as the D_{eff} of 16 μm is reached, it can reasonably be assumed that the assemblage comprises both diatoms and *Phaeocystis*. The main accessory pigment in *Phaeocystis* is 19-hexanolyoxyfucoxanthin, a derivative of fucoxanthin, a dominant light harvesting pigment in diatoms, and so it may be reasonable to model the intracellular absorption properties of individual cells with the generalised eukaryote refractive indices, but this species forms large floating colonies which result in quite different optical effects, and this cannot currently be addressed with the model. Thus, while the likely presence of *Phaeocystis* is acknowledged, it is not explicitly catered for in the modelling. This does not affect the observations on identifying changes in D_{eff} in the discussion below.

2.3.2. Benguela-Like Case Study: Addressing Pigment Variability

The assemblages modelled in the first case study address optical changes due only to biomass (i.e., concentration of Chl *a* pigment) and size (assemblage D_{eff}), as the same set of generalised Chl *a*-carotenoid refractive indices is used for all phytoplankton particles represented. However, this approach addresses only a small subset of important changes in phytoplankton assemblage type, and in the presence of variability in dominant accessory pigments, the EAP model can be set to incorporate different refractive indices as appropriate for phytoplankton displaying accessory pigments other than carotenoids.

To illustrate the effects of pigment variability, this case study simulates a transition from a high biomass *Myrionecta rubra*-dominated assemblage, to a high biomass peridinin (carotenoid)-containing dinoflagellate-dominated assemblage. *M. rubra* is a fascinating but troublesome ciliate species, and enjoys an endosymbiontic relationship with cryptophytes containing the diagnostic pigment phycoerythrin [77], and so "borrows" their characteristic red colour. *M. rubra* blooms can reach extraordinary biomass, resulting in darkly pigmented 'red tide' waters that have negative impacts both ecologically (depletion of nutrients, and the potential for anoxia as the bloom dies), as well as on the recreational use of coastal waters [77]. Again, assemblages are modelled using a Standard normal size distribution and the same values of ϵ are used as for the carotenoids. This ensures that, in this example, all assemblage changes observed are due only to pigment-related differences.

It should be made clear that the modelled transition is not intended to represent a likely ecological succession (except possibly a Lagrangian one, if a dinoflagellate bloom is advected into a previously *M. rubra*-dominated region), but rather to test what biomass and pigment differences are required for the detection of distinct optical conditions, particularly in the context of remote sensing.

2.3.3. Spectral Shape and Sensitivity Analyses

To test the sensitivity of the EAP model, a general allometric approximation of changing D_{eff} from 2 to 8 µm was chosen for this analysis, which ranges from 0.1 to 10 mg/m^3. It is recognised that this scenario does not represent all possible ecological changes, but is a reasonable approximation for a mid-range biomass diatom and dinoflagellate-dominated environment where there may be a detectable PFT signal.

3. Results and Discussion

3.1. Quantifying the Contribution of Phytoplankton to the R_{rs} Signal

Remembering that the Remote Sensing Reflectance (R_{rs}) is grossly proportional to b_b/a [78], it should be noted that, for a given D_{eff} and phytoplankton group, $b_{b\phi}/a_\phi$ will be constant for any given concentration of Chl *a* because the package effect observed with increasing Chl *a* concentration is implicit in the model (see the comparison of Bricaud and EAP a^*_ϕ for varying Chl *a* concentrations in Lain et al. [65]). However, the contribution of the phytoplankton IOPs to the total, i.e., $b_{b\phi}/a_\phi$ as a percentage of total b_b/a, will vary. The EAP model, used together with Hydrolight, allows the inspection of any component optical quantity of interest, and here, the contribution of the phytoplankton IOPs to the total IOP budget is investigated. EAP phytoplankton IOPs are used with Hydrolight to calculate a full radiative transfer solution resulting in a new theoretical quantity, $R_{rs}\phi$. This quantity is introduced as an approximate quantification of the phytoplankton contribution to the bulk R_{rs}, in order to more intuitively understand the relative optical contributions in terms of remote sensing. While acknowledging that R_{rs} is not an additive quantity, $R_{rs}\phi$ is the calculation of reflectance with only water and phytoplankton IOPs. It does not account for any optical interaction between the phytoplankton and other in-water constituents likely to be present in natural waters, such as CDOM or detrital and mineral particles. These interactions are assumed to be secondary to the contribution of phytoplankton, but have not been quantified. It is anticipated that trans-spectral effects are most likely to suffer from this type of subtractive approach, but a full photon tracing model (such as a Monte Carlo model) would be needed to ascertain this. By modelling the phytoplankton contribution to the water-leaving signal, we can assess the availability of signal for PFT retrieval.

Being able to identify the spectral regions sensitive to changes in phytoplankton assemblage (focusing on those due to change in assemblage D_{eff}) is valuable, especially to identify spectral regions which might be sufficiently independent from the ambiguity introduced by other in-water constituents. This allows the quantification of the phytoplankton signal with confidence, even where these other constituents are not well characterised. The spectral regions of maximum proportional phytoplankton signal are the ones which hold potential for detecting PFT changes from an in-water perspective, as these represent the regions of the largest phytoplankton-related signal variability as the assemblage changes.

The resulting contribution of phytoplankton to the total R_{rs} is shown in Figure 1, for typical Case 1 waters as a simple illustrative example. In this example, a_{gd} covaries with Chl a while b_{bnap} is constant. This represents the combined a_{cdom}, known to approximately covary with Chl a, and detrital absorption a_{det}, which is assumed to be small with respect to a_{cdom} [79] and approximately constant (a_{det} is neglected altogether in some IOP models e.g., Alvain et al. [9]). As CDOM does not scatter, b_{bnap} represents the scatter of only the detrital component of the non-algal constituents. Thus, it is assumed that biomass increases together with a_{cdom}, against a relatively unchanging background detrital population whose IOPs are dominated by backscatter. As the phytoplankton contribution to the IOP budget increases (i.e., generally, as biomass increases), the impact of the other constituents is proportionally less in the R_{rs}. This varies with D_{eff}, and is observable in Figure 1 to a greater degree in the R_{rs} with a smaller (nominal) D_{eff} of 2 μm as compared with a larger (nominal) D_{eff} of 12 μm to show the higher level of phytoplankton backscatter of small cells contributing to brighter R_{rs}, which is less sensitive to the addition of scattering from other sources.

For each D_{eff}, it is evident that the phytoplankton percentage contribution to the bulk R_{rs} increases with biomass. However, it can be seen that there is a dependency on D_{eff} which, when considered in the context of transitioning assemblages, is not straightforward. This observation indicates a requirement to go beyond the Case 1/Case 2 water type distinction for PFT signal analysis and applications: the differential in phytoplankton scatter as D_{eff} varies in both water types must be considered as well as variable b_{bnap} in Case 2. When it comes to retrieving information about the phytoplankton IOPs, their proportional contribution to the bulk water-leaving signal (or the total IOPs) should be considered. Figure 2 demonstrates the proportional IOP contributions for the 0.1, 1 and 10 mg/m^3 cases.

3.2. Case Study 1: Separating the Effects of Biomass from the Effects of D_{eff} Change

Figure 3 presents two distinct events which illustrate the interdependency of the size and biomass signals. Modelled R_{rs} are shown for selected adjacent stations (20 to 21 is marked A; 12 to 13 is marked B) where a nominal threshold of change detectable by satellite is reached in the blue and green spectral regions, in other words, where a change in R_{rs} would be evident on a satellite image. Given the ambiguity in the causality of the phytoplankton signal, assessing the magnitude of changes to the water-leaving signal as the in-water constituents vary will give an indication of whether there may be enough radiometric signal at TOA to even detect the change. A threshold in situ measurement resolution of 1×10^{-4} sr^{-1} [80] is taken as an indication of sensitivity to detecting change in R_{rs} by direct measurement. Given an average estimated uncertainty in satellite R_{rs} of $\pm$ 0.6 $\times$ 10^{-3} sr^{-1} across the spectrum [81], here a conservative 1×10^{-3} sr^{-1} is used to indicate a potentially detectable change in water-leaving signal from satellite. These thresholds are not definitive and are used purely for the purpose of contextualising the discussion. Both examples in Figure 3 display large changes in R_{rs}, but these are causally distinct: (A) represents a large change in Chl a concentration *and* in D_{eff}, while (B) represents a large change in Chl a concentration but a negligible change in D_{eff}.

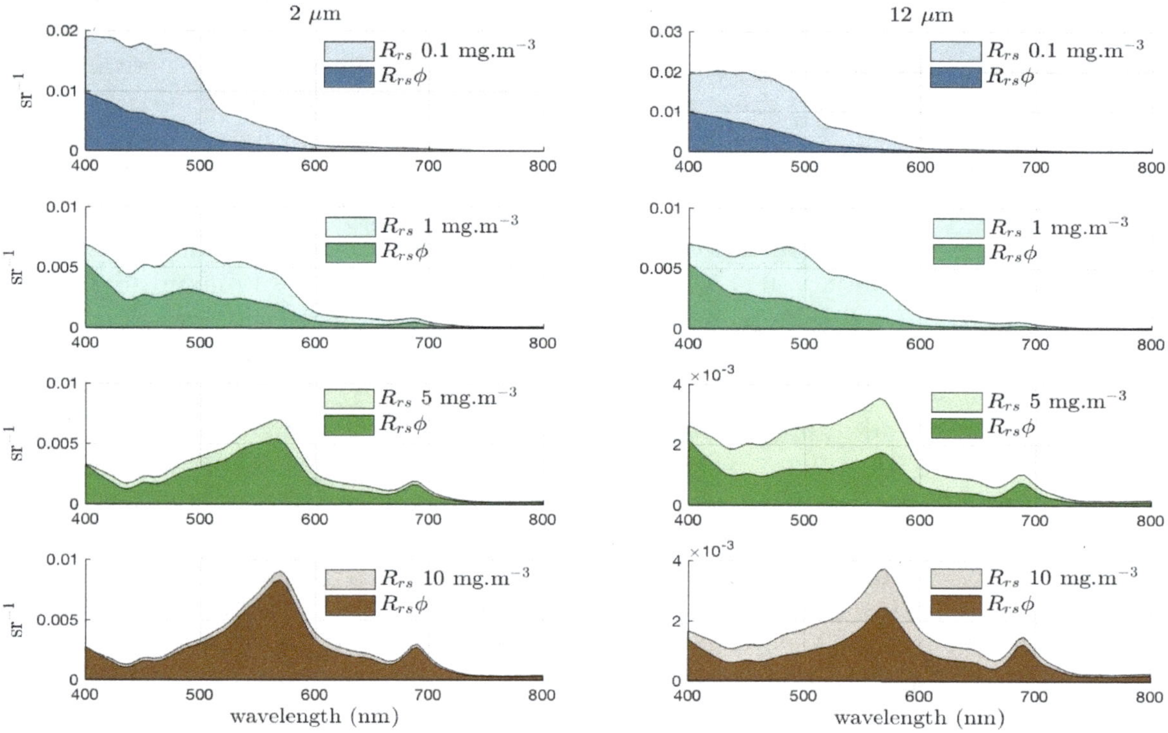

Figure 1. Relative contribution of phytoplankton to total R_{rs} (with $a_{gd}(400) = 0.07 \cdot [Chl_a]^{0.75}$, and $b_{bnap}(550) = 0.005$ m^{-1}) for increasing biomass with $D_{eff} = 2$ and 12 μm. These populations are idealised examples and not intended to represent any observed relationship between Chl a concentration and D_{eff}.

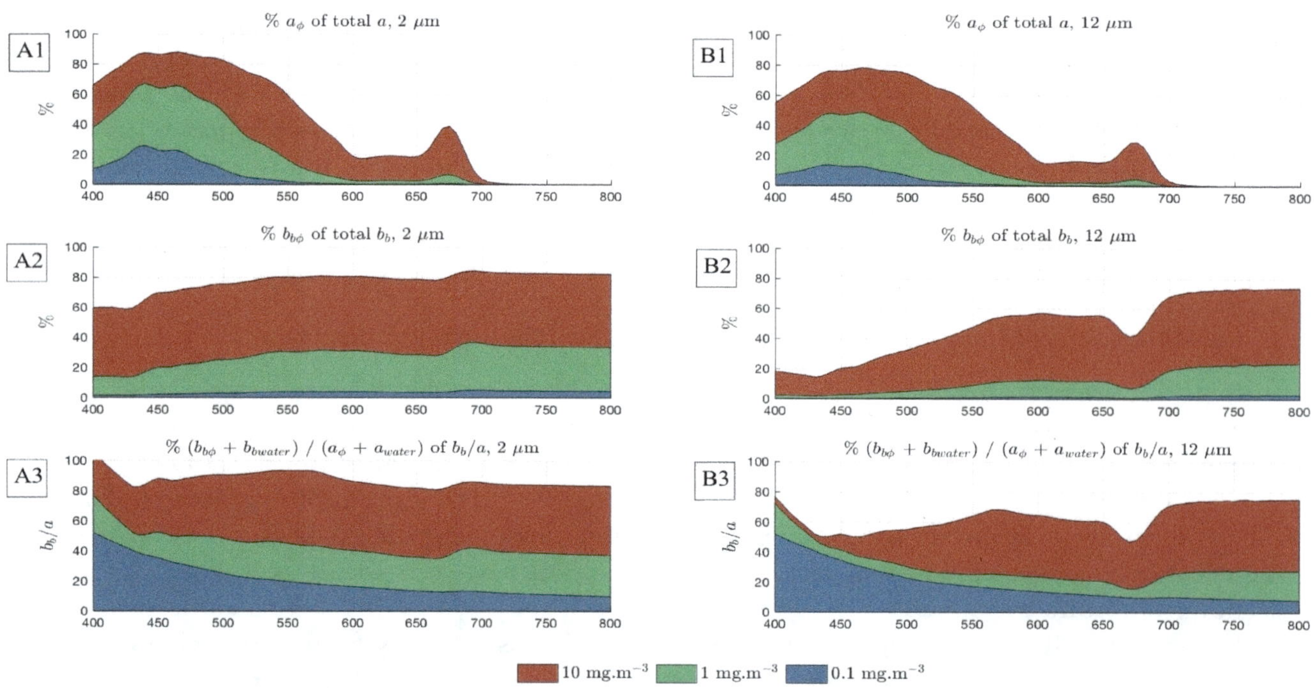

Figure 2. Example proportional phytoplankton to total Inherent Optical Property (IOP) contributions for Case 1 waters, for idealised eukaryote assemblages of 2 and 12 μm.

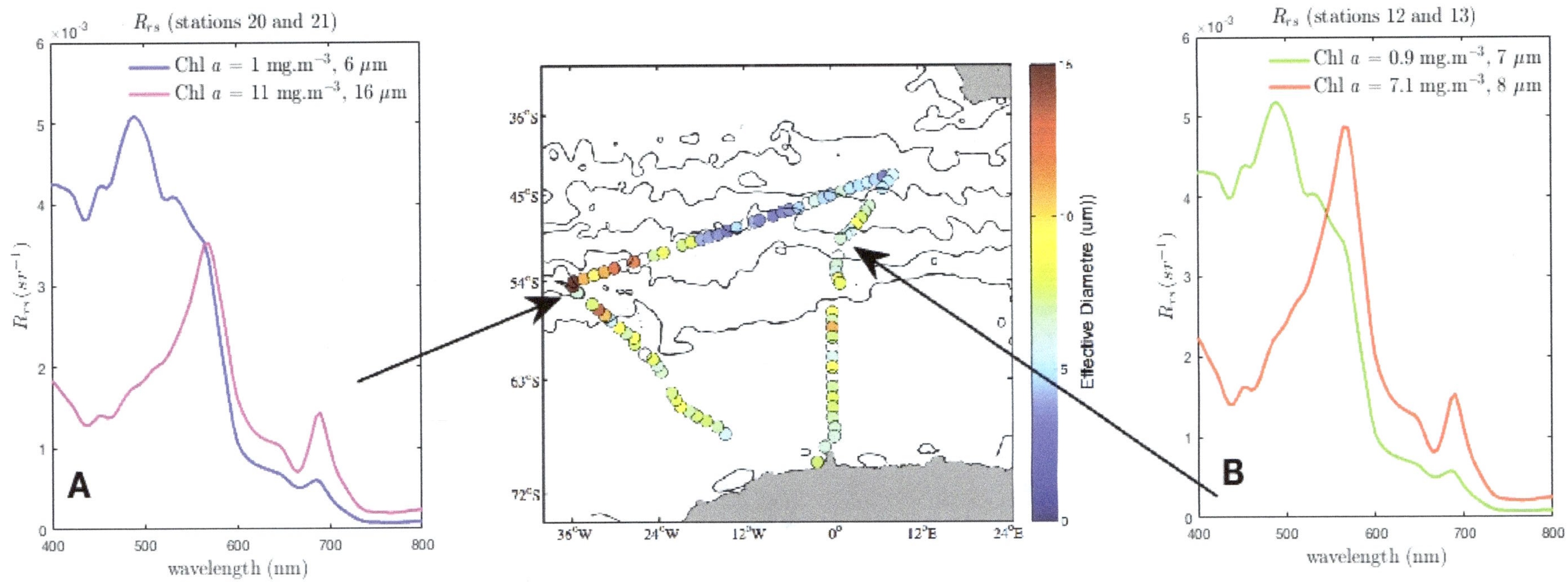

Figure 3. Modelled R_{rs} for stations 20, 21, 12 and 13 of SOSCEX III. The modelled bulk R_{rs} are calculated using Equivalent Algal Populations (EAP) generalised Chl a-carotenoid refractive indices and measured Chl a concentrations for the phytoplankton component, and include estimated $a_{gd}(\lambda)$ and $b_{bnap}(\lambda)$ contributions appropriate for this region [75,76]. Stations 20 to 21 (**A**) represent a large change in both Chl a concentration and in D_{eff}. Stations 12 to 13 (**B**) represent a large change in Chl a concentration only. The centre panel shows the measured D_{eff} for the cruise track (starting at the ice shelf on the bottom right and continuing in an anticlockwise direction.) The effective diameter image is courtesy of SANAE 55 Report [74].

Station 20 to 21 therefore represents a significant phytoplankton community shift, as large changes in both D_{eff} (from 6 to 16 μm) and Chl *a* concentration (from 1 to 11 /m^3) were recorded. To isolate this change in phytoplankton signal, the differences in $R_{rs}\phi$ for an assemblage with D_{eff} = 6 μm and an assemblage with D_{eff} = 16 μm are presented in Figure 4A for the measured range of Chl *a* concentration. Note that the large differences of about 2×10^{-3} sr^{-1} in R_{rs} in the blue, observable in Figure 3A, only appears at very low biomass in Figure 4A as this difference in R_{rs} is almost entirely due to the difference in biomass and not a change in D_{eff}. The spectral location of the most promising size-related signal for PFT retrieval is evidently dependent on biomass, and, at low biomass, it is positioned near 435 nm, while, at higher biomass, it is around 570 nm. As this is the phytoplankton-only signal, the question remains to what extent this signal is expressed in the bulk R_{rs}, when the optical impact of the non-algal constituents is also considered.

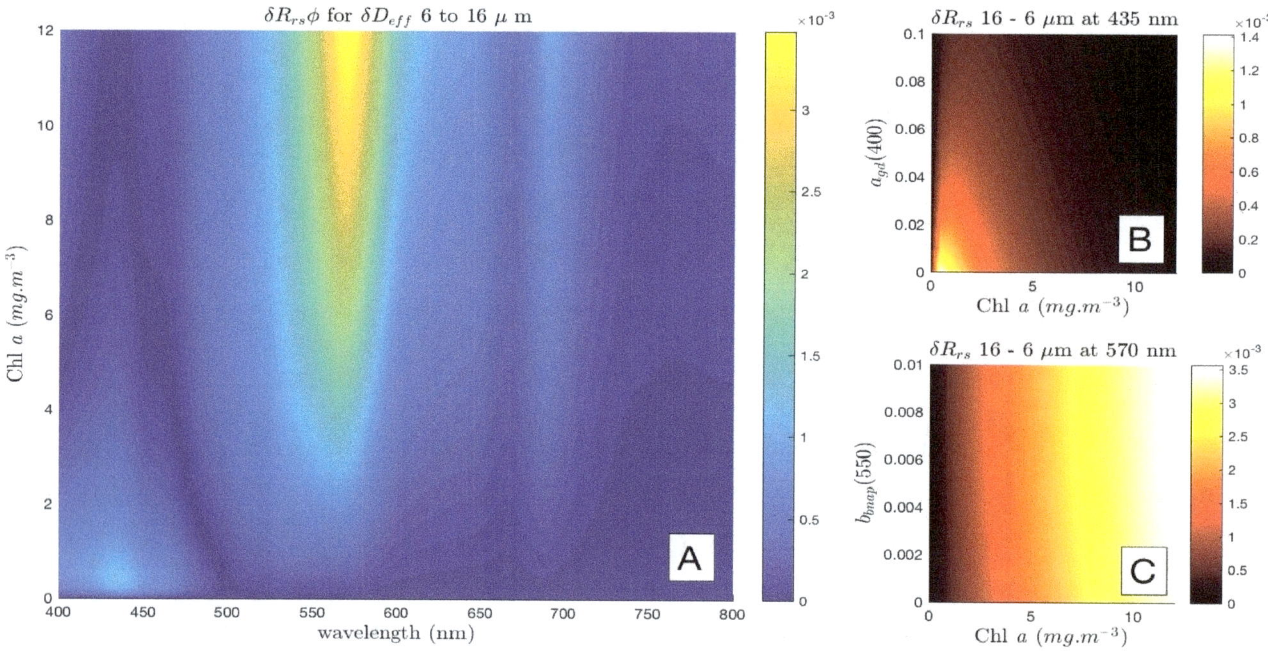

Figure 4. Southern Ocean stations 20 to 21: $\delta R_{rs}\phi$ is shown for δD_{eff} of 6 to 16 μm (**A**). The effect of $a_{gd}(\lambda)$ at 435 nm is shown in (**B**), and $b_{bnap}(\lambda)$ at 570 nm in (**C**). The units of the colour bars are sr^{-1}.

Working with the change in phytoplankton size signal identified at 435 nm (bottom left corner, Figure 4A), $a_{gd}(\lambda)$ is added at increasing concentrations to simulate a range of bulk R_{rs} at 435 nm in Figure 4B, and $b_{bnap}(\lambda)$ is likewise added incrementally at 570 nm in Figure 4C. In these plots, horizontal gradients indicate R_{rs} sensitivity primarily to the constituent on the y axis, while vertical gradients indicate that the change in R_{rs} is driven by the biomass, and is not sensitive to variability on the y axis.

Figure 4B shows that the difference in bulk R_{rs} for the given δD_{eff} is only detectable at the satellite threshold level (shown in yellow) at low biomass under very low $a_{gd}(\lambda)$ conditions. As biomass increases, increasing absorption by phytoplankton as well as by additional $a_{gd}(\lambda)$, reduces the magnitude of the water-leaving signal and renders any δD_{eff} information ambiguous. When additionally considering the brightening effect of $b_{bnap}(\lambda)$ in the blue (not quantified here), it can readily be perceived that the water-leaving signal is too complex at 435 nm to retrieve useful size information.

In Figure 4C, the relationship with $b_{bnap}(\lambda)$ at 570 nm is more straightforward. Change in R_{rs} due to δD_{eff} is detectable in the bulk R_{rs} at the satellite threshold (in red) from about 2.5 mg/m^3 upwards regardless of the $b_{bnap}(\lambda)$ contribution, at least for Case 1 type conditions. The magnitude of this signal is almost entirely biomass driven. (This is in line with the observation made by [13] that the MODIS wavebands at 531 and 551 nm are good indicators of backscatter anomalies because their magnitude is proportional to the addition or removal of particulate backscattering, and the longer wavelength band at 551 nm is less affected by variability in both $a_{gd}(\lambda)$ and phytoplankton absorption [10].)

It should be appreciated, though, that $R_{rs}\phi$ in these figures is representing the change in R_{rs} due to size *at* a particular biomass (i.e., biomass is constant while assemblage characteristics vary), effectively removing the effects of simultaneous biomass changes. Figure 5 simulates a transition from 6 to 16 D_{eff} with biomass 1 to 11 mg/m^3, where the intermediate values of both D_{eff} and Chl *a* are simply linearly interpolated. The vertical lines highlight 435 and 570 nm which were identified in Figure 4A as being the spectral regions of greatest size-driven signal. In Figure 5, while biomass and size effects combine to form large changes in $R_{rs}\phi$ in the blue, it is the smaller signal around 570 nm that contains the most size-driven change as it is not affected by biomass to the same degree. Figure 4B,C show that the signal at 435 nm is sensitive to the effects of variable a_{gd}, while the phytoplankton signal at 570 nm remains robust against variability in the non-algal optical contributions.

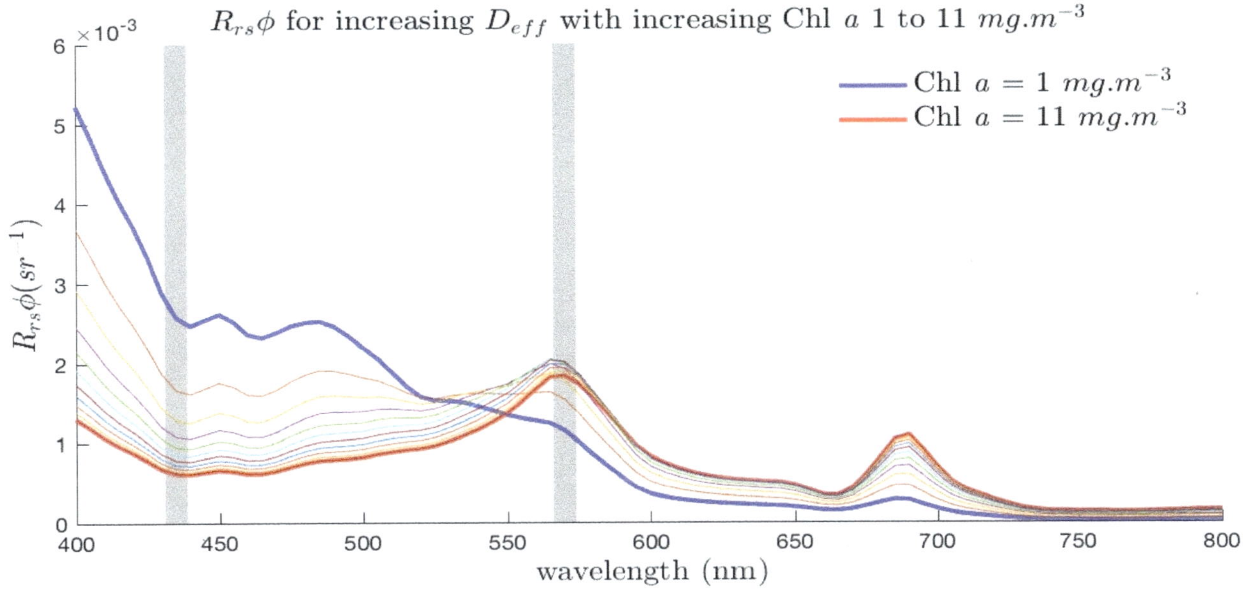

Figure 5. A simulated transition from 6 to 16 D_{eff} with biomass 1 to 11 mg/m^3. Intermediate values of D_{eff} and Chl *a* are simply linearly interpolated. The lines highlight 435 nm and 570 nm, regions of maximum size signal, which are (at 435 nm) and are not (at 570 nm) sensitive to the effects of additional optical constituents.

By contrast, stations 12 to 13 exhibit a large change in R_{rs}—seen first in Figure 4B; shown again in Figure 6A—with an increase in Chl *a* from 0.9 to 7.1 mg/m^3 but only a very small change in D_{eff} from 7 to 8 μm. This is likely, given the location in the lee of the South Sandwich Islands, to reflect a diatom bloom associated with island wake effects, due to fertilisation by terrestrial iron [82]. Tracing the signal due to this change in D_{eff} across all Chl *a* concentrations in this range in Figure 6B shows that there is a size related signal between 550 and 600 nm, but it is of an order of magnitude less than in the previous example, and so does not show potential for detection by satellite radiometry. This is illustrated further in the lower panel (C), showing the location of this signal but also that it is almost all attributable to biomass—as shown by the $R_{rs}\phi$ representing D_{eff} 7 at 7.1 mg/m^3 i.e., what the higher biomass R_{rs} would look like without the increase in effective diameter as the assemblage changes. It can be seen quite clearly from these spectra that a difference in the blue due only to this δD_{eff}, with any variability $a_{gd}(\lambda)$, would not be detectable by any means.

It should be noted that the spectral locations of maximum δD_{eff} features are a direct consequence of the spectral nature of the IOPs used in the modelling, and that both of these examples use the same Chl *a*-carotenoid refractive indices to generate the phytoplankton IOPs. In other words, as phytoplankton IOPs are adjusted to represent pigment differences, the spectral character of the assemblage change will vary. A slight migration in the exact location of the maximum available δD_{eff} signal is observable with different ranges of D_{eff}, although within the Chl a-carotenoid group it remains between 550 and 600 nm for any difference in D_{eff} between 1 and 40 μm. (This is discussed in more detail later with respect to Figure 9, and an additional figure is shown in Appendix.)

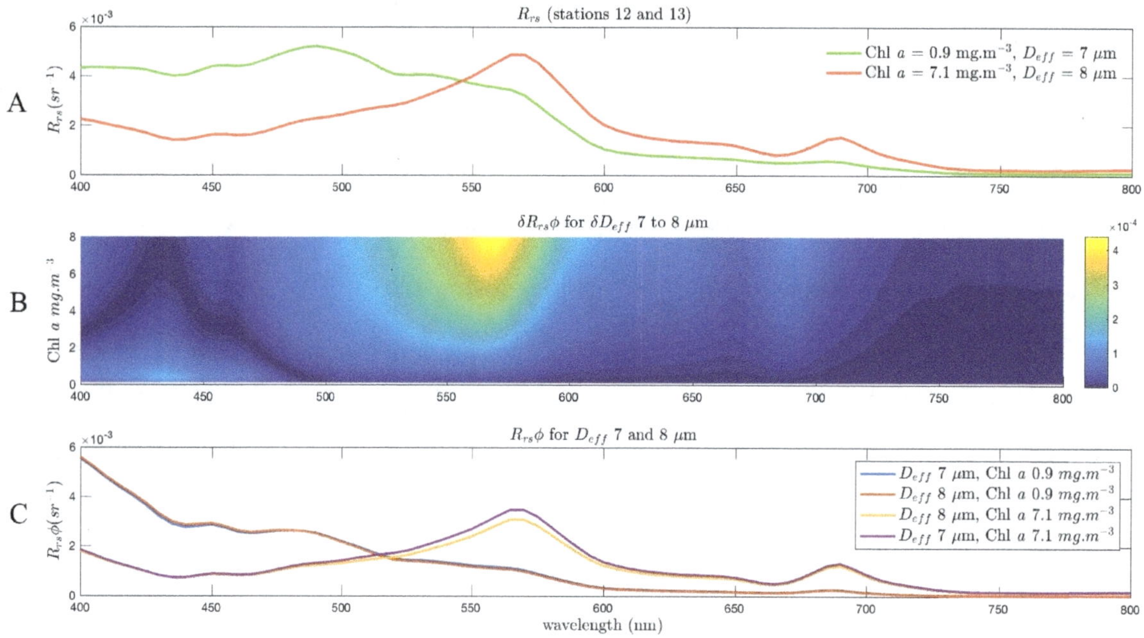

Figure 6. Modelled R_{rs} for Stations 12 and 13 (**A**), with EAP eukaryote phytoplankton IOPs, and $a_{gd}(\lambda)$ and $b_{bnap}(\lambda)$ components estimated guided by observations in [75,76], respectively; (**B**) shows $\delta R_{rs}\phi$ for this large change in Chl a concentration (1 to 7 mg/m^3) but a small δD_{eff} of 7–8 μm. The unit of the colour bar is sr^{-1}. Note that the results are one order of magnitude less than in the previous example; (**C**) shows the negligible effect on $R_{rs}\phi$ of a change in D_{eff} from 7 to 8 μm at the measured Chl a concentrations.

3.3. Case Study 2: Addressing Pigment Variability

Both the phycoerythrin-containing and peridinin-containing assemblages are modelled here, Figures 7 and 8) with D_{eff} of 12 μm, so the simulated optical changes as the assemblage changes from *M. rubra*-dominated to dinoflagellate-dominated are all due to differences in pigmentation, for any given Chl a concentration. From the log-scale R_{rs}, it is evident that the pigment-related differences in R_{rs} become larger as biomass increases. In the very high biomass blooms ($\geq$30 mg·m^{-3}) typical of the Benguela system, it is known that *M. Rubrum*—containing assemblages are identifiable from MERIS satellite imagery [83] due to the effects of the diagnostic phycoerythrin peak (at 565 nm) appearing in the 560:520 nm band ratio.

An analogous study of the sensitivity of the maximum $\delta R_{rs}\phi$ signal to non-algal constituents is made at 570 nm for the pigment-driven feature appearing at high biomass (10 mg·m^3, Figure 9). The sensitivity of pigment-driven differences to non-algal effects is in contrast to the Southern Ocean size example in that it is largely driven by variability in $b_{bnap}(\lambda)$. As biomass increases past 10 mg·m^3, the magnitude of the $\delta R_{rs}\phi$ grows as $b_{bnap}(\lambda)$ increases, showing no impact at all of biomass past 20 mg·m^3. What this means is that while significant biomass is required to detect pigment changes, past a certain upper biomass limit, the magnitude of the pigment differential signal grows proportionally as R_{rs} is augmented by non-algal scatter.

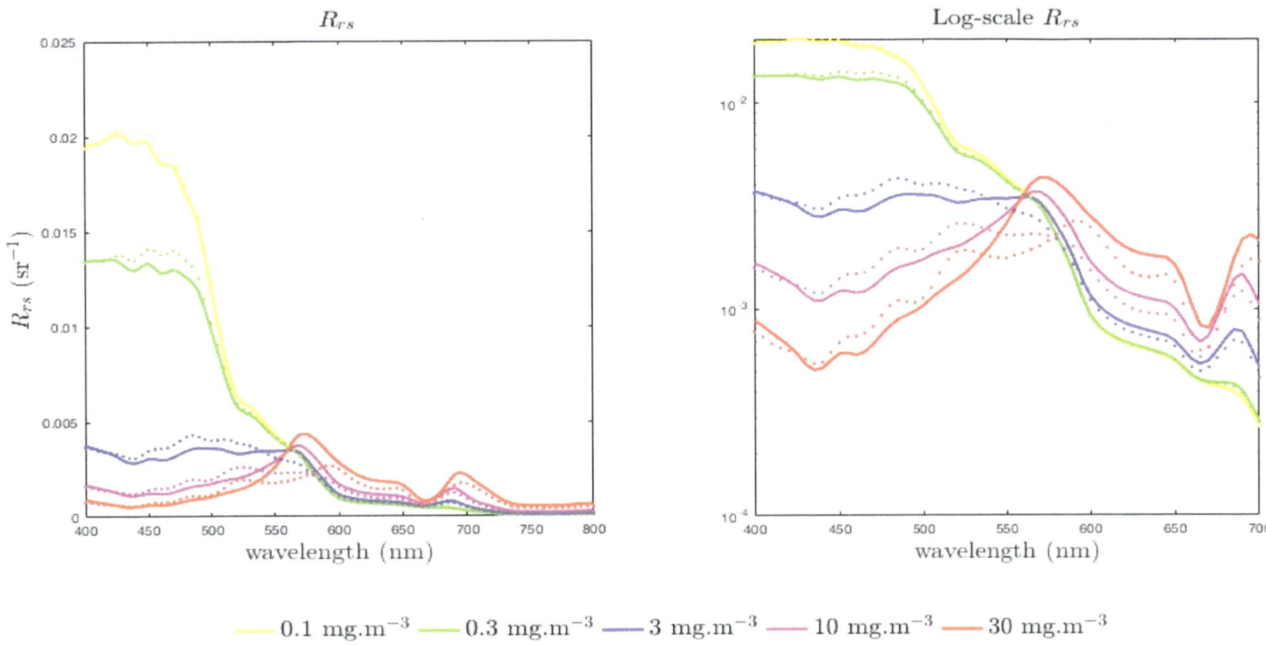

Figure 7. Benguela-like pigment-based experiment: Modelled R_{rs} shown for Chl *a*-carotenoid pigmented assemblages (solid lines) and phycoerythrin containing assemblages (dotted lines) for identical Chl *a* concentrations, at 0.1, 0.3, 3, 10 and 30 mg·m^{-3}. There is no change in D_{eff}, both are 12 µm. The non-algal optical constituents are modelled with $a_{gd}(400) = 0.07 * [Chl_a]^{0.75}$, and $b_{bnap}(550) = 0.005$ m^{-1}.

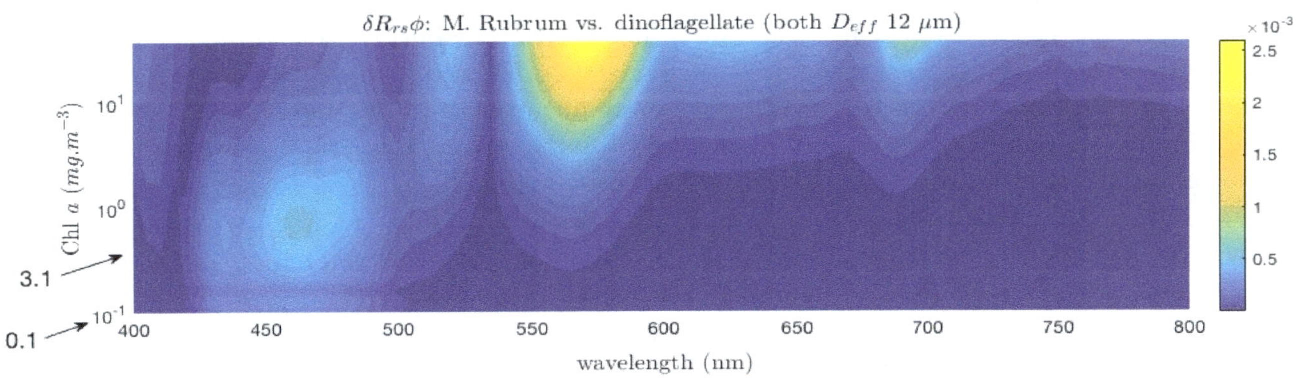

Figure 8. δR_{rs} shown for a change from a high biomass *Myrionecta rubra*-dominated assemblage, to a high biomass peridinin (carotenoid)-containing dinoflagellate-dominated assemblage. There is no change in D_{eff}.

Absorption-based pigment differences are therefore sensitive to scattering variability unless the biomass is very high, and this is particularly relevant in spectral regions affected by scattering variability due to changes in D_{eff} (Figure 4). Noting the log-scale Chl *a* axis of Figure 9, it can be observed that while the respective magnitudes of the pigment-driven (D_{eff} = 12 µm, Figure 9) and size-driven (D_{eff} from 6 to 16 µm, Figure 4) signals are comparable at the point that pigment differences appear (i.e., 10 mg·m^3), the size-driven feature is more sensitive at lower biomass—detectable at around 2 mg·m^3 for the given change in D_{eff}. This sensitivity will be affected by the size range in question and also by pigment concentrations, but it can be inferred that, generally, where size changes of this range (or larger) take place together with pigment changes, it is the size change that drives the variability in the water-leaving signal, and changes in the reflectance due to a substantial change in D_{eff} are observable at lower biomass than those due to pigment changes.

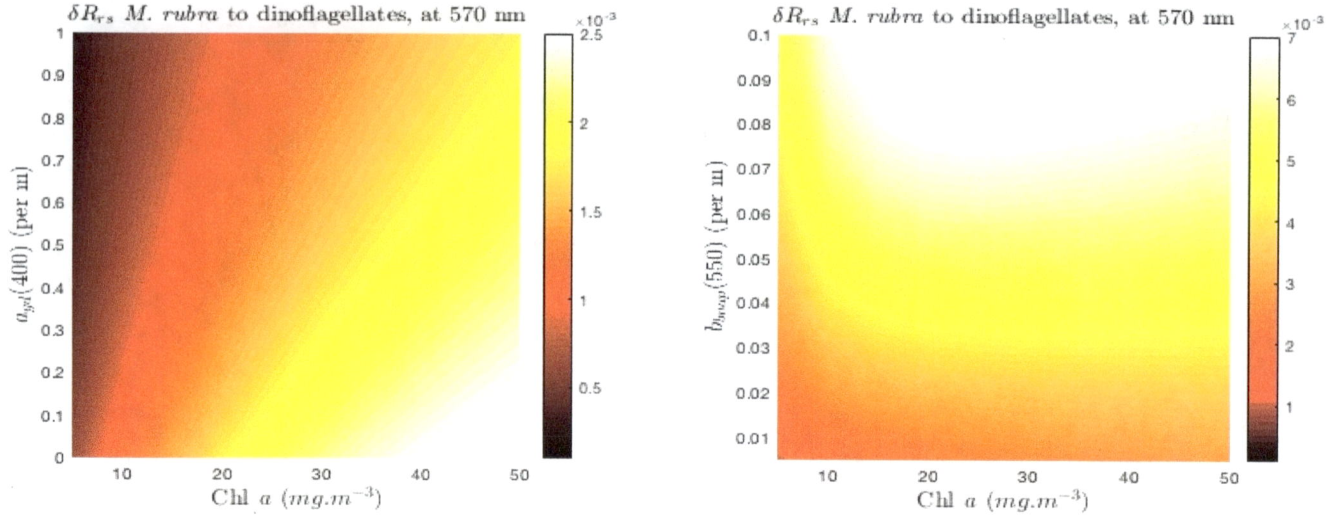

Figure 9. δR_{rs} sensitivity to a_{gd} and b_{bnap} at 570 nm, for a high biomass *Myrionecta rubra*-dominated assemblage, to a high biomass peridinin (carotenoid)-containing dinoflagellate-dominated assemblage.

This applies equally to accessory pigments other than phycoerythrin absorbing in spectral regions affected by phytoplankton scatter and/or by variability in non-algal scatter. This speaks to the importance of both the magnitude of the change in $b_b\phi(\lambda)/a\phi(\lambda)$ and the proportional contribution of $b_b\phi(\lambda)/a\phi(\lambda)$ to the total $b_b(\lambda)/a(\lambda)$ when evaluating the potential for accessing these signals from R_{rs} (see Figure 2).

This result also implies that there are cases where the δD_{eff} signal is augmented by pigment changes, for example when moving from small fucoxanthin- or peridinin-dominated cells to large phycoerythrin-dominated cells, as the optical effect of reduction in backscatter around 560–570 nm by large cells will be enhanced by additional pigment absorption in the δR_{rs} signal. However, there remains a complex optical relationship with biomass, and this effect needs to be properly accounted for in order to accurately detect the augmented signal.

The EAP approach to pigments does not address the extent to which fine spectral resolution accessory pigment absorption features persist in R_{rs}, nor their retrieval from hyperspectral radiometry. The intention with the EAP model is to demonstrate the dependence on biomass to retrieve absorption features, and the inherent signal ambiguity as the contrasting optical effects of $b_b\phi(\lambda)$ and $a\phi(\lambda)$ interact to form the phytoplankton signal within the R_{rs}. It is worth noting that hyperspectral radiometry will not overcome the inherent signal-related constraints identified in this study.

3.4. Radiometric Sensitivity of EAP Size-Based PFT Detection—Magnitude of $\delta R_{rs}\phi$

Having established that at low biomass the PFT signal in the blue is easily overwhelmed by the effects of $a_{gd}(\lambda)$ and $b_{bnap}(\lambda)$, and that pigment effects are generally secondary to those of δD_{eff}, the PFT signal due to phytoplankton scattering in the 500 to 600 nm region can be evaluated for sensitivity in terms of changes in D_{eff} and biomass. To this end, the EAP model is again coupled with Hydrolight to simulate expected variability in R_{rs} due to changes in D_{eff} with the aim of evaluating the sensitivity of the model.

Figure 10A demonstrates how the combined effects of biomass and D_{eff} interact to form the maximum available $\delta R_{rs}\phi$ signal at low biomass and small size ranges. The figure shows that this maximum lies between 520 and 570 nm—the exact wavelength varies with both size difference and biomass. The shifting position of maximum $\delta R_{rs}\phi$ is shown in Figure 10B. Increasing biomass improves the ability to trace the size-related effects, and a D_{eff} change from 2 µm up to at least 8 µm is not detectable at the threshold in oceanic conditions with Chl a < 1 mg/m^3.

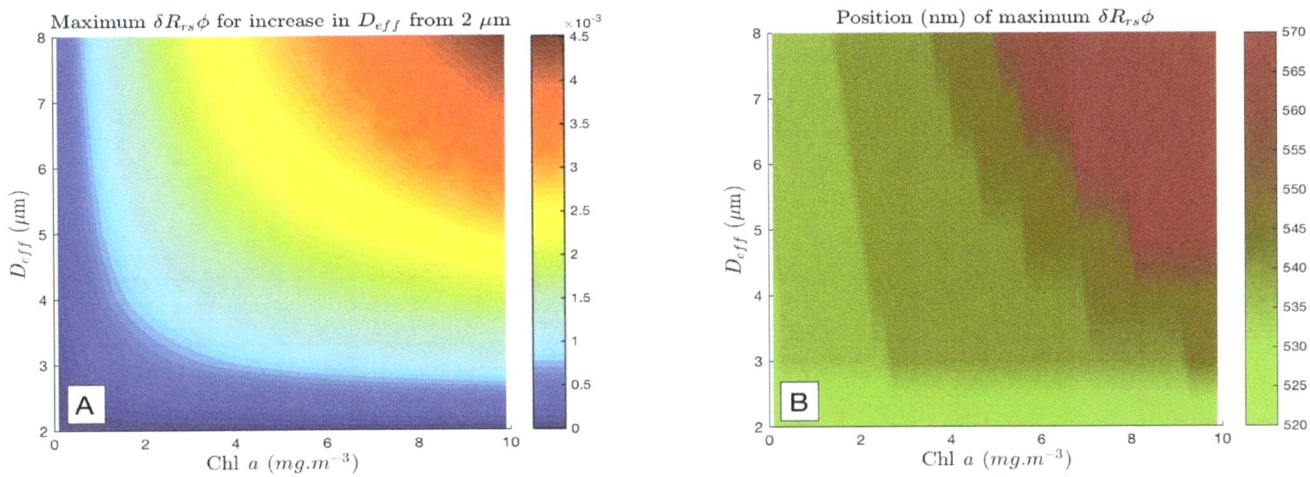

Figure 10. Maximum $\delta R_{rs}\phi$ for δD_{eff} from a starting assemblage with D_{eff} 2 µm, as Chl a varies (**A**). Note that the $\delta R_{rs}\phi$ occurs at different wavelengths from 500 to 600 nm (**B**), and this shows the maximum signal, so there is no exact wavelength information in (**A**). Using a difference of 1×10^{-3} sr^{-1} as a threshold for detection by satellite, it can be seen that, while the maximum size change here (2 to 8 µm) is not detectable with Chl $a < 1$ mg/m^3, by 10 mg/m^3, even a small change in D_{eff} results in a detectable change in R_{rs}.

Using 1×10^{-3} sr^{-1} as a threshold for detection by satellite, it can be seen that an ecologically significant shift in D_{eff} from 2 or 3 to 6 µm, such as at the onset of an oceanic bloom, looks potentially detectable from about 2 mg/m^3. By 10 mg/m^3, even a small change in D_{eff} results in a detectable change in $R_{rs}\phi$, but, as biomass falls below this, the change in D_{eff} must be increasingly large to be detected. This is consistent with inversion studies of EAP sensitivity [37]. Note that this experiment addresses only the phytoplankton-related signal, and that when attempting to identify these signals in the bulk R_{rs}, it is necessary to consider the sensitivity of the phytoplankton signal to the optical effects of the non-algal constituents. These results can be considered to show the minimum threshold for potential detection i.e., the signal is further ambiguated by non-algal optics in a real-world context.

The spectrally shifting nature of the $\delta R_{rs}\phi$ signal for oceanic PFT applications provides a strong case for hyperspectral sensors in the 520 to 570 nm wavelength region. The extent to which the $\delta R_{rs}\phi$ signal persists in fixed waveband ratios is investigated in the next section on shape sensitivity.

3.5. Spectral Shape Sensitivity of EAP Size-Based PFT Detection

To further test the sensitivity of the EAP model and the causal IOP variability in terms of identifiable changes in spectral shape from a multi-spectral perspective, $R_{rs}\phi$ ratios for 440:560 nm (blue:green), 560:665 nm (green:red) and 665:710 nm (red:NIR) wavelengths were calculated for a range of D_{eff} and biomass.

These are shown in Figure 11, representing corresponding changes in the $R_{rs}\phi$ and in the phytoplankton backscattering, for these wavelength pairs. The B:G $R_{rs}\phi$ ratio shows a strong biomass dependency and a small sensitivity to size at large sizes, for $0.5 \leq$ Chl $a \leq 4.5$ mg/m^3. The R:NIR ratio shows some sensitivity to larger sizes from about 3 mg/m^3, but this decreases as biomass increases. The G:R ratio shows a significant size-related feature for small sizes (≤ 6 µm) from biomass of about 2 mg/m^3 upwards (encircled in Figure 10). This is where a peak in the corresponding $b_{b\phi}$ ratio appears, suggesting that the large change in magnitude of the Chl a-specific backscatter $b_b^*\phi$ between small D_{eff} (Figure 12) is directly responsible for the sensitivity in the $R_{rs}\phi$ G:R ratio seen in Figure 11.

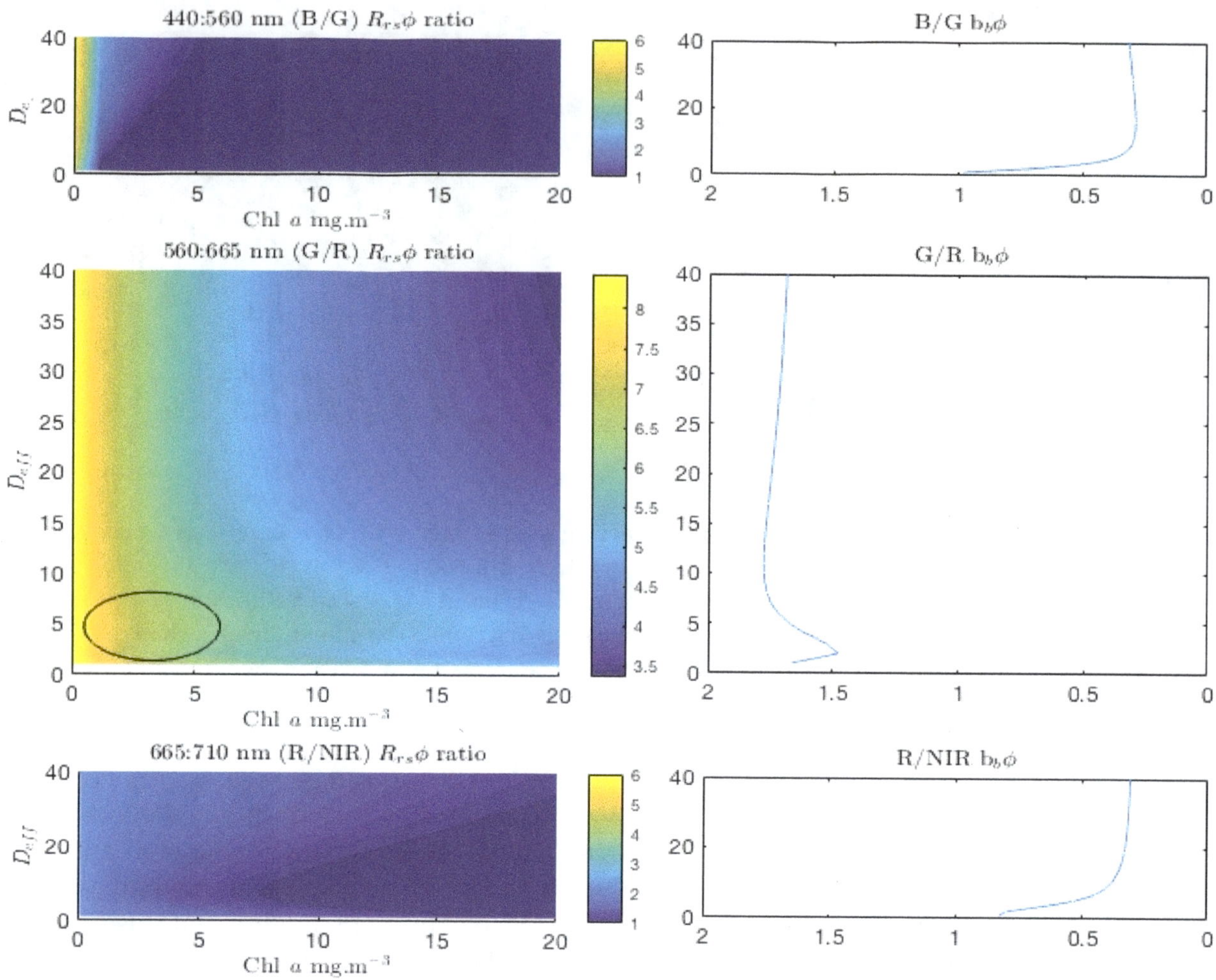

Figure 11. $R_{rs}\phi$ ratios for blue:green, green:red and red:NIR (Near Infra-Red) wavelengths as shown, for Chl a concentrations of 0.1 to 20 mg/m^3 and D_{eff} 1 to 40 µm. The B/G ratio shows a strong biomass dependency and a small sensitivity to size at large sizes, for $0.5 \leq$ Chl $a \leq 4.5$ mg/m^3. The $b_{b\phi}$ ratios all display a strong size signal at 2–4 µm, and the G/R ratio shows a corresponding size-related feature.

This is an important finding. There is a marked size dependency in all of the $b_b^*\phi$ ratios, with the greatest rate of change somewhere between D_{eff} 2 and 8 µm, but it is only in the case of the G:R ratio that the magnitude of the backscatter is sufficient for this signal to be identifiable in the $R_{rs}\phi$. Given that the radiometric signal in the blue is greatly reduced by large phytoplankton absorption and a_{gd}, and the red and NIR wavelengths are similarly affected by the absorption of water, it can be concluded that the main driver of the useable PFT signal in the green and red is phytoplankton backscatter.

Figure 13 shows the rapid increase in the proportional contribution of phytoplankton to total backscatter at 560 and 665 nm. It is known that, for typical diatom/dinoflagellate assemblages, the 560 nm region is more influenced by backscatter than by absorption. The fact that the magnitude of the total backscatter is much lower at 665 than at 560 nm, together with the strong absorption by water in this region, result in a small useable R_{rs} signal. A contribution of approximately 40% of phytoplankton to total b_b at 560 nm corresponds with the limits of detectable $\delta R_{rs}\phi$ (see Figure 2), indicating that this is the proportion at which phytoplankton backscatter starts driving the total water-leaving signal around 560 nm. Consequently, this is the minimum contribution for which some δD_{eff} information may be known. For an oceanic bloom example δD_{eff} from 2–6 µm, this threshold contribution is reached at about 2 mg/m^3, while to detect an example δD_{eff} of 10 to 20 µm in a eukaryotic succession, extremely high biomass is required. The mid-range biomass sensitivity demonstrated here presents opportunities for identifying higher resolution size classes than the 2 to 20 μm and >20 μm categories currently frequently employed [6,8,10,84]. The ability to achieve better resolution within the 2–20 μm size class is particularly desirable for marine ecosystem modelling [6].

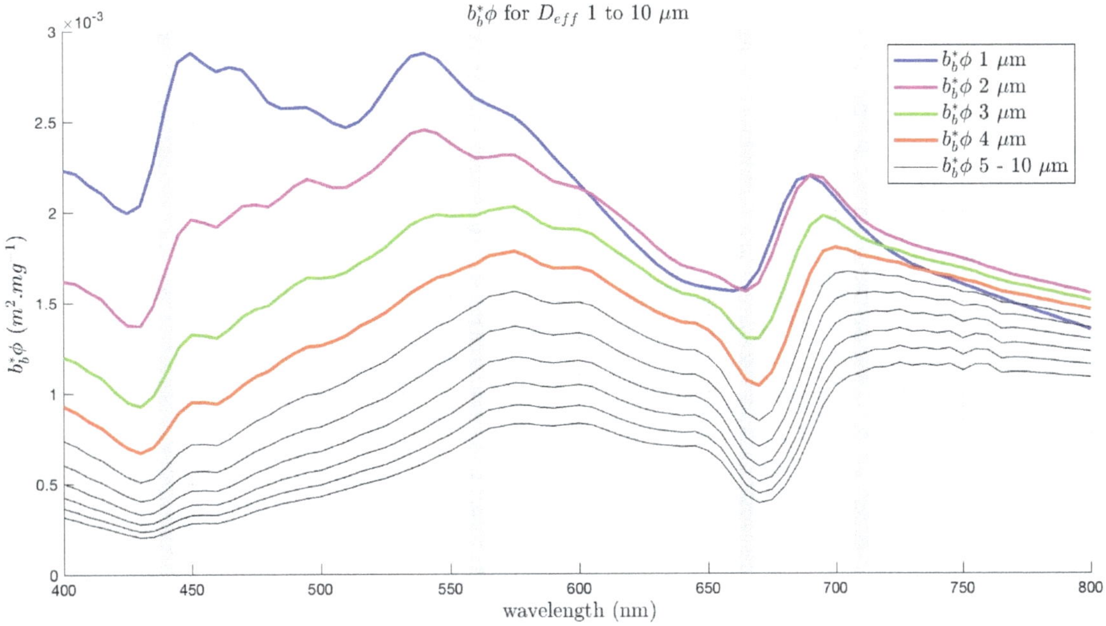

Figure 12. $b_b^*\phi$ shown for D_{eff} 1 to 10 μm. The largest differences in backscatter across the spectrum occur between 1 and 4 μm, with the exception of the overlapping of $b_b^*\phi$ in the red and NIR.

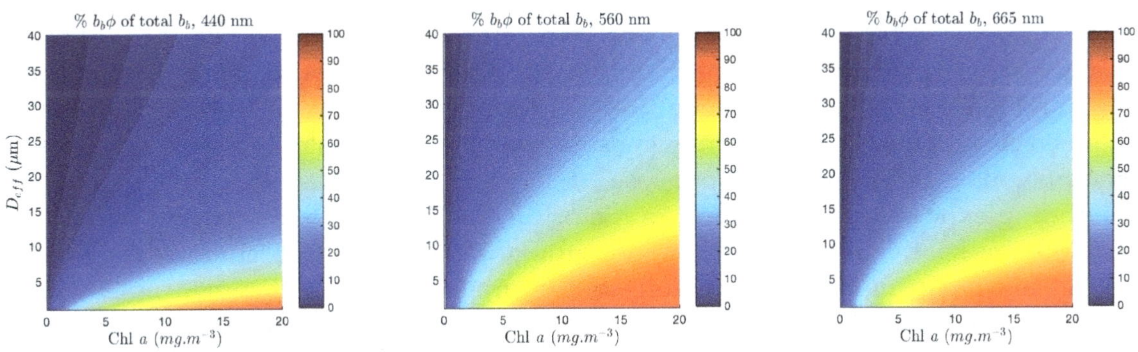

Figure 13. Percentage contribution of phytoplankton to total backscatter (including water, and with nominal $b_{bnap}(550) = 0.005$), shown for D_{eff} 1 to 40 μm and Chl a from 0.1 to 20 mg/m^3, at 440, 560 and 665 nm.

3.6. Considering Uncertainties

Particularly when considering δR_{rs} retrievals from satellite, it is important and necessary to contextualise the magnitude of the PFT signal with respect to uncertainties on the satellite radiometry. A brief study on model and associated radiometric uncertainty is available in Appendix D. An important observation is that, while the 500 to 600 nm region of the promising PFT signal may be mostly insensitive to the effects of non-algal constituents, it is also where variability in R_{rs} due to the different approaches to phytoplankton phase functions is important, emphasising the critical role of phytoplankton scatter in this signal.

4. Conclusions

The distinct causal optical effects of variations in phytoplankton biomass, mean assemblage size, pigment-related spectral variability and non-algal constituents are not easily identified, with substantial interdependency and spectral ambiguity. Consistent with previous studies [37,39], it can be seen here that ambiguity is critical in attempting to resolve the phytoplankton community structure signals.

The case studies illustrate how most of the R_{rs} signal that is due to phytoplankton is driven by biomass, an expected result. This concept underpins, after all, the primary missions of most ocean colour sensors: resolving variability in phytoplankton biomass. The study shows that quantitative consideration of the constraints of biomass and the phytoplankton contribution to the total IOP budget is required when addressing the PFT question.

The key findings include the assertions that most of the absorption driven phytoplankton signal in R_{rs} in the blue is too ambiguous to use, and that the most useful PFT signal is caused by spectral backscatter. Furthermore, the ability to assess the PFT signal against non-algal optical contributions is largely driven by biomass and the IOP budget.

Overall, spectral scattering properties of natural waters are not well characterised [85,86], and phytoplankton spectral backscattering characteristics are underexploited in terms of their impact on the water-leaving signal. The importance of better representing the angular and spectrally variable nature of phytoplankton scattering has been established [35], and it is clear that phytoplankton backscatter is at the heart of the PFT question.

The size-related PFT signal is driven by phytoplankton scattering, and spectral regions where scattering is at its most sensitive to D_{eff} show the most potential for PFT detection from the total water-leaving signal. There is most sensitivity to size-related changes in $b_{b\phi}$ in the 1–6 μm size range. Phytoplankton size-related features, most likely driven by phytoplankton absorption variablity, appear in R_{rs} around 430 nm at low biomass, and scattering-driven size-related features in the 520 to 570 nm spectral region at elevated biomass. The water-leaving signal in the blue spectral region is highly complex and ambiguous, being the result of varied and contrasting effects of absorbing and scattering characteristics of both algal and non-algal in-water constituents. Consequently, the size-related signal appearing in low biomass waters (<1 mg/m^3) may be useful only when $a_{gd}(\lambda)$ and $b_{bnap}(\lambda)$ are exactly known. Accessory pigment absorption features that persist in R_{rs} in low biomass suffer from the same vulnerability to uncertainty in the non-algal constituents. Satellite measurement uncertainty and $a_{gd}(\lambda)$ retrievals may in future be improved (e.g., with the use of radiometry in the UV), but, given current uncertainties, achieving sufficiently accurate satellite estimates of the non-algal optical components is unlikely for this purpose.

This finding exposes a vulnerability in historical approaches to phytoplankton identification and quantification based on the features of phytoplankton absorption characteristics in the blue. Satellite PFT methods using this approach all suffer from this shortcoming where assemblage-related variability is secondary to biomass effects, and where phytoplankton relationships with $a_{gd}(\lambda)$ and $b_{bnap}(\lambda)$ are not precisely known. These approaches additionally rely on implicit relationships between Chl *a* and D_{eff} which may not always hold. This work shows that, at low biomass (< 1 mg·m^3), where R_{rs} is absorption-dominated, it is unlikely that there is sufficient size- or pigment-driven PFT signal to be retrieved from satellite radiometry without making these assumptions. (Phytoplankton whose prominent absorption features are at longer wavelengths, such as phycocyanin-containing cyanobacteria, present a different case). Isolating variability in $R_{rs}\phi$ as D_{eff} and biomass vary shows that an example oceanic bloom δD_{eff} from 2 to 6 μm is only detectable at the satellite measurement threshold of 1×10^{-3} sr^{-1} when the biomass reaches about 2 mg/m^3 (Figure 10A).

Consequently, it is the size-related backscatter-driven signal in the 500 to 570 nm region, appearing at substantial biomass, that is the most useful for PFT identification from satellite radiometry as it is sufficiently insensitive to reasonable variability in both $a_{gd}(\lambda)$ and $b_{bnap}(\lambda)$ (when composed of small particles) (Figure 4). Variability in scatter due to non-algal particulate in the same size range as phytoplankton will likely ambiguate the distinctive spectral scatter of PFT changes, and this has not yet been tested. The location of the maximum $\delta R_{rs}\phi$ size feature shifts between 520 and 570 nm (Figure 10B), suggesting strongly that hyperspectral data in this region would add greater capability here. Further analysis is needed to quantify the potential advantages of hyper- over multi-spectral data with respect to this shifting maximum signal, and also with respect to the reduced SNR implicit in narrow waveband measurements.

Understanding the proportional phytoplankton contribution to the total IOP budget and the resulting water-leaving signal is central to the determination of sufficient phytoplankton-driven signal containing PFT information. The proportional 'net' contribution of phytoplankton i.e., $b_{b\phi}/a_{\phi}$ as a percentage of total b_b/a, has been identified as the driver of PFT sensitivity in the R_{rs}. Given the detectable differences in R_{rs} as size and biomass change, a proportional phytoplankton contribution of approximately 40% to the total b_b appears to a reasonable minimum threshold in terms of yielding a detectable optical change. The proportional contribution always varies with the non-algal optical constituents $a_{gd}(\lambda)$ and $b_{nap}(\lambda)$.

Despite the many sources of model uncertainty and the requirement for model validation in specific regions, these results indicate the necessity of approaching PFTs from a strongly biophysical perspective. There is a great need for better characterisation of phytoplankton community structure and improved handling of the complex spectral and angular nature of phytoplankton scattering.

The EAP model code in Matlab R2018a (The Mathworks Inc., Natick, Mass, United States) or in Python 3.7, as well as the Fortran routine for Hydrolight allowing the choice of discretised EAP phase function based on wavelength rather than backscatter fraction (see Appendix D: Uncertainties), are freely available to the community. Please contact the corresponding author.

Author Contributions: Conceptualization, L.R.L. and S.B.; Formal analysis, L.R.L.; Investigation, L.R.L.; Project administration, S.B.; Software, L.R.L.; Supervision, S.B.; Writing—original draft, L.R.L.; Writing—review & editing, S.B.

Acknowledgments: Thanks to Curtis Mobley for assistance with Hydrolight and the examiners of Lisl Robertson Lain's PhD thesis for their valuable comments. Input from two anonymous reviewers of this manuscript, as well as from Dariusz Stramski, was gratefully received.

Appendix A. Phytoplankton Assemblage Variability in the EAP Model

The successful validation of the model in very high biomass Benguela conditions [65] gives confidence in the representation of the phytoplankton component of the water-leaving signal, as it is known that, in these cases, the R_{rs} is overwhelmingly dominated by phytoplankton. It is concluded in Lain et al. [65] that it is the EAP's detailed handling of phytoplankton spectral backscatter that sets it apart from other IOP models. The core mathematics of the model are fully described in Bernard et al. [27]. A detailed study of EAP phytoplankton angular scattering and phase functions is available in Lain et al. [35].

Appendix A.1. Justification for Using Measurement-Derived Refractive Indices across Wide Size Ranges

The main light-harvesting pigments in typical diatom and dinoflagellate assemblages (fucoxanthin and peridinin, respectively)—while chemotaxonomically distinct—display the typical broad, featureless absorption spectra characteristic of carotenoids, with peaks centered around 500 nm [87] and vary well within the natural variability of phytoplankton absorption (Figure A1). They consequently have similar refractive indices [27] and so these types were combined into a generalised set of diatom/dinoflagellate IOPs, as no significant difference was found between the dinoflagellate and diatom groups in terms of their optics that could not be attributed to the respective particle sizes (see also [38,88]). This group of IOPs should correctly be referred to as Chl *a*-carotenoid IOPs.

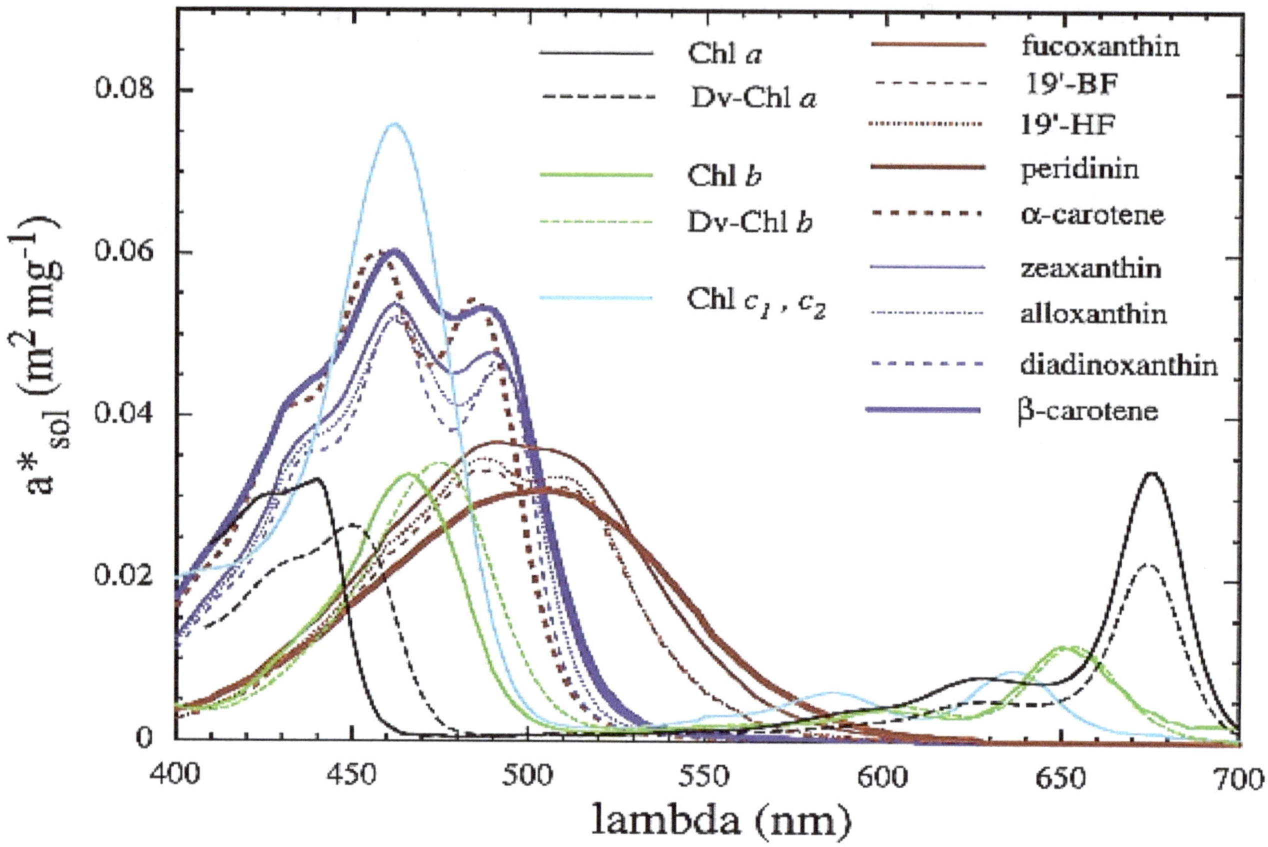

Figure A1. Pigment absorption spectra from Bricaud et al. [87], reprinted with permission from the American Geophysical Union. The broad featureless absorption spectra of fucoxanthin and peridinin peaking at around 500 nm are shown by the thin and thick brown lines, respectively.

While the measurements and refractive index derivations [89] were performed for cells of approximately 12 to 20 μm, it should be made clear that the imaginary refractive index characterises the absorption of the intracellular material and has absolutely no dependency on cell size. It can be inferred therefore that these refractive indices can be used to represent a range of phytoplankton sizes displaying dominant Chl *a* and carotenoid pigments. Southern Ocean nanophytoplankton comprise mainly diatoms and dinoflagellates, but also chlorophytes and haptophytes in smaller proportions [90,91]. The latter two phytoplankton groups are generally dominated by Chl *a* and the fucoxanthin pigment derivatives 19′-hex-fucoxanthin and 19′-but-fucoxanthin [91], which display somewhat elevated absorption peaks located at slightly shorter wavelengths with respect to fucoxanthin [87,92]. The optical influence of these derivative pigments is assumed to be negligible in the context of the case studies for two reasons.

Firstly, in the nanophytoplankton group as a whole, the fucoxanthin-derived pigments occur in far lesser concentrations than that of fucoxanthin itself [91], which is represented by the refractive indices. This is reinforced by the derivation of a nanophytoplankton group of refractive indices in non-bloom conditions.

(D_{eff} = 2 μm), whose impact on the IOPs of small cells was not sufficiently different from the diatom-dinoflagellate group to warrant its routine use for small cells. Secondly, these pigments act to increase phytoplankton absorption around the 450 nm spectral region, identified the spectral region as very vulnerable to small variability in a_{gd}, and to large satellite R_{rs} measurement uncertainty, so the conclusions regarding the spectral regions containing the most useful signatures for PFT detection still hold.

EAP sensitivity testing has indicated that increasing the intracellular Chl *a* density C_i may be appropriate for small cells, but this is not explored here.

Appendix A.2. EAP Phytoplankton IOPss

Phytoplankton-specific IOPs are presented in Figures A5 and A6 for generalised Chl *a*-carotenoid assemblages and for phycoerythrin-containing assemblages, respectively.

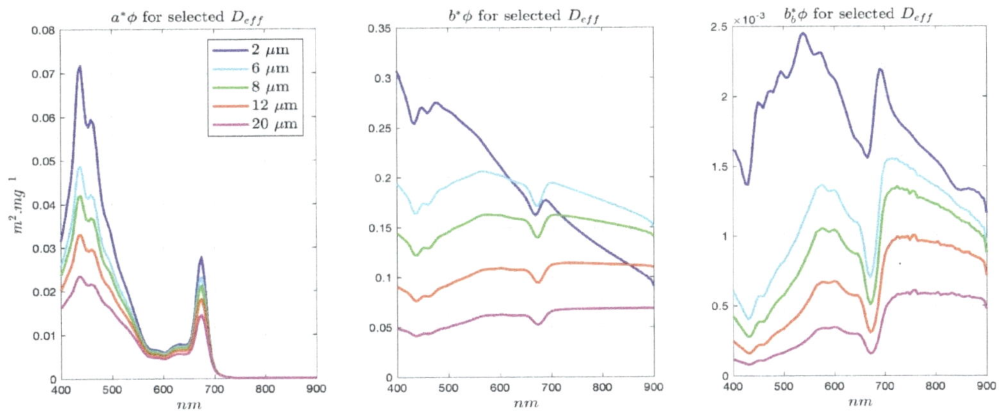

Figure A2. EAP Eukaryote Chl *a*-carotenoid-dominated IOPs for a range of assemblage D_{eff}.

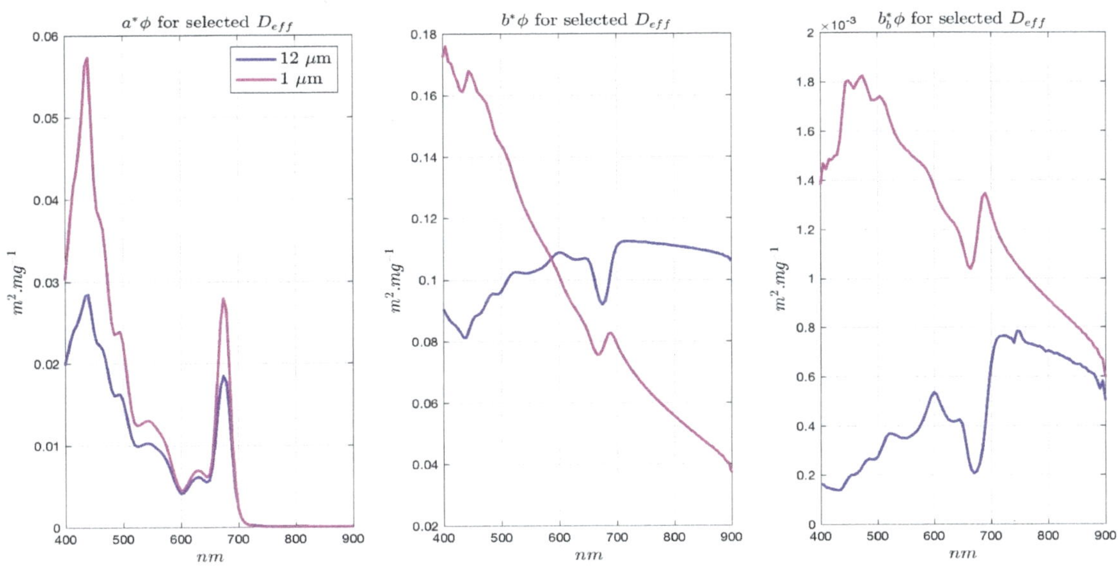

Figure A3. EAP Phycoerythrin-containing IOPs (based on *Myrionecta Rubra*), used for cryptophyte-dominated assemblages in the Benguela, and *Synechococcus* sp. in the Southern Ocean.

Appendix A.3. EAP $a_{gd}(\lambda)$ Parameterisation

A simple exponential combined gelbstoff and detrital absorption term $a_{gd}(\lambda)$ [93,94] is used as a representative of commonly occurring conditions in the Benguela:

$$a_{gd}(\lambda) = a_{gd}(400)\exp[-S(\lambda - 400)]. \tag{A1}$$

The exponential slope factor S is given a constant value of 0.012 [95]. This value, derived for the Benguela system, is not adjusted for the $a_{gd}(\lambda)$ term used in the Southern Ocean Case Studies. This is acknowledged as a source of uncertainty, but supporting literature suggests that values in the range $0.0140 \pm 0.0032\ \text{nm}^{-1}$ cater adequately for a variety of water types [93].

An observed relationship of

$$a_{gd}(400) = 0.0904\log[Chl_a] + 0.1287 \tag{A2}$$

from measurements in the Benguela is used to scale the gelbstof/detrital exponential term, and $a_{gd}(750)$ onwards is assumed to be zero. This parameterisation was derived for high biomass environments. At very low biomass (< 1 mg/m^3), the log$[Chl_a]$ term becomes negative, and so, for the Southern Ocean case studies, this parameterisation was amended to

$$a_{gd}(400) = 0.07 \cdot [Chl_a]^{0.75} \tag{A3}$$

following Alvain et al. [21], noting that the referenced parameterisation is for 440 nm and not 400, but also that the a_{gd} term is used as an approximate measure of total signal sensitivity, and so, in this sense, an absolutely accurate term is not a requirement.

Appendix A.4. EAP $b_{bnap}(\lambda)$ Parameterisation

Non-algal backscattering is modelled after Roesler and Perry [96] who describe a small particle backscattering term represented by a power law relationship (their b_{bs}, referred to as b_{bnap} in the EAP model). It has a constant spectral shape dependent only on wavelength, but variable in magnitude.

$$b_{bnap}(\lambda) = \lambda^{-1.2}. \tag{A4}$$

This is then adjusted to a selected value of $b_{bnap}(550)$, as detailed in the text.

Small particle (non-algal) scatter b_{nap} is approximated as 50 times the b_{bnap} in the Benguela examples and as 100 times the b_{bnap} in the Southern Ocean examples. This yields a non-algal particulate backscattering probability ($\tilde{b}_{bnap}$) of 0.02 (2%) and 0.01 (1%), respectively. This is assumed to be reasonable given that it has been shown that the total particulate backscattering probability $\tilde{b}_b$ varies in the range 1.2 to 3.2 % in coastal waters dominated by non-algal particles (i.e., Case 2) [97], and that generally accepted values for $\tilde{b}_b$ in Case 1 waters is around 1% [98].

Keeping the non-phytoplankton backscattering constant with Chl *a* results in a dependent but nonlinear relationship, resulting in an overall $\tilde{b}_b$ that decreases as Chl *a* increases (Figure A4), noting the spectral variability of small phytoplankton at elevated biomass.

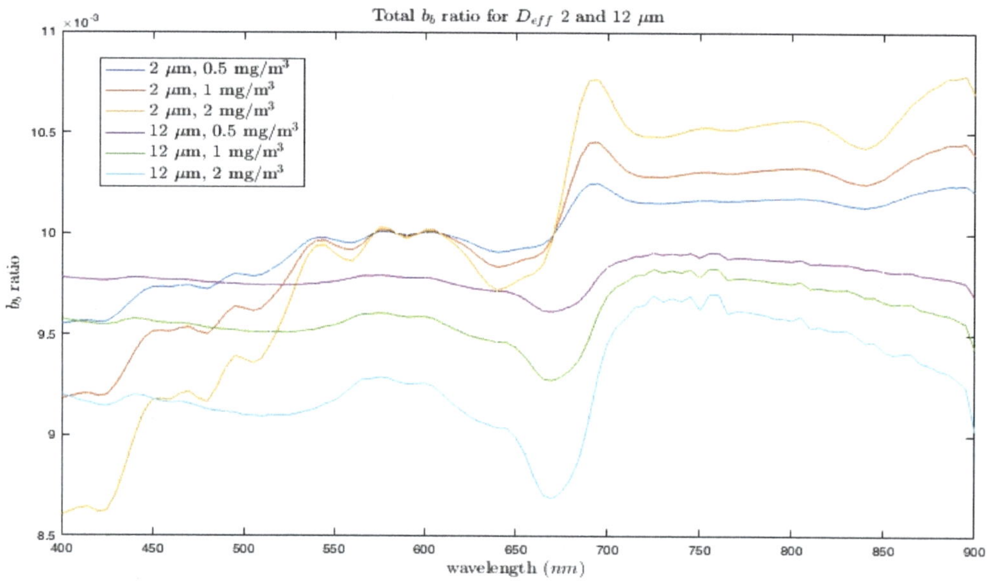

Figure A4. Bulk backscatter ratio shown for D_{eff} 2 and 12 μm, with nominal $b_{bnap}(550) = 0.01$ m^{-1} and b_{nap} as 50 times the b_{bnap}, as for a coastal environment, shown for Chl *a* of 0.5, 1.0 and 2.0 mg/m^3. The elevated backscatter ratio of coastal environments with respect to the Southern Ocean (where b_{nap} is modelled as 100 times the b_{bnap}) is attributed to the contribution of terrestrial mineral particles with a high refractive index [66,99].

Appendix B. Measurements and Modelling Parameters

Appendix B.1. Chl a

Chl *a* measurements are made using a Turner 10-AU Fluorometer (Turner Designs, San Jose, CA, USA), following Holm-Hansen et al. [100].

Appendix B.2. Model Parameters Used for Hydrolight-Ecolights

For most of the experiments, Ecolight's 2-component IOP model was used to generate $R_{rs}(\lambda)$. The "clearest natural water" IOPs were selected for Component 1 (water). IOPs for component 2 (everything else) were precomputed in Matlab from the EAP phytoplankton IOPs and additional $a_{gd}(\lambda)$ and $b_{bnap}(\lambda)$ contributions as required.

Fluorescence quantum efficiency ϕ was approximated by Chl *a* concentration:

$$<10 \text{ mg/m}^3 = 1\%$$
$$10\text{–}50 \text{ mg/m}^3 = 0.6\%$$
$$50\text{–}100 \text{ mg/m}^3 = 0.2\%$$
$$>100 \text{ mg/m}^3 = 0.1\%.$$

These values are based on MODIS ϕ_{sat} climatologies [101], and measurements [102] to characterise the reduction in ϕ as eutrophication increases.

A constant set of generalised atmospheric conditions was selected for all experiments. An annual average for solar irradiance and a solar zenith of 30° was used in lieu of time and location.

Appendix C. Position of Maximum $\delta R_{rs}\phi$

Further to Figure 9B in the main text, Figure A5 shows the position of maximum $\delta R_{rs}\phi$ for assemblage changes from 8 µm and 14 µm, respectively, for Chl *a* concentrations between 10 and 20 mg/m³. It can be seen that, at these high biomass concentrations, there is no spectral migration of the maximum $\delta R_{rs}\phi$ with biomass, and that once the maximum $\delta R_{rs}\phi$ signal reaches 570 nm, its location does not change with increasing D_{eff} for biomass up to 20 mg/m³. This has been tested up to 40 µm (not shown).

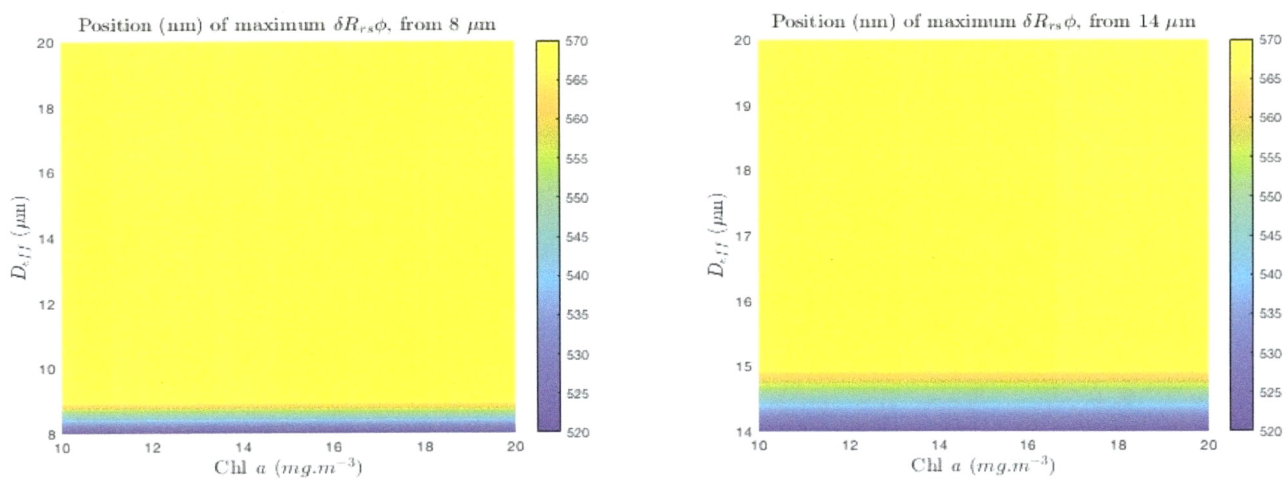

Figure A5. Spectral position of maximum $\delta R_{rs}\phi$ for assemblage changes from 8 µm and 14 µm, respectively

Appendix D. Uncertaintiess

Uncertainties in satellite radiometry are given in the main text, and model error in terms of uncertainty/variability in the phase function is described fully in [35]. There are many additional sources of uncertainty in the model (non-sphericity of phytoplankton, approximations in size distribution, Chlorophyll *a* density, to name a few), and further work is needed to quantify them appropriately. For demonstrative purposes here, given that any retrieval of size properties would be performed with the model itself, the model uncertainty is constrained to just that of the phase function variability, as this has a size implication in itself, as shown in [35].

In Figure A6, the model uncertainty is shown for $R_{rs}\phi$ against a background of total R_{rs} with nominal additional $a_{gd}(\lambda)$ and $b_{nap}(\lambda)$, together with the satellite R_{rs} measurement uncertainty. Despite the small model uncertainty on the phytoplanton signal in the blue, the huge impact of additional $a_{gd}(\lambda)$ and the large satellite radiometric uncertainty clearly show the large degree of ambiguity and potential error in the retrieval of the phytoplankton component, even if the $a_{gd}(\lambda)$ is exactly known. Satellite-derived $a_{gd}(\lambda)$ (and CDOM) products have large uncertainties: r^2 of less than 0.25 for three different $a_{gd}(\lambda)$ algorithms against in situ data [16]—noting that dependence on the atmospheric correction means that a significant level of error is propagated through the algorithms from this source, particularly in the blue.

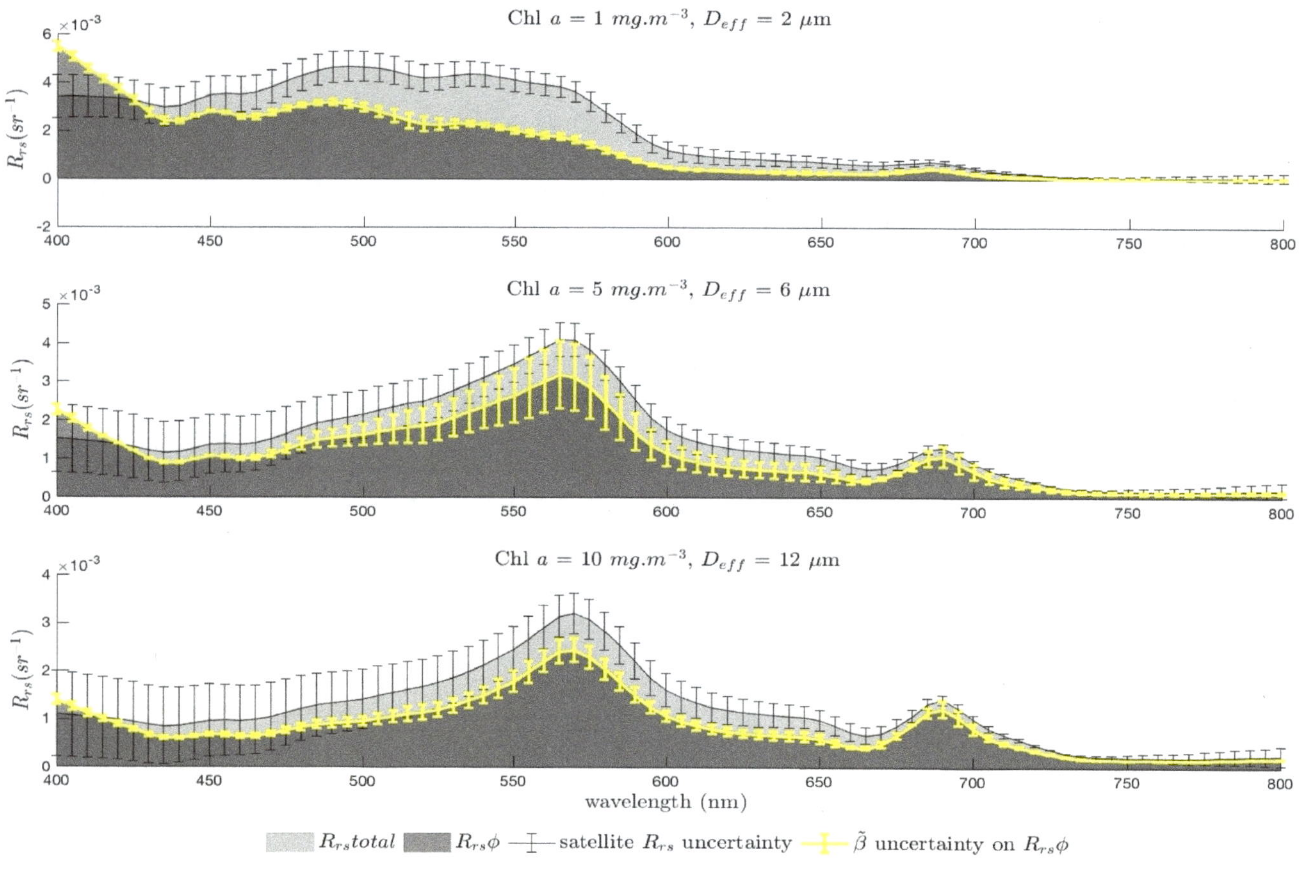

Figure A6. Total R_{rs} with satellite measurement uncertainties in the blue and red bands from [16] and linearly interpolated between them. An indication of model uncertainty on the $R_{rs}\phi$ is calculated by the spectral differences resulting from the use of a combined $b_{bp}(\lambda)$-specific Fournier Forand phase function independent of wavelength, vs. wavelength- and $b_{b\phi}(\lambda)$-dependent EAP phase functions.

The most significant spectral regions of Figure A6 in the context of this study are those where the uncertainty on $R_{rs}\phi$ overlap with the bulk satellite R_{rs} measurement uncertainty in each example. These are the spectral regions where the phytoplankton-specific signal dominates the bulk signal to the point that they are arguably indistinguishable, so these regions are particularly promising in terms of PFT detection from the bulk R_{rs}. It is also encouraging to note that the regions of maximum δD_{eff} previously identified fall within these regions, meaning that particularly close to 570 nm, the bulk signal not only closely reflects the causal phytoplankton signal but is also not very sensitive to reasonable variability in $a_{gd}(\lambda)$ and $b_{nap}(\lambda)$. However, it is an important consideration that these are also regions of large uncertainty in the size signal, and that, as D_{eff} increases, the R_{rs} expression of reduced phytoplankton scatter becomes more vulnerable to variable $b_{nap}(\lambda)$.

At low biomass, the phytoplankton signal falls well outside of the bulk measurement uncertainty, but the question of whether phytoplankton IOPs could be retrieved from the bulk depends on the resulting proportional contribution to the total. With reduced $b_b\phi/a\phi$, even small variability in the non-algal contribution to b_b/a results in signal ambiguity. In this case, the additional $a_{gd}(\lambda)$ and $b_{nap}(\lambda)$ contributions need to be exactly known, in order to be able to retrieve any PFT information.

It can also be observed in Figure A6 that the magnitude of model uncertainty is less, and the proportional contribution of phytoplankton to the bulk IOPs is greater, at wavelengths slightly shorter than those of the maximum δD_{eff} in the case studies. Thus, the spectral location of the largest observable δD_{eff} signal may not necessarily be the most revealing of PFT discrimination in terms of the associated uncertainties. A sophisticated uncertainty model would be necessary to calculate the respective advantages of reduced contribution uncertainties on a smaller signal vs. slightly larger uncertainties on a larger workable signal. It is also worth considering that, even where the bulk and $R_{rs}\phi$ signals are distinct, there are spectral regions where they are parallel i.e., maintain the same shape. It can be concluded that the phytoplankton contribution determines the spectral shape in these regions—although the uncertainty associated with a smooth $b_{bnap}(\lambda)$ curve is also not quantified here. This information could potentially be exploited to investigate PFT signal from the bulk R_{rs}.

Further work on incorporating EAP phase functions into Hydrolight has enabled the $R_{rs}\phi$ presented here to include the fluorescence term, and this is also a spectral region of a large proportional phytoplankton contribution together with small model uncertainty as calculated by the difference in approach to scattering phase functions. This region (around 685 nm) appears in the maximum δR_{rs} plots from the case studies, but has not been discussed as confidence in modelling this spectral region accurately needs to be improved with respect to natural variability in a fluorescence quantum yield and phytoplankton response to the light environment. However, it is known that this region holds further useful information on phytoplankton health [103] as well as size.

Overall, the uncertainties in both measured and modelled quantities should be considered in terms of the proportional contribution by phytoplankton. The highest proportional phytoplankton contribution to the bulk optics, and therefore the most promising signal for PFTs, occurs where elevated scatter due to biomass is complemented by the elevated scatter of small phytoplankton cells. Approaches to modelling the phase functions result in an inherent ambiguity of about 4 µm at very high biomass, but this drops with biomass and as D_{eff} increases.

References

1. Field, C.B.; Behrenfeld, M.J.; Randerson, J.T. Primary production of the biosphere: Integrating terrestrial and oceanic components. *Science* **1998**, *281*, 237–240. [CrossRef] [PubMed]
2. McClain, C.R. A decade of satellite ocean color observations. *Annu. Rev. Mar. Sci.* **2009**, *1*, 19–42. [CrossRef] [PubMed]
3. Swart, S.; Chang, N.; Fauchereau, N.; Joubert, W.; Lucas, M.; Mtshali, T.; Roychoudhury, A.; Tagliabue, A.; Thomalla, S.; Waldron, H.; et al. Southern Ocean Seasonal Cycle Experiment 2012: Seasonal scale climate and carbon cycle links. *S. Afr. J. Sci.* **2012**, *108*, 11–13. [CrossRef]
4. Thomalla, S.; Fauchereau, N.; Swart, S.; Monteiro, P. Regional scale characteristics of the seasonal cycle of chlorophyll in the Southern Ocean. *Biogeosciences* **2011**, *8*, 2849. [CrossRef]
5. Ryan-Keogh, T.J.; Thomalla, S.J.; Mtshali, T.N.; Little, H. Modelled estimates of spatial variability of iron stress in the Atlantic sector of the Southern Ocean. *Biogeosciences* **2017**, *14*, 3883–3897. [CrossRef]
6. Brewin, R.J.W.; Ciavatta, S.; Sathyendranath, S.; Jackson, T.; Tilstone, G.; Curran, K.; Airs, R.L.; Cummings, D.; Brotas, V.; Organelli, E.; et al. Uncertainty in Ocean-Color Estimates of Chlorophyll for Phytoplankton Groups. *Front. Mar. Sci.* **2017**, *4*, 104. [CrossRef]

7. Antoine, D.; d'Ortenzio, F.; Hooker, S.B.; Bécu, G.; Gentili, B.; Tailliez, D.; Scott, A.J. Assessment of uncertainty in the ocean reflectance determined by three satellite ocean color sensors (MERIS, SeaWiFS and MODIS-A) at an offshore site in the Mediterranean Sea (BOUSSOLE project). *J. Geophys. Res. Oceans* **2008**, *113*, 2156–2202. [CrossRef]
8. Sathyendranath, S.; Watts, L.; Devred, E.; Platt, T.; Caverhill, C.; Maass, H. Discrimination of diatoms from other phytoplankton using ocean-colour data. *Mar. Ecol. Prog. Ser.* **2004**, *272*, 59–68. [CrossRef]
9. Alvain, S.; Loisel, H.; Dessailly, D. Theoretical analysis of ocean color radiances anomalies and implications for phytoplankton groups detection in Case 1 waters. *Opt. Express* **2012**, *20*, 1070–1083. [CrossRef]
10. Kostadinov, T.; Siegel, D.; Maritorena, S. Retrieval of the particle size distribution from satellite ocean color observations. *J. Geophys. Res. Oceans* **2009**, *114*. [CrossRef]
11. Kostadinov, T.S. Carbon-based phytoplankton size classes retrieved via ocean color estimates of the particle size distribution. *Ocean Sci.* **2016**, *12*, 561. [CrossRef]
12. Anderson, T.R. Plankton functional type modelling: Running before we can walk? *J. Plankton Res.* **2005**, *27*, 1073–1081. [CrossRef]
13. Brown, C.A.; Huot, Y.; Werdell, P.J.; Gentili, B.; Claustre, H. The origin and global distribution of second order variability in satellite ocean color and its potential applications to algorithm development. *Remote Sens. Environ.* **2008**, *112*, 4186–4203. [CrossRef]
14. MÉlin, F.; Sclep, G.; Jackson, T.; Sathyendranath, S. Uncertainty estimates of remote sensing reflectance derived from comparison of ocean color satellite data sets. *Remote Sens. Environ.* **2016**, *177*, 107–124. [CrossRef]
15. Lee, Z.P. Remote Sensing of Inherent Optical Properties: Fundamentals, Tests of Algorithms and Applications. *Rep. Int. Ocean Colour Coord. Group* **2006**, *5*, 1–122.
16. Mélin, F.; Zibordi, G.; Berthon, J.F. Assessment of satellite ocean color products at a coastal site. *Remote Sens. Environ.* **2007**, *110*, 192–215. [CrossRef]
17. Mouw, C.B.; Hardman-Mountford, N.J.; Alvain, S.; Bracher, A.; Brewin, R.J.W.; Bricaud, A.; Ciotti, A.M.; Devred, E.; Fujiwara, A.; Hirata, T.; et al. A Consumer's Guide to Satellite Remote Sensing of Multiple Phytoplankton Groups in the Global Ocean. *Front. Mar. Sci.* **2017**, *4*, 41. [CrossRef]
18. Hirata, T.; Hardman-Mountford, N.J.; Brewin, R.J.W.; Aiken, J.; Barlow, R.; Suzuki, K.; Isada, T.; Howell, E.; Hashioka, T.; Noguchi-Aita, M.; et al. Synoptic relationships between surface Chlorophyll-a and diagnostic pigments specific to phytoplankton functional types. *Biogeosciences* **2011**, *8*, 311–327. [CrossRef]
19. Brewin, R.J.W.; Hardman-Mountford, N.J.; Lavender, S.J.; Raitsos, D.E.; Hirata, T.; Uitz, J.; Devred, E.; Bricaud, A.; Ciotti, A.; Gentili, B. An intercomparison of bio-optical techniques for detecting dominant phytoplankton size class from satellite remote sensing. *Remote Sens. Environ.* **2011**, *115*, 325–339. [CrossRef]
20. Brewin, R.J.W.; Sathyendranath, S.; Hirata, T.; Lavender, S.J.; Barciela, R.M.; Hardman-Mountford, N.J. A three-component model of phytoplankton size class for the Atlantic Ocean. *Ecol. Modell.* **2010**, *221*, 1472–1483. [CrossRef]
21. Alvain, S.; Moulin, C.; Dandonneau, Y.; Bréon, F.M. Remote sensing of phytoplankton groups in Case 1 waters from global SeaWiFS imagery. *Deep Sea Res. Part I Oceanogr. Res. Pap.* **2005**, *52*, 1989–2004. [CrossRef]
22. Alvain, S.; Moulin, C.; Dandonneau, Y.; Loisel, H. Seasonal distribution and succession of dominant phytoplankton groups in the global ocean: A satellite view. *Glob. Biogeochem. Cycles* **2008**, *22*, 1–15. [CrossRef]
23. Devred, E.; Sathyendranath, S.; Stuart, V.; Platt, T. A three component classification of phytoplankton absorption spectra: Application to ocean-color data. *Remote Sens. Environ.* **2011**, *115*, 2255–2266. [CrossRef]
24. Ciotti, A.M.; Bricaud, A. Retrievals of a size parameter for phytoplankton and spectral light absorption by colored detrital matter from water-leaving radiances at SeaWiFS channels in a continental shelf region off Brazil. *Limnol. Oceanogr. Methods* **2006**, *4*, 237–253. [CrossRef]
25. Bracher, A.; Vountas, M.; Dinter, T.; Burrows, J.; Röttgers, R.; Peeken, I. Quantitative observation of cyanobacteria and diatoms from space using PhytoDOAS on SCIAMACHY data. *Biogeosciences* **2009**, *6*, 751–764. [CrossRef]
26. Kostadinov, T.S.; Siegel, D.A.; Maritorena, S. Global variability of phytoplankton functional types from space: assessment via the particle size distribution. *Biogeosciences* **2010**, *7*, 3239–3257. [CrossRef]
27. Bernard, S.; Probyn, T.A.; Quirantes, A. Simulating the optical properties of phytoplankton cells using a two-layered spherical geometry. *Biogeosci. Discuss.* **2009**, *6*, 1497–1563. [CrossRef]

28. Uitz, J.; Stramski, D.; Reynolds, R.A.; Dubranna, J. Assessing phytoplankton community composition from hyperspectral measurements of phytoplankton absorption coefficient and remote-sensing reflectance in open-ocean environments. *Remote Sens. Environ.* **2015**, *171*, 58–74. [CrossRef]
29. Torrecilla, E.; Stramski, D.; Reynolds, R.A.; Millán-Núñez, E.; Piera, J. Cluster analysis of hyperspectral optical data for discriminating phytoplankton pigment assemblages in the open ocean. *Remote Sens. Environ.* **2011**, *115*, 2578–2593. [CrossRef]
30. Xi, H.; Hieronymi, M.; Röttgers, R.; Krasemann, H.; Qiu, Z. Hyperspectral differentiation of phytoplankton taxonomic groups: A comparison between using remote sensing reflectance and absorption spectra. *Remote Sens.* **2015**, *7*, 14781–14805. [CrossRef]
31. Sadeghi, A.; Dinter, T.; Vountas, M.; Taylor, B.; Altenburg-Soppa, M.; Bracher, A. Remote sensing of coccolithophore blooms in selected oceanic regions using the PhytoDOAS method applied to hyper-spectral satellite data. *Biogeosciences* **2012**, *9*, 2127–2143. [CrossRef]
32. Stramski, D.; Bricaud, A.; Morel, A. Modeling the inherent optical properties of the ocean based on the detailed composition of the planktonic community. *Appl. Opt.* **2001**, *40*, 2929–2945. [CrossRef] [PubMed]
33. Matsuoka, A.; Huot, Y.; Shimada, K.; Saitoh, S.I.; Babin, M. Bio-optical characteristics of the western Arctic Ocean: Implications for ocean color algorithms. *Can. J. Remote Sens.* **2007**, *33*, 503–518. [CrossRef]
34. Bernard, S.; Shillington, F.A.; Probyn, T.A. The use of equivalent size distributions of natural phytoplankton assemblages for optical modeling. *Opt. Express* **2007**, *15*, 1995–2007. [CrossRef] [PubMed]
35. Lain, L.R.; Bernard, S.; Matthews, M.W. Understanding the contribution of phytoplankton phase functions to uncertainties in the water colour signal. *Opt. Express* **2017**, *25*, A151–A165. [CrossRef] [PubMed]
36. Morel, A. Consequences of a Synechococcus bloom upon the optical properties of oceanic (case 1) waters. *Limnol. Oceanogr.* **1997**, *42*, 1746–1754. [CrossRef]
37. Evers-King, H.; Bernard, S.; Lain, L.R.; Probyn, T.A. Sensitivity in reflectance attributed to phytoplankton cell size: Forward and inverse modelling approaches. *Opt. Express* **2014**, *22*, 11536–11551. [CrossRef]
38. Organelli, E.; Nuccio, C.; Lazzara, L.; Uitz, J.; Bricaud, A.; Massi, L. On the discrimination of multiple phytoplankton groups from light absorption spectra of assemblages with mixed taxonomic composition and variable light conditions. *Appl. Opt.* **2017**, *56*, 3952. [CrossRef]
39. Defoin-Platel, M.; Chami, M. How ambiguous is the inverse problem of ocean color in coastal waters? *J. Geophys. Res. Oceans* **2007**, *112*. [CrossRef]
40. Kirk, J. A theoretical analysis of the contribution of algal cells to the attenuation of light within natural waters I. General treatment of suspensions of pigmented cells. *New Phytol.* **1975**, *75*, 11–20. [CrossRef]
41. Morel, A.; Bricaud, A. Theoretical results concerning light absorption in a discrete medium, and application to specific absorption of phytoplankton. *Deep Sea Res.* **1981**, *28*, 1375–1393. [CrossRef]
42. Sathyendranath, S.; Lazzara, L.; Prieur, L. Variations in the spectral values of specific absorption of phytoplankton. *Limnol. Oceanogr.* **1987**, *32*, 403–415. [CrossRef]
43. Bricaud, A.; Bédhomme, A.; Morel, A. Optical properties of diverse phytoplanktonic species: Experimental results and theoretical interpretation. *J. Plankton Res.* **1988**, *10*, 851–873. [CrossRef]
44. Ahn, Y.H.; Bricaud, A.; Morel, A. Light backscattering efficiency and related properties of some phytoplankters. *Deep Sea Res. Part A Oceanogr. Res. Pap.* **1992**, *39*, 1835–1855. [CrossRef]
45. Bricaud, A.; Stramski, D. Spectral absorption coefficients of living phytoplankton and nonalgal biogenous matter: A comparison between the Peru upwelling areaand the Sargasso Sea. *Limnol. Oceanogr.* **1990**, *35*, 562–582. [CrossRef]
46. Le Quéré, C.; Harrison, S.P.; Prentice, I.C.; Buitenhuis, E.T.; Aumont, O.; Bopp, L.; Claustre, H.; Cotrim Da Cunha, L.; Geider, R.; Giraud, X.; et al. Ecosystem dynamics based on plankton functional types for global ocean biogeochemistry models. *Glob. Chang. Biol.* **2005**, *11*, 2016–2040.
47. Nair, A.; Sathyendranath, S.; Platt, T.; Morales, J.; Stuart, V.; Forget, M.H.; Devred, E.; Bouman, H. Remote sensing of phytoplankton functional types. *Remote Sens. Environ.* **2008**, *112*, 3366–3375. [CrossRef]
48. IOCCG. Phytoplankton Functional Types from Space. In *Reports of the International Ocean Colour Coordinating Group*; IOCCG: Dartmouth, NS, Canada, 2014.
49. Dutkiewicz, S.; Hickman, A.; Jahn, O.; Gregg, W.; Mouw, C.; Follows, M. Capturing optically important constituents and properties in a marine biogeochemical and ecosystem model. *Biogeosciences* **2015**, *12*, 4447–4481. [CrossRef]

50. Moutier, W.; Duforet-Gaurier, L.; Thyssen, M.; Loisel, H.; Meriaux, X.; Courcot, L.; Dessailly, D.; Reve, A.H.; Gregori, G.; Alvain, S.; et al. Evolution of the scattering properties of phytoplankton cells from flow cytometry measurements. *PLoS ONE* **2017**, *12*, e0181180. [CrossRef]
51. Stramski, D.; Morel, A. Optical properties of photosynthetic picoplankton in different physiological states as affected by growth irradiance. *Deep Sea Res.* **1990**, *37*, 245–266. [CrossRef]
52. Reynolds, R.A.; Stramski, D.; Kiefer, D.A. The effect of nitrogren limitation on the absorption and scattering properties of the marine diatom Thalassiosira pseudonana. *Limnol. Oceanogr.* **1997**, *42*, 881–892. [CrossRef]
53. Stramski, D.; Sciandra, A.; Claustre, H. Effects of temperature, nitrogen, and light limitation on the optical properties of the marine diatom Thalassiosira pseudonana. *Limnol. Oceanogr.* **2002**, *47*, 392–403. [CrossRef]
54. Stramski, D.; Reynolds, R.A. Diel variations in the optical properties of a marine diatom. *Limnol. Oceanogr.* **1993**, *38*, 1347–1364. [CrossRef]
55. Stramski, D.; Shalapyonok, A.; Reynolds, R.A. Optical characterization of the oceanic unicellular cyanobacterium Synechococcus grown under a day-night cycle in natural irradiance. *J. Geophys. Res. Oceans* **1995**, *100*, 13295–13307. [CrossRef]
56. Sauer, M.J.; Roesler, C.; Werdell, P.; Barnard, A. Under the hood of satellite empirical chlorophyll a algorithms: Revealing the dependencies of maximum band ratio algorithms on inherent optical properties. *Opt. Express* **2012**, *20*, 20920–20933. [CrossRef] [PubMed]
57. Sathyendranath, S.; Brewin, R.J.; Jackson, T.; Mélin, F.; Platt, T. Ocean-colour products for climate-change studies: What are their ideal characteristics? *Remote Sens. Environ.* **2017**, *203*, 125–138. [CrossRef]
58. Mobley, C.D.; Stramski, D. Effects of microbial particles on oceanic optics: Methodology for radiative transfer modeling and example simulations. *Limnol. Oceanogr.* **1997**, *42*, 550–560. [CrossRef]
59. Stramski, D.; Mobley, C.D. Effects of microbial particles on oceanic optics: A database of single-particle optical properties. *Limnol. Oceanogr.* **1997**, *42*, 538–549. [CrossRef]
60. Morel, A.; Prieur, L. Analysis of variations in ocean color. *Limnol. Oceanogr.* **1977**, *22*, 709–722. [CrossRef]
61. Hoepffner, N.; Sathyendranath, S. Effect of pigment composition on absorption properties of phytoplankton. *Mar. Ecol. Prog. Ser.* **1991**, 11–23. [CrossRef]
62. Olson, R.; Zettler, E.; Anderson, O. Discrimination of eukaryotic phytoplankton cell types from light scatter and autofluorescence properties measured by flow cytometry. *Cytom. Part A* **1989**, *10*, 636–643. [CrossRef] [PubMed]
63. Matthews, M.W.; Bernard, S. Using a two-layered sphere model to investigate the impact of gas vacuoles on the inherent optical properties of M. aeruginosa. *Biogeosciences* **2013**, *10*, 8139–8157. [CrossRef]
64. Smith, M.E.; Lain, L.R.; Bernard, S. An optimized Chlorophyll a switching algorithm for MERIS and OLCI in phytoplankton-dominated waters. *Remote Sens. Environ.* **2018**, *215*, 217–227. [CrossRef]
65. Lain, L.R.; Bernard, S.; Evers-King, H. Biophysical modelling of phytoplankton communities from first principles using two-layered spheres: Equivalent Algal Populations (EAP) model. *Opt. Express* **2014**, *22*, 16745–16758. [CrossRef] [PubMed]
66. Stramski, D.; Boss, E.; Bogucki, D.; Voss, K. The role of seawater constituents in light backscattering in the ocean. *Prog. Oceanogr.* **2004**, *61*, 27–56. [CrossRef]
67. Toll, J.S. Causality and the Dispersion Relation: Logical Foundations. *Phys. Rev.* **1956**, *104*, 1760–1770. [CrossRef]
68. Aas, E. Refractive index of phytoplankton derived from its metabolite composition. *J. Plankton Res.* **1996**, *18*, 2223–2249. [CrossRef]
69. Bernard, S.; Probyn, T.; Barlow, R. Measured and modelled optical properties of particulate matter in the southern Benguela. *S. Afr. J. Sci.* **2001**, *97*, 410–420.
70. Stramski, D. Refractive index of planktonic cells as a measure of cellular carbon and chlorophyll a content. *Deep Sea Res. Part I Oceanogr. Res. Pap.* **1999**, *46*, 335–351. [CrossRef]
71. Johnsen, G.; Samset, O.; Granskog, L.; Sakshaug, E. In vivo absorption characteristics in 10 classes of bloom-forming phytoplankton: Taconomic characteristics and responses to photoadaptation by means of discriminant and HPLC analysis. *Mar. Ecol. Prog. Ser.* **1994**, 149–157. [CrossRef]
72. Bricaud, A.; Babin, M.; Morel, A.; Claustre, H. Variability in the chlorophyll-specific absorption coefficients of natural phytoplankton: Analysis and parameterization. *J. Geophys. Res. Oceans* **1995**, *100*, 13321–13332. [CrossRef]

73. Constable, A.J.; Melbourne-Thomas, J.; Corney, S.P.; Arrigo, K.R.; Barbraud, C.; Barnes, D.; Bindoff, N.L.; Boyd, P.W.; Brandt, A.; Costa, D.P.; et al. Climate change and Southern Ocean ecosystems I: How changes in physical habitats directly affect marine biota. *Glob. Chang. Biol.* **2014**, *20*, 3004–3025. [CrossRef] [PubMed]
74. Mtshali, T.N. *SANAE 55 Cruise Report*; South African National Antarctic Programme (SANAP): Cape Town, South Africa, 2016.
75. Del Castillo, C.E.; Miller, R.L. Horizontal and vertical distributions of colored dissolved organic matter during the Southern Ocean Gas Exchange Experiment. *J. Geophys. Res. Oceans* **2011**, *116*. [CrossRef]
76. Reynolds, R.; Stramski, D.; Mitchell, B. A chlorophyll-dependent semianalytical reflectance model derived from field measurements of absorption and backscattering coefficients within the Southern Ocean. *J. Geophys. Res. Oceans* **2001**, *106*, 7125–7138. [CrossRef]
77. Gustafson D.E., Jr.; Stoecker, D.K.; Johnson, M.D.; Van Heukelem, W.F.; Sneider, K. Cryptophyte algae are robbed of their organelles by the marine ciliate Mesodinium rubrum. *Nature* **2000**, *405*, 1049. [CrossRef] [PubMed]
78. Stramska, M.; Stramski, D.; Mitchell, B.G.; Mobley, C.D. Estimation of the absorption and backscattering coefficients from in water radiometric measurements. *Limnol. Oceanogr.* **2000**, *45*, 628–641. [CrossRef]
79. Oubelkheir, K.; Claustre, H.; Bricaud, A.; Babin, M. Partitioning total spectral absorption in phytoplankton and colored detrital material contributions. *Limnol. Oceanogr. Methods* **2007**, *5*, 384–395. [CrossRef]
80. Chami, M.; McKee, D.; Leymarie, E.; Khomenko, G. Influence of the angular shape of the volume-scattering function and multiple scattering on remote sensing reflectance. *Appl. Opt.* **2006**, *45*, 9210–9220. [CrossRef] [PubMed]
81. Gordon, H.R. Atmospheric correction of ocean color imagery in the Earth Observing System era. *J. Geophys. Res. Atmos.* **1997**, *102*, 17081–17106. [CrossRef]
82. Boyd, P.; Ellwood, M. The biogeochemical cycle of iron in the ocean. *Nat. Geosci.* **2010**, *10*, 675–682. [CrossRef]
83. Bernard, S.; Pitcher, G.; Evers-King, H.; Robertson, L.; Matthews, M.; Rabagliati, A.; Balt, C. Ocean Colour Remote Sensing of Harmful Algal Blooms in the Benguela System. In *Remote Sensing of the African Seas*; Springer: Berlin, Germany, 2014; pp. 185–203.
84. Devred, E.; Sathyendranath, S.; Stuart, V.; Maass, H.; Ulloa, O.; Platt, T. A two-component model of phytoplankton absorption in the open ocean: Theory and applications. *J. Geophys. Res. Oceans* **2006**, *111*. [CrossRef]
85. Tan, H.; Oishi, T.; Tanaka, A.; Doerffer, R. Accurate estimation of the backscattering coefficient by light scattering at two backward angles. *Appl. Opt.* **2015**, *54*, 7718–7733. [CrossRef] [PubMed]
86. Harmel, T.; Hieronymi, M.; Slade, W.; Röttgers, R.; Roullier, F.; Chami, M. Laboratory experiments for inter-comparison of three volume scattering meters to measure angular scattering properties of hydrosols. *Opt. Express* **2016**, *24*, 234–256. [CrossRef] [PubMed]
87. Bricaud, A.; Claustre, H.; Ras, J.; Oubelkheir, K. Natural variability of phytoplanktonic absorption in oceanic waters: Influence of the size structure of algal populations. *J. Geophys. Res. Oceans* **2004**, *109*. [CrossRef]
88. Dierssen, H.M.; Kudela, R.M.; Ryan, J.P. Red and black tides: Quantitative analysis of water-leaving radiance and perceived color for phytoplankton, colored dissolved organic matter, and suspended sediments. *Limnol. Oceanogr.* **2006**, *51*, 2646–2659. [CrossRef]
89. Quirantes, A.; Bernard, S. Light-scattering methods for modelling algal particles as a collection of coated and/or nonspherical scatterers. *J. Quant. Spectrosc. Radiat. Transf.* **2006**, *100*, 315–324. [CrossRef]
90. Ishikawa, A.; Wright, S.W.; van den Enden, R.; Davidson, A.T.; Marchant, H.J. Abundance, size structure and community composition of phytoplankton in the Southern Ocean in the austral summer 1999/2000. *Pol. Biosci.* **2002**, *15*, 11–26.
91. Wright, S.; Thomas, D.; Marchant, H.; Higgins, H.; Mackey, M.; Mackey, D. Analysis of phytoplankton of the Australian sector of the Southern Ocean: Comparisons of microscopy and size frequency data with interpretations of pigment HPLC data using the\'CHEMTAX\'matrix factorisation program. *Mar. Ecol. Prog. Ser.* **1996**, *144*, 285–298. [CrossRef]
92. Wright, S.W.; Jeffrey, S. Fucoxanthin pigment markers of marine phytoplankton analysed by HPLC and HPTLC. *Mar. Ecol. Prog. Ser.* **1987**, 259–266. [CrossRef]
93. Bricaud, A.; Morel, A.; Prieur, L. Absorption by dissolved organic matter of the sea (yellow substance) in the UV and visible domains. *Limnol. Oceanogr.* **1981**, *26*, 43–53. [CrossRef]

94. Roesler, C.S.; Perry, M.J.; Carder, K.L. Modeling in situ phytoplankton absorption from total absorption spectra in productive inland marine waters. *Limnol. Oceanogr.* **1989**, *34*, 1510–1523. [CrossRef]
95. Bernard, S.; Probyn, T.A.; Shillington, F.A. Towards the validation of SeaWiFS in southern African waters: The effects of gelbstoff. *S. Afr. J. Mar. Sci.* **1998**, *19*, 15–25. [CrossRef]
96. Roesler, C.S.; Perry, M.J. In situ phytoplankton absorption, fluorescence emission, and particulate backscattering spectra determined from reflectance. *J. Geophys. Res.* **1995**, *100*, 13279–13294. [CrossRef]
97. Chami, M.; Shybanov, E.B.; Khomenko, G.A.; Lee, M.E.G.; Martynov, O.V.; Korotaev, G.K. Spectral variation of the volume scattering function measured over the full range of scattering angles in a coastal environment. *Appl. Opt.* **2006**, *45*, 3605–3619. [CrossRef] [PubMed]
98. Twardowski, M.S.; Boss, E.; Macdonald, J.B.; Pegau, W.S.; Barnard, A.H.; Zaneveld, J.R.V. Model for estimating bulk refractive index from the optical backscattering ratio and the implications for understanding particle composition in case I and case II waters. *J. Geophys. Res* **2001**, *106*, 14129–14142. [CrossRef]
99. Boss, E.; Pegau, W.; Lee, M.; Twardowski, M.; Shybanov, E.; Korotaev, G.; Baratange, F. Particulate backscattering ratio at LEO 15 and its use to study particle composition and distribution. *J. Geophys. Res. Oceans* **2004**, *109*. [CrossRef]
100. Holm-Hansen, O.; Lorenzen, C.J.; Holmes, R.W.; Strickland, J.D. Fluorometric determination of chlorophyll. *ICES J. Mar. Sci.* **1965**, *30*, 3–15. [CrossRef]
101. Behrenfeld, M.J.; Westberry, T.K.; Boss, E. Satellite-detected fluorescence reveals global physiology of ocean phytoplankton. *Biogeosciences* **2009**, 779–794. [CrossRef]
102. Ostrowska, M.; Woźniak, B.; Dera, J. Modelled quantum yields and energy efficiency of fluorescence photosynthesis and heat production by phytoplankton in the World Ocean. *Oceanologia* **2012**, *54*, 565–610. [CrossRef]
103. Greene, R.M.; Geider, R.J.; Kolber, Z.; Falkowski, P.G. Iron-induced changes in light harvesting and photochemical energy conversion processes in eukaryotic marine algae. *Plant Physiol.* **1992**, *100*, 565–575. [CrossRef]

10

Assessing Fluorescent Organic Matter in Natural Waters: Towards In Situ Excitation–Emission Matrix Spectroscopy

Oliver Zielinski [1,*], Nick Rüssmeier [1,2], Oliver D. Ferdinand [1], Mario L. Miranda [1,3] and Jochen Wollschläger [1,4]

1 Center for Marine Sensors, Institute for Chemistry and Biology of the Marine Environment, University Oldenburg, 26382 Wilhelmshaven, Germany; nick.ruessmeier@uol.de (N.R.); oliver.ferdinand@uol.de (O.D.F.); mario.luis.miranda.montenegro@uol.de (M.L.M.); jochen.wollschlaeger@uol.de (J.W.)
2 OFFIS e.V. Institute for Information Technology, 26121 Oldenburg, Germany
3 Laboratory of Air and Water Quality, University of Panama, Panamá 4, Panamá
4 Helmholtz-Zentrum Geesthacht, Institute of Coastal Research, Max-Planck-Str. 1, 21502 Geesthacht
* Correspondence: oliver.zielinski@uol.de

Featured Application: This proof-of-concept work of in situ EEMS might find applications on different autonomous platforms in biogeochemical observing systems for the marine environment.

Abstract: Natural organic matter (NOM) is a key parameter in aquatic biogeochemical processes. Part of the NOM pool exhibits optical properties, namely absorption and fluorescence. The latter is frequently utilized in laboratory measurements of its dissolved fraction (fluorescent dissolved organic matter, FDOM) through excitation–emission matrix spectroscopy (EEMS). We present the design and field application of a novel EEMS sensor system applicable in situ, the 'Kallemeter'. Observations are based on a field campaign, starting in Norwegian coastal waters entering the Trondheimsfjord. Comparison against the bulk fluorescence of two commercial FDOM sensors exhibited a good correspondence of the different methods and the ability to resolve gradients and dynamics along the transect. Complementary laboratory EEM spectra measurements of surface water samples and their subsequent PARAFAC analysis revealed three dominant components while the 'Kallemeter' EEMS sensor system was able to produce reasonable EEM spectra in high DOM concentrated water bodies, yet high noise levels must be addressed in order to provide comparable PARAFAC components. Achievements and limitations of this proof-of-concept are discussed providing guidance towards full in situ EEMS measurements to resolve rapid changes and processes in natural waters based on the assessment of spectral properties. Their combination with multiwavelength FDOM sensors onboard autonomous platforms will enhance our capacities in observing biogeochemical processes in the marine environment in spatiotemporal and spectral dimensions.

Keywords: natural organic matter; DOM; FDOM; CDOM; Gelbstoff; EEMS; PARAFAC; marine sensors; Kallemeter; FerryBox; Trondheimsfjord; Norway

1. Introduction

The investigation of natural organic matter (NOM) and especially its dissolved fraction (DOM) is of high relevance in aquatic sciences. Studies of NOM and DOM include their sinks and sources as part of the global carbon cycle [1–4], its stability and degradability by abiotic and biotic processes [5,6], as well as natural and anthropogenic ecosystem effects e.g., from hydrocarbons [7]. Complementary to laboratory approaches, the need to measure DOM directly in the water column (in situ) initiated the development of field-applicable optical sensors that can be attached to autonomous vehicles or ship-based platforms [8]. Addressing the optical properties, namely absorption and fluorescence, of certain DOM fractions in situ sensors enable a high-resolution assessment of these ecosystem-relevant parameters or proxies on different observation platforms [9,10]. The absorption of light from fractions of the DOM pool is a long known feature, which was initially investigated by the German chemist Kurt Kalle in the early 20th century [11]. Observing an increased absorption of light towards the blue and ultraviolet (UV) part of the spectrum, he denoted this fraction of DOM as 'Gelbstoff' or 'yellow substance' [12,13], apparently inspired by the yellowish color of the samples taken from the Baltic Sea. Nowadays, this fraction is known as colored or chromophoric dissolved organic matter, short CDOM [14,15]. The measurement principle for CDOM absorption (an inherent optical property, or IOP, of natural water bodies) is based on the reduction of light intensity from a light source to a detector over a certain distance. It typically requires a filtered water sample, since only the dissolved and not the particulate fraction is of interest, and the latter could introduce errors due to light scattering. While in situ filtration is technically possible and practiced e.g., in underway systems [16], it is rather sophisticated and therefore not realized in today's commercially submersible CDOM sensors [17]. For a feasible, but still technically complex, approach to derive CDOM absorption in subsea or underway applications, the scattering loss of photons is suppressed by reflective tubes [18] or a reflecting cavity [19–21]. Yet in the majority of today's observing platforms and ship operated sensor systems, CDOM absorption measurements are not implemented.

The most common sensors within CTD-samplers, FerryBox systems, autonomous underwater vehicles (AUVs), gliders, biogeochemical Argo floats and alike are CDOM fluorometers. Applying a UV light source and detecting an emitted fluorescence signal at a higher wavelength (lower energy), these sensors utilize a principle that Kalle once described as 'skyblue fluorescence' [12,22]. In addition to these single-channel sensors, a few sensor systems commercially exist that enable a selection of several wavelengths pairs [17]. Recently, a matrix fluorescence sensor was presented that can sense a set of 4 detection wavelengths for three different excitation wavelengths, thus spanning a matrix of 12 combinations [23]. While CDOM fluorescence is technically speaking a correct term, it can be mixed up with CDOM (by definition an absorption property) and reports or commercial product descriptions sometimes tend to incorrectly shorten measurement devices as 'CDOM sensors'. To reflect the fact that the fluorescing part of the DOM is a subfraction of the absorbing fraction of DOM, we herein use FDOM as a designator for fluorescent DOM, as common in recent literature [9,24–26]. Laboratory measurement of FDOM is typically performed with UV–VIS spectrofluorometers [27,28]. They enable a free selection of excitation and detection wavelengths and thus different forms of results: (a) emission spectra (with a fixed excitation wavelength); (b) excitation spectra (with a fixed emission wavelength); (c) synchronous scans (excitation and emission changing at the same rate); and (d) excitation–emission matrices (EEM). The latter provides a full scan of a range of excitation wavelengths, performing for each of these excitation wavelengths a full emission scan. The resulting matrix of fluorescence intensities, sometimes referred to as 'optical fingerprint', is reflecting all fluorescing components and its peaks are associated with different components from the DOM pool [14,29]. A classical labeling of the fluorophores in DOM identifies five main peaks A, C, B, T, and M. Those peaks are extensively used to identify the main composition and origin of FDOM. Previous reports indicate the correlation between microbial presence and freshly produced DOM (peaks M, B and T), while allochthonous DOM has been associated to the peaks A and C [29]. EEM spectra are frequently used in recent literature [27,28] and several parameters are calculated from them, including the humification index [30], biological index [31], and recently produced material index [32]. A statistical analysis of EEM spectra is often performed by parallel factor analysis (PARAFAC), a method that identifies the underlying components from a set of spectra [33,34]. EEM spectroscopy (EEMS) and its spectral analysis can be considered the state-of-the-art in FDOM assessment, however, to the best of our knowledge, no submersible sensor exists that enables the high-resolution (in terms of wavelengths intervals) measurement of excitation–emission matrix spectra with in situ equipment.

Herein, we will present (i) design and application of a novel in situ EEMS sensor system (in memory of Kurt Kalle named the 'Kallemeter') that enables full optical fingerprints in a submersible design, and compare these results with (ii) flow-through data from two commercial FDOM sensors, and (iii) spectral components identified in surface water samples from a laboratory UV–VIS spectrofluorometer, based on a field campaign in the Trondheimsfjord, Norway.

2. Materials and Methods

The following section starts with a detailed description of the design of the novel 'Kallemeter' in situ EEMS sensor system, including initial laboratory measurements to illustrate its performance. To enable a comparison of bulk fluorescence detected, two commercial FDOM sensors will briefly be introduced, both operated in a commercially available flow-through sensor system (FerryBox, -4H-JENA engineering, Jena, Germany). This will be followed by a description of the laboratory methods applied on discrete water samples, and the subsequently used statistical method to derive the components of the EEM spectra observed. Finally, we will provide information about the study site itself.

2.1. Design of an In Situ Capable EEMS Sensor System

The newly-developed submersible EEMS sensor system provides comprehensive excitation–emission matrices for in situ fluorescence measurements of organic matter in natural waters. This subchapter presents its design, beginning with a representation of the operation mode in Section 2.1.1. Followed by this, a description of the technical implementation and the system design is given in Section 2.1.2. Section 2.1.3 presents the evaluation of the system performance.

2.1.1. Submersible EEMS Sensor System and Modes of Operation

Investigations on fluorescent organic matter can benefit from spatiotemporal information that enable a comprehensive interpretation of the dataset in context of environmental conditions. The 'Kallemeter' EEMS sensor system supports in situ explorations with three common modes of operation for field studies: vertical profiling, underway sensing (e.g., installed in a moonpool of a vessel), and moored operation (e.g., integrated to a subsea platform). Figure 1 shows the deployment of the EEMS sensor system in profiling mode (Figure 1a) and the deployment in a moonpool of a research vessel for in situ underway investigations (Figure 1b,c). The submersible design offers automated data acquisition from water samples to a maximum water depth of 200 m in a freely selectable time interval, without the support of an operator. Once installed, autonomous long-term investigations from days to weeks are possible, depending on biofouling [35] of the optical sensor unit and sediment load of the water filtration unit.

2.1.2. EEMS Sensor System Design and Technical Implementation

The entire EEMS sensor system is designed for a stand-alone operation as an underwater unit that only needs to be supplied with electrical power from an external power source. In this way, a moored operation with an external 12 V DC battery pack can also be realized. In case of shipborne installations like profiling or underway measurements, the system can be connected to an on-board unit via a 200 m seaworthy cable, as illustrated in Figure 2a. The on-board unit provides 130 V DC power supply and data transmission to the 'Kallemeter' via Ethernet. This gives the opportunity to connect an external PC for remote control of the underwater unit, automatic data backup, or real-time visualization of the in situ measurements. In addition to the in situ fluorescence spectrometer unit, the 'Kallemeter' was equipped with a turbidity meter (STM/bh, Seapoint Sensors Inc., Exeter, NH, USA) and a CTD (Conductivity, Temperature, Depth) (miniCTD probe, ASD Sensortechnik GmbH, Trappenkamp, Germany). The 'Kallemeter' housing is made of a solid PEEK (polyether ether ketone) tube with steel plates at its ends, which also support the dissipation of thermal energy to the aquatic environment through internally forced convection, as this is important to provide long-term stability of the measurement system (light source and optical measurement devices). The EEMS sensor system is controlled through an integrated Microsoft Windows 7 PC, digital I/O converters, and LabVIEW software (version 2013, National Instruments, Austin, TX, USA), using a custom-made control program (named 'EEMsea'). The latter allows for comprehensive adjustments of the measurement procedures, like EEM wavelengths or control of the sampling interval. All EEM datasets are stored in a comma-separated value (CSV) file format and include system- and sampling-relevant metadata in the header (e.g., sampling-identifier, internal temperature, and coefficients), which are assigned to the standardized file structure from PerkinElmer Luminescence Spectrometers ([Info], [Setup], [SampleInfo], [SampleData]; PerkinElmer, Waltham, MA, USA). Therefore, postprocessing and statistical analysis of EEM spectra can easily be applied by standard third party software or Matlab Toolboxes (The MathWorks, Natick, MA, USA) e.g., for PARAFAC [27].

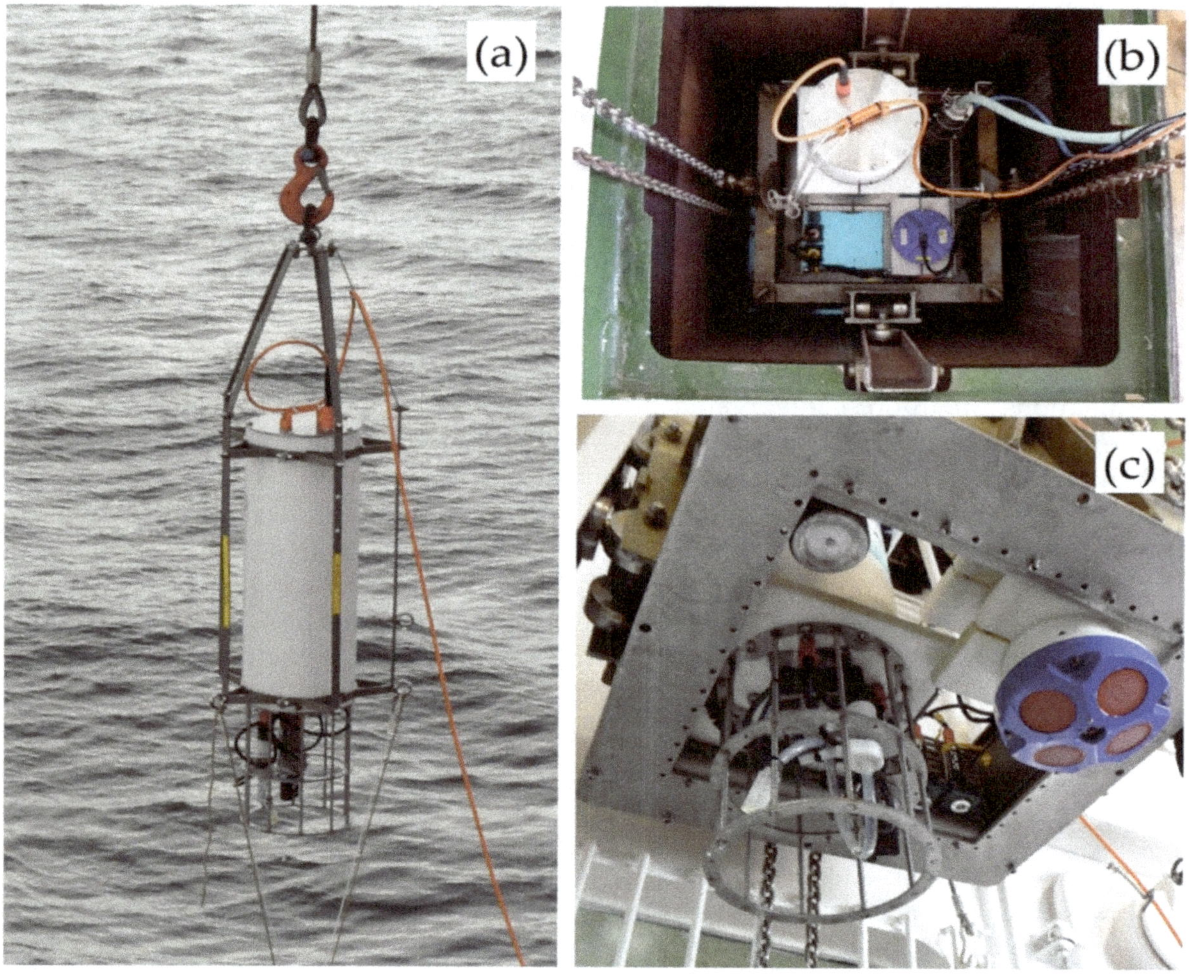

Figure 1. (**a**) Excitation–emission matrix spectroscopy (EEMS) sensor system in profiling mode on a winch of a research vessel. (**b**) EEMS sensor system, installation for in situ underway observations in the moonpool of the research vessel, top-view. Close to the EEMS sensor system, an acoustic Doppler current profiler (ADCP) as well as the water intake pump for the FerryBox system were installed in the moonpool-frame. (**c**) View from below.

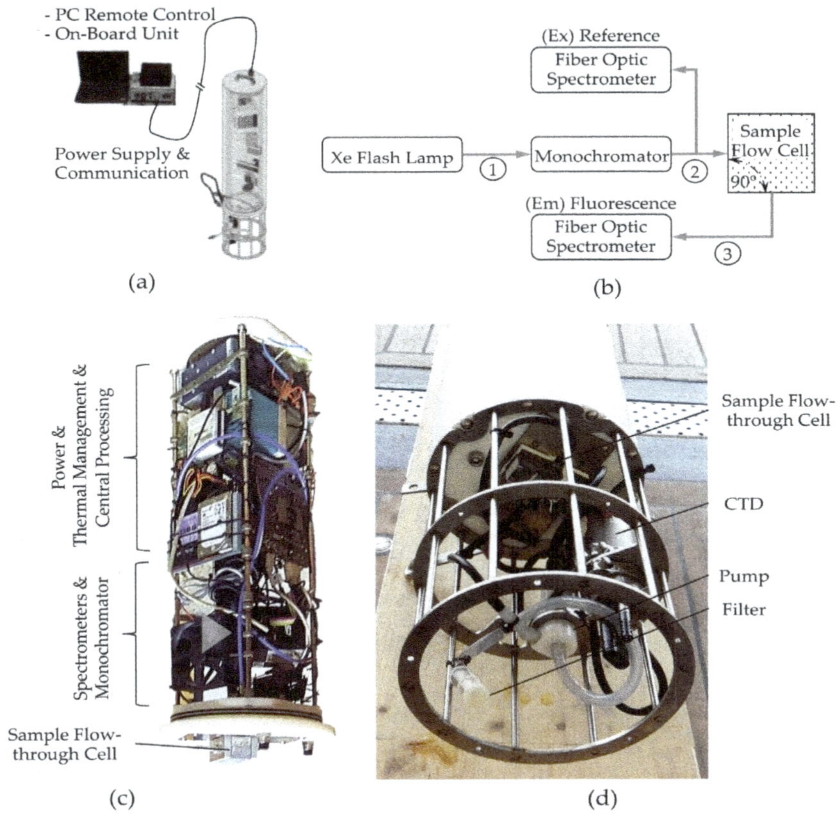

Figure 2. System components of the submersible EEMS sensor system. (**a**) System connected to an on-board unit and remote PC-control (optional). (**b**) Schematic structure of excitation–emission matrices (EEM) fluorescence spectrometer hardware and optical components (connected by optical fibers 1, 2, 3). (**c**) Internal close-up with electronics and optical components. (**d**) Close-up of the outer components of the EEMS sensor system.

The entire EEM fluorescence spectrometer design, which is presented in Figure 2b, consists of the following fluorescence measurement components: In the fluorescence excitation strand, a Xenon flash lamp (PX-2, Ocean Optics, Largo, FL, USA) is used as a broadband light source, which is coupled to a monochromator (CM 110, Spectral Products, Putnam, CT, USA) via an optical fiber (1). The output of the monochromator is coupled via an optical Y-fiber (2) to the reference spectrometer (USB2000+, Ocean Optics, Largo, FL, USA). Through this, a reference measurement of the excitation spectrum is realized. In parallel, the optical Y-fiber (2) connects the monochromator output to a flow-through sample cell in which the sample medium is located. The fluorescence emission strand consists of the detection spectrometer (QE-Pro, Ocean Optics, Largo, FL, USA) coupled via an optical fiber (3) to the sample flow-through cell. In the following, details of the fluorescence measuring system and its components are presented.

Fluorescence Excitation Light Source

Some in situ sensor designs use Xenon flash (or pulsed) lamps due to their high energy output in the ultraviolet (UV) and visible regions of the light spectrum as well as their long lifespan [36]. For these reasons, the 'Kallemeter' also utilizes a Xenon flash lamp (PX-2, Ocean Optics, Largo, FL, USA), which emits pulsed light with a continuous spectrum in a fluorescence relevant excitation wavelength range from 220 nm to 750 nm. A pulse power of maximum 45 microjoules/pulse (with 5 µs pulse duration at 1/3 height of pulse) and a lifetime of 10^9 pulses are expected for this 'expendable part' (estimated 230 days continuous operation @ 50 Hz pulse rate). A stable output from pulse to pulse is guaranteed for up to 100 Hz repetition rate, so the flash delay results in 10 milliseconds. Since the total intensity typically decreases slowly over the operating period [37], a measurement of the excitation spectrum is implemented to provide metadata of the EEMS sensor system status (presented in the subsection "Quality status of the excitation wavelengths").

Control of the Fluorescence Excitation Band

A monochromator (Type CM 110, Spectral Products, Putnam, CT, USA) is used to separate a narrow excitation band from the spectrum of the Xenon flash lamp. A full width at half maximum (FWHM) effective bandwidth of 10 nm is realized (therefore no slits are used) and the excitation wavelengths are selectable from 200 nm to 750 nm in 1 nm steps (with ruled grating AG1200-00300-303). The monochromator provides a coupling and decoupling angle of 14.8° and results to 0.13 numerical aperture (NA).

Quality Status of the Excitation Wavelengths

A wavelength-calibrated reference spectrometer (USB2000+, Ocean Optics, Largo, FL, USA) controls the quality status of the excitation. The internally used grid provides a spectral sensitivity range from 200 nm to 800 nm with a fixed slit width of 100 μm, and a possible integration time from 1 millisecond to 65 s. Reference data are recorded in parallel to the fluorescence measurements but are not included in the calculation of the EEM spectra. Instead, they only serve as metadata for monitoring the quality of the excitation light source state over the operation time.

Water Sampling Flow-Through Cell

Various conceptual instrument designs and principles for fluorescence detection hardware are available for different applications: right angle detection, intersecting cones, flow-through, and fiber optic designs [36]. The 'Kallemeter' sample flow-through cell design combines right angle detection (the commonly used 90° emission–excitation setup in luminescence spectrometry) with fiber optics and intersecting cones in a closed flow-through design, which is illustrated in Figure 3a. In addition, the measuring cell makes use of well-directed multireflective intersections of the optical fiber aperture to achieve multiple excitations and detection cone overlaps. Thus, the optical path within the water sample is enhanced and an effective increase of factor 2.5 of the fluorescence signal for the highest signal-to-noise ratio can be achieved (Appendix A, Table A1). Two mirror elements are used for this purpose, which are arranged opposite to the optical fiber terminations in a way that no direct fluorescence excitation light can enter in the optical fiber of the detector strand. A concave mirror (CM127-025-F01 - Ø1/2" UV Enhanced Al-Coated, f = 25.0, Thorlabs, Newton, NJ, USA) is used in (opposite) combination to the excitation cone. With a focal length of 25 mm, the focus point of the selected concave mirrors is almost at the end of the optical fibers distance of 22 mm. A planar mirror (PF05-03-F01 - Ø1/2" UV Enhanced Al-Coated, Thorlabs, Newton, NJ, USA) is used in (opposite) combination to the emission cone of the optical fibers.

As the sample cell forms the contact point between the optical parts and the water sample, it is installed in a separate housing at the bottom of the EEMS sensor system (see Figure 2c,d). Thus, a decoupling of the fluid-carrying and pressurized parts from the sensitive internal structures of the measurement unit is achieved and a simple sealing and pressure tightness can be realized. The optical interfaces of the sample cell, which are in contact with a seawater sample, are covered with superior transparent quartz glass windows (#45-310. TECHSPEC, Ø 15 mm Uncoated, 1λ Fused Silica Window, Edmund Optics, Barrington, NJ, USA). In order to ensure biogeochemical inertness of the sample cell, a specific adhesive (LS2-6140, NuSil Technology, Carpinteria, CA, USA) was selected through in situ tests as a suitable material for connecting the sample cell components and quartz windows. Even after several days in aqueous solution, this adhesive did not emit any fluorescent substances.

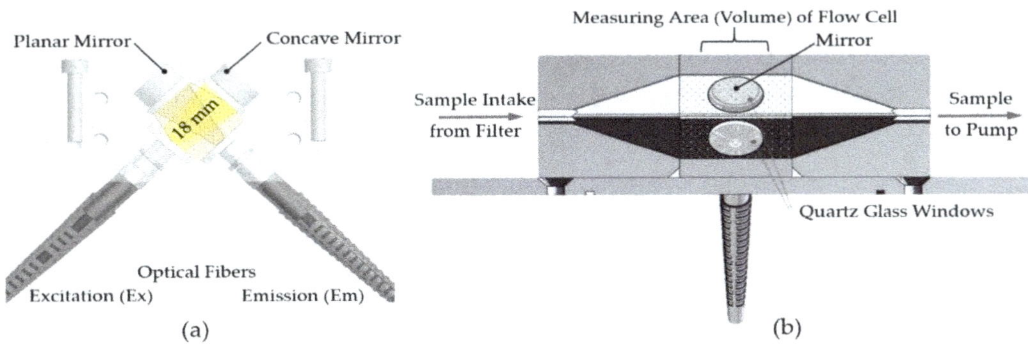

Figure 3. Details of the EEMS sensor system sample flow-through cell. (**a**) Measuring area-section with right angle arrangement of the excitation and emission optical fibers including the concave and planar mirror, respectively. (**b**) Cut-section through the flow-through cell with media intake and quartz glass windows.

The flow-through geometry type results in a greater efficiency in signal output compared to an open-faced design. Thus it is a good choice for waters with a wide range of NOM concentrations and low numbers of particles, as the sensor geometry (sample volume) is well defined [36]. A closed flow-through cell design, as illustrated by the cut-section in Figure 3b, provides shading of the optical inner parts from ambient light. This is a further benefit for an in situ application as it reduces the effect of biofouling [38]. During the sampling process, the closed sample cell is flushed with the water sample. The inlet side of a mini submersible pump for profiling applications (SBE 5M-1, Sea-Bird Scientific, Bellevue, WA, USA) is connected to the water outlet of the sample cell, while a reusable 50-μm nylon screen filtration element (type series F20 filter housing, Wolftechnik Filtersysteme, Germany) is connected to the sample cell water inlet via opaque hoses. This prevents an uncontrolled exchange between the ambient water and the water sample volume during an EEM spectrum measurement. This is particularly important for in situ underway operations when the 'Kallemeter' moves through different water bodies during an ongoing EEM measurement. Note that the setup presented herein was not equipped with a 0.2 μm or 0.45 μm filter, as recommended to separate the dissolved from the particulate fraction of organic matter. This is also the case for common bulk fluorescence sensors, as used in this study for comparison. However, as a pumped flow-through system, the 'Kallemeter' is generally capable of such a filtration unit as used e.g., for in situ spectral absorption measurements of CDOM [17].

Main Fluorescence Detector

The main fluorescence detector unit of the 'Kallemeter' EEMS sensor system is a wavelength-calibrated CCD digital spectrometer with a high dynamic range of ~85.000:1 (QE-Pro, Ocean Optics, Largo, FL, USA). The fluorescence emitted by the water sample is transmitted via an optical fiber, which is installed at the sample flow-through cell. The spectrometer built-in grating enables the detection in a full range of emission wavelengths from 200 nm to 950 nm, which corresponds to the detection range of commonly used laboratory luminescence spectrometers (e.g., the LS55, PerkinElmer, Waltham, MA, USA). The slit of the detector spectrometer has a width of 100 μm. Long-term stability and noise reduction is achieved by internal Peltier cooling of the detector to −10°C, as all fluorescence measurements signals are subject to alterations as a function of temperature. Temperature changes can arise either from self-heating of the instrument or from changes in ambient temperature of the deployment environment. Integration times of 8 milliseconds up to 60 min are selectable to enable stable measurements for environments where sample signal intensity is small. The spectrometer includes further a second order wavelength filter, preventing detection of Rayleigh and Raman scattering of higher orders.

The chosen detector spectrometer offers automatic dark measurements without external shutter units, which is used to obtain related metadata to correct for instrument response variables. As a routine by the 'EEMsea' software, a dark count measurement is carried out once initially per EEM spectrum measuring cycle and is automatically subtracted from the fluorescence data. Investigations of wavelength assigned intensity dark count measurements are presented in Appendix A, Figure A1 through a long-term measurement test (116 h) exhibiting stable operating conditions with no drift or significant invariance over time.

Optical Fibers

Optical fibers in the system setup are solarization-resistant, as ultraviolet radiation below 260 nm degrades transmission in silica fibers, also known as UV degradation. The custom-made designs, as illustrated in Figure 4a, provide maximum fiber diameters (80 single fibers 200 μm core diameter, Ø 2.6 mm bunch, LOPTEK, Berlin, Germany) and achieve a high transmission (96%) in all relevant excitation wavelengths (labeled (1) in Figure 2b). The numerical aperture is NA = 0.22. Figure 4b shows the optical Y-fiber and the single 80 μm fiber to the reference spectrometer that measures the reference excitation spectrum (labeled (2) in Figure 2b). This constellation prevents super-saturation or even damage of the reference spectrometer. For the fluorescence emission (detection) strand, a conventional optical fiber is used (QP1000-2-UV–Vis, Ocean Optics, Largo, FL, USA), connecting the sample flow-through cell to the main fiber optic spectrometer (labeled (3) in Figure 2b).

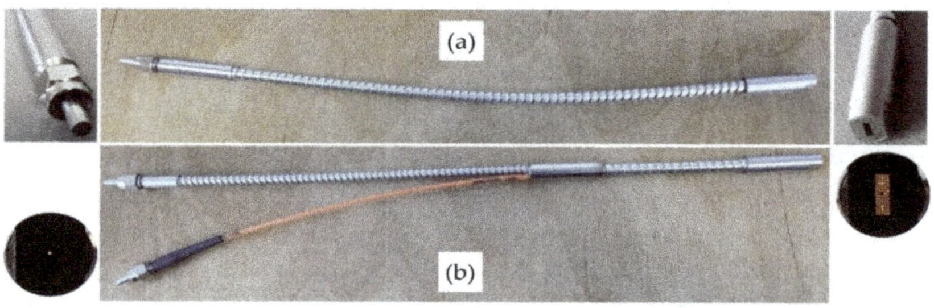

Figure 4. (**a**) Optical fiber coupling of the Xenon flash lamp with the monochromator (labeled (1) in Figure 2b). (**b**) Optical Y-fiber coupling of the monochromator with the reference spectrometer and the water sample flow-through cell (labeled (2) in Figure 2b).

2.1.3. EEMS Sensor System Fluorescence Performance Evaluation

As the complete 'Kallemeter' EEMS sensor system is based on a new design, the system performance for wavelength calibration, fluorescence intensity and signal-to-noise ratio was examined prior to field application. Considering the main focus of this article being the field application, the most important facts and a compilation of regularly practiced calibration methods for the in situ EEMS sensor system will be presented in the following section. All settings for the following measurements are listed in detail in Table A2 of the Appendix A.

The general procedure to calibrate an in situ fluorometer consists of (i) precalibration with tests of pressure, mechanical, and electronic stability as well as precision, (ii) signal output calibration to measure the dark and maximum counts, (iii) internal temperature calibration, (iv) determination and record of offset values with purified water at different temperatures, and (v) manufacturer calibration to obtain the scale factor of the fluorometer [36]. Calibration should be repeated in regular intervals, because environmental conditions can vary and lamp as well as spectrometer performance can degrade over time. Table 1 summarizes the EEMS sensor system specifications.

Signal-To-Noise Ratio

The signal-to-noise ratio (S/N) of the Raman band of tap water is easy to realize and a widely used method to quantify the sensitivity of a spectrofluorometer [39,40]. The S/N (where S denotes signal intensity and N noise intensity) of the Raman band is calculated with the following equation.

$$S/N = (S - N)/\sqrt{N}$$

For the 'Kallemeter' a S/N of 34 was determined for the Raman band measurement of tap water as presented in Figure A2.

Table 1. Overview of the 'Kallemeter' EEMS sensor system, fluorescence hardware specifications.

Parameter	Specification
Light source	Pulsed Xenon flash, 9.9 Watts
Excitation range	220–750 nm
Excitation bandpass	10 nm (FWHM)
Excitation monochromator	Czerny–Turner design, 110 mm focal length
Excitation wavelength precision / accuracy	0.2 nm/±0.6 nm
Emission range	200–950 nm
Emission bandpass	None (full set of wavelengths)
Emission detection spectrograph	(QE-Pro), Fixed, 101 mm focal length
Emission detector	TE-cooled CCD fiber optic spectrometer
Emission integration time	8 ms to 60 min
Excitation reference measurement	(USB2000+), CCD fiber optic spectrometer
Sensitivity	Water-Raman S/N 34

Wavelength Calibration

The wavelength calibration of the EEMS sensor system needs no special adjustments by the user, as both spectrometers used are calibrated by the manufacturers and therefore are supplied with a set of calibration coefficients. For verification we performed a laboratory check with a reference standard for molecular fluorescence spectrophotometry (Compound 610; excitation: 440 nm; emission: 480 nm; nominal concentration of 1×10^{-6} M, Starna Scientific Ltd., Ilford, UK), data not shown. Such a measurement allows the evaluation of the EEMS optical setup over time and corresponding correction methods for the emission wavelengths.

Intensity Calibration with Suwannee River Standard

Since 1982, representatives of the International Humic Substance Society (IHSS) have isolated reference fulvic acids, humic acids, and natural organic matter (NOM) from the Suwannee River in southeastern Georgia, USA. Suwannee River has a remarkable low variance in its elemental composition and is therefore a commonly accepted standard. It has a long history as reference material for multidisciplinary research areas, including the marine community [41]. For the fluorescence measurement of the 'Kallemeter', a sulfuric acid dilution of Suwannee River (sample charge number 2S101H) was used with concentrations of 2.5 ppm and 25 ppm. The EEM spectra of the two concentrations of Suwannee River standard are shown in Figure 5. Receiving reasonable signal intensity in both samples indicated sufficient signal intensity for nominal concentrations at marine study sites.

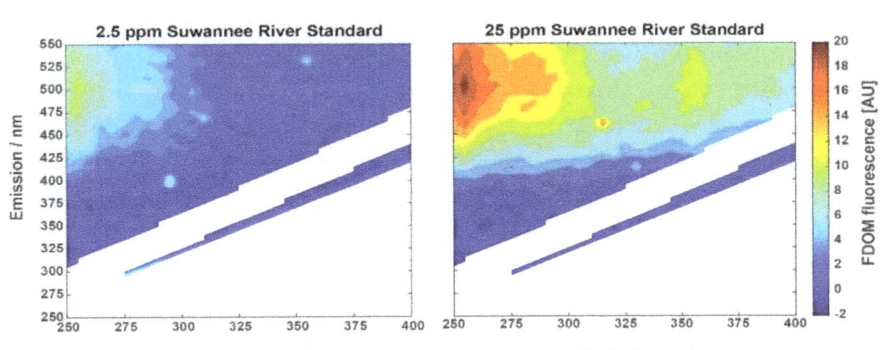

Figure 5. EEM spectra for two concentrations of Suwannee River standard. Excitation wavelengths were used for the x-axis, while emission wavelengths were applied for the y-axis [42]. Color bar (z-axis) denotes relative fluorescence intensity.

2.2. FerryBox-Based Organic Matter Fluorescence Measurements

An important step in the evaluation of the 'Kallemeter' is the comparison with other existing sensors and approaches, here with two commercially available sensors, the Cyclops-7-U (Turner Designs, San Jose, CA, USA) and the MatrixFlu-UV (TriOS, Rastede, Germany) mounted in a FerryBox system (4H-JENA engineering, Jena, Germany) installed on the ship during the research cruise (see Section 2.3). Furthermore a LS-55 spectrofluorometer (PerkinElmer, Waltham, MA, USA) was used as a common standard for laboratory measurements (see Section 2.4), providing complete EEM spectra directly comparable with the 'Kallemeter'. The FDOM sensors either provide only certain parts of the full EEM (MatrixFlu-UV) or only bulk fluorescence data (Cyclops-7-U).

As described previously [43], there are two versions of the MatrixFlu sensor, one for the visible wavelengths, with a focus on phytoplankton, and one for the ultraviolet, dedicated to FDOM components. The latter is used here and referred to as MatrixFlu-UV. The fundamental idea is to measure the typical peaks of organic matter fluorescence, as described by Coble [29], instead of assessing the whole excitation and emission spectrum, thus enabling a rapid detection and a high temporal resolution. This section describes only the sensor's main features, as its technical design has been published elsewhere [23,44].

The MatrixFlu-UV sensor consists of an excitation light source with distinct wavelengths combined with a set of detectors collecting a near field emission of a fluorescence signal at certain wavelengths directly in the water column (open sensor interface). Each combination of an excitation and emission wavelength results in a different fluorescence channel. The optical part contains a single semiconductor device emitting light at three different wavelengths (254 nm, 280 nm, and 320 nm) on the same optical axes. Four single semiconductor detectors are arranged around it collecting the fluorescent signal in an angle of approx. 170° to the optical axes. Different wavebands are selected via customizable narrow bandwidth filters (here 280 nm, 360 nm, 460 nm, and 540 nm). By doing this, all possible combinations result in twelve different channels.

FDOM components referred to in this work are mainly associated to (a) terrestrial or old marine humic-like substances (peak C), (b) "freshly produced" marine humic-like substances (peak M), and protein-like Tryptophan substances (peak T) [45,46]. Humic components show a further related excitation maxima for UV light of higher energy (lower wavelength) but same emission position or range (peak A) [45]. Data from all MatrixFlu-UV channels are collected simultaneously. For the comparison with the 'Kallemeter' and the other bulk fluorescence sensor (described below), Channel 11 was extracted and evaluated, corresponding to peak C (EX/EM 320 nm/460 nm). Data was averaged on every minute values and smoothed by a moving mean filter (window size 13). Factory calibration and further information about the sensor performance are available in [23].

The Cyclops-7-U sensor estimates bulk FDOM concentration by measuring the fluorescence emission at 470 nm (60 nm peak width) for an excitation at 325 nm (120 nm bandwidth filter [47]). Thus, for a comparison with the data from this sensor, the 'Kallemeter' in situ EEM signal in the area corresponding to an excitation of 265 to 385 nm and an emission of 440 to 500 nm have been integrated. This provides a 'Kallemeter' derived bulk fluorescence signal that corresponds to the excitation and emission ranges given in the specifications of the Cyclops-7-U sensor.

2.3. FerryBox Installation and Other Oceanographic Parameters Measured

The water intake of the FerryBox was realized in the moonpool of the research vessel (in approx. 4 m depth) close to the measuring head of the submersed 'Kallemeter' to ensure a comparability of the measurements between the three types of instruments. Besides the FDOM data, readings of standard oceanographic parameters, such as temperature, salinity, and chlorophyll *a* fluorescence were also obtained. All FerryBox data were averaged on minutely values and smoothed by a moving mean filter (window size 11).

2.4. FDOM Laboratory Water Samples

Sampling from the near surface water, here defined as the samples collected between 0.1 and 2.0 m, was obtained with a submersible pump system and stored in 5 L white plastic bottles, at all stations. Samples were filtered in the laboratory immediately after sampling, and stored at 4 °C in previously rinsed glass bottles. CDOM absorption and EEM spectra measurements were completed within 24 h after sampling.

Samples were filtered with a glass syringe device through GF/F Nucleopore filters (0.2 µm, GE Healthcare Whatman, Piscataway Township, NJ, USA). CDOM absorbance spectra were obtained between 220 and 650 nm at 5 nm intervals on a Shimadzu UV-2550PC UV–VIS spectrophotometer (Shimadzu, Kyoto, Japan) with 10 cm quartz cuvettes. All measurements were performed at room temperature using freshly produced ultrapurified water (UPW; Milli-Q 185plus water purification system, Merck Millipore, Burlington, MA, USA) as reference. CDOM absorption was calculated by multiplying absorbance by 2.303/r, where r is the cuvette length (in meter). FDOM EEM spectra measurements were conducted at room temperature on a LS55 spectrofluorometer (PerkinElmer, Waltham, MA, USA) equipped with a Xenon flash lamp, pulsed at line frequency of 60 Hz. The excitation (EX) and emission (EM) scanning ranges were set to 200 to 400 nm (5 nm interval) and 220 to 550 nm (0.5 nm interval), respectively. The scanning speed was 1200 nm min^{-1}, the EX and EM slit was set to 10 nm, and the detector gain was at medium range (775 V). Daily measurements of UPW were used as references. Water Raman scatter peaks were eliminated by subtracting the EEM spectra of UPW blanks. Inner filter effect and Rayleigh scatter peaks were corrected using the drEEM toolbox [33]. Lamp variations were normalized to the Raman peak integral (EX: 350 nm) using freshly produced UPW, therefore producing normalized intensities in Raman units (RU) [39]. EEM spectra were resized due to high S/N in certain regions. Therefore, excitation wavelengths below 250 nm and emission wavelengths below 300 and above 550 nm were excluded from the subsequent PARAFAC modeling process.

PARAFAC modeling was performed with the drEEM toolbox (ver.0.2.0) in MATLAB R2015b (The MathWorks, Natick, MA, USA) according to a method described previously [33], employing the N-way toolbox as the engine of the PARAFAC algorithm [34]. A three-component model was validated by split-half analysis, random initialization analysis, and residuals analysis. The model explained up to 99.95% of the variability within the lab-based LS55 spectrofluorometer EEM dataset.

A similar PARAFAC analysis was applied on the Kallemeter-based EEM spectra along cruise track between station 11 and 15. Correction for inner filter effects was realized by utilizing the measured absorption spectra from the nearest station and using Kallemeter blank measurements from an initial laboratory test with purified water. Four out of 48 spectra were excluded from the analysis, due to high residual error and leverage load, validating a two-component model that explained up to 61% of the variability within the dataset through split-half analysis, random initialization analysis, and residuals analysis. A third component could not be detected.

2.5. Study Site

In order to assess the performance of the 'Kallemeter' in comparison with the MatrixFlu-UV and the FerryBox-mounted Cyclops-7-U fluorescence sensors in the field, the data obtained during an expedition (HE491) with research vessel R/V Heincke from the coastal sea into the Trondheimsfjord (Norway) were analyzed. Figure 6 shows a map of the study site and the cruise track investigated. The instruments were in parallel operation from 19th to 25th of July 2017. The Trondheimsfjord is the third-largest fjord in Norway with a length of 140 km, a volume of 235 km^3, and a surface area of 1420 km^2. It is divided in three basins and has a seaward sill depth of 195 m [48]. It is characterized by comparably high CDOM concentrations as well as vertical and horizontal gradients [49], making it an ideal test site.

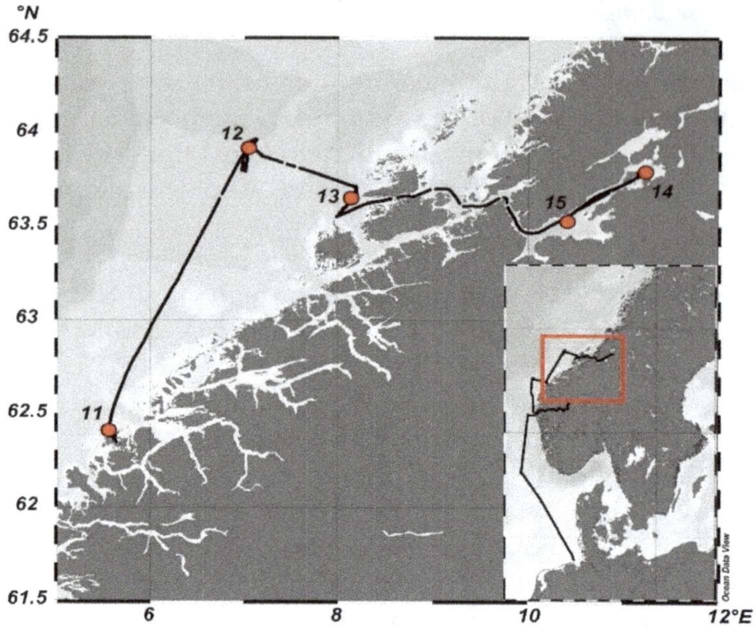

Figure 6. Overview of the study area (insert). Map shows Trondheimsfjord and its adjacent coastal sea with cruise track (solid black line) and stations (red dots) of R/V Heincke expedition HE491.

3. Results

The FerryBox data is well suited to provide a first overview of the overall fjord hydrography. The FerryBox dataset started on 19 July in the Norwegian coastal sea (station 11) heading first northwards (station 12) and then into the Trondheimsfjord (station 13), through the mid-fjord section and towards the innermost part of the fjord (station 14). From there, the transect went back to the mid-section and ended at station 15 on 25 July. There is a considerable difference in all parameters measured between the area outside and within the fjord (Figure 7), indicating the presence of two water bodies with different properties. With the exception of salinity, values of all parameters are higher within the fjord. Salinity is highest towards the ocean side, with values of more than 32, rapidly decreasing inside the fjord towards the mid-section exhibiting values of around 24. Temperature increases slightly from around 14 °C in the outer section towards 15 to 17 °C in the mid- and inner-part of the Trondheimsfjord. Chlorophyll-a fluorescence (here provided in AU, as the sensor has not been specifically calibrated for chl-a before the cruise) shows a gradual increase from the seaward part of the fjord to a maximum at the end of the fjord. FDOM from the Cyclops-7-U fluorometer (provided in AU) shows, inverse to salinity, a four-fold increase and steep gradient from the seaward side of the fjord towards the mid- and inner-part.

Comparing the 'Kallemeter' bulk fluorescence signal extracted from the EEM spectra against the bulk fluorescence signal of the MatrixFlu-UV channel 11 and the FerryBox inbuilt Cyclops-7-U FDOM sensor data (Figure 8), it can be seen that all three signals showed a steep increase on 23 July, when entering Trondheimsfjord. MatrixFlu-UV exhibited a maximum at the end of that day, followed by a decrease on 24 July towards a steady level around half of the signal range. FDOM bulk fluorescence from the FerryBox inbuilt Cyclops-7-U sensor increased in the same steep way towards a higher level on 23 July. Signal intensity stayed on this level with only a slight decrease in the morning of 24 July. 'Kallemeter' bulk fluorescence signal, as indicated by red stars, exhibited the same dynamical behavior as the Cyclops-7-U. A linear regression analysis was performed (Figure 9) of the extracted 'Kallemeter' bulk fluorescence against the Cyclops-7-U sensor (coefficient of determination $R^2 = 0.96$) as well as the corresponding channel of the MatrixFlu-UV ($R^2 = 0.88$), both attached to the flow through system.

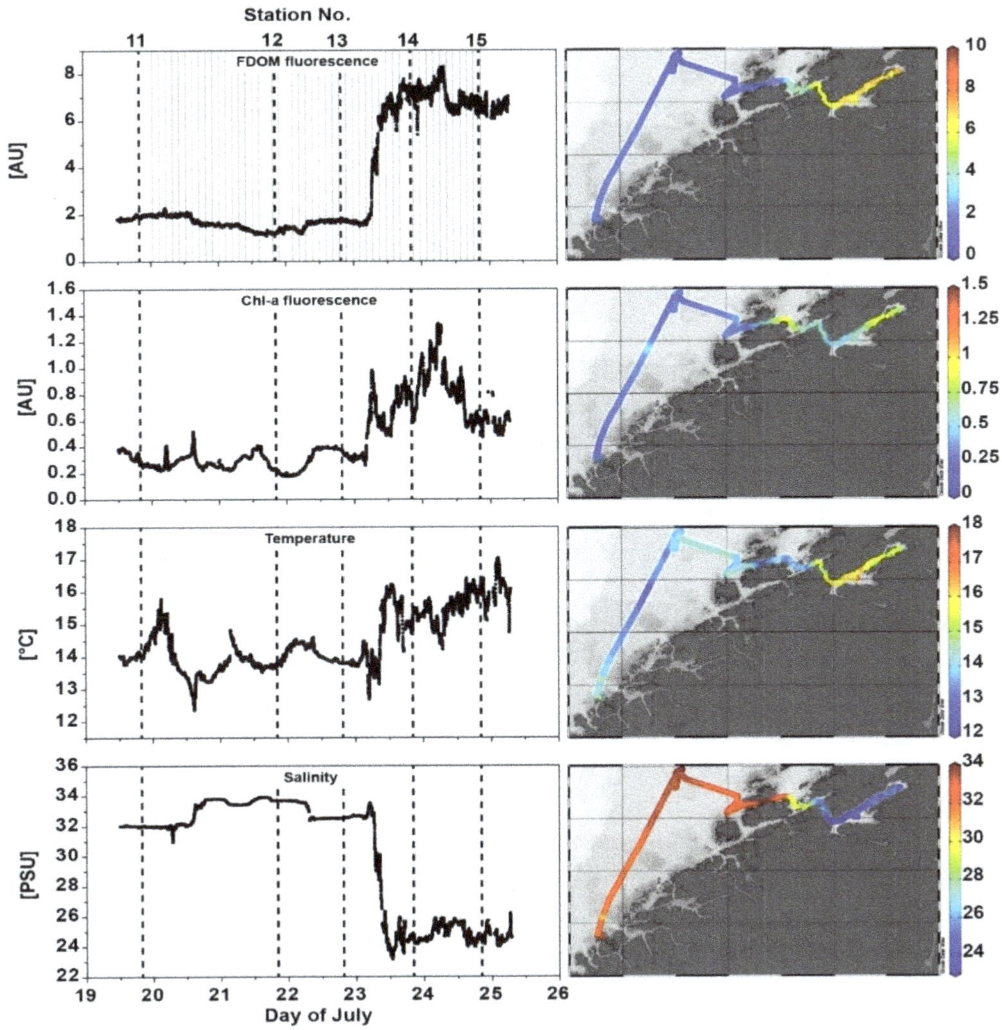

Figure 7. Underway oceanographic data measured by the FerryBox. From top to bottom: Fluorescent dissolved organic matter (FDOM) from the Cyclops-7-U bulk fluorometer, chlorophyll a fluorescence, temperature, and salinity. The left panels display the data as time series, the corresponding right ones as a map plot. Dashed lines indicate the stations where water samples were collected. Gray lines in the top-left panel indicate availability of the in situ Kallemeter-based EEM spectra (N = 48).

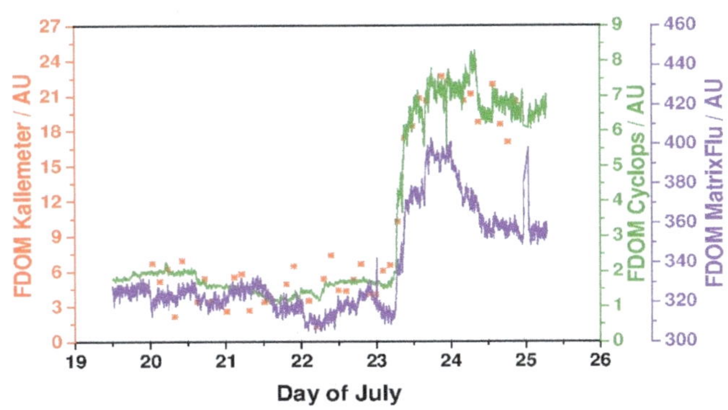

Figure 8. Comparative view on the extracted underway measurements from the MatrixFlu-UV (violet), the FerryBox FDOM Cyclops-7-U sensor (green) and the extracted bulk fluorescence signal from the EEM spectra (red stars), obtained along a transect from the Norwegian coastal sea into the Trondheimsfjord.

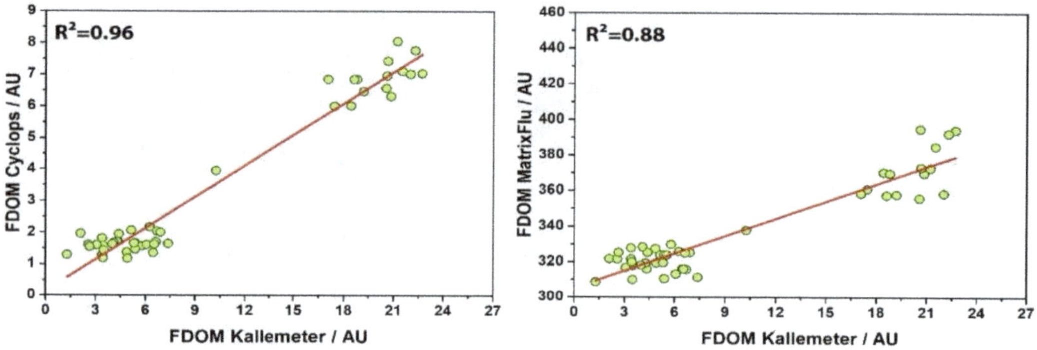

Figure 9. Comparison of extracted 'Kallemeter' bulk fluorescence signal against FerryBox FDOM Cyclops-7-U sensor (**left panel**) as well as the corresponding channel of the MatrixFlu-UV (**right panel**). Red lines represent the results of a linear regression.

In situ underway EEM measurements, measured by the 'Kallemeter' in the moonpool of R/V Heincke, provided a total of 48 spectra in 5 days (20 to 25 July). All fluorescence information and metadata were continuously recorded from open water into the Trondheimsfjord. A comparison of normalized EEM spectra of the lab-based LS55 fluorometer with the corresponding nearest normalized Kallemeter-based EEM spectra at stations 12–15 (Figure 10; station 11 had no close match-up within a time frame of ±1 h) shows good similarity for stations 14 and 15, where signal intensities are 5-times higher as compared to stations 12 and 13 (compare Figure 8). These lower concentrations of fluorescent organic matter at stations 12 and 13 were only resolved from the apparently more sensitive laboratory LS55 spectrofluorometer, thus illustrating the current limits of the 'Kallemeter' with respect to low concentrations and the effects of noise. Please note that we applied normalization to the EEM spectra to enable a comparison of spectra signatures measured, independent from the signal strength. A non-normalized set of 'Kallemeter' EEM spectra is provided in the appendix (Figure A3) showing for every fourth dataset (a) an EEM spectra with EX wavelengths 230–550 nm, $\Delta\lambda = 10$ nm, integration time of 60 s and two averaged scans, slit = 10 nm, EM wavelengths 189–990 nm, and (b) an emission scan with EM wavelengths 189–990 nm for 250 nm EX wavelength. These Kallemeter-based EEM spectra show an increase in fluorescence intensities inside Trondheimsfjord as compared to the coastal water conditions in the EM wavelength range from 400 to 550 nm.

PARAFAC analysis of the lab-based LS55 spectrofluorometer EEM spectra from water samples gathered at different depths demonstrated the presence of three main components (C1, C2, and C3) in the water samples (Figure 11, top panel). These components were characterized by their excitation and emission pairs (EX/EM) as follows. Component (C1) corresponding to terrestrial humic-like peaks A_C and C, component (C2) corresponding to marine humic-like peaks A_M and M, and component (C3) corresponding to protein-like Tryptophan peak T, as suggested by Coble [29]. The spectral composition of the recorded lab-based LS55 spectrofluorometer EEM spectra (provided in Table 2) shows that the fluorescent intensity of the peaks C1, C2, and C3 remains constant between the open ocean stations (ST11 and ST12) and the outer section of the fjord (ST13). Once the mid- and inner-part of the fjord was reached, we recorded a sharp increment (by a factor of 5) in the fluorescent intensity for the components C1 and C2 (ST14 and ST15), while C3 stays nearly constant.

Performing PARAFAC analysis of the Kallemeter-based EEM spectra along the transect resulted in a model with two components validated (Figure 11, bottom panel). These components, even though influenced by the noisy EEM spectra that suited as model dataset, can be related according to their excitation/emission pairs to the presence of terrestrial humic-like peaks A_C and C, and marine humic-like peaks A_M and M. Although reasonable EEMS where produced by the 'Kallemeter', the presence of noise and suspended materials in the samples must be addressed in the future, in order to minimize the observed displacement on the peak's position.

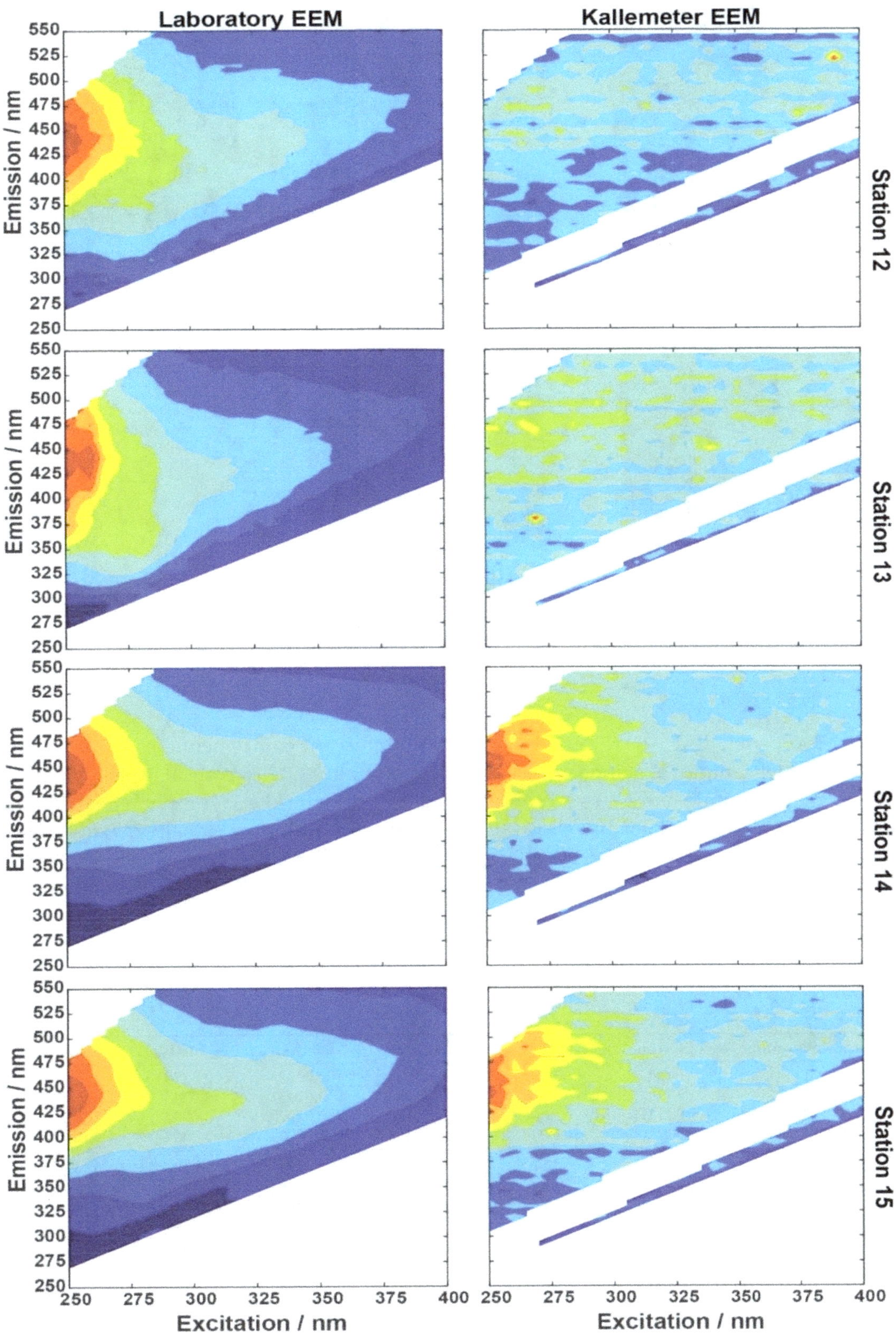

Figure 10. Normalized EEM spectra from lab-based water sample analysis with the LS55 laboratory spectrofluorometer (**left**) and nearby (match-up of ±1 h) in situ measurements with the Kallemeter (**right**) for stations 12–15 of R/V Heincke expedition HE491 along a transect from the coastal ocean into the Trondheimsfjord. Notice that FDOM intensity in stations 14 and 15 are 5-times higher compared to stations 12 and 13.

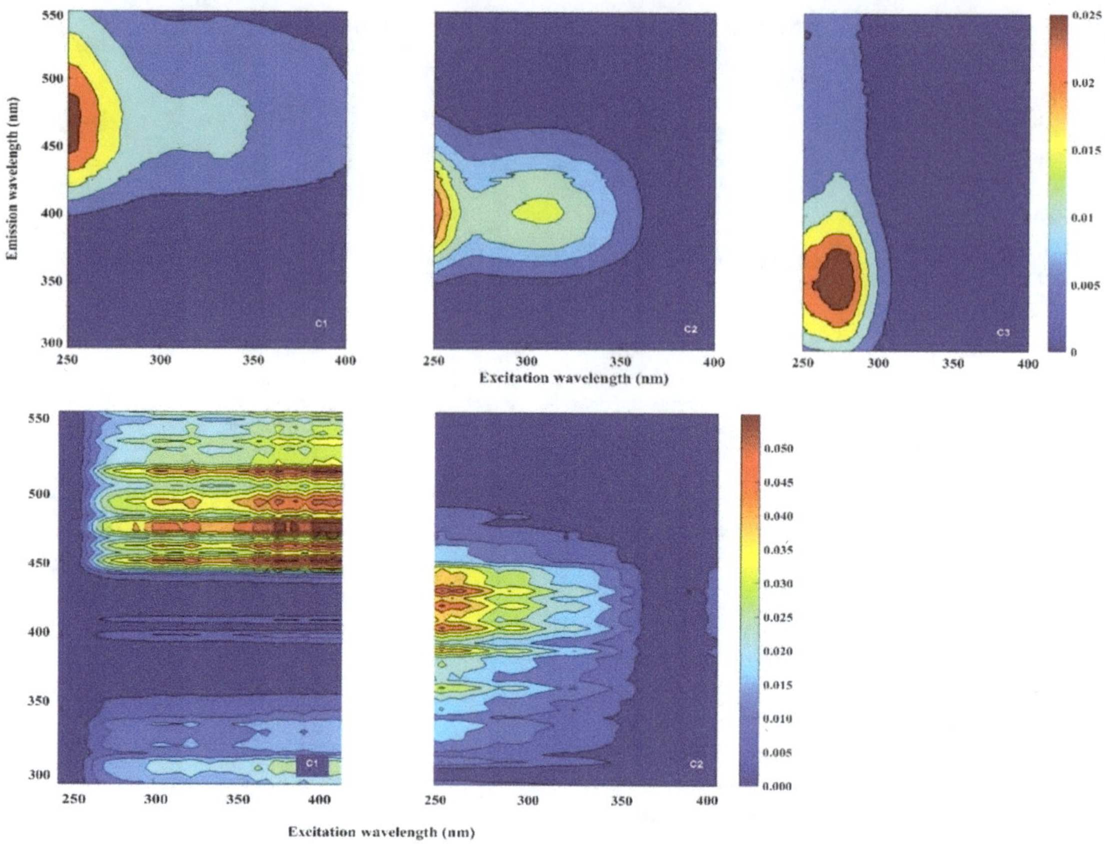

Figure 11. Top panel: PARAFAC-derived principal fluorescent components from lab-based EEM spectra from water samples at stations 11–15. From left to right: (C1) Terrestrial humic-like peaks A_C and C; (C2) marine humic-like peaks A_M and M; and (C3) protein-like Tryptophan peak T, all peaks denoted after Coble [29]. Color bar shows Raman units (RU). Bottom panel: PARAFAC derived principal fluorescent components from Kallemeter-based EEM spectra along the transect.

Table 2. Spectroscopic composition of the lab-based LS55 spectrofluorometer EEM spectra analyzed water samples in Trondheimsfjord. C1, C2, and C3 represent the intensity of PARAFAC components in Raman units.

Sample Number	Station	Depth (m)	C1 (RU)	C2 (RU)	C3 (RU)	Date
1	ST11	0.10	0.0364	0.0302	0.0238	20 July 2017
2	ST11	0.50	0.0349	0.0298	0.0243	
3	ST11	1.50	0.0354	0.0297	0.0283	
4	ST11	2.00	0.0331	0.0289	0.0304	
5	ST12	0.10	0.0252	0.0194	0.0162	22 July 2017
6	ST12	0.50	0.0231	0.0177	0.0106	
7	ST12	1.50	0.0226	0.0174	0.0127	
8	ST12	2.00	0.0227	0.0177	0.0130	
9	ST13	0.10	0.0355	0.0278	0.0214	23 July 2017
10	ST13	0.50	0.0355	0.0295	0.0287	
11	ST13	1.50	0.0323	0.0255	0.0173	
12	ST13	2.00	0.0322	0.0263	0.0191	
13	ST14	0.10	0.1772	0.1054	0.0363	24 July 2017
14	ST14	0.50	0.1745	0.1017	0.0247	
15	ST14	1.50	0.1763	0.1035	0.0248	
16	ST14	2.00	0.1749	0.1028	0.0263	
17	ST15	0.10	0.1471	0.0904	0.0378	25 July 2017
18	ST15	0.50	0.1493	0.0899	0.0233	
19	ST15	1.50	0.1543	0.0928	0.0241	
20	ST15	2.00	0.1530	0.0919	0.0232	

4. Discussion, Limitations, and Conclusions

The main objective of this study was to present a novel in situ EEMS sensor system—the 'Kallemeter'—enabling full optical fingerprints in a submersible design. We presented the design of this system within the methods Section 2.1 and illustrated its application based on laboratory samples of diluted Suwannee River standards with 2.5 ppm and 25 ppm, respectively. Fluorescence exhibited around EX: 240 nm and EM: 450 nm was corresponding to terrestrial humic-like peak C substances, as expected for Suwannee River standards, with increasing intensity and additional visible contributions extending to higher excitation wavelengths [50,51]. The assessment of the 'Kallemeter' EEMS sensor systems' signal-to-noise ratio (S/N) based on the Raman peak of water in a tap-water sample, provided an S/N of 34. Compared to commercially available laboratory spectrofluorometers that exhibit S/N of 500 or higher the 'Kallemeter' is much more limited in resolving low signal intensities [39]. Part of this is based on the different optical components implemented, e.g., a fiber optic spectrometer compared to a monochromator plus photomultiplier setup for the LS55 spectrofluorometer, still considered as the gold-standard for high sensitivity applications. Also the choice of the excitation light was influenced by power- and heat-considerations and therefore restricted to <10 Watts, where other laboratory instruments like the Aqualog (Horiba, Kyoto, Japan) are equipped with up to 150 Watt Xenon flash lamps [52]. The strongest influence can be assumed from the application of fiber optics within the sensor, resulting in losses while coupling light in and out as well as transmitting it in the fiber. The use of fiber optics is a result of the compact design that a submersible housing requires—a potential redesign of the 'Kallemeter' will take this into account and try to partially avoid the use of optical fibers through direct light paths.

The field application of the 'Kallemeter' was realized in the moonpool of R/V Heincke on transect from Norwegian coastal waters into the Trondheimsfjord in July 2017. The system was successfully operated for five days in a fully submerged installation. EEM spectra derived from this application showed very low fluorescence intensities superimposed by noise components for all coastal sea measurements and the outer fjord area, while mid- and inner-fjord measurements exhibited clear broad signals of approximately EX: 250–300 nm and EM: 400–500 nm. These observations are consistent with the spectral component intensities derived from laboratory UV–VIS spectrofluorometer (LS55, PerkinElmer, Waltham, MA, USA) measurements of surface water samples. With respect to components C1 and C2 derived from PARAFACS analysis these were ranging for the outer fjord between 0.02 and 0.04 Raman Units, while mid- and inner-fjord values ranged between 0.09 and 0.18 RU. The direct comparison of normalized EEM spectra from the lab-based spectrofluorometer and the 'Kallemeter' (Figure 10) illustrated both, the capability of the 'Kallemeter' to provide full EEM spectra if fluorescent organic matter concentrations are sufficiently high (stations 14 and 15), and the limitation to resolve low concentrations against the background noise as encountered at stations 12 and 13 in coastal waters. The latter is a result of the lower S/N and sensitivity, discussed above. The application of the 'Kallemeter' EEMS sensor system in its current form therefore seems to be limited towards areas with an increased amount of fluorescent organic matter, typically encountered in near coastal areas, estuaries and inland waters. Furthermore a study of the impact of noise or signal degradation applied on EEM spectra that serve as PARAFAC model input would be of interest to assess general limitations, however that is beyond the scope of this study.

The PARAFAC analysis of the Kallemeter-based EEM spectra enabled us to model two components, corresponding to terrestrial humic-like peaks A_C and C, and marine humic-like peaks A_M and M. This was compared to a PARAFAC analysis of the LS55 spectrofluorometer EEM spectra from near surface water samples at the distinct stations that successfully validated three components. While lab-based components 1 and 2 resemble the in situ Kallemeter components (even though limited by high noise influence), the third component, namely the protein-like Tryptophan peak T, could not be reproduced from the 'Kallemeter' data. Since peak T could be quantified in the laboratory measurements to an average intensity of 0.023 Raman units without any signal increase in the Trondheimsfjord, we consider this as a current limitation for the 'Kallemeter'.

The ability to perform a PARAFAC analysis on automatically measured in situ EEM spectra is a major achievement and, to the best of our knowledge, the first of its kind. We encountered a couple of challenges to that end, that we provide here as current limitations but also avenues for future improvements. (A) Since we did not filter the water entering the measurement cell with 0.2 μm (only applying 50 μm to keep coarse material out) the signal observed is originating from dissolved, colloidal and particulate fractions of organic matter. For low amounts of particulate material this might not be a problem and indeed all in situ FDOM bulk fluorescence sensors (including the two used in this study for comparison) are also measuring unfiltered samples. However, for direct comparison with laboratory measured (filtered) FDOM EEM spectra this can be considered an obstacle. To overcome this, the installation of a filter unit at the inlet, containing an appropriate 0.2 μm filter can be easily realized in future applications of the 'Kallemeter'. It can be speculated that part of the noisiness of the spectra and PARAFAC components can be associated to the influence of particulate material. (B) All EEM spectra need to be corrected for the inner filter effect (IFE). This typically requires a parallel measurement of the absorption spectra of the filtered water sample over the same wavelength range. The 'Kallemeter' itself is not suited to perform this absorption measurement, therefore we applied absorption spectra from water samples analyzed with a laboratory spectrophotometer. These were only available at the five stations investigated and not at the 48 measurement locations of the 'Kallemeter'. Assigning absorption spectra from samples that were gathered up to 24 h earlier or later is introducing uncertainties, even though we took care to respect the different oceanographic gradients encountered in the matching procedure. A much better way to account for this need of parallel absorption measurements will be to use automated in situ capable methods [53]. (C) Applying mirrors in the sample flow through cell, we increased the fluorescence signal intensity by a factor of 2.5, however IFE corrections applied to correct EEM spectra do not consider these and are therefore overcorrecting. Luciani et al. [54] showed how mirrors change the fluorescence signal and can be at the same time utilized to correct for IFE, if measurements with and without mirrors are performed on the same sample. This could be another methodological improvement to overcome constraints mentioned above.

Comparing the results from the bulk fluorescence signal of the 'Kallemeter' (calculated from the EEM spectra) with the flow through data from the commercially available FDOM sensors (a) inside the FerryBox (Cyclops-7-U, Turner Design, San Jose, CA, USA) and (b) the channel 11 from the recently developed matrix fluorescence sensor (MatrixFlu-UV, TriOS, Rastede, Germany), a good correspondence could be observed. The maximum exhibited on 23 July and the following decrease in the 'Kallemeter' fluorescence intensity (compare Figure 8) seems to be better reproduced by the Cyclops-7-U FDOM sensor. The corresponding MatrixFlu-UV channel is able to reproduce the same steep increase, however shows a signal reduction afterwards, not represented by the other methods applied in this study. As this study reports one of the first field datasets achieved with the MatrixFlu-UV sensor we cannot judge on this readings but see a clear need to investigate the sensors responses.

The MatrixFlu-UV sensor was investigated herein only with respect to one of its 12 channels representing a main spectral component in the investigated environment. It is therefore herein utilized as single-channel sensor and its further value in rapid quasi-EEM spectra measurements will be investigated in future works. Other multichannel approaches exist, with two [55] or three [56] combinations technically built up from single EX/EM combinations. A recent study of Nordic seas FDOM intensities [57] utilized a novel three-channel fluorometer (WETStar, Sea-Bird Scientific, Bellevue, WA, USA). This system excites the water sample in a flow-through cell with two LEDs at 280 nm and 310 nm. Two detectors measure the emission at 350 nm and 450 nm, allowing for a combination of channels in specific peak areas: peak C (with EX/EM 310 nm/450 nm), peak A_C

(with EX/EM 280 nm/450 nm), and peak T (with EX/EM 280 nm/350 nm). All applications mentioned above underline the potential of multichannel FDOM measurements for specific components dominant in natural water bodies. The technological step from multichannel sensing to real excitation–emission matrix spectroscopy with high spectral resolution in both, the excitation and the emission sides, is huge, since it requires complete different technologies. An interim step with six (up to 12) distinct LED excitation wavelengths and emission spectra with 7.5-nm spacing (named 'LEDIF') was presented as part of a field-deployable optical system [58].

The 'Kallemeter' represents, to the best of our knowledge, the first submersible EEMS sensor system that performs full optical fingerprints. While the detection side is comparable to the LEDIF approach, using a fiber optical spectrometer, excitation was realized in our system by a Xenon flash lamp and an embedded controlled monochromator. This enabled us to achieve 10-nm steps in the excitation wavelengths selection (and finer steps are technically feasible); however, it also slows down the measurement process to approx. 30 min per individual sample. This is in the same order of magnitude as the LS55 (PerkinElmer, Waltham, MA, USA) UV–VIS laboratory spectrofluorometer used in this study. Faster laboratory systems exist nowadays (e.g., Aqualog, Horiba, Kyoto, Japan) that enable EEM spectra measurements below 5 min that require, however, further technical improvements (e.g., a TE cooled back-illuminated CCD fluorescence detector, and a high power Xenon flash lamp). Modern ocean observing strategies combine sensors with high spatiotemporal resolution but limited information depth with more sophisticated sensors, that provide a high depth of information but are limited in their spatial and temporal resolution [8]. The 'Kallemeter' is one of those highly sophisticated sensors and will likely, as others in that field [59,60], improve in its operational specifications, i.e., size, weight, and power. Multichannel fluorescence sensors for FDOM are already smaller, lightweight, and less power consuming. Their application on different autonomous platforms has been demonstrated [17,55,58]. Thus, they are the linkage towards resolving rapid changes and processes. An assessment of fluorescent organic matter in high spatiotemporal and spectral resolution is therefore technically feasible and likely to become standard in future biogeochemical ocean observatories.

Author Contributions: Conceptualization, O.Z.; Methodology, O.Z., N.R., O.D.F., M.L.M., and J.W.; Validation and analysis, O.Z., N.R., O.D.F., M.L.M., and J.W.; Field investigations, N.R, O.D.F., M.L.M., and J.W.; Project Administration, O.Z. and N.R.; Funding Acquisition and Resources, O.Z.; All authors contributed to the writing and revision process.

Acknowledgments: The authors are thankful to the master and crew onboard the R/V Heincke HE491 campaign. The cruise was supported by the Alfred Wegener Institute for Polar and Marine Research (AWI) and the German Federal Ministry of Education and Research (BMBF). Sincere thanks to Anna Friedrichs and Kathrin Dietrich for assistance in data collection and laboratory analysis of water samples. Additional thanks to Beke Tietjen for supporting the preparation of Figures 8 and 9 in their early form, Ursel Gerken for checking spelling and grammar, and Julian Beck for constructing the flow-through cuvette of the 'Kallemeter'. MatrixFlu-UV was built and provided by TriOS Mess- und Datentechnik GmbH, Rastede, Germany. FerryBox was kindly provided by Wilhelm Petersen, Institute of Coastal Research at the Helmholtz-Zentrum Geesthacht, Germany. Several components of the 'Kallemeter' were kindly provided by Axel Bochert and Jan Boelmann, University of Applied Science, Bremerhaven, Germany. The feedback from the participants of the WOMS 2018 workshop in Toulon (France), two anonymous reviewers, and Dariusz Stramski as editor is gratefully acknowledged.

Appendix A

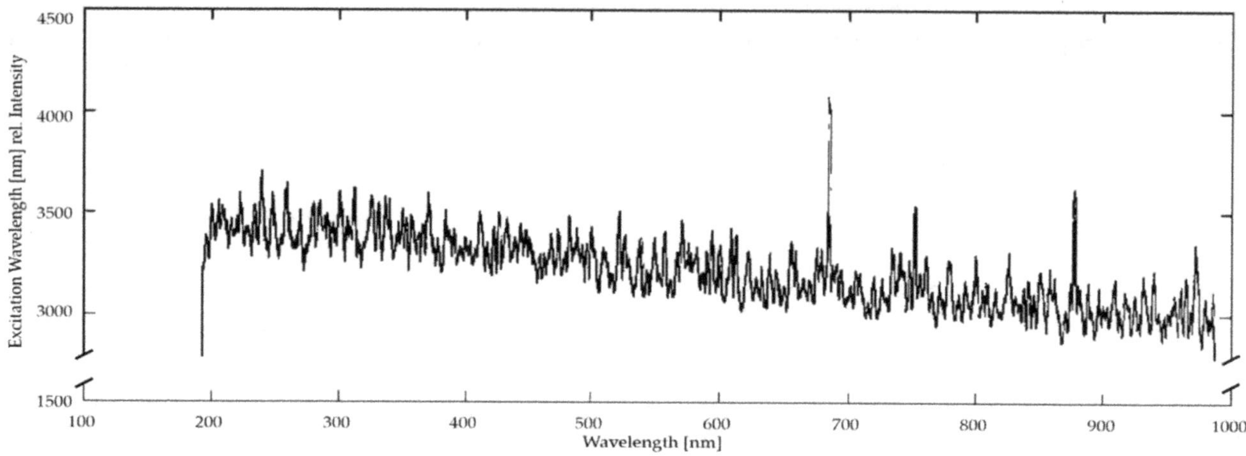

Figure A1. Dark counts of the 'Kallemeter' fluorescence detection spectrometer, measured by a long-term in situ test of 50 full EEM spectra (presented as an overlay) over a period of 116 h exhibiting a constant slight intensity gradient against the ordinate (wavelength).

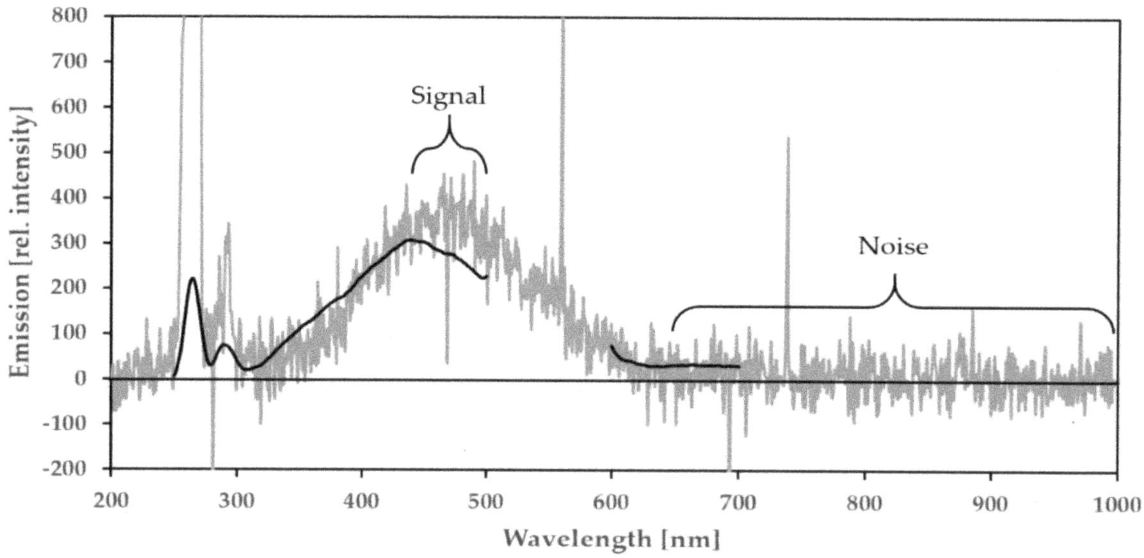

Figure A2. Signal to noise measurement of tap water with the 'Kallemeter' fluorescence detection spectrometer (grey line, integration time 300 s), and the LS55 laboratory spectrofluorometer (black line, 2nd maxima masked), both excitation wavelengths are 260 nm.

Table A1. Comparison of different mirror combinations within the measuring cell. Shown: Mirror on (opposite) excitation and emission side, maximum value of the fluorescence signal, signal-to-noise ratio (S/N). Measured: Compound 610 from Starna Scientific Ltd. (Ilford, UK), flash lamp frequency: 100 Hz, excitation wavelength 440 nm, integration time: 5 s.

Excitation Side	Emission Side	Max [counts]	S/N
Concave	Concave	678	156
Concave	Planar	571	157
Planar	Concave	650	187
Planar	Planar	570	145
Absent	Absent	346	94

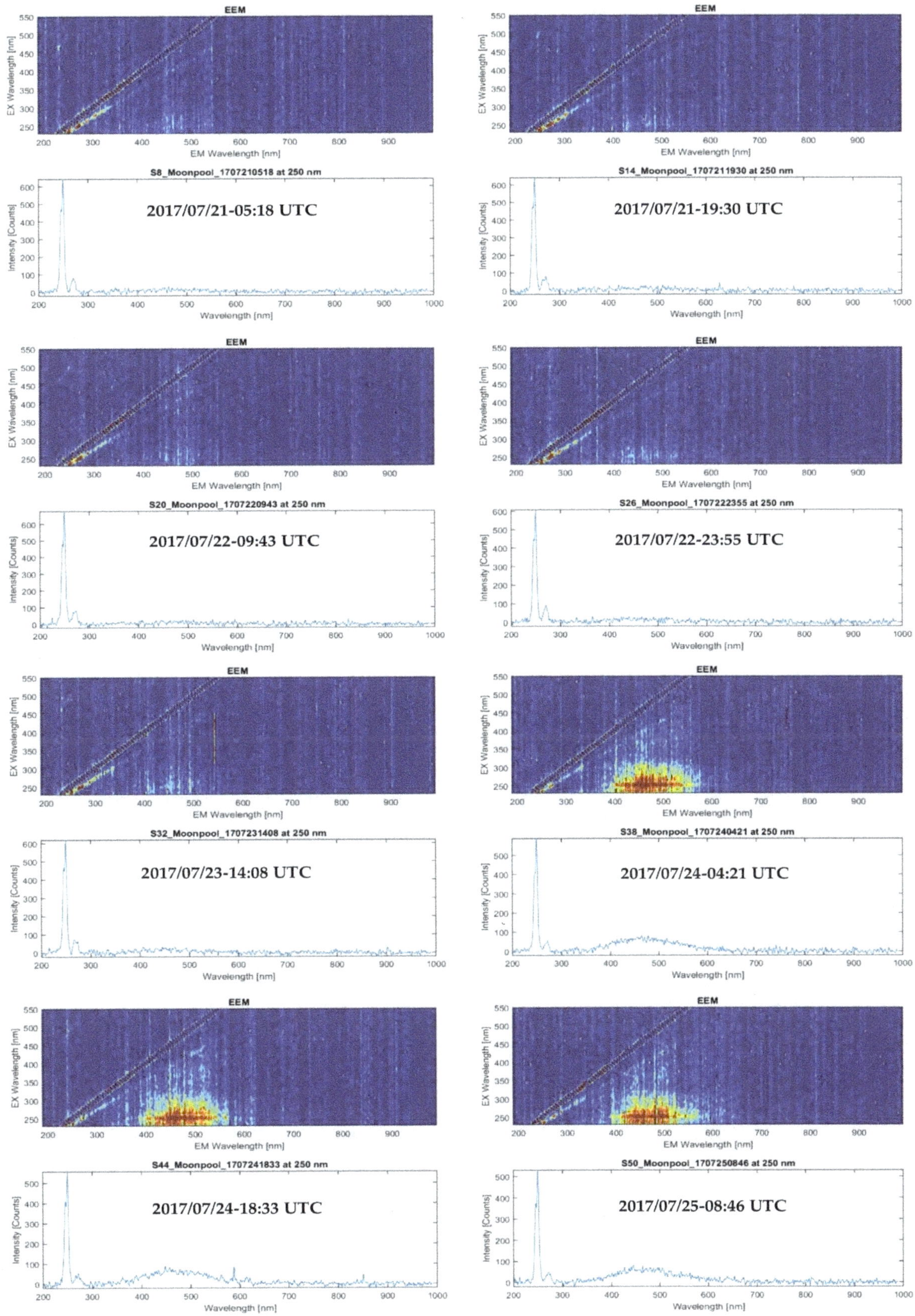

Figure A3. In situ underway measurements of EEM spectra, obtained from the moonpool of R/V Heincke (HE491) along transect from the coastal ocean into the Trondheimsfjord.

Table A2. EEMS measurement parameters for intensity calibration procedures.

Parameter	Settings
Excitation start	200 nm
Scan steps	5 nm
Excitation end	450 nm
Dark correction	True
Dark noise multi scan	20
CM110 offset correction	None
CM110 slits	None
Integration time	6 s
Scans to average	10
Box car	0
Trigger mode	0
Detector temperature	−10 °C
PX-2 connect	Reference
Multiple strobe	On
Flash delay	10 milliseconds
Warm-up	None

References

1. Mostofa, K.; Vione, D.; Mottaleb, M.; Yoshioka, T. (Eds.) *Photobiogeochemistry of Organic Matter: Principles and Practices in Water Environments*, 1st ed.; Springer: Heidelberg, Germany, 2013.
2. Dittmar, T. Reasons behind the long-term stability of marine organic matter. In *The Biogeochemistry of Marine Dissolved Organic Matter*, 2nd ed.; Hansell, D.A., Carlson, C.A., Eds.; Academic Press: Boston, MA, USA, 2015; pp. 369–388.
3. Ridgwell, A.; Arndt, S. Why Dissolved Organics Matter: DOC in Ancient Oceans and Past Climate Change. In *The Biogeochemistry of Marine Dissolved Organic Matter*, 2nd ed.; Hansell, D.A., Carlson, C.A., Eds.; Academic Press: Boston, MA, USA, 2015; pp. 1–20.
4. Watson, J.; Zielinski, O. *Subsea Optics and Imaging*; Woodhead Publishing Limited: Cambridge, UK, 2013.
5. Miranda, M.L.; Mustaffa, N.I.H.; Robinson, T.B.; Stolle, C.; Ribas-Ribas, M.; Wurl, O.; Zielinski, O. Influence of solar radiation on biogeochemical parameters and fluorescent dissolved organic matter (FDOM) in the sea surface microlayer of the southern coastal North Sea. *Elem. Sci. Anth.* **2018**, *6*, 15. [CrossRef]
6. Parlanti, E.; Wörz, K.; Geoffroy, L.; Lamotte, M. Dissolved organic matter fluorescence spectroscopy as a tool to estimate biological activity in a coastal zone submitted to anthropogenic inputs. *Org. Geochem.* **2000**, *31*, 1765–1781. [CrossRef]
7. Drozdowska, V.; Freda, W.; Baszanowska, E.; Rudz, K.; Darecki, M.; Heldt, J.T.; Toczek, H. Spectral properties of natural and oil polluted Baltic seawater—Results of measurements and modeling. *Eur. Phys. J. Spec. Top.* **2013**, *222*, 1–14. [CrossRef]
8. Zielinski, O.; Busch, J.A.; Cembella, A.D.; Daly, K.L.; Engelbrektsson, J.; Hannides, A.K.; Schmidt, H. Detecting marine hazardous substances and organisms: Sensors for pollutants, toxins, and pathogens. *Ocean Sci.* **2009**, *5*, 329–349. [CrossRef]
9. Friedrichs, A.; Ferdinand, O.; Miranda, M.L.; Zielinski, O. SmartFluo goes FDOM: Advancement of the DIY fluorometer approach towards UV excitation. In Proceedings of the OCEANS 2017, Aberdeen, UK, 19–22 June 2017.
10. Wymore, A.S.; Potter, J.; Rodríguez-Cardona, B.; McDowell, W.H. Using In-Situ Optical Sensors to Understand the Biogeochemistry of Dissolved Organic Matter Across a Stream Network. *Water Resour. Res.* **2018**, *54*, 2949–2958. [CrossRef]
11. Kalle, K. Zum Probleme der Meereswasserfarbe. *Ann. Hydrol. Mar. Mitt.* **1938**, *66*, 1–13.
12. Kirk, J.T.O. *Light and Photosynthesis in Aquatic Ecosystems*, 3rd ed.; Cambridge University Press: New York, NY, USA, 2011; p. 638.
13. Kowalczuk, P.; Ston-Egiert, J.; Cooper, W.J.; Whitehead, R.F.; Durako, M.J. Characterization of chromophoric dissolved organic matter (CDOM) in the Baltic Sea by excitation emission matrix fluorescence spectroscopy. *Mar. Chem.* **2005**, *96*, 273–292. [CrossRef]

14. Coble, P. Marine optical biogeochemistry: The chemistry of ocean color. *Chem. Rev.* **2007**, *107*, 402–418. [CrossRef]
15. Para, J.; Coble, P.G.; Charrière, B.; Tedetti, M.; Fontana, C.; Sempéré, R. Fluorescense and absorption properties of chromophoric dissolved organic matter (CDOM) in coastal surface waters of the northwestern Mediterranean Sea, influence of the Rhône river. *Biogeosciences* **2010**, *7*, 4083–4103. [CrossRef]
16. Schüβler, U.; Kremling, K. A pumping system for underway sampling of dissolved and particulate trace elements in near-surface waters. *Deep. Sea Res. Part I Oceanogr. Res. Pap.* **1993**, *40*, 257–266. [CrossRef]
17. Moore, C.; Barnand, A.; Fietzek, P.; Lewis, M.; Sosik, H.; White, S.; Zielinski, O. Optical tools for ocean monitoring and research. *Ocean Sci.* **2009**, *5*, 661–684. [CrossRef]
18. Zaneveld, J.R.V.; Kitchen, J.C.; Moore, C.C. Scattering error correction of reflection-tube absorption meters. In Proceedings of the Ocean Optics XII, Bergen, Norway, 26 October 1994.
19. Tassan, S.; Ferrari, G.M. Variability of light absorption by aquatic particles in the near-infrared spectral region. *Appl. Opt.* **2003**, *42*, 4802–4810. [CrossRef] [PubMed]
20. Fry, E.S.; Kattawar, G.W.; Pope, R.M. Integrating cavity absorption meter. *Appl. Opt.* **1992**, *31*, 2055–2065. [CrossRef] [PubMed]
21. Röttgers, R.; Doerffer, R. Measurements of optical absorption by chromophoric dissolved organic matter using a point-source integrating-cavity absorption meter. *Limnol. Oceanogr. Methods* **2007**, *5*, 126–135. [CrossRef]
22. Determann, S.; Lobbes, J.M.; Reuter, R.; Rullkötter, J. Ultraviolet fluorescence excitation and emission spectroscopy of marine algae and bacteria. *Mar. Chem.* **1998**, *62*, 137–156. [CrossRef]
23. Ferdinand, O.; Friedrichs, A.; Miranda, M.L.; Voß, D.; Zielinski, O. Next generation fluorescence sensor with multiple excitation and emission wavelengths—NeXOS MatrixFlu-UV. In Proceedings of the OCEANS 2017, Aberdeen, UK, 19–22 June 2017.
24. Fukuzaki, K.; Imai, I.; Fukushima, K.; Ishii, K.-I.; Sawayama, S.; Yoshioka, T. Fluorescent characteristics of dissolved organic matter produced by bloom-forming coastal phytoplankton. *J. Plankton Res.* **2014**, *36*, 685–694. [CrossRef]
25. Lee, E.; Yoo, G.; Jeong, Y.; Kim, K.; Park, J.; Oh, N. Comparison of UV–VIS and FDOM sensors for in situ monitoring of stream DOC concentrations. *Biogeosciences* **2015**, *12*, 3109–3118. [CrossRef]
26. Mustaffa, N.I.H.; Ribas-Ribas, M.; Wurl, O. High-resolution variability of the enrichment of fluorescence dissolved organic carbon in the sea surface microlayer of an upwelling region. *Elem. Sci. Anth.* **2017**, *5*, 52. [CrossRef]
27. Kowalczuk, P.; Tilstone, G.H.; Zabłocka, M.; Röttgers, R.; Thomas, R. Composition of dissolved organic matter along an Atlantic Meridional Transect from fluorescence spectroscopy and Parallel Factor Analysis. *Mar. Chem.* **2013**, *157*, 170–184. [CrossRef]
28. Miranda, M.L.; Trozjuck, A.; Voss, D.; Gassmann, S.; Zielinski, O. Spectroscopic evidence of anthropogenic compounds extraction from polymers by fluorescent dissolved organic matter in natural water. *J. Eur. Opt. Soc. Rapid Publ.* **2016**, *11*, 16014. [CrossRef]
29. Coble, P. Characterization of marine and terrestrial DOM in seawater using excitation-emission matrix spectroscopy. *Mar. Chem.* **1996**, *51*, 325–346. [CrossRef]
30. Zsolnay, A.; Baigar, E.; Jimenez, M.; Steinweg, B.; Saccomandi, F. Differentiating with fluorescence spectroscopy the sources of dissolved organic matter in soils subjected to drying. *Chemosphere* **1999**, *38*, 45–50. [CrossRef]
31. Huguet, A.; Vacher, L.; Relexans, S.; Saubusse, S.; Froidefond, J.M.; Parlanti, E. Properties of fluorescent dissolved organic matter in the Gironde Estuary. *Org. Geochem.* **2009**, *40*, 706–719. [CrossRef]
32. Drozdowska, V.; Kowalczuk, P.; Jozefowicz, M. Spectrofluorometric characteristics of fluorescent dissolved organic matter in a surface microlayer in the Southern Baltic coastal waters. *J. Eur. Opt. Soc. Rapid Publ.* **2015**, *10*, 15050. [CrossRef]
33. Murphy, K.R.; Stedmon, C.A.; Graeber, D.; Bro, R. Fluorescense spectroscopy and multi-way techniques. Parafac. *Anal. Methods* **2013**, *5*, 6557–6566. [CrossRef]
34. Bro, R. PARAFAC. Tutorial and applications. *Chemom. Intell. Lab. Syst.* **1997**, *38*, 149–171. [CrossRef]
35. Delauney, L.; Compere, C.; Lehaitre, M. Biofouling protection for marine environmental sensors. *Ocean Sci.* **2010**, *6*, 503–511. [CrossRef]

36. Coble, P.; Spencer, G.; Baker, A.; Darren, M. (Eds.) *Aquatic Organic Matter Fluorescence*; Cambridge University Press: Cambridge, UK, 2014.

37. Ocean Optics. *PX-2 Pulsed Xenon Lamp Stability*; Engineering Note 110-00000-000-04-0704; Ocean Optics: Largo, FL, USA, 2002.

38. Manov, D.V.; Chang, G.C.; Dickey, T.D. Methods for reducing biofouling of morred optical sensors. *J. Atmos. Ocean. Technol.* **2004**, *21*, 958–968. [CrossRef]

39. Lawaetz, A.J.; Stedmon, C.A. Fluorescence Intensity Calibration Using the Raman Scatter Peak of Water. *Appl. Spectrosc.* **2009**, *63*, 936–940. [CrossRef] [PubMed]

40. Heibati, M.; Stedmon, C.A.; Stenroth, K.; Rauch, S.; Toljander, J.; Säve-Söderbergh, M.; Murphy, K.R. Assessment of drinking water quality at the tap using fluorescence spectroscopy. *Water Res.* **2017**, *125*, 1–10. [CrossRef] [PubMed]

41. Perdue, E.M. Standard and Reference Samples of Humic Acids, Fulvic Acids, and Natural Organic Matter from the Suwannee River, Georgia: Thirty Years of Isolation and Characterization. In *Functions of Natural Organic Matter in Changing Environment*; Springer: Heidelberg, Germany, 2013.

42. Stedmon, C.A.; Markager, S. Resolving the variability in dissolved organic matter fluorescence in a temperate estuary and its catchment using PARAFAC analysis. *Limnol. Oceanogr.* **2005**, *50*, 686–697. [CrossRef]

43. Pearlman, J.; Zielinski, O. A new generation of optical systems for ocean monitoring—Matrix fluorescence for multifunctional ocean sensing. *Sea Technol.* **2017**, *2*, 30–33.

44. Delory, E.; Castro, A.; Waldmann, C.; Rolin, J.F.; Woerther, P.; Gille, J.; Del Rio, J.; Zielinski, O.; Golmen, L.; Hareide, N.R.; et al. NeXOS development plans in ocean optics, acoustics and observing systems interoperability. In Proceedings of the 2014 IEEE Sensor Systems for a Changing Ocean (SSCO), Brest, France, 13–17 October 2014.

45. Coble, P. Colored dissolved organic matter in seawater. In *Subsea Optics and Imaging*; Watson, J., Zielinski, O., Eds.; Woodhead: Cambridge, UK, 2013; pp. 98–118.

46. Stedmon, C.A.; Markager, S.; Bro, R. Tracing dissolved organic matter in aquatic environments using a new approach to fluorescence spectroscopy. *Mar. Chem.* **2003**, *82*, 239–254. [CrossRef]

47. Turner Designs. *Optical Specification Guide for Cyclops Submersible Sensors*; Engineering Note 998-2181 Rev. U; Turner Designs: San Jose, CA, USA, 2018.

48. Jacobson, P. Physical Oceanography of the Trondheimsfjord. *Geophys. Astrophys. Fluid Dyn.* **1983**, *26*, 3–26. [CrossRef]

49. Mascarenhas, V.J.; Voß, D.; Wollschlaeger, J.; Zielinski, O. Fjord light regime: Bio-optical variability, absorption budget, and hyperspectral light availability in Sognefjord and Trondheimsfjord, Norway. *J. Geophys. Res. Oceans* **2017**, *122*, 3828–3847. [CrossRef]

50. Carstea, E. Fluorescence Spectroscopy as a Potential Tool for In-Situ Monitoring of Dissolved Organic Matter in Surface Water Systems. In *Water Pollution*; InTech: London, UK, 2012; pp. 47–68.

51. Goldberg, M.; Wiener, E. Fluorescence Measurements of the Volume, Shape, and Fluorophore Composition of Fulvic Acid from the Suwannee River. In *Humic Substances in the Suwannee River, Georgia; Interactions, Properties, and Proposed Structures*; Averett, R.C., Leenheer, A., McKnight, M., Thorn, A., Eds.; United States Geological Survey Water-Supply Paper 2373; USGS: Reston, VA, USA, 1994; pp. 100–113.

52. Gilmore, A.M.; Cohen, S.M. Analysis of chromophoric dissolved organic matter in water by EEMs with HORIBA-Jobin Yvon fluorescence instrument called "Aqualog". *Readout* **2013**, *41*, 19–24.

53. Wollschläger, J.; Voß, D.; Zielinski, O.; Petersen, W. In situ observations of biological and environmental parameters by means of optics-development of next-generation ocean sensors with special focus on an integrating cavity approach. *IEEE J. Ocean. Eng.* **2016**, *41*, 753–762. [CrossRef]

54. Luciani, X.; Redon, R.; Mounier, S. How to correct inner filter effects altering 3D fluorescence spectra by using a mirrored cell. *Chemom. Intell. Lab. Syst.* **2013**, *126*, 91–99. [CrossRef]

55. Cyr, F.; Tedetti, M.; Besson, F.; Beguery, L.; Doglioli, A.M.; Petrenko, A.A.; Goutx, M. A New Glider-Compatible Optical Sensor for Dissolved Organic Matter Measurements: Test Case from the NW Mediterranean Sea. *Front. Mar. Sci.* **2017**, *4*, 89. [CrossRef]

56. Mayerfeld, P. Fluorometers: Integration experiences with unmanned vehicles. In Proceedings of the OCEANS 2017, Anchorage, AK, USA, 18–21 September 2017.

57. Makarewicz, A.; Kowalczuk, P.; Sagan, S.; Granskog, M.A.; Pavlov, A.K.; Zdun, A.; Borzycka, K.; Zabłocka, M. Characteristics of chromophoric and fluorescent dissolved organic matter in the Nordic Seas. *Ocean Sci.* **2018**, *14*, 543–562. [CrossRef]
58. Ng, C.L.; Senft-Grupp, S.; Hemond, H.F. A multi-platform optical sensor for in situ sensing of water chemistry. *Limnol. Oceanogr. Methods* **2012**, *10*, 978–990. [CrossRef]
59. Herfort, L.; Seaton, C.; Wilkin, M.; Roman, B.; Preston, C.M.; Marin, R., III; Seitz, K.; Smith, M.W.; Haynes, V.; Scholin, C.A.; et al. Use of continuous, real-time observations and model simulations to achieve autonomous, adaptive sampling of microbial processes with a robotic sampler. *Limnol. Oceanogr. Methods* **2016**, *14*, 50–67. [CrossRef]
60. Olson, R.; Shalapyonok, A.; Kalb, D.; Graves, S.; Sosik, H. Imaging FlowCytobot modified for high throughput by in-line acoustic focusing of sample particles. *Limnol. Oceanogr. Methods* **2017**, *15*, 867–874. [CrossRef]

PERMISSIONS

All chapters in this book were first published in MDPI; hereby published with permission under the Creative Commons Attribution License or equivalent. Every chapter published in this book has been scrutinized by our experts. Their significance has been extensively debated. The topics covered herein carry significant findings which will fuel the growth of the discipline. They may even be implemented as practical applications or may be referred to as a beginning point for another development.

The contributors of this book come from diverse backgrounds, making this book a truly international effort. This book will bring forth new frontiers with its revolutionizing research information and detailed analysis of the nascent developments around the world.

We would like to thank all the contributing authors for lending their expertise to make the book truly unique. They have played a crucial role in the development of this book. Without their invaluable contributions this book wouldn't have been possible. They have made vital efforts to compile up to date information on the varied aspects of this subject to make this book a valuable addition to the collection of many professionals and students.

This book was conceptualized with the vision of imparting up-to-date information and advanced data in this field. To ensure the same, a matchless editorial board was set up. Every individual on the board went through rigorous rounds of assessment to prove their worth. After which they invested a large part of their time researching and compiling the most relevant data for our readers.

The editorial board has been involved in producing this book since its inception. They have spent rigorous hours researching and exploring the diverse topics which have resulted in the successful publishing of this book. They have passed on their knowledge of decades through this book. To expedite this challenging task, the publisher supported the team at every step. A small team of assistant editors was also appointed to further simplify the editing procedure and attain best results for the readers.

Apart from the editorial board, the designing team has also invested a significant amount of their time in understanding the subject and creating the most relevant covers. They scrutinized every image to scout for the most suitable representation of the subject and create an appropriate cover for the book.

The publishing team has been an ardent support to the editorial, designing and production team. Their endless efforts to recruit the best for this project, has resulted in the accomplishment of this book. They are a veteran in the field of academics and their pool of knowledge is as vast as their experience in printing. Their expertise and guidance has proved useful at every step. Their uncompromising quality standards have made this book an exceptional effort. Their encouragement from time to time has been an inspiration for everyone.

The publisher and the editorial board hope that this book will prove to be a valuable piece of knowledge for researchers, students, practitioners and scholars across the globe.

LIST OF CONTRIBUTORS

Rosalia Santoleri, Simone Colella, Mario Benincasa and Vittorio Ernesto Brando
Istituto di Scienze Marine (CNR-ISMAR), Via Fosso del Cavaliere 100, 00133 Rome, Italy

Marco Bracaglia
Istituto di Scienze Marine (CNR-ISMAR), Via Fosso del Cavaliere 100, 00133 Rome, Italy
Dipartimento di Scienze e Tecnologie, Università degli Studi di Napoli Parthenope, Via Amm. F. Acton 38, 80133, Naples, Italy

Marco Bellacicco and Salvatore Marullo
Climate Modelling Laboratory, Italian National Agency for New Technologies, Energy and Sustainable Economic Development (ENEA), Via E. Fermi 45, 00044 Frascati, Italy
Consiglio Nazionale delle Ricerche (CNR), Istituto di Scienze Marine (ISMAR), Via Fosso del Cavaliere, 100 Rome, Italy

Jaime Pitarch, Emanuele Organelli and Gianluca Volpe
Consiglio Nazionale delle Ricerche (CNR), Istituto di Scienze Marine (ISMAR), Via Fosso del Cavaliere, 100 Rome, Italy

Victor Martinez-Vicente
Plymouth Marine Laboratory (PML), Prospect Place, The Hoe, Plymouth PL1 3DH, UK

Arthur C. R. Gleason, Kenneth J. Voss and Howard R. Gordon
Physics Department, University of Miami, Coral Gables, FL 33146, USA

Michael S. Twardowski
Harbor Branch Oceanographic Institute, Fort Pierce, FL 34946, USA

Jean-François Berthon
Joint Research Centre of the European Commission, 21027 Ispra, Italy

Xiaodong Zhang
Department of Earth System Science and Policy, University of North Dakota, Grand Forks, ND 58202, USA

Lianbo Hu
Department of Earth System Science and Policy, University of North Dakota, Grand Forks, ND 58202, USA
Ocean Remote Sensing Institute, Ocean University of China, Qingdao 266001, China

Dirk Aurin
NASA Goddard Space Flight Center (USRA), Greenbelt, MD 20771, USA

Antonio Mannino
NASA Goddard Space Flight Center, Greenbelt, MD 20771, USA

David J. Lary
University of Texas at Dallas, Richardson, TX 75080, USA

Linhai Li and Dariusz Stramski
Scripps Institution of Oceanography, University of California San Diego, La Jolla, CA 92093-0238, USA

Mirosław Darecki
Institute of Oceanology, Polish Academy of Sciences, 81-712 Sopot, Poland

Emmanuel Boss
University of Maine, Orono, ME 04469, USA

Christopher R. Sherwood
U.S. Geological Survey, Woods Hole, MA 02543, USA

Paul Hill and Tim Milligan
Dalhousie University, 1355 Oxford Street, Halifax, NS B3H 4R2, Canada

Michael Twardowski
Harbor Branch Oceanographic Institute, Florida Atlantic University, Ft. Pierce, FL 34946, USA
Sunstone Scientific LLC, Vero Beach, FL 32963, USA

Alberto Tonizzo
Sunstone Scientific LLC, Vero Beach, FL 32963, USA

Lisl Robertson Lain
Department of Oceanography, University of Cape Town, Cape Town 7700, South Africa

Stewart Bernard
Department of Oceanography, University of Cape Town, Cape Town 7700, South Africa
NRE, Centre for Scientific and Industrial Research (CSIR), Cape Town 7700, South Africa

Oliver Zielinski and Oliver D. Ferdinand
Center for Marine Sensors, Institute for Chemistry and Biology of the Marine Environment, University Oldenburg, 26382 Wilhelmshaven, Germany

Nick Rüssmeier
Center for Marine Sensors, Institute for Chemistry and Biology of the Marine Environment, University Oldenburg, 26382 Wilhelmshaven, Germany
OFFIS e.V. Institute for Information Technology, 26121 Oldenburg, Germany

Mario L. Miranda
Center for Marine Sensors, Institute for Chemistry and Biology of the Marine Environment, University Oldenburg, 26382 Wilhelmshaven, Germany
Laboratory of Air and Water Quality, University of Panama, Panamá 4, Panamá

Jochen Wollschläger
Center for Marine Sensors, Institute for Chemistry and Biology of the Marine Environment, University Oldenburg, 26382 Wilhelmshaven, Germany
Helmholtz-Zentrum Geesthacht, Institute of Coastal Research, Max-Planck-Str. 1, 21502 Geesthacht

Index

www.ingramcontent.com/pod-product-compliance
Lightning Source LLC
LaVergne TN
LVHW061510220926
848485LV00017B/792